World Epidemics

*A Cultural Chronology of Disease
from Prehistory to the Era of Zika*

SECOND EDITION

MARY ELLEN SNODGRASS

McFarland & Company, Inc., Publishers
Jefferson, North Carolina

LIBRARY OF CONGRESS CATALOGUING-IN-PUBLICATION DATA

Names: Snodgrass, Mary Ellen, author.
Title: World epidemics : a cultural chronology of disease from prehistory
to the era of Zika / Mary Ellen Snodgrass.
Description: Second edition. | Jefferson, North Carolina : McFarland &
Company, Inc., Publishers, 2017 | Revised edition of: World epidemics :
a cultural chronology of disease from prehistory to the era of SARS.
c2003. | Includes bibliographical references and index.
Identifiers: LCCN 2017043128 | ISBN 9781476671246
(softcover : acid free paper) ∞
Subjects: LCSH: Epidemics—History. | Epidemics—Social aspects—History.
Classification: LCC RA649 .S65 2017 | DDC 614.4—dc23
LC record available at https://lccn.loc.gov/2017043128

BRITISH LIBRARY CATALOGUING DATA ARE AVAILABLE

ISBN (print) 978-1-4766-7124-6
ISBN (ebook) 978-1-4766-3106-6

Front cover image © 2017 iStock

Printed in the United States of America

McFarland & Company, Inc., Publishers
Box 611, Jefferson, North Carolina 28640
www.mcfarlandpub.com

For the professionals
who keep my family well:

Dr. Peter Bradshaw
Dr. Bruce Carlton
Dr. Daniel Couture
Dr. Elizabeth Cressy
Dr. John Davis
Dr. Ronald Key
Dr. Jerry Pruitt
Dr. Jessica Tate

Acknowledgments

Many thanks to the following people for their help with this book: Beth Bradshaw and Martin Otts, reference librarians, Patrick Beaver Library, Hickory, North Carolina; Diane Brannon, R.N., Hickory, North Carolina; and Mark Schumacher, reference librarian, Jackson Library, University of North Carolina–Greensboro. Special thanks to Alexander Street Press for the use of their superb databases.

Table of Contents

The misery of the people is the most fertile mother of diseases.
—Johann Peter Frank, director-general
of public health in Austrian Lombardy,
in a speech delivered on May 5, 1790

The living, the dying, and the dead huddled together in one mass. Some unfortunates in the most disgusting state of smallpox, distressingly ill with ophthalmia, a few perfectly blind, others living skeletons, with difficulty crawled from below unable to bear the weight of their miserable bodies. Mothers with young infants hanging at their breasts unable to give them a drop of nourishment. How they had brought them thus far appeared astonishing. All were perfectly naked. Their limbs were excoriated from lying on the hard plank for so long a period. On going below, the stench was insupportable. How beings could breathe such an atmosphere and live appeared incredible.

—Logbook, British warship Fawn,
after seizure of a slaver bound
for Brazil, February 1841

Many epidemics are remembered, though few details are known.
—J. Stanley Gardiner, author, 1898

Preface

This encyclopedia invites the reader, writer, historian, researcher, health professional, teacher, librarian, journalist, and general reader to learn about a wide range of world epidemics and to contemplate the lives of earnest and dedicated men and women who, from prehistory to the present, have advanced the understanding of what makes people sick.

The text offers an easy-to-use source of information arranged by estimated time spans and exact dates. Entries of various lengths summarize the data available on incidents of contagion across the globe. Lengthy histories are broken up by date and linked with cross-references to round out events that fail to fit into neat time frames— for example, the spread of smallpox over the Western Hemisphere, the experience of sailors with scurvy, and the advance of AIDS and Zika worldwide. Enhancing historical data are citations from personal and public documents, a map (on page 4) of resurgent vector-borne disease, comparative charts of types of infection and results of scourges, and rough estimates of numbers of people affected by each epidemic.

Research materials include personal letters and diaries, annual summaries of the projects of missions and religious houses, health statistics from cities and the World Health Organization, and the published findings of scientific experimentation and clinical trials in *Lancet*, the *Journal of the American Medical Association*, and the *British Medical Journal*. To the researcher beginning a probe of world epidemics, three invaluable electronic databases are Alexander Street's *Early Encounters in North America*, the online *New England Journal of Medicine*, and ABC-Clio's two-part online database comprised of *America: History and Life*, which covers learned articles about North America, and *Historical Abstracts*, which presents parallel information worldwide from 1450 to the present. For more current data, the Gale Group's *SIRS Research* database, the London *Times Digital Archive*, and *CPI.Q* (Canadian periodicals) provide information from hundreds of newspapers and periodicals.

Readers will find information on healers such as Galen and Benjamin Rush, researchers like Robert Koch and Joseph Goldberger, published works on epidemics such as John Snow's pinpointing of public waterworks as reservoirs of cholera, particulars of single diseases such as hookworm or West Nile fever, and the health histories of a geographic locale or people, for example, tribes of Papua New Guinea, native Hawaiians, the Sadlermiut Eskimo, or the Cree of midwestern Canada. Several other aids accompany the text.

A glossary clarifies 116 essential terms, such as patient zero, zoonosis, pathogen, sequelae, paleopathology, World Health Organization, and Médecins sans Frontières (Doctors without Borders), an international humanitarian task force against disease.

Appendix A lists individual diseases by their proper and informal names, e.g., spotted fever/ typhus and dumdum fever/leishmaniasis; the Latinate names of such pathogens as *Escherichia coli* and *encephalitis lethargica*; dates and places of early research on each disease such as Wilhelm Burgdorfer's analysis of Lyme disease in the U.S. in 1998; and the manner of infection, whether by polluted water, infected air, insect or animal bite, or contact with human effluvia.

Appendix B names historic writings on disease and presents them in alphabetical order. It also includes translations of titles, e.g., Cyprian's *De Mortalitate* (On Mortality) and Paul Louis Simond's *La Propagation de la Peste* (The Spread of Plague).

Appendix C presents the writings by date,

1

beginning in antiquity with the works of Sushruta, Hippocrates, and Areteus to the most current writings on SARS and MERS.

Appendix D presents writings by such prominent medical writers as Hideyo Noguchi, Robert Koch, Girolamo Fracastorius, Rhazes of Baghdad, and Eugen Fischer, designer of a system of eugenics that influenced Adolf Hitler's human experimentation on "expendable" Gypsies, mental defectives, and Jews.

The bibliography presents research sources in two forms: a general bibliography lists materials covering more than one disease or summarizing the whole span of epidemiology; and a second bibliography presents titles under 88 headings of specific diseases, beginning with acrodynia and AIDS and concluding with yaws and yellow fever. The topics include a subheading for bioterrorism.

An exhaustive index lists people, places, beliefs, infections, titles, historical events, and medical milestones along with the specific dates of entries (for example, Metís, Cairo, Mormonism, Listeria, Syphilis sive Morbus Gallicus [syphilis or French disease] [1530], Boer War, and Salk vaccine). Generous cross-referencing clarifies details such as parrot fever as a folk name for psittacosis, specific examples of nursing theory, significant types of food poisoning, and puerperal fever as an element of women's history.

Several books were especially informative. I could not have gained the insight to compile this text without these valuable current sources:

Elizabeth A. Fenn's *Pox Americana: The Great Smallpox Epidemic of 1775–82* (2001); George C. Cohen's *Encyclopedia of Plague and Pestilence* (2001); Norman Cantor's *In the Wake of the Plague: The Black Death and the World It Made* (2001); E. Margaret Crawford's *The Invisible Enemy: A Natural History of Viruses* (2000);

Robin Marantz Henig's *A Dancing Matrix: How Science Confronts Emerging Viruses* (1993); Alan Kraut's *Silent Travelers: Germs, Genes, and the "Immigrant Menace"* (1994); William Naphy and Andrew Spicer's *The Black Death and the History of Plagues, 1345–1730* (2000); Thomas M. Daniel and Frederick C. Robbins's *Polio* (1997); Katherine Ott's *Fevered Lives: Tuberculosis in American Culture Since 1870* (1996); Andrew Cliff, Peter Haggett, and M. R. Smallman-Raynor's *Island Epidemics* (2000); Pete Davies's *The Devil's Flu* (2000); Rosemary Horrox's *The Black Death* (2000); and Micheal Clodfelter's *Warfare and Armed Conflicts* (2002).

Also valuable were several of my own reference works: *Historical Encyclopedia of Nursing* (2000), *World Food* (2012), *Who's Who in the Middle Ages* (2013), *Settlers of the American West* (2015), *The Civil War Era and Reconstruction* (2015), *American Women Speak* (2016), and *Colonial Women's Art* (2017).

Of particular help were current publications, news sources, and academic journals, notably, Reuters, the BBC News, *National Geographic*, *Nature*, *The Wall Street Journal*, *Medical History*, *African Historical Studies*, *American Journal of Public Health*, *Canadian Historical Review*, *Archaeology*, and online newspapers from around the world, including the *China Daily*, *Al-Ahram Weekly*, and the *Sydney Sun Herald*. I am also indebted to numerous Internet sites, particularly "Plagues and Peoples: The Columbian Exchange," "The Condition of the Working Class in England," "Epidemic Disease," "Science in Al-Andalus," "A History of Western Medicine and Surgery," "Outbreak of Resistant Foodborne Illness," "Under the Knife: A History of Medicine," "Women in Medicine," and "Women's Philanthropic and Charitable Work."

Introduction

Disease is a fearful unknown within the human condition. It is not by chance that, for all time, the Four Horsemen of the Apocalypse have galloped at will across the plane of world affairs. Without a backward glance, Famine, War, and Pestilence have left a trail of hapless victims for Death to finish off—from warriors, peasants, foundlings, laborers, and slaves to rajahs, chiefs, princes, and potentates.

The image of the notorious quartet of killers is more than a dramatic symbol. History bears out the connection between human mischance and outbreaks of sickness. Thus, the poorest and unluckiest of peoples have borne the brunt of a tumbling house of cards, beginning with failed harvests and progressing to poverty, starvation, political dissension, violence, and catastrophe. The paradigm of human miseries has burdened a wide range of eras: the fall of Rome, the Crusades, Christopher Columbus's expeditions to the Caribbean, the coming of the *coureurs de bois* to the Great Lakes and Mississippi Valley, the tragic failure of the Napoleonic wars, the colonization of South Africa and Oceania, the annual Islamic hadj to Mecca, and the transportation of slaves shackled in ships' holds on the middle passage from the Bight of Benin. The events varied in scope and purpose, but all ended in mass graves.

The proportion of wellness to suffering has altered with the cadence of the Four Horses' gait: long periods of health and prosperity have preceded unforetold terror from the Black Death, deadly scourges among the Indians of New England and Yucatán, the London Plague of 1665, scurvy among crews of naval expeditions to the Pacific, seasonal cholera among Pakistani and Ceylonese, trachoma contracted among quarantined immigrants at Ellis Island, and terrifying Ebola in the villages of the Congo.

In the 1910s, World War I had not drawn to a close before Spanish influenza swept the globe, filling the lungs of doughboys and civilians alike with septic fluids and turning their faces and extremities blue within hours of the first symptom. Similarly, the post–World War II generation of victors looked toward good times, but instead met erratic outbreaks of typhus, syphilis, and polio, the mystifying crippler that suffocated the youngest and weakest and confined survivors to iron lungs, rocking beds, or a lifetime of leg braces and wheelchairs. In the decades after the bombings of Hiroshima and Nagasaki in August 1945, leukemia, breast cancer, and thyroid carcinoma engulfed Japan's survivors of the world's only deliberate radiation poisoning of civilians as an act of war.

The record of these historic pestilences is a curious blend of fact, legend, and supposition. Pre-literate people made their marks on papyri and Egyptian tomb walls, Peruvian pottery, Eskimo oral tradition and Asian story-talk, Hindu and Polynesian ritual, and the winter counts of North America's Plains Indians. In literate lands, witnesses recorded the plight of the sick and dying in almost every form of creative writing— hymns and prayers, Buddhist and Koranic scripture, witty aphorisms and verse, letters and journals, plays, novels, posters, and *gonfaloni*, the plague banners that marked religious processions intended to alter God's thinking on who should live and who should perish. These human responses shared space with the cold, immutable data of science and city and state death rolls in preserving for all time the outcome of historic eras, good and bad.

A chronology of epidemics offers the reader a panoramic overview of patterns in human behavior. Repeatedly, villages and towns have faced so many dead and dying that overwrought survivors have abandoned the amenities of burial, the comforts of deathbed confession and farewell, even the decencies of food and nurse-care owed to family, servants, neighbors, and friends. Likewise diverted from their aims, battalion leaders, priests, health officials, and civic authorities abandoned long-range goals to rescue the hapless from dysentery, leishmaniasis, dengue fever, and typhus. In nobler scenarios, individuals have martyred themselves to succor and treat the sick, compose bequests of the dying, and nurture orphans. During outbreaks of the Yellow Jack in Philadelphia, New Orleans, Savannah, and Charleston, selfless physicians faced more patients in a day than they would normally treat in several lifetimes. West African health professionals faced certain death by succoring the dying in the throes of ebola.

The flip side of the best of times reminds the student of history that layers of civility, piety, and compassion can melt away at the first toll of the leper's bell, the digging of common burial pits, and the nightly rumble of the dead-cart. As white patches of leprosy or the buboes of plague enveloped Europeans and as smallpox felled settlers of New England, grim sermonizers pointed fingers at the morally lax and blamed them for calling down certain death on themselves and the righteous. In retaliation against ill fate, the ignorant felt justified in charging Gypsies, Jews, Indians, black slaves, Asian immigrants, refugees, and homeless wanderers as sources of diphtheria, smallpox, yellow fever, and plague. Advancing from name-calling, timorous locals reshaped fear into anger, displacing their terror and grief as merciless outcries, rioting, and barbarity. The most virulent people seized, dispossessed, and tormented innocent prey before ripping them apart or burning them alive in their dwellings. The most shameless victimizers stopped the vengeful juggernaut long enough to enrich themselves by looting homes and stripping corpses.

Just as the worst of times have created hellholes of suffering on Africa's coast, in the British prison ships during the American Revolution, and among immigrants to Canada, huddled throngs of Bengali, and transportees to Australia, the Zeitgeist has offered opportunity to history's most honored altruists: nurse Florence Nightingale at Scutari; herbalist Mary Jane Seacole at the Crimean front; nutritionist Alexis Soyer during the Irish Potato Famine; researcher Hideyo Noguchi in Japan; Jesuit fathers and Sisters of Charity in Quebec; Mother Teresa among India's outcasts; Father Damien among the lepers sequestered at Kalawao, Molokai; Qui Xi at immunization clinics in China; Rose Kaplan's nurse corps in Jerusalem and Alexandria, Egypt; and Albert and Hélène Breslau Schweitzer at Lambaréné in Gabon.

From the challenge of epidemics have come brave inoculators who have risked mayhem to protect the healthy, sanitarians engineering more wholesome living conditions, and volunteers like Clara Maass, who died in Havana after exposing herself to mosquito bites to pinpoint the insect vectors of yellow fever. From the puzzle of bacillus and virus have sprung Sir David Bruce and Lady Elizabeth Steele-Bruce's innovative methods of ending the scourge of sleeping sickness in Africa, the World Health Organization's victory over smallpox and polio, and Carlos Juan Finlay and William Crawford Gorgas's efforts to rid the air of mosquitoes that inflicted malaria and yellow fever on the laborers and engineers who cut a canal across Panama.

Taken as a unit, the history of epidemics refuses to lodge neatly within the medical realm. Episodes of the battle against disease align the demographics of illness with a variety of memorabilia: the children's song "Ring Around a Rosy"; census figures on Assyrian steles; changes in Puritan headstone art; the formulation of eau de cologne; the coining of English money as tokens of the king's touch against scrofula; and the impersonal greetings of Ugandans, who abandoned handshakes and hugs to halt the spread of Ebola. Fear of contagion has altered bureaucratic treatment of intravenous drug users and HIV carriers in Cuba, the burial style of the Tlingit, funerary feasting among South Pacific cannibals, and criteria for military nurses. The prevention and treatment of disease forced U.S. southerners to acknowledge widespread hookworm, caused medical and dental staffs to wear masks and rubber gloves to prevent HIV, made prison provisioners give up polished rice to end outbreaks of pellagra, and turned English sailors into limeys. Disease permeates language with myriad judgmental names for syphilis, the abbreviation "E coli" for a common food poisoning, the "white man's grave" for Africa, the Japanese *itai-itai* ("ouch-ouch") disease for an affliction of joints, the historic Legionnaire's disease for a mysterious outbreak during a Philadelphia convention, the "eight-day" sickness among doomed newborns on Saint Gilda Island, and "typhoid Mary" as a general term for anyone who haphazardly spreads contagion to the unwary.

The study of epidemics also recognizes some familiar figures in new roles—George Washington, Lady Mary Wortley Montagu, and Benjamin Franklin as promoters of variolation, Noah Webster as an expert on contagion, Louisa May Alcott and Walt Whitman as wartime medics, and English sea captain James Cook and American artist George Catlin as eyewitnesses to the infection of virgin-soil populations.

As a glimpse of humankind at cotside and graveside, the study of epidemics displays emotion at its keenest and logic at its most pragmatic. The individual incidents of infections—among measles victims on a single island like Fiji or Iceland, or among flu victims over whole continents—evidence the ability of human beings to adapt to nature at its most perverse. In propitiating gods, administering herbs, vaccinating the uninfected, or bidding farewell to a hopeless case, survivors' responses reveal individual values and belief systems as well as the outlook of the times, whether sanguine, indifferent, or resigned. Taken as a whole, the experience of epidemics insists that humanity treasure life as fragile and fleeting.

The Chronology

1,500,000 B.C.E. Deposits of new bone on the remains of *Homo erectus* from the Middle Pleistocene Era in Kenya attest to the presence of the yaws bacteria, a chronic tropical infection of bones, cartilage, and skin causing disfiguring lesions.

500,000 B.C.E. New bone deposits on remains of a *Homo erectus* femur at Venosa, Italy, suggest that humans migrating north from Africa brought yaws bacteria with them.

10,000 B.C.E. Smallpox epidemics first slew humans in northeastern Africa along river deltas inhabited by the first farmers.

8000 B.C.E. Malaria, perhaps the biggest killer in history, dates to Africa around 10,000 B.C.E. Within two millennia, it progressed in virulence with the settlement of humans and the development of agrarian lifestyles. Living in close-knit communities in the Middle East may have encouraged breeding grounds for insects in pools and ponds.

4000 B.C.E. In Egypt, people feared the growing threat of leprosy, an incurable and highly contagious disease. To ward off infection, families and some physicians shunned victims. Lepers became virtual pariahs to society and formed their own colonies to provide mutual support to fellow sufferers.

3700 B.C.E. A hieroglyph from Memphis pictured bone formations in the skeleton of the Egyptian temple priest Ruma, the gatekeeper at Astarte's temple. Anthropologists concluded that withering in the leg indicated crippling caused by polio. The image is the earliest known depiction of the disease.

3180 B.C.E. In Egypt, the first recorded epidemic appeared in *Aigyptiaka* (*A History of Egypt*) by Manetho of Sebennytos, a scribe and priest-chronicler at Heliopolis in 280 B.C.E. He listed the pharaohs in order from legendary origins to 323 B.C.E. and stated that the reign of Mempses coincided with a terrible pestilence.

ca. 3000 B.C.E. The smallpox virus, a derivative of the original orthopoxvirus that emerged among animals of the Central African rain forests, evolved into a human scourge. Spreading northeast, it became endemic among large populations inhabiting the Nile and infected traders and soldiers who passed through the cosmopolitan ports of North Africa.

ca. 3000 B.C.E. Examination of Nesperehân, a high priest of Amon at Thebes under the Pharaoh Ramses, confirmed the presence of spinal tuberculosis in Egypt. Paleopathologists identified the disease from the presence of a spinal abscess, evidence of chronic tuberculosis. *See also* **ca. 1000 C.E.**

2600 B.C.E. The Chinese recognized beriberi as a disease and documented cases in medical texts compiled around 2000 B.C.E.

1770 B.C.E. At Mari in Mesopotamia, King Zimri-Lim had his scribe post a warning to Queen Shiptu concerning a woman named Nanna, who suffered from infectious lesions. The king's orders, inscribed on a clay tablet in Akkadian, forbade drinking from a common cup or using a seat or a bed after Nanna had touched it. His discussion of Nanna's lesions indicates a sophistication about *mustahhizu* (contagious disease).

1600 B.C.E. In Egypt, mummies at the Giza pyramids bore evidence of osteosarcoma or bone cancer.

1580 B.C.E. As revealed by Danish physician Ove Hamburger, a stele from the 18th Egyptian dynasty pictured a male figure, possibly a priest, suspected of suffering paralytic infantile polio. Clues to his condition include a withered, shortened right leg, foot in an equine position, and a walking stick on which he leaned.

1500 B.C.E. Egyptian traders and cameleers carried smallpox on overland routes through Arabia to trading centers in India.

1495 B.C.E. According to biblical history in Exodus 7–13, during the rule of a pharaoh, Egypt suffered a series of natural disasters—blood-red river water, fish kills, frogs, dust storms, a cattle disease, lice, flies, boils, hail, crop failure, and locusts. Scholars surmise that the fifth plague may have been anthrax, which killed camels, cattle, sheep, and oxen. Afterward, a dramatically lethal plague struck Egyptians, but did not harm the Israelites, who may have developed an immunity to the disease. Whatever the circumstances, the enslaved Israelites, led by brothers Moses and Aaron, used the opportunity to flee bondage. *See also* **March 1603.**

In Leviticus, a book of rules for God's people to observe, Jehovah delivers a holiness code that warns the Israelites of punishment for disloyalty:

> I also will do this unto you; I will even appoint over you terror, *schanhepheth* [consumption], and the burning ague, that shall consume the eyes, and cause sorrow of heart: and ye shall sow your seed in vain, for your enemies shall eat it [Leviticus 26:16].

In the priestly tradition of upbraiding those weak in faith, the harsh exhortation clearly draws upon common knowledge of fevers and tuberculosis and possibly trachoma, a granulation that damages the eyes.

From this same era, biblical mention of a "fiery serpent" in Numbers 21:6, 8 and Deuteronomy 8:15 suggests the guinea worm, the cause of dracunculiasis, a parasitic infestation caused by contaminated drinking water. Reference to the worms along the Red Sea implies that Egyptians and their Hebrew slaves suffered dracunculiasis during the Exodus. Examination of a mummy from the Manchester Egyptian Mummy Project disclosed a calcified worm. Additional references to the parasite come from poetry by Vasistha found in the Rig Veda (ca. 1700–1400 B.C.E.), the oldest and most revered Indian scripture.

1480 B.C.E. As chronicled in Numbers 11:32–35, a plague broke out among the Israelites after Moses and his brother Aaron led them from the Sinai to Kadesh, possibly Wadi Murrah about 30 miles northeast of Sinai, where they remained for 35 years. During a protracted struggle between the willful, disobedient Israelites and the law of Yahweh, the refugees wearied of Yahweh's gift of manna from heaven. In place of simple desert meals, they lusted for the varieties of meat they had eaten during their enslavement to the Egyptian pharaoh. Yahweh told Moses that he would supply them with so much meat for a month that they would sicken from their food orgy.

As promised, Moses observed a heavenly wind bringing quail from the shores of the Gulf of Aqaba on their annual spring migration from Africa to Europe. When the exhausted birds fell to earth a day's journey from the camp perimeter, the gluttonous Israelites left their tents, gathered the birds, and began eating. According to scripture, "Ere it was chewed, the wrath of the Lord was kindled against the people, and the Lord smote the people with a very great plague" (Numbers 11:33–35). After the burial of the victims outside the camp, the Lord named the unidentified spot Kibroth-hattavah, meaning "graves of the longing or greedy." The chastened Israelites, restored to their original obedience around the holy Ark of the Covenant, unified around their leader and moved northeast to Hazeroth.

1400 B.C.E. In India, an ancient Hindu myth, "The Sacrifice of Daksha," referred to leprosy as *kushtha*. The story is found in the sacred Atharva Veda (Lore of the Fire Priests), attributed to the mythic Dhanvantari, the first physician and teacher of healing, around 800 B.C.E. One folkloric charm commands:

> Born by night art thou, O plant, dark, black, sable.
> Do thou, that art rich in colour, stain this leprosy, and the gray spots!
> The leprosy and the gray spots drive away from here.
> May thy native colour settle upon thee;
> The white spots cause to fly away!
> Sable is thy hiding-place, sable thy dwelling-place, sable art thou,
> O plant: drive away from here the speckled spots! ["Atharva Veda"].

Although the descriptions seem to refer to Hansen's disease, the term may name a family of dermatological ills. Additional commentary on infectious disease named coughs, diarrhea, paralysis, *apatik* (scrofula), *balasa* (degenerative bone and joint disease), *takman* (fever), and *rajaya-*

kshma (tuberculosis), which caused frequent relapses.

1346 B.C.E. During a war with Egypt, the Hittites of Anatolia suffered a smallpox epidemic, the first recorded in history. After the death of King Suppiluliumas I and his heir, Arnuwandas, the Hittites did not recover their former power. The typical ritual to rid the people of plague was to drive a ram down the road toward the enemy to transfer contagion. Thus, soldiers rid themselves of the pox and put the source of infection to good use as a biological weapon.

1300 B.C.E. In India, the *Ordinances of Manu* warned young people not to choose mates from families infected with tuberculosis. A special prayer marked the disease, thought to riddle the lower castes:

> Oh Fever, with thy Brother Consumption, with thy Sister Cough, go to the people below [Daniel 1997, p. 13].

1250 B.C.E. To rid the realm of malformed and grossly diseased people, Ramses II ordered 80,000 Egyptian lepers out of the public domain and into a settlement village on the rim of the Sahara Desert.

1200 B.C.E. The examination of a pair of mummies from Egypt's 20th dynasty disclosed eggs of the schistosome, which may have infected humans from as early as 3,000 B.C.E. The discoverer, French physician Marc Armand Ruffer, Professor of Bacteriology at Cairo Medical School, invented the field of paleopathology and wrote *Studies in Paleopathology of Egypt* (1921). He concluded that schistosomiasis, also called snail fever or bilharzia, may have been common in ancient Egypt. At Deir el-Bahari, he also located evidence of smallpox and presented extensive studies of anthracosis, arteriosclerosis, meningitis, plague, pneumonia, staphylococcus infection, and tuberculosis. Further investigation pointed to cholera, cirrhosis of the liver, dysentery, leprosy, malaria, tonsillitis, and typhoid fever.

1193 B.C.E. In the Valley of the Kings at Thebes during the 19th Dynasty, the mummy of the Pharaoh Siptah (Akhenre Setepenre) of Egypt went to the grave with a withered foot and shortened leg crippled by polio, or possibly cerebral palsy. The decedent expired suddenly at age 20 in the sixth year of his reign before his sarcophagus was complete. His tomb remained undisturbed until Egyptologist Sir Grafton Elliot Smith excavated it on August 29, 1905. Forensic study determined that polio was endemic in ancient Egypt.

1190 B.C.E. As described in the opening lines of Homer's *Iliad* (ca. 850 B.C.E.), a plague weakened Agamemnon's army in the ninth year of a ten-year war against the Trojans. The text of the Greek epic connects the outbreak to an act of impiety, the commander-in-chief's refusal to release Chryseis to her father, the Trojan priest Chryses. Chryses makes a moving plea to Apollo, god of healing:

> Hear me, god of the silver bow. Ruler over Chryse and Cilla and Tenedos, O Mouse-god. If ever I burned fat offerings to you, give me my desire. Let the Greeks pay through your arrows for my tears [Homer (tr. Richards) 1950, p. 34].

Homer describes the virulent disease as an onslaught of arrows from the divine bow and makes no mention of victim recovery. Because Chryses refers to rats, historians deduce that the pestilence might have been bubonic plague. Other possibilities are anthrax, a zoonotic that could have struck both animal and human victims, and dysentery, a common pestilence to armies living in close, unsanitary conditions.

The pestilence extends over nine days and blankets the camp "like the night" (ibid.). It strikes first their pack mules and guard dogs before advancing to Greek troops. So many die "till the fires of the dead were burning night and day" (ibid.). The mighty warrior Achilles intervenes on the tenth day and calls on Calchas, a seer who had led the Greek convoy to Troy.

After Agamemnon atones by sending twenty sailors to row Chryseis out to the ship that would take her home, he orders his followers to cleanse themselves and their belongings in seawater and to propitiate Apollo with fragrant burnt offerings, perhaps to quell the stench of disease and incinerated corpses. Because Agamemnon is rough with old Chryses, Apollo extends the epidemic. The connection between Agamemnon's arrogance and effrontery to a holy man are typical of Greek literature, which is predicated on themes of personal faults and their influence on world affairs.

ca. 1157 B.C.E. Pharaoh Ramses V may be the first identified victim of smallpox. His burial site in the temple at Dayr al-Bahri near the Valley of the Kings in western Thebes was the discovery of French Egyptologist Victor Loret on March 9, 1898. Examination of the young king's mummy from Egypt's 20th Dynasty revealed pockmarks from pustules on his face and torso. The damage to his skin suggests that he died around age 35 to 40 of an infectious disease that was probably endemic in ancient Egypt.

1141 B.C.E. According to I Samuel 5–6, the Philistines attacked the Israelites at the battle of Ebenezer in central Samaria and stole the Ark of the Covenant, the repository of sacred power. The Philistines transported their booty to Ashdod, one of their chief cities, where the people suffered emerods, an unidentified disease that struck them in their "secret parts," possibly their lymph glands. After seven months of illness, the thieves returned the ark on a cattle-drawn cart to its traditional owners at Beth-shemesh west of Jerusalem.

Upon the ark's arrival, 70 Israelites—inflated in scripture for the sake of the telling to 50,070—died upon coming in contact with the contagion borne by the cart. One Oxford clinician, William Porter MacArthur, an expert on tropical disease in the mid–1900s, proposed a diagnosis of bubonic plague because of its emergence as swellings in the groin. Because of their rapid demise an alternate diagnosis is that pneumonic plague killed the Israelites.

1122 B.C.E. After smallpox traveled from Egypt through India, Hindus called upon Mata Sitala (Mother Cool One), a kindly, maternal goddess, to save them from the scourge. The Huns brought the disease east to China during the reign of the Emperor Tcheou, when people named it "venom from the mother's breast."

1017 B.C.E. As confirmed in II Samuel 24 and I Chronicles 21, in the time of King David, 70,000 Israelites died of an epidemic that census takers may have spread in fulfillment of their duties counting the tribes from Dan north of the Sea of Galilee to Beersheba west of the Dead Sea.

790 B.C.E. As explained by the biographer Plutarch's *Lives* (115), plague struck Rome and spread to Laurentum in Latium southeast of the harbor of Ostia. The Camerians to the northeast used the opportunity to attack Rome, but Romulus, Rome's founder and first king, surprised them and killed 6,000 before seizing their stronghold.

710 B.C.E. Rome suffered an epidemic in the eighth year of the reign of Numa Pompilius, the king who followed Romulus. After the god Jupiter dropped a bronze shield from the sky, Numa suppressed the disease by enshrining the divine gift in the place where it fell and by having the vestal virgins sprinkle the water of the sacred spring to cleanse their temple, which was sacred to Vesta, goddess of the hearth.

701 B.C.E. When King Sennacherib of Assyria sent his army against Judah and took 46 cities, Hezekiah, the 13th king of Judah after David, prayed to Yahweh for divine intervention. By morning, according to II Kings 19:35 and Isaiah 37:36, around 185,000 enemy lay dead from the onslaught of God's angel, possibly through dysentery or bubonic plague. Confirming this historic event are three sources: the translation of an account on the doorjamb of Sennacherib's throne room at Nineveh, made by Sir Austin Henry Layard in 1847; the Sennacherib prism at the University of Chicago's Oriental Institute; and the Taylor Prism, a baked clay stele from around 689 B.C.E. bearing the Assyrian version of the event.

Conflicting accounts by the Greek historian Herodotus and the Judaeo-Roman Jewish chronicler Josephus obscure the historic facts, which suggest that a rodent-borne plague beset Assyrian troops. Herodotus explains in Book II: 141 of *The Persian Wars* (ca. 428 B.C.E.) that, when the Assyrians marched on Egypt's King Sethos at Pelusium on the Nile delta, a plague of mice overran the Assyrian camp and chewed the leather bow strings, shield straps, and arrow cases. As told by Herodotus, the defenseless soldiers "abandoned their position and suffered severe losses during their retreat" (1961, p. 158). The historian adds that a stone statue of Sethos survives at Hephaestus's temple picturing the king with a mouse in his hand.

In Josephus's version, found in Book X:141 of *Antiquities of the Jews* (94), the biblical angel of death is altered into an epidemic:

> Now when Sennacherib was returning from his Egyptian war to Jerusalem, he found his army under Rabshakeh his general in danger, for God had sent a pestilential distemper upon his army; and on the very first night of the siege, a hundred fourscore and five thousand, with their captains

and generals, were destroyed. So the king was in a great dread and in a terrible agony at this calamity; and being in great fear for his whole army, he fled with the rest of his forces to his own kingdom, and to his city Nineveh [1960, p. 213].

The situation worsened when the king arrived home to civil unrest. In his weakened condition, he was unable to fight off Adrammelech and Seraser, his older sons, who slew him in his temple. The Assyrians rebelled against the usurpers, driving them into Armenia.

ca. 668 B.C.E. As described in the library of King Ashurbanipal at Nineveh, at the pinnacle of Assyrian conquests and cultural attainment, the dracunculiasis infestation may have entered Mesopotamia. It probably arrived in Egyptian prisoners transported to Assyria around the beginning of the seventh century B.C.E.

ca. 642 B.C.E. According to Livy, Rome's revered mythographer and historian, a pestilence assailed the city in the time of Tullus Hostilius, the third king of Rome. The king at first castigated soldiers who were too sick for duty until he himself contracted the ailment. He consulted the writings of Numa, his predecessor, and found a description of secret rites honoring Jupiter Elicius or "the omen summoner," to whom Numa had raised an altar on the Aventine Hill. Because Tullus failed to perform the ceremony properly, lightning struck him and burned his house.

600 B.C.E. The first Greek doctor, Aesculapius (also Asclepius or Asklepios), was the son of Apollo, god of healing. Skilled in pharmacology, herbalism, and surgery, Aesculapius became a revered cult figure worshipped at the Delphic Oracle, where people terrified of plagues begged to be spared. They also propitiated the healer at his temples in Cos, Epidaurus, and Pergamum, where Aesculapian healers practiced. *See also* **431 B.C.E.; 292 B.C.E.**

ca. 600 B.C.E. Leprosy first appeared in Indian medical literature. The father of surgery, Sushruta, a respected physician and medical writer, described the disease in the text of *Sushruta Samhita*, a classic compendium of Ayurvedic medicine. He also prescribed rehydration and a liquid diet of rice gruel to revive and nourish cholera victims. Of tuberculosis, he could only say that it was incurable. Doctors who treated tubercular patients risked their reputation for skill and knowledge.

ca. 580 B.C.E. After the art students of the Cretan sculptor Daedalus settled at Sicyon on the northeastern Peloponnesus, they abandoned their art academy for a journey to Turkey. Because a plague broke out at Sicyon, citizens consulted the Pythia, the oracle of Delphi, who informed them that the plague was a punishment on the people because the sculptor left their temple friezes unfinished. As dramatized by the Roman encyclopedist Pliny the Elder, the Sicyons wooed Daedalus back to his work with prayer and bonuses.

480 B.C.E. In Chapter VIII of *The Histories*, the Greek traveler and chronicler Herodotus described an outbreak of dysentery among the 800,000 Persian soldiers of Xerxes after his loss to the Greeks at the battle of Salamis, history's first decisive sea battle. In a forced march the way they had come through Thessaly to the original embarkation point across the Hellespont, the 300,000 Persian survivors lived off the land, devouring grain, grass, and the bark and leaves of trees. Along the evacuation route through Paeonia and Macedonia, Xerxes left the dead and dying in the care of local people. Upon arrival at Abydos on the northwestern tip of Turkey, his men ate and drank too greedily and suffered more deaths before the Persians could reach Sardis to the southeast.

452 B.C.E. Dionysius of Halicarnassus, a Greek chronicler of the first century B.C.E., described an epidemic that killed most of Rome's slaves and half the citizens. The weakened populace failed to keep up with cremation of the dead and jettisoned bodies into the sewer and into the Tiber River. Healthy Romans sickened from the odor and from drinking polluted water from the river, which also spread contagion to nearby shepherds. The Romans propitiated their gods to no avail.

When the nearby Aequians took advantage of the situation to menace Rome, the city's ambassadors petitioned allies to fend off an attack. However, neighboring tribes also contracted the disease as far away as the Sabines and Volsci. Because harvests went unreaped, famine worsened the Romans' condition. Livy, Rome's most prominent historian, spoke of the loss of Rome's leading citizens—consuls Sextus Quintilius and Spurius Furius, as well as senators and tribunes.

431 B.C.E. The first volume of Greek historian Thucydides's six-book *The Peloponnesian War* characterizes the disorganized response of Athenians to an unidentified fever that raged for two years and took one-third of the populace. The disease, which originated in Ethiopia and passed to Piraeus and Athens through Egypt and Libya, weakened soldiers under command of Archidamus and prefaced the Athenian loss to Sparta. Thucydides summarized symptoms—cough, fetid breath, fever, hoarseness, inflamed eyes, sneezing, sore throat, and vomiting followed by convulsions, an outbreak of pustules, exhaustion, and severe diarrhea.

Nothing seemed to stall the epidemic. Of the patients of frustrated caregivers, Thucydides said:

> Some died from want of care, but so did others who were receiving the greatest attention. No single remedy was established as a specific; for what did good to one did harm to another.... More often the sick and the dying were tended by the pitying care of those who had recovered, because they knew the course of the disease and were themselves free from apprehension [Thucydides 1963, p. 75].

He reported speculation that the disease may have resulted from well water poisoned by Spartan assassins, but left the final judgment on the source of contagion to more learned professionals.

The epidemic left Athens in pitiable condition. Without adequate burial parties, the dead piled up in the streets and in fountains and temples. The scourge was so loathsome that scavenger birds and animals shunned diseased corpses. Those carrion-eaters that even touched the rotting flesh died immediately.

In terms of mores, Thucydides made observations about human nature in plague time. In the midst of hysteria, he found Athenians spending their savings haphazardly enjoying temporal pleasures that death might soon deprive them of. His despairing countrymen flouted law, respect for the gods, and personal honor. People ceased supplicating deities and stopped consulting oracles. They abandoned traditional burial ritual and interred corpses where they could, even in the gravesites belonging to others. Similar discourtesy to corpses on crematory fires resulted in the indiscriminate heaping of other dead on the pyre.

Thucydides was the first chronicler to recognize that those who survived outbreaks of the disease remained immune for life. He observed that some survivors mistakenly believed that their resistance extended to protection from all contagion. When the epidemic returned at the end of 427 B.C.E., it contributed to the downfall of Athens. Medical historians debated the identity of the Thucydidean pestilence, considering bubonic plague, smallpox, typhoid fever, and typhus. The detailed list of symptoms also strongly allies with the course of measles in adults. Other possibilities include ergotism from grain blighted with the fungus *Claviceps purpurea*. Because the traditional deities had failed to halt the catastrophe, in 420 B.C.E., the Greeks initiated the cult of Aesculapius, god of healing.

429 B.C.E. After Pericles dispatched a naval convoy to capture a Spartan post at Potidaea (south of modern Thessaloniki), so many Athenian sailors fell ill on board the ship that the flotilla turned back. Historians have debated the cause of the outbreak, considering bubonic plague, scarlet fever, smallpox, and typhus.

405 B.C.E. During the Punic Wars, when the Carthaginian general Himilco marched on Syracusa in Sicily, to the west at Acragas, his army contracted a devastating disease, possibly measles. In the words of the historian Livy:

> The nursing of the sick and contact with them spread the disease, so that either those who had caught it died neglected and abandoned, or else they carried off with them those who were waiting on them and nursing them, and who had thus become infected. Deaths and funerals were a daily spectacle; on all sides, day and night, were heard the wailings for the dead ["Ancient Medicine"].

The scourge killed Himilco's cousin, leaving the general in charge. He had no choice but to seek peace with Dionysius the Elder, leader of the Syracusans. The disease followed the soldiers home and spread among Carthaginian citizens and their neighbors.

Another outbreak in 397 B.C.E. brought a subsequent military disaster on Himilco's forces as they again assaulted Syracusa. Contributing to contagion was hot, dry weather and the crowding of the insurgents in a steamy fen. Dionysius the Elder, who successfully held off the siege, blamed the impious Carthaginians for plundering temples and incurring the gods' punishment in the form of disease.

winter ca. 400 B.C.E. In Book I of *The Epidemics* (ca. 310 B.C.E.), Hippocrates, the father of medicine, or one of his interns described *phthisis* (tuberculosis) as a scourge of young adults. He thought it "the most considerable of the diseases which then prevailed, and the only one which proved fatal to many persons" (Daniel 1997, p. 18). The text warned of fever, persistent sweating of the upper torso, a hard cough to bring up thick phlegm, and a general wasting of the body's energies. The philosopher Plato, Hippocrates's contemporary, considered TB victims hopeless cases whom doctors and the state should not attempt to treat.

Book VI recorded "the cough of Perinthus," an event datable to winter around 400 B.C.E. Among the Thracians of Perinthus on the Sea of Marmora, the serious coughing disease brought on pneumonia and other complications. Historians surmise from the symptoms of paralyzed palate and angina that Hippocrates witnessed an epidemic of diphtheria, influenza, or pertussis.

365 B.C.E. As described by Paulus Orosius of Braga, Spain, the first Christian historian and author of the seven-volume *Historiarum Adversus Paganos* (History against the Pagans) (417), a wasting disease attacked Rome and struck in waves over a long period. The sickness reached along the Rhine to Gallic and German tribes, weakening the northern resistance of Roman imperialism.

334 B.C.E. When malaria moved north from Ethiopia and Egypt, it assailed Greece. Alexander the Great used the epidemic to his advantage and conquered the area. The disease turned on his Macedonian forces along their route toward India and may have been the cause of his own serious illness at Tarsus in September 333 B.C.E. Exacerbating his ill health were frequent late-night bouts of drinking. He died of protracted fever and abdominal pain in central Babylon (modern Iraq) on June 10, 323 B.C.E.

ca. 310 B.C.E. In *The Epidemics* (ca. 310 B.C.E.), Hippocrates, prominent diagnostician and the father of classical medical arts, first described outbreaks of mumps at the island of Thasos south of Macedonia as a broad swelling about the ears that sometimes inflamed the testicles. In describing a hemorrhagic fever, he outlined unusual weather conditions. Medical historians have posed bubonic plague, relapsing fever, and typhus as the source of the outbreak.

300 B.C.E. An epidemic of malaria at the Greek colony of Paestum in Campania south of Naples, Italy, derived from stagnant irrigation pools that formed bogs. Because of the region's bad reputation for *mal aria* (bad air), travelers and settlers avoided it. Abandoned to nature, Paestum receded into oblivion until highway builders rediscovered it in 1752.

ca. 300 B.C.E. An Italian urn housed in Paris at the Louvre depicts an adult male leaning on a stick. His misshapen legs bear evidence of polio.

292 B.C.E. As characterized by Livy, foremost Roman historian, an epidemic swept Rome and its environs. On priestly advice, the people sought the statue of Aesculapius housed at Epidaurus. The envoys assumed that a snake found on their ship embodied the healing god's spirit. Ovid, author of the *Metamorphoses* (8 B.C.E.), described how the snake climbed the mast as though looking for a resting place. When the snake left the ship on an island in the Tiber River and restored the city to health, the grateful Romans raised a temple and cult to Aesculapius.

fall 218 B.C.E. When Hannibal invaded Rome, he led his Carthaginians across the Alps along with Balearic slingers, Gallic spear wielders, Iberian heavy infantry, Libyan foot soldiers, Numidian light cavalry, and African javelin throwers perched atop elephants. The massed force carried malaria into Rome's outskirts, producing endemic disease among rural Italians. In terror of the scourge, so many outsiders trekked to Rome to propitiate Febris, the fever goddess, that they required the building of two additional worship centers.

ca. 250 B.C.E. Yüeh Ling, author of *Monthly Ordinances*, spoke of *chiai nio* (tuberculosis), which caused patients to cough up blood. Chinese doctors treated consumptives with moxibustion, the burning of mugwort leaves on sensitive places on the patient's body.

212 B.C.E. According to the historian Livy, during the Second Punic War, the Roman army encountered a serious outbreak, probably influenza, as soldiers assaulted the Carthaginian stronghold at Syracusa, Sicily. Both sides suffered the loss of foot soldiers and officers. Poet Tiberius Catius Silius Italicus, author of the 17-book *Punica* (ca. 70), characterized the onset of sickness. General Marcus Claudius Marcellus evaded the pestilence by marching his men away from mosquito-infested lowlands and into Syracusa, which fell to Rome.

ca. 176 B.C.E. Shunyu Yi, a noted Chinese doctor, examined maidservants of royalty for disease. In one, he detected *shang phi* (consumption), a disease that killed her within six months.

62 B.C.E. On return from Asia Minor, the legions of Pompey the Great carried leprosy from the realm of Mithridates (modern northern Turkey) to Rome, the first time that Europeans had encountered the scourge. Roman occupational troops in Gaul, Spain, Germany, and Britain spread the infection, which they called *elephas*. It later acquired the Greek designation *lepros* (scaly). It wasn't until 30 that Aulus Cornelius Celsus, a writer of Alexandrian medicine, described the disease in detail.

27 B.C.E. An observer of Rome's endemic malaria, poet Marcus Terentius Varro, author of *Rerum Rusticarum* (Country Matters) (27 B.C.E.), warned readers to avoid building on a river bank or marshy ground, which spread a miasma to eyes, nose, and mouth that endangered life.

ca. 15 C.E. The Greek geographer and historian Strabo of Pontus and later of Rome wrote about disease in his encyclopedic *Geographica* (ca. 20). He correctly surmised that plague struck Greece during an increase in rodent populations. Over the first century, the pandemic spread east and south to Egypt, Libya, and Syria.

ca. 95 Philostratus's *Life of Apollonius of Tyana* (ca. 190) recorded that plague assailed Ephesus about the time of the Emperor Domitian's death, and the Ephesians appealed to the Turkish magician Apollonius for help. He summoned citizens to the theater before a statue of Apollo and chose a wretched beggar as a public sacrifice. Philostratus claimed that, after the people stoned the old man to death, they removed the stones and found that he had disappeared, leaving a rabid hunting hound crushed to death in his stead.

100 Pottery from Peru and Ecuador indicated the presence of facial ulceration, scarring, and malformation of the mucous membranes among the pre–Inca. The cause was epidemic leishmaniasis, a protozoan skin disorder carried by female phlebotomine sandflies. In more severe form, it struck the liver, lymphatic system, and spleen, causing lesions, anemia, and death.

100 Alexandrian anatomist Rufus of Ephesus, a Greek-speaking physician during the reign of the Roman Emperor Trajan from 98 to 117, summarized elements of bubonic plague that swept over Egypt, Libya, and Syria. As preserved in the 70-book *Collections of Oribasius Sardianus* (ca. 365), compiled by the Byzantine encyclopedist who treated the Emperor Julian, Rufus established a name for accurate disease pathologies and produced numerous treatises, 58 of which were translated into Arabic.

110 A Greek clinician, Areteus (or Aretaeus) the Cappadocian, who practiced in Alexandria, apparently had many tubercular patients. From experience, he wrote that the people most likely to contract the disease were thin and narrow-chested with prominent throats. He remarked that near the end of their suffering, patients hemorrhaged from the lungs, but felt no pain.

In *On Cholera* (ca. 110), Areteus spoke of sweating, weak pulse, a collection of matter in the stomach, and propulsive vomiting coinciding with a discharge of fetid excrement. He added that patients died piteously of spasm, dry heaves, and suffocation. In *On Ulcerations about the Tonsils* (ca. 110), he characterized the fever, inflamed neck, and loose teeth in victims of diphtheria.

160 The Han Empire collapsed after northern insurgents carried "barbarian boils," or bubonic plague, into China.

165 In the fifth year of the Parthian War, a disease, possibly smallpox, struck the soldiers of the Roman general Avidius Cassius while they attacked Syria at Seleucia (central Iraq) on the Tigris River. When legions returned from a campaign against Armenia in Parthia (modern Iran), they infected Rome. The next year, the pestilence, called the Antonine Plague or the Plague of Galen, weakened Rome at the rate of 10,000 infections per day. The disease killed up to 2,000 daily, reaching a total estimated from 3.5 million to seven million, thus contributing significantly to the decline of the empire.

The scourge spread over southern Europe, up

the Rhine River, and into Asia Minor. In spring 169, at Altinum in northeastern Italy, the contagion killed the Emperor Lucius Verus, co-ruler with philosopher-emperor Marcus Aurelius. The epidemic gave Goth and Vandal insurgents an opportunity to threaten Aquileia to the northeast. The epidemic emerged at Venetia (Venice) in 166. To save the empire, Marcus Aurelius summoned the Greek doctor and medical writer Galen of Pergamum, author of *De Alimentorum Facultatibus (Concerning the Workings of Digestion)* (ca. 180), and made him court physician. To supplement the sickly army, the emperor sold personal goods to pay gladiators, slaves, and German and Scythian mercenaries to defend Rome. By rewarding his alien soldiers with land, he initiated a weakening of Roman control of the empire and a demand from outsiders for Roman citizenship. On March 17, 180, the emperor himself succumbed to the disease at his military headquarters in Vienna, ending a golden age and leaving in charge his inept son Commodus.

251 An epidemic, possibly measles, called the Plague of Cyprian, bore the name of Saint Cyprian, bishop of Carthage, the first recorder of symptoms. The outbreak began in Ethiopia and radiated north to Egypt's Nile Delta, where it killed 67 percent of Alexandria's populace. The pestilence dogged the Roman Empire for nearly two decades, killing up to 5,000 daily. In his sermon *De Mortalitate* (On Mortality) (252), Cyprian mused:

> This mortality is a bane to the Jews and pagans and enemies of Christ, to the servants of God it is a salutary departure. As to the fact that without any discrimination in the human race the just are dying with the unjust, it is not for you to think that the destruction is a common one for both the evil and the good [McNeill 1976, p. 136].

The decline in the work force reduced the army, weakened the economy, and led to the debasement of coinage from a lack of laborers to work the silver mines.

For fifteen years, the disease killed Romans, including the Emperor Claudius Gothicus in 270. To flesh out the legions, Roman officers enlisted mercenaries. The government put the poor to work building arches, baths, temples, and theaters. Bolstered by Cyprian's sermon, the church used public panic as an opportunity to convert huge numbers of pagans to Christianity by prom-

ising life after death. The response of grieving survivors to the disease may have established black as the color of mourning.

310 Maximinus II, the Roman viceroy of Egypt and Syria, faced famine and plague during a drought compounded by war with Armenia. As church historian Eusebius of Caesarea, author of the ten-volume *Ecclesiastical History* (325), explained, the wasting disease forced people to beg for food and fight with stray dogs for scraps. Some families mourned multiple members at a single funeral. The pestilence contributed to the number of people abandoning paganism and turning to Christianity.

312 Following bouts with famine and locusts, the people of central and northern China suffered an epidemic that left their country a wasteland. The 100 taxpayers of Sheni were reduced to only two.

ca. 320 Chinese medical writer Ho Kung characterized an epidemic causing pustules filled with yellowish fluid. The devastating skin disorder disfigured survivors with purplish scars that remained discolored for a year. Historians deduce that he was describing smallpox or perhaps measles.

350 At Edessa in northern Greece, Saint Ephrem the Syrian, an influential Jesuit preacher and hymn writer, distributed food, organized an ambulance service, and supervised a 300-bed community hospital treating victims of plague. When famine and contagion worsened, he exhausted his strength, sickened, and died on June 9, 363.

370 Saint Basil the Great of Caesarea south of Haifa, Israel, the father of Eastern monasticism, became the first caregiver to establish a treatment center for lepers. He founded a hospice, the earliest of Christian lazarettos, on the Mediterranean coast of Palestine and received patients who survived as wandering outcasts from society. His contemporary, Zodicus of Constantinople, contracted the disease and built a leprosarium that received all classes of sufferers, even Constantina, the daughter of the emperor Constantius.

390 At Porto, a harbor town west of Rome, Fabiola, a wealthy Roman matron, originated the concept of the public hospice in the Western world. She sheltered sick pilgrims and the poor and outcast ailing from life-threatening diseases. To assure the safety of patients, she traveled the roads, retrieving those who fell by the way. During the Crusades, the concept of the hospice spread across the Eastern Mediterranean.

400s On long sea voyages, the Chinese battled seasickness and scurvy among ship's crews by supplementing their diet with fresh ginger, which they grew in shipboard pots.

410 When Alaric led the Visigoths against Rome, the bogs of Campania harbored malaria-carrying insects, which caused his sudden death as his forces prepared to invade Sicily. Medical historians surmise that a deadly strain of the disease may have contributed to the fall of Imperial Rome. The atmosphere worsened for local people as the municipal machinery collapsed, leaving foul ditches undrained and fountains and waterways stagnant.

432 At Armagh, Saint Patrick, Ireland's patron saint and minister to the poor, opened his home to itinerant lepers. He amazed local people by embracing lepers and personally baptized the sick and treated their lesions.

446 As described by theologian and chronicler the Venerable Bede in *Historia Ecclesiastica Gentis Anglorum* (*Ecclesiastical History of the English People*) (713), the Saxons of England experienced a lull in constant wars with the Celts and Picts and grew lazy and wasteful. A plague struck, bringing what Bede interpreted as God's judgment against sinful, luxury-loving backsliders.

452 After smallpox engulfed Provence, France, Bishop Nicaise (or Nicasius), the builder of Rheims Cathedral, became the patron saint of the diseased and protector of the afflicted. He recovered from the pox by rubbing holy oil on his skin.

470 In Ireland, Saint Bridget of Kildare opened the first of three convents to wandering lepers. Men and women who admired her benevolence joined her order. A contemporary, Saint Scholastica, Saint Benedict's twin, performed the same service for plague victims in Monte Cassino, Italy, where she educated nuns in nurse care.

481 In France, Anne of Clovis initiated touch ceremonies for the king's evil or scrofula, a serious tubercular condition of the glands of the neck and throat. Her pious gestures and prayers for the sick were thought to bring comfort or a complete cure.

ca. 500 Cholera began to reach epidemic numbers in Asia, the Middle East, and the Eastern Mediterranean.

537 According to the *Annales Cambriae* (*Annals of Wales*) (954), after the battle of Camlann, in which Mordred mortally wounded King Arthur, a plague struck Britain and Ireland.

541 The first recorded outbreak of bubonic plague was the world's first great pandemic. Called the *Mortalitas Magna* (*Great Death*) or the Plague of Justinian, it began in Arabia and Pelusium on the Nile delta and inundated Syria, Persia, and Palestine with sudden outbreaks of fever followed by collapse, emergence of buboes, delirium, vomiting of blood, and death. The pestilence reached Constantinople by the spring of 542 and lurked around the eastern Mediterranean until the 760s. *See also* 1320.

The historian and city prefect Procopius of Caesarea, an eyewitness to the four-month catastrophe, estimated that disease killed up to ten thousand daily and wiped out 40 percent of the capital city before devastating Alexandria, Palestine, and Syria. In all, the pestilence destroyed up to one quarter of the population of the eastern Mediterranean. The Emperor Justinian, who survived infection, rewarded with food and cash the heroes of the epidemic, those immune citizens who removed and disposed of bodies. While the military and commerce came to a standstill, the burden of so many corpses forced health authorities to push some out to sea on barges and to wall up others in towers. The strain on the emperor and his treasury ended his plans to reinvigorate the Roman Empire.

543 The first outbreak of bubonic plague reached France, according to Bishop Gregory of Tours, author of the ten-volume *Historia Francorum* (*History of the Franks*) (ca. 591). During a serious outbreak at Arles, an angel appeared promising that Bishop Gall of Clermont would survive for eight years.

547 Plague spread from Egypt to the British Isles after 541. The chronicler Saint Gildas, son of a chief who became a monk at Clyde, Wales, described the first plague epidemic in Britain in *De Excido et Conquestu Britanniae* (*On the Ruin* *and Conquest of the Britains*) (ca. 560). The pestilence killed Saint Finnian of Clonard and also Maelgwn, king of Gwynedd, after he fled contagion and locked himself into the chapel at Rhos. According to the *Annales Cambriae* (*Annals of Wales*) (954), "Thus they say 'the long sleep of Maelgwn in the court of Rhos.' Then was the yellow plague" (*Annales Cambriae*). As described in Ruaidhri O'Luinin's *Annals of Ulster* (1489), the *Cron Chonaill* (yellow pestilence) that struck Ireland in 548 may have been relapsing fever.

552 The arrival of the first Buddhist missionary from Korea brought measles and smallpox to Japan, a virgin-soil population.

565 A disciple of Saint Patrick, the charismatic healer Saint Columba of Gartan, Ireland, set up religious stations at Kell, Derry, and Durrow. Banished to the Scottish island of Iona, he built a leprosarium, where he spooned *aquavit* and *usquebeatha*, the original Irish whisky, into patients as stimulants.

566 As described in the *De Gestis Langobardorum* (*History of the Lombards*) (ca. 770) of Paul the Deacon, a Benedictine fabulist and historian during Charlemagne's reign, bubonic plague assailed Liguria in southwestern Europe. As far away as Bavaria, the depleted region fell silent as whole families died, leaving livestock and fowl untended and crops unharvested.

570 The Koran cites a miracle, the attack at Yemen and Mecca by Christian Ethiopian forces mounted on elephants. To protect the holy Ka'abah shrine from destruction, Allah dispatched birds, which pelted the insurgents with rocks. As explained in the legend, each stone bore the name of the victim it would hit. The blows, which could pierce helmets, created pestilential pustules that killed the Ethiopian Christian emperor, 'Abraha (or 'Abreha) al-Ashram, during his hasty retreat to Sana'a, Yemen. Medical historians interpreted the story of the Elephant War Epidemic as a fanciful description of measles or smallpox, which the Ethiopian chronicler El Hameesy corroborated.

571 Bishop Gregory of Tours, author of the ten-volume *Historia Francorum* (*History of the Franks*) (ca. 591), described an unspecified plague in Auvergne, Bourges, Chalon, Dijon, and Lyons. It was so widespread that it forced survivors to inter corpses in plague pits. The hero of the epidemic,

Father Cato, remained on duty to comfort the sick, say mass, and bury the dead until his own death. Another victim, Bishop Cautinus, returned after fleeing the region and died the Friday before Easter.

August 580 Bishop Gregory of Tours, author of *Historia Francorum* (*History of the Franks*) (ca. 591), characterized the onset of dysentery in France during a civil war. The disease struck King Chilperic I and his young son as well as Prince Chlodobert. Queen Fredegunda, pretending to avenge her sons' deaths, plotted the assassination of her husband's children by his first wife, Audovera, and also arranged the murder of Chilperic's brother Sigebert. When the epidemic felled Queen Austrechild, she swore her husband, King Guntram of Burgundy, to a vengeful pact to assassinate the doctors who failed to save her.

583 In response to the number of lepers living among healthy citizens, the Council of Lyon, the first Catholic convention held in France, condemned those with leprous lesions to remain apart from the rest of society on pain of death.

585 A scourge, probably smallpox, arrived in Japan from Korea, afflicting a majority of the citizenry. During the rule of the Emperor Bidatsu, the outbreak coincided with the Soga clan's expansion of Buddhism from Korea through ritual, design, literature, and sculpture. Because of the disease, Japanese skeptics believed that the kami, Shinto divinities, sickened them because they had embraced a new faith. Rioters stripped and flogged clerics and nuns, torched the Sogas' holy dwellings, and tossed statues of Buddha into a moat.

In the aftermath, during an upsurge of religiosity among the Japanese, the emperor had the temples replaced. Shortly before he died of smallpox, he halted the proselytizing of Shintoists to the new faith. After the death of Bidatsu's son and heir Yomei as well as numerous clan leaders, they were succeeded by the emperor's 11-year-old grandson, Crown Prince Shotoku, the eventual co-regent with his mother, the Empress Suiko. A promoter of Confucianism and Buddhism, he became a devotee of Buddha and, in the city of Nara, ordered the casting of the Shakyamuni Buddha and enshrined it in the Gango-ji temple.

588 According to Bishop Gregory of Tours, author of *Historia Francorum* (*History of the*

Franks) (ca. 591), the plague struck Marseilles after a ship brought the disease from Spain. People who purchased goods from the traders contracted the disease, which struck rapidly and spread throughout the harbor town. Of the king, Gregory reported in Chapter 9:

> Like some good bishop providing the remedies by which the wounds of a common sinner might be healed, King Guntram ordered the entire people to assemble in church, and Rogations to be celebrated there with great devotion. He then commanded that they should eat and drink nothing else but barley bread and pure water, and that all should be regular in keeping the vigils. His orders were obeyed. For three days his own alms were greater than usual, and he seemed so anxious about all his people that he might well have been taken for one of our Lord's bishops rather than for a king ["Britain and Europe"].

Meanwhile, Bishop Theodore set a poor example of cowardice by immuring himself in prayer at the church of the Holy Victor during the two-month siege. After a lull, the disease resumed its virulence and spread north up the Rhone River to Avignon and Viviers, followed a year later with an outbreak at Nantes.

589–590 Following heavy flooding of the Tiber River, a resurgence of plague at Verona, Ravenna, and Rome killed Pope Pelagius II on February 7, 590. The pestilence yielded about the time that his successor, Pope Gregory the Great or Saint Gregory I, ordered a seven-part litany and headed a prayerful procession of the devout through the streets of Rome. The people credited a heavily armed Archangel Michael at the head of the entourage with their salvation. The raising of a statue to his intervention on Hadrian's Tomb, now called the Castel Sant' Angelo, created a lasting and highly visible monument to Christian miracles.

590 Ergotism caused an outbreak of Saint Anthony's Fire in France. It may have derived from eating bread made of rye infested with the ergot fungus. Called "mad grain," this major food source could produce a burning sensation, numbness, convulsions, hallucinations, psychosis, gangrene, and death.

599 A new cycle of bubonic plague in Italy and Provence left 85 percent of the population alive, but reeling from demographic and economic loss as merchants abandoned trade routes. Barbarian invaders used this period of military weakness as an opportunity for raids. When fanatic Islamic hordes rushed into Persia and Byzantium to take advantage of governmental disruption, the Eastern Roman Empire called home forces stationed in the western end of the Mediterranean to strengthen the home guard. As the agrarian populace withdrew from the Negev, irrigation networks went to ruin, weeds reclaimed orchards and fields, and the desert spread over once arable land.

610 Bubonic plague appears to have emerged in China at Kwangtung northwest of Hong Kong. Because infection was rare farther inland, historians deduced that it arrived from the coast, perhaps from contact with merchant vessels from the Mediterranean.

616 Khosrau II, the tyrant of Persia, brought smallpox to Jerusalem during a massacre of Christians. As he pursued his conquests in North Africa, he spread the scourge to Egypt.

639 When plague struck 'Amwas-Emmaus in Syria northwest of Jerusalem, the name "Bir et-Ta'un" (*The Well of the Plague*) commemorated a terrifying illness that followed a devastating famine. Residents still recovering from hunger fled from a pestilence they interpreted as a sign of judgment day. According to legend, Caliph 'Umar ibn al-Khattab, ruler of 'Amwas, arrived at the edge of the city when news of epidemic reached him. He pondered a Koranic injunction to stay in a city during an epidemic, but to avoid entering a city beset by disease, a wise proverb that presaged the Renaissance concept of quarantine.

Avoidance of 'Amwas determined its destiny. As the plague overran the Islamic enclaves of Syria and Palestine, it killed General Abu Ubaidah, 25,000 of his soldiers, and Yazid Abu Sufyan, governor of Syria. Palestinian authorities chose a new headquarters at Ramleh or Ramla, which served as a stronghold and Arab capital of Israel for 300 years. The city of 'Amwas remained neglected and barren until the arrival of the Crusaders in June 1099, when they built a fort and restored an abandoned church on their last stop on the march to Jerusalem.

644 In the eighth year of his reign, King Rothari of Lombardy, a noted lawgiver, seized all lepers and imprisoned them in an abandoned structure outside of Milan. To survive, the sick depended on parcels of food and supplies delivered by relatives and friends.

647 With the arrival of diseased Arab Muslims, smallpox struck the Phoenician colony of Tripoli (modern Libya) and advanced as the invaders marched across the North African coast.

653 According to an addendum to the fourth-century writings of Chinese physician Ho Kung, a disfiguring disease, possibly smallpox or measles, invaded east across China. One remedy involved mixing boiled mallow with garlic as a dressing for rice. The medical historian explained that, because the sickness emerged when Chinese soldiers fought insurgents at Nan-yang northwest of Shanghai, the disease was called the barbarian pox.

664 A wave of pandemic bubonic plague swept over Europe for two decades. In his *Ecclesiastical History* (713), the Venerable Bede, a Saxon historian at Jarrow in northeastern England, described the advance of disease from southern Britain to Northumbria and west to Ireland, where the *Liber Landavensis* (*Llandaff Register*) (1132) incorrectly identified it as *pestis flava* (yellow plague), another name for relapsing fever.

680 When plague spread from Rome over Italy, according to Paul the Deacon's *Historia Romana* (*Roman History*) (ca. 770), peasants began seeing miraculous signs and wonders. Survivors venerated Saint Sebastian, a third-century martyr of Roman militarism who guarded them against disease. The origin of the cult was the removal of Sebastian's remains from Rome to the church of San Pietro in Vincoli at Pavia. During the holy procession, the epidemic abated.

683 Saint Giles (or Saint Aegidius) of Athens, the lame patron of cripples, built a hermitage on the Rhone River near Nimes, France, and treated victims of endemic leprosy.

ca. 700 A Nazca boy around ten years of age died of tuberculosis of the spine and major organs. His mummy, located at Hacienda Agua Salada, Peru, attested to the presence of the disease among indigenous people. Paleopathologists determined that the child required a special seat because of paralysis and wizening of his lower limbs. Earlier proofs of the disease in Central America exist at the Museo Popol-Vuh in Guatemala from around 600 B.C.E. to 300 B.C.E.

700s The *tokusei* (virtuous acts of government), decrees issued by Japanese sovereigns, averted human calamities such as disease. Rulers issued the benevolent decrees canceling debt or granting amnesty to increase their reputation for virtue, to end famine or epidemic, or to cure sickness in the imperial house.

710 Smallpox reached Spain after Tariq ibn Ziyad, the Muslim governor of Tangier, and his Moors attacked the Goths at Gibraltar, killing King Roderick, the last Visigoth king, in the first year of his reign. The success of Saracen forces carried their campaign and smallpox north to the Pyrenees.

732 When Saracen troops fought their way to Tours, Charles Martel, the romanticized "hammerer" and founder of the Carolingian dynasty, defeated them, but did not stop the advance of measles and smallpox in France.

735 A three-year bout of smallpox migrated from China and Korea to feudal Japan during the reign of the Emperor Shomu (or Seimu Tienno), when typhoons, earthquakes, and drought worsened conditions for peasants. Hearsay blamed the outbreak on one capsized fisherman who washed up in Korea, where he contracted the disease. As chronicled in *Ishinho* (982), a 30-volume medical history compiled from Chinese texts by the Buddhist physician Yasuhori Tanbo (or Tamhu), Chinese authorities quarantined victims to hospital wards. Shamans performed ceremonies before the epidemic god and distributed *majinai* (magic charms) inscribed on *mokkan* (wooden tablets), wooden dolls to represent patients, and clay pots painted with human faces as means of dispelling contagion. A third of the Japanese died, including a half million at Nara. Among the dead were Muchimaro, Fusasaki, Umakai, and Maro, the four brothers of the ruling Fujiwara family. The imperial government lacked laborers and reconsidered exemptions for Buddhist priests.

A religious fervor fueled greater devotion to Buddhism, for whom devotees, whole villages, and pilgrims offered up *gankake* (or *gandate*), prayers or petitions to a Shinto or Buddhist deity to beg protection from contagion. To stave off further epidemics, the Emperor Shomu propitiated Buddha to end the epidemic by erecting pagodas, an abbey, and nunneries in each province. At the imperial palace, he built the Buddhist

Todaiji temple, the grandest religious project of the Nara period, containing a gilded bronze likeness of the Vairochana Buddha or Great Buddha along with display areas for valued paintings, glassware, jewelry, textiles, and musical instruments from across Asia. Outside the building, he positioned the Shosoin (or Seisoyin), a museum of medicinal herbs. He also constructed the Kokubunji temple complex, comprised of gate, main hall, lecture hall, scriptorium, belfry, monastery, and seven-storied pagoda.

746 The emergence of bubonic plague at Constantinople, derived from Syria, according to the *Chronographia* (ca. 810) of the monk Theophanes, but passed over shipping routes across North Africa to Sicily, Italy, and Greece. It arrived last to the Hellespont by a meandering port-to-port route because the Arabs had cut off overland trade to the area. The sickness devastated locals, causing them to speculate on cruciform shapes that appeared on personal garments and church vestments. The dying imagined that murderers stalked their houses; burial squads resorted to digging graves in any open field, grove, vineyard, or empty reservoir. To restore the traditions and ideals of the Hellenic realm, Constantine V imported newcomers from Greece and the Aegean Isles to settle vacant plots of farmland and transferred his commercial center from the Golden Horn to the Marmara coast.

mid–700s St. Walpurga (or Walburga) of Heidenheim, a Wessex princess from Devonshire, England, established a six-bed hospice at the double monastery at Hildesheim in southern Bavaria to minister to plague victims. By her example, nursing became a Christian priority.

751 North of Paris, Pepin the Short, king of the Franks, remanded lepers to areas apart from healthy society. He also prohibited marriage for lepers. The mates of lepers were allowed to seek divorce.

762 The first of a series of coastal outbreaks of epidemic disease struck China and killed half of the citizenry of Shantung on the eastern coast.

790 A smallpox outbreak that began in China proved so deadly to Nara, Japan, that it afflicted all citizens below the age of 30 and caused a high rate of death. In 814 a re-emergence of the disease spread from China or Korea via the harbor of Dazaifu in the south, killing 50 percent of the populace.

850 Yaws is evident in remains of early inhabitants of the Mariana Islands in the Pacific.

857 According to the anonymous author of the *Annales Xantenses* of 847, a major outbreak of Saint Anthony's fire or ergotism in Germany, the first documented epidemic, struck Rhineland peasants. Brought on by spoiled grain, the scourge killed thousands from gangrenous ergotism to the extremities. The onset began with swollen blisters and advanced to corrupt flesh, loss of limbs, convulsions, and death.

869 Japan's Gion (also Gion Goryoe or Gion'e) festival honored Gavagriva, a health deity and guardian of the Jetavana monastery in India on the border of Nepal. The festival originated during an epidemic that struck Kyoto in summer following the annual spring flooding. To propitiate the gods, devotees erected 66 *hoko* (spears) symbolizing 66 provinces and prayed for deliverance. The ritual, such as that sponsored by Kyoto's Yasaka Shrine, featured costumes and parade floats to house deities on earth. The celebration remained standard until the Onin War of 1467. In 1600 during the Edo period, a merchant guild revived the custom.

900 Late Woodland Indians of the Great Lakes, St. Lawrence River Valley, Appalachias, and Mississippi River Valley suffered endemic blastomycosis, a fungal infection that targeted teens and young adults with vertebral lesions. The bone and lung disease flourished as tribes shifted from hunting and gathering to cultivation of gardens in acidic soil.

915 During a golden age, epidemic smallpox in Japan provoked terror after it struck the Emperor Daigo, a progressive ruler and beloved arts patron. To contain the disease, he freed his subjects from taxes, forced labor, and unfulfilled pledges. The devout called for prayer and cleansing rituals. He survived 15 more years and died on October 23, 930.

925 An eminent Islamic doctor and chemist, Rhazes of Baghdad, Iraq, studied contagion from the Greek texts of Hippocrates, Thucydides, and Galen. Rhazes produced an original monograph, *De Variolis et Morbillis Commentarius* (*On Smallpox and Measles*) (910), the first authoritative description of two persistent plagues that targeted children, particularly males. His observation of pale flesh, edema, sweating, nosebleed, and reddened eyes accompanied description of pustules. He followed with a more detailed clinical handbook, *The Compendium* (ca. 925), which summarized patient histories that confirmed his diagnoses.

936 King Athelstane, England's first *Rex Totius Britanniae* (*King of All Britain*), founded a leprosarium at York. The hospice offered food, shelter, and palliatives, but promised no cures.

947 Smallpox returned to Japan, afflicting the Emperor Murakami in the first year of his reign. It also infected his predecessor, Suzaku, who abdicated in 946 and died on September 6, 952. At provincial temples, religious officials offered public prayers for the emperors in the Shinto tradition. Murakami lived another 20 years and died on July 5, 967.

951 Kuya (or Koya), a son of royalty and charismatic Buddhist mendicant monk, earned the popular name *Ichi no Shonin* or *Ichi no Hijiri* (*Saint of the Marketplace*). To end an epidemic in Kyoto, Japan, he erected an image of the Eleven-Headed Kannon at Saikoji in Nagano and introduced the *nenbutsu* folk dances, which involved prayers and the sounding of gongs and drums. The sacred performance also served to bring rain during a drought. Popular wisdom called for *umeboshi* (pickled plums) eaten with tea as a palliative for sickness.

974 According to the medical text *Ishinho* (982), by Yasuhori Tanbo, Japanese doctors instituted the first pesthouses and isolation wards during a smallpox epidemic. They also employed red banners in patient rooms and red wrappings on the sick.

994 A year-long pestilence—possibly smallpox—swept Japan, killing over half the population, including many nobles and ministers. Within four years, the country fell under another scourge of measles. The disease did not spare any social class. Its virulence and its return in 1025 to a new generation of sufferers are topics of the *Eiga Monogatari* (*A Tale of Flowering Fortune*) (ca. 1028), a story sequence of aristocratic life begun by a female courtier, Akazome Emon, and completed by another writer.

ca. 1000 King Olaf II of Norway, a noted lawgiver and evangelist, excommunicated all lepers in his realm and exiled them from healthy communities. His hard-handed decree both disenfranchised the sick and deprived them of mates.

ca. 1000 Epidemic tuberculosis claimed lives in the upper midwest of the Mississippi Valley during the rise of the Cahokia state and its sister colonies. Anthropologists documented tubercular lesions on many Oneota skeletons south of Lake Michigan and surmised that for each diseased skeleton, there were around 300 victims of the disease who may have died before damage appeared on bones. No one knows how the disease emerged among Native Americans.

1000 In Africa, Asia, and Europe, pandemic leprosy so decimated the population for the next two centuries that the healthy demonized and banished victims as though they were bearers of evil. *See also* 1106. The First Crusade advanced *Mycobacterium leprae* east to the Byzantine Empire. Monastic orders applied compassion to the unique social ill by building leprosaria and treating patients humanely within quarantined space. Health authorities in Denmark, Sweden, Norway, France, Italy, Germany, and England immured the sick in lazar houses and buried victims in mass graves. In 1220, Archbishop Henry of Dublin fought contagion by erecting a quarantine compound for victims arriving by sea. The disease raged into a pandemic to 1400 before abating, but it did not reach the Western Hemisphere until the arrival of Columbus in 1492. Sixteenth-century European slave traders introduced the bacteria to West Africa.

early 1000s In the time of Arab physician and healer Avicenna of Kazakhstan, author of *Khitab ash-Shifa* (*Healing of the Soul*), a medical text, and *Qanun fi at-tibb* (*The Canon of Medicine*), a classic compendium, Persians deliberately exposed their children to cattle infected with cowpox. The transmission of the mild disease created antibodies that protected the children permanently from smallpox.

1039 During an epidemic of Saint Anthony's fire or ergotism, the philanthropist Gaston de la Valloire founded a hospital at Vienne, France, to receive victims. Greatly affected were nursing infants, children, and invalids, particularly in rural areas where people regularly ate rye bread, the source of the malady.

1059 Philip I of France became the first king to touch people suffering from scrofula, called the king's evil in England or Saint Marcoul's Evil in France. Philip intoned the phrase, "Le roi te touche, Dieu te guérit" ["The king touches thee, God heals thee."] ("Le Roi et l'Etat"). According to the doctrine of the divine right of kings, Philip exuded a holy power that he could direct to the sick. However, his healing power disappeared because of his dissolute lifestyle. To recover the healing touch, he made a pilgrimage to Corbeny in northeastern France to the shrine of Saint Marcoul, who was Abbot of Nanteuil until 552 C.E. Philip resumed healing hordes of sufferers at a church, cloisters, and park, where the Grand Almoner passed a coin to each seeker as a royal token and talisman against illness.

1070 Lanfranc of Pavia, the administrator of the Canterbury see, responded to the growing pandemic of leprosy in England by establishing a leprosarium at Harbledown.

summer 1081 During the holy war between the papacy and the Holy Roman Empire, disease became the unknown quantity in a perpetual shift of the balance of power. As Pope Gregory VII holed up at Castel Sant' Angelo, the tomb of Hadrian that evolved into Rome's impregnable papal stronghold, the Holy Roman Emperor Henry IV laid siege. Imperial troops fell in large numbers from summer heat and a parallel onslaught of dysentery, malaria, and typhoid. Into 1083, Henry continued to strike at the autocratic Gregory, who excommunicated him for secular offenses to the church and for attempting to supplant the pope with antipope Clement III. The Norman warrior Robert Guiscard, who liberated Gregory from self-imprisonment at the city fort, died on July 17, 1085, from plague contracted at the Greek isle of Cephalonia.

August 1083 The first recorded epidemic of typhus occurred near Salerno, Italy, at Trinità della Cava, a Benedictine abbey. Over a period of two months, the monks suffered fever and swellings. A parallel outbreak of symptoms struck Brescia. *See also* 1489.

1095 At isolation centers throughout Europe, the monastic nursing brothers of the Order of Antonines, founded by Gaston de Dauphiné, specialized in the care of victims of ergotism or erysipelas, a streptococcal infection. The Antonians headquartered at the first hospital, erected near the Church of Saint Anthony at Saint-Didier de la Mothe outside Grenoble, France. The true cure of the painful convulsions, tingling, and numbness of ergotism derived from journeying outside the victims' home district and eating grains in other areas that were not infested with the ergot fungus.

Into the late 1100s, travelers afflicted with ergot poisoning (or Saint Anthony's fire) sheltered at a hospice and at Saint-Antoine-de-Viennois, France, where Antonite monks received and cared for them. To sanctify their hostel, they transported the relics of Saint Anthony, a fourth-century Egyptian hermit, to the premises. Poets and artists including 16th-century Dutch artist Hieronymus Bosch associated Saint Anthony with the disease at 360 other reception centers in Central and Western Europe. In 1512–16, the Alsatian-French artist Matthias Grünewald painted an altarpiece at the hospital chapel of an Antonite center in Isenheim, Germany.

In 1797, Pope Pius VI suppressed the order of the Hospitallers of Saint Anthony, primarily because the disease had waned. However, in Sologne, France, ergotism remained a danger until 1821, when the replacement of rye with potatoes rid the diet of the cause of the burning pain. The disease continued to plague Europeans into the time of tsarist Russia, where a high death rate resulted from spoiled grain in the diet.

1096 At Prague, Czechoslovakia, Thaddeus von Hajek (or Hagecius), personal physician to the Emperor Rudolph II of Bohemia, recorded in his chronicle a medieval typhus epidemic, which he identified by head pain. *See also* 1489.

summer 1098 During the First Crusade, an epidemic of typhoid fever overwhelmed Christian soldiers at Antioch, Syria, killing 100,000 during a twelve-week hiatus and wiping out 2,500 German mercenaries. Worsening the situation were hunger, damp, and bitter winter weather, the causes of numerous desertions. After the Christians captured Jerusalem, a second outbreak of malaria, scurvy, or typhoid weakened survivors. Historian William of Tyre, author of the 23-volume *Historia Hieroslymitana* (*History of Jerusalem*) (1184), gave no details of the sickness, which struck at Marra, Syria. Angered at their lot, the men torched the town as a gesture to Raymond of Toulouse of their despair. The next year, the army foundered after over 70 percent of the horses died of an equine epidemic.

1100s At the abbey of Saint Ponziano outside Spoleto, Italy, votive tablets from the crypt describe the ancient cult of Saint Senzia, who protected residents from malaria during the rise of bog waters when the Tessino River flooded.

1100s Excavations in Phoenix, Arizona, at Salt River Hohokam sites produced skeletal remains suggesting serious nutritional deficiencies over several centuries, beginning with the twelfth century. According to paleopathologists, as game and wild foods became scarce during floods, droughts, and crop failures, children at the Grand Canal Ruins, Casa Buena, and La Ciudad suffered rickets and limited growth because of malnutrition in infancy and early childhood. In decline, the Casas Grandes society disintegrated. The Hohokam merged with the O'odham around 1450. Similarly malformed were skeletons from Zuñi societies at Hawikku and nearby Heshotauthla in west central New Mexico, and images of handicapped human figures on hieroglyphs of the Maya, Aztec, and Inca and on pipes made by Adena-Hopewell artisans of the Mississippi and Ohio valleys.

1106 In St. Omer, France, a wealthy altruist named Vumrad volunteered to replace an inconvenient lazar house with a new dwelling on church land farther from the populace to house male lepers. To assure the sick their privacy and to protect the healthy, an official edict declared that the property would have its own priest and church and a burial ground restricted to lepers and their servants.

1117 A year before her death, Queen Matilda or Good Queen Maud, wife of Henry I, followed the example of Saint Giles of Athens, patron of outcasts, by erecting a leprosarium, St. Giles Hospital, that survived outside London for 400 years. The hospital served during a surge in leprosy cases in England. In token of charity, the lepers reciprocated by extending a cup to prisoners as they plodded from Tyburn prison to the gallows.

1147 At the beginning of the Second Crusade, among the pilgrims and troops of the Emperor Louis VII, a pestilence—possibly bubonic plague, dysentery, or typhoid fever—struck at Adalia in southwestern Anatolia, severely crippling Christian forces and infecting local peasants. According to monk-chronicler Odo (or Eudes) of Deuil, royal chaplain and author of the three-volume *La Croisade de Louis VII, Roi de France* (*The Crusade of Louis VII, King of France*) (ca. 1160), Turkish forces easily overcame their attackers and murdered in flight those despairing French infantrymen fleeing the city.

ca. 1150 The Japanese instituted the *Natsu Matsuri* (summer festival), an urban ritual of dancing, fireworks, and parade floats intended to ward off plague and pestilence.

1155 At the battle of Tortona southwest of Milan, Holy Roman Emperor Frederick I Barbarossa on his first expedition to Italy contaminated the Guelph enemy's water supply by tossing in rotting corpses of men and animals. The act was a common bioterrorist strategy for spreading sickness throughout a besieged countryside.

August 2, 1167 At Rome before engaging Pope Alexander III's Sicilian troops, the imperial army of Frederick I Barbarossa incurred huge numbers of incapacitated soldiers from an unidentified disease. The outbreak followed a heavy summer storm that spilled sewage from drains into the open. Historians surmise that the disease was malaria or typhus. As a result of life-threatening sickness, which struck Frederick himself and killed his counselors and officers, the insurgents retreated into Lombardy.

1180 While atoning for the murder of Thomas à Becket, King Henry II of England supported the establishment of a Carthusian abbey in Witham, Somerset, where Saint Hugh of Lincoln cared for lepers.

summer 1188 Giraldus Cambrensis (also Gerald de Barri or Gerallt Cymro), a Welsh chronicler and author of *Itinerarium Cambriae* (*On Itineraries of Wales*) (1191), summarized an outbreak of spontaneous frenzied dance or dance mania that he witnessed in Brecknockshire in south central Wales at the church of Saint Almedha. Some participants tore off their clothes and frolicked naked; others laughed, wept, jerked, and twitched distractedly, stamped their feet, and rolled on the ground. According to some sources, most of the dancers were women. Observers named the epidemic hysteria Saint

Vitus's dance after the patron saint of epileptics. It was a form of tarantism, an episode of frenetic shaking, delusions of saints and angels, and convulsions resulting from the real or imaginary bite of the tarantula. The scientific name, choreomania, derives from the Greek for "dance mania" (Bartholomew, 2000). Alternate diagnostic names, Sydenham's chorea and rheumatic chorea, refer to involuntary spasms in young children.

early 1190 During the Third Crusade, 95 percent of Europe's 100,000 soldiers either died of hunger or plague or deserted to escape certain death. At Antioch in southern Anatolia, German troops suffered the greatest loss. Into 1191 at Acre, an Israeli seaport, pilgrim troops fell to famine and scurvy at the rate of up to 200 per day. Their adversaries, Saladin's Muslim army, were less severely crippled by disease.

Among the dead in January 1190 was Duke Ferdinand III of Swabia. In addition to many soldiers, both Philip II of France and Richard the Lion-Hearted contracted fever. In April 1191, the English flotilla of 25 ships brought enough food to revive bodies and spirits while the French protected the port. After protracted sufferings, in July, the Christian coalition managed to overcome the Muslims.

1197 St. John of Math organized the Order of Trinitarians in Paris to treat domestic epidemics as well as to succor victims of combat wounds. Called "Friars of the Ass" for their lowly mounts, the almoners canvassed France and Spain on behalf of the sick and aided Christians incarcerated in Muslim jails.

1200s In Bengal and across northern India, Hindu peasants turned increasingly to Mata Sitala (*Mother Cool One*) as the goddess of smallpox and other infectious diseases. Propitiation of Sitala involved naturalistic treatment of the disease. Patients who contracted the pox merited reverence for carrying divinity in them as she covered their bodies with her *khel* (sport) or her *lila* (story). Out of respect to Sitala, bestower of smallpox, the sick bore the misery as a token of honor.

Medical specialists administered purgatives to remove impurities and massaged ailing bodies with cooling *neem pattha* (leaves) and a paste of flour and turmeric. Healers treated fever with cold drinks and such foods as plantain, cold rice, and yogurt. The patient avoided warmth, including the heat generated by exercise and sexual activity. Emerging buboes required lancing with thorns to relieve pressure and pain. In an act of sacred celebration, the *tikadar* (inoculator) injected effluvia into the sick to weaken the infection. When patients recovered, they prayed to Mata Sitala, carried her image in procession, cooled it in the Hooghly River, and presented gifts of flowers, coconut, and rice. *See also* **1122 B.C.E., 1877.**

1203 At Neumarkt, Saint Hedwig of Bavaria opened a convent, school, and leprosarium for women that remained active for 35 years. In this same period, when some 19,000 hospices for lepers were in operation in Europe, Fachtna of Kilcolgan, Ireland, received lepers at a guesthouse. Later in the century, outbreaks of leprosy peaked across the continent. Community lazar-houses or lazarettes, the compulsory isolation sanitariums at which municipal officials confined lepers, developed into almshouses or general hospitals receiving victims of other diseases and afflictions.

November 1218 In the second year of the Fifth Crusade, launched by Pope Honorius III, Christian forces in Egypt led by Cardinal Pelagius faced the outbreak of dysentery. After pressing on to the port city of Damietta along the Nile Delta in December, 20 percent of their number died from scurvy and exhaustion during heavy downpours that flooded their camp. Inside the city, Muslim citizens also suffered from an epidemic of granular conjunctivitis or trachoma. The army of Sultan al-Malik al-Kamil decamped in February 1219, leaving the way open for Christians to take the city. Inside, Christian parties found only 3,000 of 80,000 inhabitants alive and only 100 still healthy. Corpses lay in the streets, where dogs gnawed them.

October 3, 1226 Saint Francis of Assisi, a preacher among Italy's poor, died of tuberculosis at age 44 at Portiuncula, Assisi. Historians assume that his volunteerism with the Franciscan *fratres minor* (minor brethren) for 15 years situated him among lepers, homeless, imprisoned, and outcast people, from whom he contracted TB, the disease of the impoverished.

September 1227 At the outset of the Sixth Crusade, when the Emperor Frederick II of Swabia was set to embark from Brindisi, Italy, for the Holy Land, his forces contracted endemic malaria. Both Frederick and the Landgrave of Thuringia caught the fever. In anger at the failed expedition leadership, Pope Gregory IX excommunicated Frederick for failing to complete the mission.

1237 On the road between Arnstadt and Erfurt in east central Germany, an outbreak of manic dancing, jumping, and leaping seized one hundred children, many of whom died of their frenzy. Others were stricken with convulsions and incurable tremors. Christian apologists interpreted the psychological anomaly, called tarantism, as satanic possession.

1243 Cambodia's King Indravarman II (also identified as Yasovarman or Jayavarman VII) died around age 23 of leprosy, which was endemic in Southeast Asia. According to Asian folklore collected by the Buddhist Institute, a Khmer legend claimed that the king contracted the disease from venom that spewed out of a serpent he was wrestling. The fight was a mythic interpretation of predatory disease. His subjects revered leprosy as a divine curse that ennobled royalty. In Angkor, a statue seven meters high at the Terrace of the Leper King preserves the king's link with the bacterium.

Easter 1249 Historian Jean de Joinville, author of *Histoire de Saint Louis* (*Life of Saint Louis*) (1309), described how King Louis IX of France led 25,000 soldiers to Egypt on the Seventh Crusade. The expeditioners encountered so many corpses bottlenecked on the Nile River at Mansourah that the king had porters bury the Christian dead and send the Saracen corpses out to sea on the current. The contagion arising from so much corruption produced a severe outbreak of typhoid fever, complicated by hunger, dysentery, and scurvy, which the biographer also suffered.

At Easter, barber-surgeons sliced away corrupt gum tissue from the afflicted. The king ordered an evacuation by sea from Damietta for his ailing troops, who rapidly surrendered to the enemy, led by Fakhr ad-Din. After the battle of Fariskur, Louis, who was soon in Mamluk custody, displayed symptoms of both scurvy and dysentery. Recognizing Louis's extraordinary charisma, the caliph freed him from his chains on the promise that he would abandon Damietta and remit a ransom of 800,000 gold coins. Muslims slew the sick soldiers whom Louis left behind. The next year, the king earned sainthood in part for succoring

lepers every three months and teaching basic nursing techniques to staff who washed the sick and soothed their lesions with ointment.

August 25, 1270 As described by biographer Jean de Joinville, author of *Histoire de Saint Louis* (*Life of Saint Louis*) (1309), on the way to Tunis during the Eighth Crusade, many soldiers in the unsanitary camp suffered from typhoid fever and dysentery. King Louis IX of France, a veteran of two crusades, died from fever along with his eldest son, Jean Tristan, on August 3. The French retrieved the beloved king's heart and bones and displayed them at the church of Saint Denis.

June 17, 1278 A sudden onset of the dancing plague or tarantism seized Dutch pedestrians crossing the Maas River bridge at Utrecht, Holland. Some 200 frenzied participants would not or could not stop gyrating. When the bridge buckled, they fell into the river. Many drowned; those who survived were conveyed to a chapel of Saint Vitus, who supposedly rid them of mania. Christian interpretation explains their deaths as divine intervention, directed to the bridge by a passing cleric.

late 1200s Arab physicians in Alexandria, Baghdad, Cairo, and Damascus segregated in contagion wards the patients who suffered from fevers and trachoma. In a separate sector, staff supervised wards treating wounds and the ills of pregnant women and infants.

1300 At Mombasa, the major port of Kenya in eastern Africa, smallpox invaded a virgin-soil population after it spread from the crews of Arabian, Chinese, and Indian trading vessels.

1300 In Flanders, government agents controlled leprosy through stringent measures. Positive diagnoses, made by members of religious orders, preceded the assignment of patients to leper hospitals or to medical ghettos. To live in Flemish society, lepers had to wear large hats and ring a bell to ward off citizens who might contract their affliction. In 1538, physicians took the place of monks and priests in making diagnoses.

1300s Sleeping sickness, a disease spread by the bite of the tsetse fly, was first rampant in the northern reaches of the Niger basin. When a version of the disease struck horses and cattle, it was called *nagana*.

1320 The Black Death, the second great cycle of bubonic plague after the Justinian plague of 541 C.E., originated in the Gobi desert in the Mongol Empire. It spread over China, killing 65 percent of the population. It migrated southwest to India and advanced on the Eurasian steppes.

June 21, 1321 Philip V of France made leprosy a treasonous offense by connecting infection with an alleged grand conspiracy with the Moorish regime of Granada, the Sultan of Babylon, and Jews to poison the wells of Christians. The source of the myth was a series of failed harvests, which left French peasants in debt to Jewish money-lenders. The grand inquisitor, Bernard Gui of Toulouse, presided over tribunals and compiled a manual of torments. By royal order, lepers were burned at the stake or boarded up in their homes and burned. The law required that they be interred without religious blessing and their estates confiscated by the government.

To rid Europe of lepers, mobs burned amply endowed leprosaria from Lausanne to the Carcassonne, looted the properties, and participated in anti–Semitic pogroms. Bishop Jacques Fournier superintended the execution of thousands. In 1338, when Fournier was named Pope Benedict XII, he denounced the persecution of lepers and rescinded the charge of treason against them.

late spring 1326 An outbreak of plague across southwestern Europe prompted Aragonian-Jewish medical author Abraham ben David Caslari of Besalu, Spain, to compose two fever texts. One, *Ma'mar Be-qadahot Divriot U-minei Qadahot* (*A Treatise on Pestilential Fevers and Other Kinds of Fevers*) (1326), noted that "fevers were lethal; they would not pass away for ten days, and many would die of them; and the fevers were incessant, with much fainting and distress" [French 1998, p. 10].

1327 St. Roch (also Roche or Rock) of Montpellier, France, a survivor of bubonic plague, became a patron of cholera and plague victims. On a pilgrimage to Italy, he set up a treatment regimen at Piacenza southeast of Milan, where he contracted plague, but recovered. Legend claims that a dog belonging to the nobleman Gothardus bore the saint bread and nursed him back to health.

Roch continued his ministrations at hospitals he founded in Aquapendente, Cesaria, Mantua, Parma, and Rome. Victims declared that he blessed the sick and caused miraculous cures. After an ignoble imprisonment and death in his hometown on August 16, prelates revived his lore during epidemics in Belgium, France, Germany, Italy, and Spain. They protected houses with the letters V.S.R. for "Vive Sanctus Roch" (Live Holy Roch). Artists depict Saint Roch dressed in pilgrim's hat, cloak, and sandals with the telltale bubo evident on his left thigh. In 1780, Jacques Louis David painted the saint beseeching the Virgin Mary and her infant son for aid to plague victims, who lie around him in various states of torment.

1332 A great medieval idealist, Magnus II Eriksson, the last of the Folkung kings, assumed power at age 16 as Magnus VII of Norway. While attempting to merge the Danish Provinces of Skåne and Blekinge with Norway, he battled an epidemic of pneumonic plague, which reduced his subjects to one-third their original number. The catastrophic loss obliged him to share power with his sons, Haakon and Erik.

1345 The Black Death, which originated in Asia with a death toll of 25 million, did not infiltrate Europe until it reached the Crimea on the north shore of the Black Sea in 1345. In 1347, the contagion advanced on Italy through Genoese merchants who arrived at the Straits of Messina. The ship was sailed by a ghost crew, most of whom were either dying or dead. First called the Black Death in Sweden in 1555, the disease raged over much of the globe until 1772.

Records indicate the path of the great pestilence and the estimated death toll in some locales where census figures attest to loss (see tables at 1656, December 19, 1665). Overall, 1,000 villages vanished as 25 million Europeans perished. *See also* **July 24, 2001**.

Those infected with bubonic plague quickly began aching and sweating, collapsing, and tossing fitfully on their pallets. The first mortal symptoms were loss of body fluids through vomiting, coughing up blood, and diarrhea. In the third stage, patients produced black patches and buboes or swellings of lymph glands under arms and in the groin, the preface to hysteria and coma. The course of the disease was so swift that attendants sometimes died before their patients expired. Because of ignorance of bacterial infection, doctors and nurses ignored the threat of running cesspools and infestations of vermin in alleyways and thatched roofs. Health advice, such as the list distributed by medical professors at the University of Paris in 1348, offered ridiculous regimens forbidding meat, fish, fat, and olive oil. The dicta encouraged only light exercise and the cooking of vegetables in rainwater.

1346 Edward III of England banished lepers from London to prevent spread of contagion from their lesions and breath, particularly the infecting of the genitals of whores.

summer 1347 Arriving from Russia, the Black Death, an epidemic of the *Yersinia pestis* (or *Pasteurella pestis*) bacterium, entered the eastern Mediterranean with the forces of Malik Ashraf, who collapsed as they besieged the area outside Baghdad, Iraq. Journeying along with pilgrims, travelers, and traders, for three years, the pestilence overran Aleppo, Antioch, and Damascus, Syria, where 1,000 died per day. The devout called on God in public prayer and interred their loved ones at mass funerals and burials. Survivors abandoned cities and farms and tried to dash ahead of the contagion without realizing that they carried it with them.

Plague ravaged Armenia, India, and Mesopotamia and struck the Black Sea area at the Genoese city of Kaffa (now Feodosia) in Crimea, where it killed 85,000. While Kipchak Khan Janibeg (or Yanibeg) directed his Tatar forces at the city in a year-long siege, Kaffans barely survived. In 1347, however, plague invaded the insurgents, forcing Janibeg to relent. His final act of vengeance was barbarous—the Mongol General Gabriel De Mussis catapulted over Kaffa's walls the plague-riddled corpses of dead soldiers, which Kaffans tossed into the sea. The Genoese panicked and launched four ships for Italy, carrying pestilence into the peninsula. Only a quarter of Venetians survived; Pisa escaped with only 30 percent of its populace.

The disease swept over Egypt, halting fishing from Damietta harbor and decimating Alexandria as up to 200 died per day from pneumonic plague. At Cairo the daily toll was 300 daily for a total of 7,000, some from septicemic plague, which infected the circulatory system, but caused no buboes. The onslaught interfered with traditional Islamic burial customs as morbidity spread along the Nile delta and west into Libya and Tunisia.

Traveling in many directions over trade routes from the Mongolian steppes throughout Asia, the disease inundated Europe from trading vessels and, over four centuries, recurred every 17 to 25 years. The cycle favored cities where the common *Rattus rattus* thrived, but developed erratically over the populace, skipping some villages and obliterating others. Worst hit were prisons, ships, abbeys, and walled communities. England lost 1.5 million or 40.5 percent of its 3.7 million citizens. The onslaught ended in 1720 at Marseilles, France.

Plague moved by water to Constantinople and Alexandria. In Cyprus, the onset of plague coincided with an earthquake, which wiped out the fishing and olive-growing industries. In panic, islanders murdered Arab servants and hurried to high ground. Legend reveals that refugees fell dead from an infectious wind, a metaphor for the contagion they carried with them in flight from the diseased coastal area.

early October 1347 At Messina, Sicily, port authorities allowed trading vessels access to wharves. Franciscan historian Michael Platiensis (or Michael of Piazza), who described the situation in *Account of the Plague* (1357), claimed that twelve Genoese vessels of unknown origin arrived carrying crew and passengers already dead or dying of a pestilence that had begun in India. Plague-carrying rats swarmed onto the land. Platiensis marveled that "In their bones [the arrivals] bore so virulent a disease that anyone who only spoke to them was seized by a mortal illness and in no manner could evade death" ("Michael Platiensis").

Platiensis blamed human-to-human spread of infection and described eruptions in thighs and upper arms, followed by violent vomiting of blood for three days preceding death. He noted that lawyers refused to attend the dying to notarize wills and clerics would not hear confessions. Those fleeing to Catania called for deliverance from holy Agatha, protector of victims of catastrophe. Queen Joanna I of Sicily and her son Don Federigo arrived to observe the hysteria.

November 1347 Catanians rejected a plea from Messina to dispatch Saint Agatha's holy relics to their town. The Archbishop of Catania mediated the crisis by dipping the relics in water and carrying the water to Messina. Citizens of Messina began a pilgrimage to Catania to venerate the sacred objects firsthand, spreading contagion as they processed. Pilgrims died in the fields or collapsed in hospitals. Attendants buried them in trenches outside Messina's city walls.

The Sicilian pestilence flourished until April 1348, killing even the heroic Archbishop of Catania and Duke Giovanni. Too late, enforcers of health regulations directed the contagion-bearing ships into the Mediterranean. In November, one of the death ships traveled to Marseilles. Others arrived in Corsica and Sardinia. From the coast, the disease infiltrated Portugal and Spain.

December 1347 Venetians suffered a severe catastrophe from plague, which reduced the population by around 75 percent. Officials took immediate action against contagion from the port area by quarantining docking vessels for 40 days, selecting barren islands as burial grounds, and setting burial standards at five feet. Surprisingly, Milan was less able to prohibit contagion, yet suffered a death rate of only 15 percent. The Milanese government appears to have protected the healthy by walling up homes containing the sick and provisioning them indirectly by placing food, medicines, and supplies in baskets lowered on ropes from the upper floors.

January 1348 Flemish historian Pierre-Jean de Smet's *Breve Chronicon Clerici Anonymi* (*Brief History of Unnamed Clerics*) (ca. 1850) explained that a trio of galleys harboring at Genoa brought bubonic plague to town. Too late, the Genoese drove them from the port, sending them down the Italian coast toward Sicily with their cache of bacilli. When the disease spread over southern Europe, at Dubrovnik, Croatia, widespread epidemic encouraged packs of wolves to devour the living, who were weak and still reeling from terror and loss.

January 9, 1348 At Basel, angry Swiss Christians blamed Jews for spreading plague. Mobs burned area rabbis and 600 Jews at the stake and left the remains unburied. A general banishment of Jews resulted in expulsion of survivors, the adaptation of their synagogue into a church, and forcible baptism of 140 Jewish children.

March 1348 At Florence, which the disease engulfed from March to September 1348, estimates of the dead ranged from 60,000 to 100,000. In spring, people began falling ill of a disease for which physicians had no tincture or cure. Female patients gave up their prudery and allowed examination and treatment by male doctors. The

healthy sequestered themselves in communes or retreated to country estates, ate and drank in moderation, and spoke with no one who might bear contagion. The servant class dried up, leaving the wealthy to fend for themselves. Those who walked in public carried flowers, herbs, and spices at their noses to cover the odor of decay.

In the absence of pallbearers drawn from family and friends, a professional band of *becchini* (gravediggers), whom the Italian poet Giovanni Boccaccio referred to as low-class corpse-carriers, performed the rites for money, sometimes hefting multiple corpses of families on one bier. When the civic structure collapsed, lawlessness overcame the city. Much feared for their contagion and cruelty, the black-hooded *becchini* were enlisted from the poor and criminals. They took license with victims, killing those dying of plague to hurry them to their graves, then robbing their estates of jewelry and cash.

Local historian Marchione di Coppo Stefani, author of the *Cronica Fiorentina* (*Chronicle of Florence*) (ca. 1382) wrote of the piteous condition of victims abandoned by their families:

> Many died unseen. So they remained in their beds until they stank. And the neighbors, if there were any, having smelled the stench, placed them in a shroud and sent them for burial. The house remained open and yet there was no one daring enough to touch anything because it seemed that things remained poisoned and that whoever used them picked up the illness [Stefani].

He commented that the city came to a virtual standstill: "All the shops were shut, taverns closed; only the apothecaries and the churches remained open. If you went outside, you found almost no one" (ibid.).

Boccaccio, who was appointed as a Florentine ambassador in 1348, spoke of the plague in the *Decamerone* (*The Decameron*) (1353), a compendium of classic *fabliaux* and short stories. In the introduction, the author commented on efforts by civil authorities to prohibit plague-carriers from entering the city and to supplicate God for deliverance from sin, which priests considered the cause of the pestilence. Boccaccio described four categories of reaction to the epidemic:

- The first group of people revelled in the streets in sybaritic abandon. In desperation, they gave up attempts to ward off infection by joining social groups at taverns and inns.

- Others withdrew into a grim seclusion.
- A third group avoided either extreme to remain cautiously active.
- Those who could afford travel fled to the countryside.

According to historian Michael Levey, author of *Florence, a Portrait* (1996), Florentines lost their faith in prayer and medicine. Grimly realistic, they diverted their minds from suffering and death with fun and satisfaction of physical appetites.

Throughout Italy, mayors shut in whole districts to separate the dying from the healthy. In Pistoia, a Tuscan city of 11,000 outside Florence, city authorities issued edicts after the plague arrived in spring 1348. Their demands were stringent:

- a prohibition of citizens from traveling and placement of a guard on the city
- a ban on importation of cloth, bed linens, or garments bearing contagion
- placement of bodies in wood caskets nailed shut to prevent odor and contagion
- a ban on the transportation of the dead from the city
- a standard setting the depth for graves at 2.5 *braccia* (67.5 inches)
- an order that mourners refrain from contact with corpses and from holding funeral feasts
- a ban on the hiring of professional mourners and musicians
- a prohibition of the purchase of new clothes for mourning to all except widows of the deceased
- an end to the wailing for the dead and the ringing of the cathedral's bells as a means of avoiding panic in the living when they realized how many had died.

Additional ordinances controlled odor from meat markets and tanneries. Such emergency measures became common over the next 16 years, when the Black Death halved the population of Europe and the eastern Mediterranean.

March 20, 1348 Lorenzo de Monaci's *Chronicon de Rebus Venetis* (*Venetian Chronicle*, 1758) attested that Doge Andrea Dandolo and the Venetian council set up a panel of three aristocrats, who prepared barge service to cemetery spaces at Saint Erasmo and on the island of San Marco. Statutes released inmates from debtors' prison, halted immigration, and stopped mendicants from currying sympathy by displaying corpses. At the harbor, authorities set up quarantine quarters on Nazarethum island and threatened to burn ships violating harbor regulations. For remaining faithful to his post during the scourge, Venetian health inspector Francesco di Roma received a pension of 25 ducats per year.

April 1348 Pedro IV of Aragon organized health care workers to protect Majorcans from plague. Nonetheless, within four weeks, some 15,000 died, leaving only 20 percent of the populace alive. The army was so depleted that, in May, the island soldiers begged the king for military aid to ward off the Muslims of Tunis. A year later, Minorca had suffered so great a loss to plague that its governor petitioned Majorca for help.

April 2, 1348 At the end of an eight-week peak in deaths, a London cemetery had received 2,000 burials at the rate of 34 a day. By June, the rate of deaths in London rose to 290 and held steady at that number until September.

April 27, 1348 A letter written by a priest to recipients in Bruges, Belgium, reported on Avignon's pitiful condition, with over 7,000 French homes boarded up and residents flowing from the city. Over the previous 44 days, 11,000 had been buried in a graveyard near Our Lady of Miracles, with more local victims interred at Saint Anthony's Hospital and elsewhere. To relieve the problem, Pope Clement VI blessed the Rhone River to consecrate the overrun of bodies that authorities dumped into the water.

The papal see at Avignon lost 70 priests and nine cardinals, leaving a few overworked prelates to conduct holy sacraments. Lay members began hearing confession. In desperation, Pope Clement VI forgave the sins of all plague victims, but he placed as a burden on Christians the edict that they must accept pestilence as God's punishment of the sinful. The pope himself, on the advice of his doctor, Guy de Chauliac, sat between two hot fires and wore a magic emerald ring to counter contagion.

May 2, 1348 When plague appeared at Pistoia, Italy, the city council issued ordinances banning the return of Pistoians from Pisa or Lucca, visits from plague-infested areas, and the importation of any linens, woolens, or corpses. Laws regulated food sales and the behavior of mourners at funerals as well as the location and depth of graves. To prevent hysteria, the city fathers halted the spread of news by bell-ringers, trumpeters, and town criers. The council amended the rules on June 4 by banning grave-digging by any but Pistoia's 16 appointed laborers.

June 1348 Plague made its way over the Tyrol to Bavaria in North Germany and west to Almeria, Barcelona, and Cervera, Spain. In 1493, the *Nuremberg Chronicle* reported on the disease: "Pestis lugubris et miseranda hoc anno MCCCXLVIII incepit" "*A woeful and wretched pestilence began in the year 1348*" (Naphy and Spicer 2000, p. 34). At Almeria, Spain, a Muslim doctor and medical historian, ibn Khatimah, blamed Beni Danna for contracting the first case and noted that symptoms included coughing up blood. The disease was so virulent among Arab soldiers that they failed to engage the Christian troops of Castile and pondered whether they should abandon Islam and convert to Christianity to avert infection. Muslims were so overawed by the disease that they named it "The Great Destruction in the Year of Annihilation."

June 2, 1348 At Siena, devastation paralleled that in other plague-ravaged locales. Agnolo di Tura del Grasso, author of *Cronica Senese (Sienese History)* (1348), summarized the loss of 80,000 in seven months:

> It is impossible for the human tongue to recount the awful truth.... Father abandoned child, wife husband, one brother another.... And in many places in Siena great pits were dug and piled deep with the multitude of dead [Kolata 1999, p. 39].

Stupefied by the immensity of suffering and loss, he reported that he buried his five children with his own hands. He feared that the end of the world had come.

Siena's inner workings came to a halt. Judges suspended civil court for three months or until the plague relented. The city council, which operated at two-thirds its usual manpower, shut down legalized gaming. The oil and wool business ceased. The church lost seven cardinals and was forced to accept laymen in place of fallen priests and friars. Parishes padded their coffers with legacies and donations, the guilt offerings or propitiations from those who could afford to bribe God. The town was so sapped by death that it courted newcomers with reduced taxes.

summer 1348 A French Carmelite monk, Jean de Venette, who completed the historiography begun by Friar Guillaume (or William) of Nangis, a Benedictine chronicler at Saint Denis, summarized that the plague

> began among the unbelievers, came to Italy, and then crossing the Alps reached Avignon, where it attacked several cardinals and took from them their whole household. Then it spread, unforeseen, to France, through Gascony and Spain, little by little, from town to town, from village to village, from house to house, and finally from person to person. It even crossed over to Germany, though it was not so bad there as with us [Newhall 1953, p. 49].

At the Hôtel-Dieu in Paris, the nuns continued to treat plague victims without fear of dying from contagion. Daily, 500 cadavers awaited burial at the cemetery of Saint Innocent, included the nursing sisters themselves who died of plague.

September 15, 1348 At Berne, Chillon, and Zurich, Switzerland, mobs threatened Jews for spreading plague. Those imprisoned at the Castle of Chillon on Lake Geneva were tortured to make them admit poisoning wells in Venice and Neustadt. The first victim, a doctor named Balavignus, a sanitarian at Strasbourg, drew suspicion because he exterminated rats and, thus, protected local Jews from contagion. On the rack, he claimed that Rabbi Jacob of Toledo sent him an egg containing an infectious powder to toss into wells at Thonon and Vivey. Another Jew, Banditonon of Neustadt, confessed to a similar poisoning of a Cartuet well. The next day, a third, Mansiono of Neustadt, admitted poisoning the waters of Evian. Zurich issued a ban on Jews on September 21. On September 26, Pope Clement VI contradicted anti–Semites by pointing out that Jews, like all Europeans, were suffering from bubonic plague.

October 1348 Diarists and journalists contributed personal experience with the way in which cities and families confronted certain death. At Siena, artisans stopped work on the cathedral. Before falling dead at his desk, banker and historian Matteo Villani added ten books to the *Chronica Universale (Universal History)* (1348), begun by his brother Giovanni Villani, who also died of plague. Matteo foresaw an apocalypse and declared that disease evidenced God's intent to exterminate humankind. He predicted that survivors would realize that God had saved them by divine grace and that the living would become humbler and better Christians by avoiding sin and embracing benevolence.

At the University of Paris, King Philip VI convened 49 health workers to compose the Paris Consilium. Expert diagnosticians predicted that no one would ever discover the cause of pestilence. Among educated guesses were descriptions of comets and other astral portents and anomalies. Most damning were lunar and solar eclipses preceding the conjunction of the planets Jupiter, Mars, and Saturn in the sign of Aquarius, an assessment based on Hippocrates's *The Epidemics* (ca. 310 B. C.) and Aristotle's *Meteorology* (ca. 350 B. C.). This death-dealing celestial alignment occurred on March 20, 1345, generating noxious fumes and putrefaction that rotted the heart and other organs, a diagnosis based on the work of physician and pharmacologist Avicenna of Kazakhstan, the author of *Qanun fi at-tibb* (*The Canon of Medicine*) (ca. 1030).

November 22, 1348 As Germans panicked at numerous deaths from plague, anti–Jewish sentiment overran Augsburg, Munich, and Wurzburg. Burnings in Solothurn, Stuttgart, and Zofingen spread to Burren, Landsberg, Lindau, and Memmingten in December, and, in January, to Freiburg, Speyer, and Ulm. Residents of Speyer rid their town of contagion by sealing Jewish corpses in wine casks and tossing them into the Rhine River. The persecutions continued at Dresden, Eisenach, Gotha, and Strasbourg in February and, the next month, at Baden, Erfurt, and Worms. At Toledo, Spain, the rumor that Jews initiated an anti–Christian plot blamed Rabbi Peyret of Chambry, Savoy, as the mastermind who sent assassins to France, Italy, and Switzerland.

throughout 1348 As a result of the Black Death, doctors remained in practice by treating patients at a distance and refusing to enter peasant hovels. The most prominent surgeon of the day, Guy de Chauliac, personal physician of Pope Clement VI, compiled *Capitulum Singulaire* (*A Singular Chapter*) (1350), a medical history that described contagion in Avignon. He noted that so many fatalities caused people to hate doctors, who could do nothing to prevent plague or relieve its symptoms. Consequently, the physicians stopped practicing and denounced their profession because they felt helpless against certain death.

To the north, ships lolled at the wind's will in the English Channel after all on board died. As farm workers were felled in their sheds and fields, crops went unharvested, cows unmilked. Some goats, pigs, and chickens reverted to the wild. Sheep, which were vulnerable to the disease, died in huge numbers, notably one pasture where 5,000 lay rotten and stinking. As the populace shrank, 200,000 villages disappeared from the map of Europe.

Among the famous killed by the plague in 1348 were Florentine historian Giovanni Villani, Florentine painter Giovanni Daddi, law professor Giovanni d'Andrea of the University of Bologna, Florentine painter Giotto's best art students, Sienese painter Ambrogio Lorenzetti and his family, the parents of Tunisian historian ibn Khaldun, Norwich Cathedral builder William de Ramsey, queen's chaplain Robert de Eglesfield of Oxford, poet Richard Rolle, the brothers of Magnus II of Sweden, royal surgeon Roger de Heyton, Abbot Simon de Bircheston of Westminster, and, on April 6, 1348, Laura de Noves of Avignon, the beloved of the poet Petrarch. Added to the list were Bishop Foulque de Chanac, Bonne de Luxembourg, Jeanne de Bourgogne, Jeanne of Navarre, Archbishop John Stratford of Canterbury, Lord Mayor John Pulteney of London, Governor John Montgomery of Calais, Bishop John de Pratis of Tournai, Belgium, Abbot Michael Mentmore of Saint Albans along with 47 monks, and Joan, daughter of Edward III. One victim, hygiene teacher and Hebrew translator Gentile da Foligno of Padua, had compiled a plague treatise, *Concilium contra Pestilenciam* (*Advice on the Plague*) (1348), which moved him to report the fear of doctors that "anyone who approaches [the plague's] dwelling shall die" (French 1998, p. 9).

The first professionals to suffer near annihilation were the brothers in monasteries, who opened their infirmaries to victims and buried the dead in hallowed grounds. At Rochester, as witnessed by Friar William of Dene, the bishop lacked a staff. In lieu of his household of priest, squires, clerics, pages, and servants, all dead, he placed in charge of duties eight women—four nuns and four novices. The Abbey of Meaux, Yorkshire, sank to ten residents, 20 percent of its original headcount. At Louth Park Abbey, a Cistercian chronicler noted that the plague, like the flood in Noah's day, took all people, good and bad and of all cultures.

Chronicler Henry Knighton, a Leicestershire priest, observed that the plague coincided with a "great murrain of sheep everywhere in the kingdom," killing 5,000 animals (Carey 1987, p. 48). He commented that men whose wives died of plague volunteered to perform priestly duties, even if they were illiterate and had no knowledge of scripture. Because of the "foul deth of Engelond," the Scots gathered in Selkirk Forest to invade northern England, but fell victim to the same epidemic (ibid.). Rapidly, they lost 5,000 men. As they turned toward home, the English followed on their heels and ravaged the weakened troops. Knighton claims that many buildings fell to ruins across the land as villages and hamlets dried up from lack of habitation.

Although record keeping was inadequate in Ireland, Brother John Clyn, a Franciscan at Carrick-on-Suir, Tipperary, and compiler of a world history, recorded a subjective evaluation of the epidemic in *Annales Hibperniae* (*The Annals of Ireland*) (1349). At Kilkenny, where his fellow monks had already died, he wrote on his parchment:

> Plague stripped cities, castles and towns of their inhabitants so thoroughly that there was scarce anyone left alive in them. The pestilence was so contagious that those who touched the dead or the sick were immediately affected themselves and died ["The Spread of the Plague"].

Noting that plague killed adult and young family members and servants in the ordinary household as well as the aristocracy, bishops, priests, and pilgrims, he feared that all would die, leaving no one to carry on Christ's mission against Satan on earth. Clyn's work ended in mid-text as he, too, succumbed to plague. In 1998, Robert Farber emulated Clyn's composition of doom by writing of his own awareness of mortality from AIDS in *Western Blot #14*.

In 1350 an unknown English writer scratched in the stones of Ashwell church tower, Hertfordshire, their dismay at the huge loss of life in a year that was "miseranda, ferox violenta" (wretched, wild, and violent) (Yonge 1864). Germany's clergy lost 35 percent; England's lost half, including three archbishops of Canterbury sequentially from May 1348 to August 1349. Of Dominican friars in Montpellier, France, only seven out of 140 survived; in Avignon, Carcassone, and Marseilles, whole houses perished.

The Alexian Brothers or Cellites, who bore the name of Alexis, a fifth-century saint and nurse, got their start in 1348 under the administration of Tobias of Mechelen, Belgium. At the height of the plague, brothers went door to door to pour healing curatives and hot soup. Without realizing the danger of personal contact, they spread infection from their tunics and cassocks. Hopeless people prayed, propitiated relics and talismens, practiced astral healing, and paid the last of their money for worthless herbal mixtures. The only nurse care involved bleeding and purging, the standard regimen of the day for all infection and a quick death for patients already weak with suffering. Some sought relief through mustard plasters, cautery, and the lancing of buboes, which triggered shock and near instant death.

In England, Flanders, France, Germany, Holland, Hungary, and Poland, the Brotherhood of Flagellants, a violent masochistic order of male and female zealots originating in eastern Europe, organized group penance and scourging to drive out sin. They carried spurious letters written by Jesus Christ and dropped from heaven urging serious acts of penance before the world came to a cataclysmic end. Processing two by two from town to town, hundreds of cowled figures bore Christian banners into areas where parishioners fearful of the plague tolled bells to welcome them. Benedictine chronicler Thomas Walsingham of Norfolk commented that the movement lacked the pope's permission. On October 20, 1349, Pope Clement VI banned the unauthorized exhibition entirely. However, interference was perilous: local prelates who tried to stop the march risked death.

In the opinion of Carmelite historian Jean de Venette, the mania suited the spirit of the plague-ravaged era. Stripping to the waist, the lay brethren encircled the sick and fell to the ground in the posture of the crucified to receive multiple lashings from the master flagellant. Chronicler Henry of Hereford, a Westphalian member of the Dominican friary at Minden, Germany, recorded that individuals produced metal-tipped leather flails and applied them to their torsos while the master called for prayer. The brothers epitomized the Christ-like man of sorrows, the bearer of the world's sins. They made no attempt to soothe, rinse, or treat their own torn flesh, which poured blood and ragged bits of flesh. Zealous onlookers soaked cloths in the gore to smear on their faces or to create healing nostrums.

Supporting the frenzy twice daily were mournful verses of a standard hymn chanted by onlookers as the ritual orgy lengthened into the night for a period of 33 days and eight hours, symbolizing the life of Christ, who died at age 33. For a fee, others could join the self-righteous throng if they promised to forgo bathing and shaving, beds and clean clothes, and sexual intimacy. Upon entry to the order, new brethren had to unburden themselves of all sins committed since age seven.

1349 In England, Bishop William of Edynton foresaw that the epidemic would spread from France to England and Scotland. He warned:

> The plague kills more viciously than a two-edged sword. Nobody dares enter any town, castle or village where it has struck. Everyone flees them in terror as they would the lair of a savage beast. There is an awful silence in such places; no merry music, no laughter. They have become dens of terror, like a wilderness [Day 1989, p. 6].

His prediction was correct. According to Galfridus de Baker, a monk at Malmesbury west of London, and to the chronicle of the Grey Friars of Lynn northeast of Cambridge, English troopers returned to Melcombe Regis, the harbor outside Weymouth, Dorset, on June 24 from combat in Gascony and bore the scourge inland. London alone sank to half its former headcount. Saint Andrews lost 24 priests, who exposed themselves to contagion while administering extreme unction to the dying. In all, Scotland's population decreased by two-thirds. As confirmed in legend, a standing stone commemorates a village that died out but for one elderly Scotswoman. On her own, she loaded a donkey cart with the corpses of villagers and buried them in a nearby field.

At Vienna, a total of 32 Austrian physicians and surgeons diagnosed ills and treated plague victims. Civil regulations required the appointment of a *magister sanitatio* (sanitation officer), the clearing of junk and refuse from basements and streets, the filling in of pits and cesspools, and the closing of public markets, baths, and bordellos, which attracted tourists and visiting merchants. Guards at city gates ousted wanderers, beggars, migrant workers, and street vendors. Compliant citizens leashed their dogs, marked the houses of the sick, buried their dead outside the city walls, and burned all infected clothing.

During the few months that disease threatened, the populace rose up against Jews and massacred them for causing the epidemic. The same scenario of ignorance and scapegoating that had thrived the previous year recurred as the hate-filled drove out or wiped out Jews in 350 German municipalities, notably Breslau, Nuremberg, and Rothenburg. At Frankfurt, some Jews immolated themselves in their homes rather than be expelled from the city. Similar pogroms killed Jews wholesale in Italy, Poland, and Spain.

Herman Gigas, a Franconian Franciscan monk, composed a chronicle on the deaths of European Jews. He remarked on rapid decline and death of patients by the second or third day and the emptying of towns and cities of people. He reported on alleged causes of the scourge:

> Some say it was brought about by the corruption of the air; others that the Jews planned to wipe out all the Christians with poison and had poisoned wells and springs everywhere. And many Jews confessed as much under torture: that they had bred spiders and toads in pots and pans, and had obtained poison from overseas…. God, the lord of vengeance, has not suffered the malice of the Jews to go unpunished [French 1998, p. 6].

Gigas added that over most of Germany, Christians condemned Jews to death without offering them an opportunity to prove their innocence of conspiracy.

The rest of Europe incurred a mounting death toll. Some 500 died daily of plague in Pisa in 1349. The number reached 600 per day in Venice and Vienna and 800 per day in Paris. When the epidemic slowed, the population of Paris had sunk to half and Florence to 20 percent of its former number, in part because the famine of 1347 had weakened so many. Of the citizenry of Hamburg and Bremen, two-thirds were dead. The busy port of Marseilles recovered from a massive loss, which mounted to 16,000 citizens dead in one month. Less resilient cities—Arras, Carcassone, Laon, Montpellier, Rheims, and Rouen—never rebounded to their previous prosperity. At Burgundy, the loss was heaviest on the church, which dispatched priests to succor the sick and dying, and in the vineyards, where skilled vintners died, leaving few in charge. At 14 wineries, there were no laborers to prune and cultivate.

North Africa, too, suffered its bout of plague. As described in the *At-Ta'rif bi-Ibn Khaldun (Autobiography of ibn Khaldun)* (1375), composed by an Arab historian born in Tunis, the sufferings of Muslim communities paralleled those of European Christians. Along the Maghrib, where Arabs dispossessed Berbers, plague swallowed up people and wiped out dynasties. As properties and cities declined and clans weakened, the Tunisian historian Khaldun feared that the disease threatened all civilization. He surmised that God decreed oblivion as a preface to the creation of a new, less iniquitous race. One significant survivor, ibn Abu Madyan of Salé, Morocco, disproved suspicion about the all-compassing lethal air by remaining indoors with his family until the plague had run its course.

January 1349 In England, erection of Winchester Cathedral halted because of the deaths of artisans and day laborers, leaving Bishop William Edendon no choice but to abandon his building schedule. At Saint Swithun's shrine at Winchester, an unruly bunch attacked a friar in the act of burying the dead and forced him to enlarge holy grounds for the receipt of plague victims. Ralph of Shrewsbury, Bishop of Bath and Wells, informed his see that bubonic plague had sapped the diocese and robbed parishes of parsons and priests, some of whom may have fled. He urged those who were unshriven to confess their sins to laymen, or in the extremes of need, to laywomen, the last possible choice to confer sanctity.

January 22, 1349 In a second eruption of anti–Semitic violence at Speyer, Germany, mobs destroyed the Jewish community and killed inhabitants. To spare their children from being proselytized with baptism, Jewish mothers threw babies into the persecutors' fires and then immolated themselves in the flames. Survivors either fled to Heidelberg or converted to Christianity. Because the city claimed all Jewish real estate, even the cemetery, the prospect of enrichment heightened the zeal of pogrom leaders.

early 1349 On her way from Finsta, Sweden, to Rome to celebrate the holy year of 1350, proclaimed by Pope Clement VI, Saint Bridget of Uppland concluded that the pious could suppress plague if they gave up extravagant dress, donated generously to benevolences, and honored the Holy Trinity with a monthly mass. Until her death in 1373, she remained in Rome to establish the Bridgettines (or Birgittines) and a hospice to treat the sick. She was so revered in times of illness that the touch of her hand was said to restore the suffering to health.

February 9, 1349 In Strasbourg, an anti–Semitic cabal of laborers forced the resignation of the more liberal Mayor Peter Swarber and two counselors. Blaming Jews for disseminating plague, the plotters rounded up 2,000 five days later. Before burning the lot, zealots spared only those who converted and children who could be forced into baptism. The mob looted the homes of those killed in the conflagration.

As described by chronicler Jacob von Konigshofen, author of *Cremation of the Strasbourg Jews* (1349), the immolation took place on a wood scaffold in the Jewish cemetery. By summer, even those Jews baptized as Christians came under suspicion of spreading plague, which had crept into the city. On September 12, 1349, the Emperor Charles IV pardoned the Strasbourg mob for the monstrous treatment of Jewish citizens. Similar persecutions re-emerged at Augsburg, Frankfurt, Nuremberg, and Regensburg.

spring 1349 The outbreak of plague in Hereford, England, derived from surrounding counties. For six months, the disease sickened West Midlanders. To combat contagion, they avoided joining social groups, ceased observing holy days and sacred feasts, and turned to the nurse care of their kin. Because many priests were either terrified of contagion or deceased, healthy clerics worked overtime at visiting parishioners and performing funerals.

Hereford's church vacancies filled and opened once more as the newly hired clergy died. The total number of bestowals of religious duties rose from fewer than ten annually to 160 in 1349, when there were 56 deaths among priests. The numbers speak of a year of serious turmoil:

> In July alone, 19 died. By 1360, religious service stabilized once more to reflect the numbers of the mid–1340s: a total of 7 were lost, 4 by death, 1 by resignation and 2 from other causes [Dohar 1995, p. 42].

More difficult for hard-pressed clergymen was the challenge of raising spirits and bolstering faith during the scourge. In the Frome deanery alone, staff presided over 158 funerals for plague victims. At Leominster, there were insufficient monks to farm their lands. The town of Yeye disappeared altogether. In an era of diminished income, church offerings were reduced to less than a living stipend for priests, who petitioned cathedral clergy for pastoral aid and succor. Because of mounting poverty, Benedictine brothers at Wenlock Abbey received a diocesan pardon of 200 marks to cover the loss of tenants.

For the common laborer, the Black Death had a salutary effect for survivors. People who had to toil for a living incurred infections more easily than those who sequestered themselves in castles and manors. Because of an immediate shortage of workers, wages rose in response to the competition for skilled help. Another shift was the increase in the importance of English, the language of the working class, and of English education in the classroom. The ambitious fled cottages and rural homes to better themselves. Those remaining behind bore the burden of heavy responsibility. From the spirit of discontent emerged the Peasants' Revolt of 1381.

May 1349 A London vessel bearing wool to Scandinavia also carried plague. Residents of Bergen, Norway, noted that the vessel drifted onto the shore because the crew shrank to cadavers. The people attempted to quarantine themselves from plague at Tuysededal, but all died except for one child. The girl, named Rype, reverted to a feral lifestyle. Meanwhile, plague drifted on to Prussia and northwest to Greenland and Iceland.

July 1349 Plague ravaged Frankfurt, Germany, at the rate of 1,000 per day at its height. In retaliation against scapegoat Jews, the Brotherhood of Flagellants initiated a murderous pogrom in the Jewish sector. In August, the anti–Semitic pogroms revived at Cologne and Mainz.

December 1349 Plague struck Cologne and killed thousands at Erfurt, Bremen, Mainz, and Munster. Torture of Jews as poisoners of wells decimated Jewish communities at Basel, Cologne, Strasbourg, Worms, and Zurich and inflamed Flemish persecutors against Jews in Flanders.

March 26, 1350 As the plague sped over Britain's outer isles, Greenland, Iceland, and Scandinavia, it wrought sudden death, dismaying King Magnus II of Sweden, who believed that God punished human sin with disease. The first European ruler to die of plague was Alfonso XI of Castile, who fell on March 26, 1350, at Gibraltar while facing combat against Arabs. *See 1435.*

1351 When the Black Death reached eastern Russia, it killed Theognostos, the patriarch of the Russian Orthodox Church from the Greek isle of Chios, as well as the Grand Duke of Moscow before moving into the Ukraine. Contagion circled south to strike Crimea, where it had first infected Europe in 1346.

1357 Domenico di Leonardo Buoninsegni's *Istoria Fiorentina* (*History of Florence*) (1580) chronicled how the great Italian city experienced an outbreak of a disease that he named "influenza di freddo (influence of cold)." The term refers to the "influence" of the stars on respiratory ills.

late 1350s As a result of mass death during the Black Death, according to French chronicler and Carmelite friar Jean de Venette, the social order tipped dangerously near chaos. Throughout Europe, commodities, food, and the labor of tradespeople and artisans more than doubled in price before spiraling downward a century later. Many were unemployed, including the *ecorcheurs* (flayers), the marauding, leaderless soldiers who turned to crime and brigandage. Kings went into debt as taxes rose. Citizens grew greedy and covetous. They hoarded coins, pressed lawsuits, and fought over wills and inheritance of lands and properties.

Famine and wars flourished, as did maligning of Jews, Gypsies, and other targetable cultures. To no avail, Pope Clement VI again condemned specious blame against single groups of people; Charles IV of Bohemia, Rupert von der Pfalz, Pedro IV of Aragon, and Duke Albert of Austria also attempted to avert pogroms. Despite the pope's issuance of bulls against violence, the result of smoldering racial prejudice within trade guilds was the near-genocide of Jews in Germany, in part because Jews appeared to withstand infection better than Christians.

1361 A new wave of bubonic plague, called the *pestis secunda*, devastated Europe. It primarily struck cities, killing 20 percent of the population. At Edinburgh, where the death count was one-third of the city, nobles suffered the greatest loss. The *Scotichronicon* (*Scots History*) (1387) compiled by John of Fordun, a priest at Aberdeen, noted: "Men shrank from it so much that, through fear of contagion, sons, fleeing as from the face of leprosy or from an adder, dared not go and see their parents in the throes of death" ("The Black Death"). Those Scots owning country homes fled the city and sheltered in isolation until the contagion passed.

In England, King Edward III outraged his people by cornering the market on masons and builders for the construction of Windsor Castle.

Chronicler Ranulf Higden, author of *Polychronion* (1363), remarked that the king's project coincided with a fall in the number of skilled laborers, many of whom died of plague. Although the men were paid royal wages, they felt the outrage of citizens needing their services. To keep the king's staff from decamping, the sheriff of Yorkshire distributed red hats and uniforms to make them easily identifiable.

1365 When plague overran Swedish communities, diarist Marienna Jensdatter, wife of Asbjørn Pedersen Marsvin of Ystad, Scania, explained that the death of husbands elevated women in patriarchal society. Many Swedish widows headed households and took over their husbands' businesses.

1366 English historian Raphael Holinshed, author of *Chronicles of England, Scotland, and Ireland* (1577), described how a scourge returned to England, killing people of all ages. To differentiate it from bubonic plague, he named the pathogen "small pockes."

1369 Europe suffered the *pestis tertia*, a third reinfection of bubonic plague, which raged for two years and killed up to 15 percent of the populace. The hurling of remains into the Thames River created a noxious stench.

1372 To treat victims of epidemics, Catherine Benincasa, proclaimed Saint Catherine of Siena, Italy's patron saint, set up a nurse corps called the Order of Saint Dominic or the Caterinati. They specialized in the care of incurables, particularly lepers and plague victims. From her hospice at Siena emerged the Hospital of Santa Maria della Scala, one of Europe's oldest medical centers.

January 17, 1374 To halt contagion when bubonic plague threatened a recurrence, Bernabo Visconti of Reggio, Lord of Milan, Italy, enacted arbitrary sanitary laws. He commanded that plague victims be carried into the countryside to recover or die and that anyone nursing the sick should withdraw from contact with others for ten days. Bernabo further ordered priests to report all illness or suffer burning at the stake for noncompliance. Anyone introducing disease to Reggio risked forfeiture of possessions.

September 11, 1374 Throughout Austria, Bohemia, Carinthia, France, Holland, Hungary, and Poland, the greatest and most chronicled episode of tarantism or epidemic dance occurred. Jean d'Outremeuse, author of *Le Geste de Liege* (*Events at Liege*) (ca. 1388), witnessed the phenomenon in September and remarked:

> There came from the north of Liege … a company of persons who all danced continually. There were linked with clothes, and they jumped and leaped…. They called loudly on Saint John the Baptist and fiercely clapped their hands [Bartholomew 2000].

Other histories describe the dancers as heathen, immodest, and shameless for their nudity and blatant sexuality. Radulpho de Rivo considered them satanic in their blend of religious and carnal behaviors. The singing and gyrations continued for days, halting only for rest, food, and drink.

At Aix-la-Chapelle in West Germany, as noted by Peter of Herental, author of *Vita Gregorii XI* (*Biography of Gregory XI*), delusional pilgrims from as far away as France broke into hysteric dancing, called the dancing mania, also Saint Vitus's dance or Saint John's dance for the midsummer festival in the saint's honor. Along the Moselle and Rhine rivers, dancing throngs twitched, postured, and tumbled in trancelike ecstasy. The number surpassed 500 at Cologne, where a mass epidemic consisted primarily of women. Participants suffered severe chest pain, which they attributed to demons requiring exorcism, and developed epileptic seizures that imitated the release of devils.

1383 During epidemics, health officials in Marseilles, a vulnerable French seaport, began quarantining the city and its docks for periods up to 40 days each. The French *quarante* (40) was the source of the public health term "quarantine."

1400 The re-emergence of plague in Iberia revived fears of the Black Death, which had thrived, then ebbed only a half century before. Gonzalo de Meno, the archbishop of Seville, withdrew to the country, but could not elude death from infection. In Lucca, Italy, apothecary and herbalist Giovanni Sercambi, who was born in 1348, grew up in a tradition of plague lore until the pathogen killed him in 1424. He wrote *novelle* (stories) in 1374 about a traveling party in flight from a plague who told stories along their journey about Italy. The framework influenced Geoffrey Chaucer's *Canterbury Tales* (1385). In 1400, Sercambi also compiled a city history, which he illustrated with ink sketches. He produced images of the Black Death that depicted contagion as a corruption of air and soil poured out by a symbolic Angel of Death in the form of arrows.

1402 When bubonic plague struck Iceland, the nation was already reeling from starvation, natural disasters, and the loss of trade goods from Norway. The disease was so relentless that it killed half the populace, annihilating entire families. After losses to the servant class, upper-level society had to fend for itself. So many farmers died that the government forced fishers to put in time working livestock, plowing, and harvesting. With fewer offspring to inherit family plots, acreage passed to the open market and into the ownership of the church.

July 1415 During the siege of Harfleur, France, dysentery, a common soldier's disease, overran the English Army. Hundreds died within two weeks. By the end of the siege on September 22, around 2,000 lay dead of dysentery; 2,000 more returned home wretchedly sick with diarrhea and dehydration. Among the victims were the Earl of Suffolk and Richard Courtenay, the Bishop of Norwich and a favorite adviser and companion of the ambitious King Henry V. The king was attending to the dying bishop on September 15 and immediately ordered Courtenay's interment under the high altar at Westminster Abbey. Seven years later, the king also succumbed to dysentery at Vincennes in eastern Paris.

May 1417 A return of bubonic plague to Florence, Italy, reduced the population from 40,000 by one-tenth to 36,000. Collapsing at the rate of 500 per day, the victims tended to be members of the laboring class rather than the rising merchant class.

1418 At Strasbourg, in Belgium, and along the lower Rhine, an outbreak of dancing plague or tarantism, a manic group hysteria, afflicted hordes of participants. Sympathetic people conducted the dancers to the chapels of Saint Vitus at Zabern and Rotestein, where priests cured them with ritual derived from the piety of Saint Vitus, patron saint of those afflicted with epilepsy and mania.

June 1425 Florentines, particularly slum children, complained of *pondi* (dysentery), which ravaged the Italian city for four months, killing 120.

1430 After waves of bubonic plague struck western Europe, the population fell to perhaps 20 million, a quarter of the number that had lived in 1290. When the disease emerged in Florence in January, it flourished in a crowded urban setting, killing many children into mid-year. Only 16 percent of those infected by the disease survived.

1432 When plague assailed Lisbon, Portugal, fleeing peasants carried infection throughout the land. Many thousands were struck down in the normal acts of the day—dining at the table, walking the streets, shopping, and worshipping. The city lost health workers and priests to the disease. So many corpses littered the streets that dogs devoured their flesh and carried contagion to more people. Monsignor Andre Dias of the Monastery of Saint Dominic urged hysterical citizens to perform acts of piety and placed on the bodies of victims drops of holy water and cards with the name of Jesus written on them. *See also* **1435**.

1434 In Castile, the Spanish King Juan II battled the endemic "Jewish leprosy" by enacting new sanitary regulations and by founding royal hospices that sheltered lepers. To enter the facilities, patients deeded their possessions and property to the king. He also suppressed a plot to murder and rob Jews who converted to Christianity.

1435 Plague infiltrated much of Portugal and raged for over three years. The disease killed King Duarte I in the sixth year of his reign, reputedly from handling infected mail. Ironically, he succeeded King João I, who had died of the disease in 1433. The sudden loss of Duarte left the Portuguese leaderless during squabbles between the Queen Mother and the Infante Dom Pedro, who sought to control the six-year-old heir, Duarte's grandson, Alfonso V the African.

1451 During seven episodes of bubonic plague between 1451 and 1503, Sisteron in south central France lost two-thirds of its population. The ignorant blamed black magic and corralled and executed suspected witches. At Brignoles, enforcers of health regulations took a more sensible approach in questioning travelers and rejecting those coming from plague-ridden areas. The least rational blame for contagion targeted lepers, whom the ignorant accused of working as agents of Jews to spread disease among Christians, a form of scapegoating and persecution that re-emerged during stressful times.

mid-summer 1451 A plague epidemic in Milan, which peaked from mid-summer to September 1451, failed to yield to local public health initiatives of exiling transients and plague victims west along the great canal to a villa at Cusago. To house the overflow of patients, health workers erected the Locus Montanee, a pesthouse at Milan's southwest edge. They lodged other plague victims in abbeys, temporary sheds, and abandoned hospices. Near the end of the scourge in 1452, Giovanni Catelano became a respected doctor at the Ospedale Maggiore and *sanità* (public health) official. He observed the lower-class sick and recorded his findings in the Milanese *Necrologi* (*Death Registers*), a testimony to symptoms and course of infection. *See also* **1468** and **1483.**

1463 Bavaria suffered a lethal four-year siege of plague that struck Augsburg, Munich, and Nuremburg and killed 8,500 in Regensburg alone. Among the dead were 16 monks at the convent of Saint Emmeran, all killed within four weeks.

1464 Visual representations of disease appeared on the *gonfaloni*, plague banners used in Italian religious processions as a form of prayer or supplication and as a call to penitence and devotion to the Virgin Mary, the chief intercessory figure to scourge victims. Painter Benedetto Bonfigli of Perugia immortalized the onset of plague in his city with "Gonfalone di San Francesco al Prato" ["Banner of Saint Francis of the Field,"] a pictorial banner designed for the church of San Francesco di Perugia. The pose characterizes the power of the Virgin Mary, who receives the prayers of eight saints and stretches her arms and cloak to protect the faithful from pestilence.

The symbolic scene below the main figure pictures a winged demon with rotting flesh who raises bow and arrows against a spear-wielding angel. In a peripheral image, the Holy Family depicted outside the city walls flees unharmed under divine protection. In 1471, another allegorical picture by Bonfigli blamed Christ, who hurls lightning at Perugia, which saints and the Virgin Mary defend.

August 1466 Around 40,000 Parisians died of a three-month epidemic of plague. The superstitious accused Jews and lepers of spreading infection. To end the misery and allay panic, pious citizens resorted to religious processions, public chanting, and displays of plague banners and religious icons.

spring 1468 During a mild plague epidemic in Milan that coincided with famine, Giovanni Catelano, a devoted and hard-working public health official, was forced to defend his professional judgments because 22-year-old Duke Galeazzo Maria Sforza feared that a full accounting of infections would ruin his large and expensive summer wedding. Health commissioner Hector Marchese defended Catelano and others who were unable to give immediate diagnosis of ambiguous symptoms and thus isolate the sick before the spread of contagion. Because Catelano pursued details of disease from relatives and other physicians, he was accused of robbing the dead and dying. After his housekeeper's death from plague, the doctor had to declare himself a possible carrier of the pathogen.

That fall, Duke Galeazzo Maria Sforza proved relentless in tracking down cases of disease. His agents located a youth who fled the center of Milan and died on the outskirts. By constructing a permanent pesthouse, the duke intended to rid the city of festering contagion.

spring 1477 A re-emergence of bubonic plague and typhus ravaged northern Italy, killing over 80,000. Spreading from Milan to Brescia, Mantua, and Venice, where 30,000 died, the plague terrified peasants and prelates and left doctors helpless to alter its course. The scourge advanced along the Adriatic coast of Croatia to Dalmatia the following year. *See also* 1489.

1479 While safely sheltered in Crepole, Italy, chronicler Francesco Pagni described the arrival of plague to Pescia, Tuscany, where 1,487 died. In typical style, he opened his document with *Ad Divam Virginem* (*To the Divine Virgin*), a propitiatory address to the Virgin Mary, intercessor for the suffering.

1483 A three-year pestilence in Milan appears to have been an epidemic of typhoid fever. One medical author, Giuliana Albini, carefully differentiated the symptoms from the city's previous onslaughts of bubonic plague. At its height in August 1485, the disease produced a mortality rate of 40 percent, primarily women and children. As late as October 8, 1485, when 51 new cases were identified, the infection still ravaged the city. On December 8, 1485, Duke Galeazzo Maria Sforza received notice that health inspectors were supervising city sectors, purging, purifying, and burning possessions, bedding, and garments to rid residences of contagion. In his 31st year as public health doctor, Giovanni Catelano was still on the job studying symptoms and examining corpses to give a true accounting of disease.

1485 To end a plague epidemic, the people of Vicenza, Italy, built the votive Church of San Rocco as a supplication to heaven. They dedicated the premises to Saint Roch (or Rock), the protector saint of pilgrims and plague victims.

late August 1485 The *sudor Anglicus* or English sweating sickness (also called bloody fever and day sweat) emerged after Henry VII's triumph at the battle of Bosworth in central England, which established his claim to the throne. The disease produced headaches, muscle pain, fever, labored breathing, rash, and unusual perspiration primarily in young victims living in England and France. An observer, historian Polydore Vergilio, author of the 27-book *Anglica Historia* (*English History*) (ca. 1537), described terrible thirst and foul body odor of the afflicted. Outbreaks recurred at London and Chester in 1508 and at Cambridge, London, and Oxford in 1517. Medical historians have compared the symptoms to scarlet fever, plague, hantavirus, and influenza.

1488 Construction of a vast pesthouse, the Lazaretto di San Gregorio of Milan, applied the architecture of Carthusian monasteries to isolating contagion. As a means of containing infection in Italy, staff assigned patients to 288 private rooms on the outer edge of a rectangular courtyard. When patient load rose to 16,000 in 1630, health workers erected temporary shelters in the center.

1489 Typhus, a wartime disease called camp fever, began during the Crusades, but did not reach epidemic proportions in western Europe until 1489, when it ravaged Spain during war between the troops of Ferdinand and Isabella and the Muslims of Granada. Spanish troop casualties totaling 20,000 were 6.6 times higher for typhus than for combat wounds. Speculation varied as to the cause of contagion. Some authorities blamed Spanish soldiers arriving from Cyprus after battling Ottoman forces. Other medical experts suspected shipboard rats imported from Asia.

1492 As described in the *Nuremberg Chronicle* (1493), the German author, Hartmann Schedel, observed a diphtheria epidemic and summarized its characteristics.

1493 Plague killed 80 percent of the citizens of Genoa, Italy. Saint Catherine of Genoa, a hospital director and author of *Dialogues on the Soul and the Body* (1510), survived infection. In recovery, she admitted plague victims to her medical center, Saint Laurence Hospital. She and her husband, Giuliano Adorno, a Franciscan tertiary, treated the dying until Giuliano also died from plague. She treated his mistress and Giuliano's illegitimate children, who also suffered the scourge. For Catherine's devotion to hospital work, she became the patron of nurses.

1493 As described in the *Treatise on the Serpentine Malady* (ca. 1520), the author, ship's surgeon Rodrigo Ruy Díaz de Isla, treated Christopher Columbus's crew for syphilis, a new disease striking Europeans. The infection passed from the Arawak to sailors via sexual contact. Of its newness, Diaz de Isla said:

> I myself name it the serpentine sickness, of the island of Hispaniola. For just as that animal is hideous, fearsome and terrible, in like wise is the malady ... which ulcerates and corrupts the flesh, breaks and destroys the bones, and cuts and shrinks the tendons [Gugliotta 2000].

After a lengthy incubation that could last for decades, syphilis reached its tertiary stage, spreading tumors and irreversibly damaging skin, bones, heart, and nervous system. It corrupted tissue and caused arthritis, heart disease, insanity, and death.

1493 The first European doctor to reach the Western Hemisphere, Diego Alvarez Chanca of Seville, royal physician to Ferdinand and Isabella of Spain and the first professional doctor to practice in the Western Hemisphere, battled an outbreak of malaria that struck Christopher Columbus and his crew. It was the first major onslaught of disease after initial European contact with Native Americans.

January 30, 1493 Swine influenza struck the Indians of Hispaniola or Santo Domingo in the Antilles, nearly annihilating the native race. As the first New World chronicler, Diego Alvarez Chanca described the flora and fauna of Hispaniola, including Taíno and Carib culture, the value of the chili pepper to native diet and healing, and the importation of pigs from the Canary Islands to Hispaniola. The animals spread swine flu in the town of La Isabela, the first Spanish settlement in the Americas, killing hundreds of islanders. There were so many dead Indians that their corpses accrued in fetid piles, rendering the area unfit for habitation.

Chanca explained that Christopher Columbus caught the acute infectious disease on December 9, 1493, and required three months recuperation. One-third of his 1,500 men sickened and died. In a letter to Ferdinand and Isabella, Columbus wrote, "The cause of the ailments so common among us, is the sustenance, and the waters, and airs, because we have noticed that they affect everyone, though few are in danger" (Guerra 1988, p. 307). He added that Chanca, the official fleet surgeon of the second voyage to the Caribbean, stayed busy treating the sick.

Chronicler Pietro Martire d'Anghiera's seven-volume De Orbe Novo (On the New World) (1494–1530) reported on the high death rate among expeditioners and Indians and added that the replacement of the dead with Lucayan Indians, a branch of the Arawak from the Bahamas, caused the newcomers to die. Of their loss, Martire remarked, "The number of such poor devils has diminished extraordinarily; many affirm that once they were 1,200,000. How many are they now? I am frightened to think about it" (ibid., p. 308). In the estimation of Spanish legal scholar Juan de Solorzano Pereira, "The breath of other people killed them" (ibid., p. 319).

Affirming the loss was a Dominican friar, Bartolomé de las Casas, author of Historia de las Indias (History of the Indies) (1532). In "Memorial on Remedies for the Indies" (1518), he mourned that "of the 1,000,000 souls there were in Hispaniola the Christians have left but 8,000 or 9,000, the rest being dead" (ibid., pp. 309–310). He also reported that a similar death rate reduced the native population of Cuba, Jamaica, the Lesser Antilles, and Puerto Rico. Chronicler Gonzalo Fernandez de Oviedo y Valdéz corroborated the sickness in Natural History of the Indies (1526) and added that the recovering Spaniards were "like citron and the color of saffron or jaundice" (ibid., p. 313).

1494 Called the great pox to distinguish it from smallpox, syphilis was introduced to Europe by Indian captives aboard Christopher Columbus's ships. A Roman physician to Pope Julius II, Giovanni de Vigo (also Jean, John, or Juan di Vigo) of Rapallo, characterized in Arte Chirurgia Copiosa (On the Complete Surgical Art) (1514) how syphilis arrived in Europe with the return of Columbus from the Indes. He explained:

> In the year 1494 ... there appeared ... throughout almost all Italy, a malady of a nature unknown to that time.... The contagion which gives rise to it comes particularly from coitus; that is, sexual commerce of a healthy man with a sick woman or the contrary [Major 1932, p. 28].

In de Vigo's opinion, the disease followed the troops of Charles VIII of France to war in Naples. In fall 1494, his hired soldiers, including some of the men who sailed with Christopher Columbus, overwhelmed Neapolitan forces.

Prostitutes servicing men on both sides of the conflict spread syphilis. By the next spring, the French, pox-ridden and tottering with joint pain, had to fall back, dispersing the infection to women they encountered along the way. As mercenaries demobilized, the disease made its way to Switzerland, Germany, Holland, the British Isles, Hungary, Poland, Russia, Africa, and Asia. Four years after the war, Charles himself died of the disease. Until the Emperor Maximilian himself was stricken, he concluded that syphilis was divine punishment of blasphemers. A blame game produced alternate names—mala napoletana (Neapolitan disease), mal de Naples (the Naples disease), and morbus gallicus (French disease). As the scourge spread east, similar attributions accused the Spanish, Polish, Germans, and Turks of harboring syphilis.

De Vigo expanded on the appearance of rigid pimples on the penis or vulva as the first symptom and described the transmission of syphilis to the genitals from copulation with "an unclene woman." De Vigo particularized symptoms:

> lytle pushes of blewe colour, otherwhyles of blacke, sometyme of whytyshe, wyth a certayn hardnes aboute the same, whych pustules could not be healed by medicine applyed with in or wythout, but that they wold enbrace the hole bodye, wyth ulceration of the genetall partes, euer returnyng agayne after they were healed,

chiefly in the ioyntes [joints], in the armes, under the knees, & in the foreheade, and welnye spredde through all the body, & yet at thys tyme they begyn euen so, but it is not so contagious as it was at the begynnyng [Carr].

1494 After a 92-year abeyance, plague beset Iceland again, making its chief kill in the remote West Fjords. As revealed in folklore, the pestilence reached the island in the form of blue fabric, possibly an historic link to an English merchant vessel. The shape-shifting cloth took on a life of its own and flitted into Hvalfjord as a blue bird, then altered itself once more into undulant mist, a sinister presence that reminded the people of sea fog. Survivors abandoned the region to locate available acreage to the north.

1495 When African slaves began cultivating sugar on the first West Indies plantations, they introduced smallpox to native islanders. Simultaneously, yellow fever imported by the expeditionary force led by Christopher Columbus flourished in virgin-soil territory. On the island of Santo Domingo, the death rate was 80 percent. At Caonabo, the Spanish triumphed in war, but shared the misery of natives stricken with yellow fever. To protect his men, Columbus had to abandon the site and establish a new headquarters before returning to Spain on March 10, 1496. *See also* June 27, 1498.

1496 The Guanche, natives of the Canary Islands who first encountered the Spanish in 1402, lost their traditional homeland to European invaders. During the final influx of Spanish to Tenerife in 1496, in a few weeks, an unidentified disease the outsiders called "Guanche drowsiness" killed hundreds of natives within weeks. The original 80,000 Guanche were so riddled by European diseases that they capitulated to the Castilians, who enslaved them and drove them off their land. Like other indigenous peoples, the Guanche virtually vanished from history. By 1600, historians declared them extinct.

1496 As stigma attached to victims of syphilis, humanist author Theodericus Ulsenius (also Dirk van Ulsen), a Nuremberg physician, issued his advice to the city fathers on prevention and treatment. He published a poem, *Vaticinium in Epidemicam Scabiem* ("Revelation on Epidemic Scabies") (1496), and *Cura Mali Franci* (*Cure of the French Pox*) (1496), an anthology of 50 aphorisms intended to instruct professional health workers. His writings incorporated the astrological and philosophical leanings of the period.

1497 Vasco da Gama's voyagers carried syphilis from Europe to the Indian Ocean, where it spread to India in 1498 and moved east to China and Japan. On his long journey around the Cape of Good Hope the following year, over 62 percent of his crew died of scurvy, a vitamin C deficiency that killed some one million sailors by 1800—more than died from combat, shipwreck, or contagious disease.

1497 Five years after Christopher Columbus's first voyage to the New World, Niccolo Leonicen, who taught at the medical college at Ferrara, Italy, authored *Libellus de Morbo Gallico* (*A Pamphlet on the French Disease*) (1497). In reference to the sudden appearance of syphilis in Europe, he wrote:

> A disease of an unusual nature has invaded Italy & many other regions. In the beginning pustules are on the private parts, soon on the whole body and frequently located on the face itself besides causing great hideousness as well as a great deal of pain. Moreover to this disease the physicians of our time do not yet give a name, but is called by the common name of French disease, as if this contagion were imported from France into Italy or because Italy was invaded at the same time both by the disease itself and the armies of the French [Major 1932, p. 16].

A contemporary, Gaspar Torella, the personal doctor of Pope Alexander VI, summarized the epidemic in Rome and called the new disease pudendagra. At Ferrara, Sebastiano dall'Aquila (also Sebastianus Aquilanus) contributed to serious local debate on the causes of the disease.

April 21, 1497 When syphilis began infecting the Scots at Aberdeen, the city fathers banned prostitution with the edict that "all the light women of the town should desist from their sins of venery" [Carr]. Those who violated the law risked branding on the cheek.

September 1497 During the reign of James IV of Scotland, an unidentified epidemic was so repugnant that the Edinburgh council enacted the Grandgore Act, banishing victims of "grandgore" (large sore or syphilis) off Leith to the mile-long island of Inchkeith in the Firth of Forth. In token of the plague's virulence, poet Robert Henrysone, called the "schoolmaster of Dunfermline," composed an 88-line propitiatory poem,

"Ane Prayer for the Pest" (ca. 1498). Repeated throughout was the cry, "Preserve us fra this perrelus pestilence" (Henrysone, 1997).

1498 Francisco López de Villalobos, a Spanish physician educated at the University of Salamanca, published *Medical Works* (1498), a poetic summary of the emergence of syphilis in Europe. He credited sexual adventures, drunkenness, and overeating as the sources of infection.

1498 News of bubonic plague so terrified Scots that, according to the *Records of the Burgh of Edinburgh*, laws forbade contact with infected villages or their belongings. Punishments ranged from exile and the burning of all belongings to the death penalty. As plague encroached on Edinburgh, more regulations forbade entering the city except by the main gate past a guardpost. Laws prohibited the roaming of children beyond their homes and required the burning of fabric imported from England. Civic authorities closed schools and taverns, inspected all imported food and merchandise, and imposed a 10 p.m. curfew.

By 1499, the spread of the disease to Haddington and Peebles required a new regulation banning from Edinburgh all travelers from those cities. Additional statutes required the execution of free-roaming dogs and pigs and the placement of unruly children in stocks for scourging. Teachers who continued to hold classes were exiled. Anyone opening a trade booth faced confiscation of goods. A head tax of 12 shillings for nobles and five shillings for peasants paid for street cleaners, who carried a white wand of office to denote their importance. Lodged in isolation at the Hospital of Saint Mary Wynd on High Street and avoided by citizens, these workers washed and smoked infected houses.

By 1500, Edinburgh's council shut down trade at the Lawnmarket and targeted servants with branding and exile if they took or bought clothing without their masters' approval. Citizens importing goods illegally faced lopping off of their hands for males and branding on the cheek for females. People living in infected houses incurred a 12-day quarantine. Belongings removed from infected houses required washing and smoking and an eight-day period of drying in fresh air. Anyone violating the regimen faced the burning of property. *See also* **1529, May 1530, September 1568.**

June 27, 1498 When Christopher Columbus made a third voyage to the New World, he halted his fleet at the Cape Verde Islands for provisioning. He cut short his stay on July 4 because of rampant yellow fever, which began infecting his men. The arrival of the fleet at La Isabela, Santo Domingo, may have caused the first yellow fever epidemic in the New World.

June 1499 As revealed by historian Raphael Holinshed's *Chronicles of England, Scotland, and Ireland* (1577), plague besieged England, striking London in June 1499 and extending into the first weeks of 1500. Between 20,000 and 30,000 died in the outbreak. Enlightened people abandoned church-led scapegoating of Jews, prostitutes, homosexuals, and Gypsies or Rom as causes of the plague and began to develop theories of biological contagion.

early 1500s Early in the 16th century, Saint Jerome Emiliani set up a receiving home in Venice for children orphaned by plague. He himself survived the disease and continued his efforts to aid the afflicted at a hospital he built at Verona in 1518.

1502 Juan Almenor, a Venetian physician and author of *De Morbo Gallico* (*On the French Disease*) (1502), noted that the upsurge in syphilis, a previously unknown ailment, earned the Italians' spite and the name Gallicus. Almenor added, "[It] should now be named Patursa, ... a disease filthie and Saturnall. It is a filthie disease, because it maketh women to be esteemed unchast and irreligious" [Major 1932, pp. 22–24]. He explained its source as twofold: "The first is the only influence or corruption of the aire, from whence we must charitably thinke, that it infected those which were religious. The second is conversation, as by kissing and sucking, as appeareth in children, or by carnall copulation" (ibid.). *See also* **1519.**

As a result of the epidemic, syphilis affected royalty as well as the commoner. In Poland, the disease compromised the aristocracy and the military. Both the Jagiellonian and Vasa dynasties died out from infertility caused by genital infection. In Germany, William II of Hesse weakened from syphilis, which altered his rule and threatened his lineage. As his health worsened in 1506, he began to secure rule for his infant son. In the German king's last three years, his mind declined to such an extent that a regency council ruled in his stead.

Just as victims of the Black Death scapegoated Jews as carriers of infection and deliberate spreaders of toxins, early 16th–century Gentiles once more passed the blame for venereal disease on to Jews. Anti-Jewish stereotypes that flourished in the Middle Ages perpetuated myths of unhygienic skin vulnerable to leprosy and syphilis. Added to the victimization of Jews, Hebraic males were accused of rabid sexual appetites and high incidence of venereal disease.

1502 Nicolás de Ovando, the first permanent governor of Hispaniola and notorious brutalizer of the Taíno, colonized Santo Domingo. To retain a labor force to work the mines, he petitioned Spanish monarchs Ferdinand and Isabella to replace dying natives with cargoes of West African slaves. As a result, most of his 2,500 followers succumbed to yellow fever, which was endemic in Africa.

1505 During a decade of petty kingdom wars, armies tramped up and down the Italian boot, spreading typhus. Commenting on the advancing contagion was the Veronese physician and medical writer Girolamo Fracastorius (also Jerome or Hieronymus Fracastor or Fracastoro), a graduate of the University of Padua. In *De Contagione et Contagiosis Morbis* (*On Contagion and Contagious Diseases*) (1546), he attributed infection to contacts with Cyprus and other Mediterranean isles.

1506 The arrival of the Portuguese to coastal Newfoundland introduced smallpox, tuberculosis, and other European diseases to the Beothuk, the original "red men," named for the red pigment they smeared on their skin. They became one of the first indigenous New World nations to disappear as a result of epidemics. The last Beothuk, Nancy Shawanadith, died at Victoria Hospital in Saint John's, Newfoundland, in 1829.

1507 Abuse of native slave labor in the silver and gold mines of Hispaniola and Cuba forced the Spanish overlords to introduce West African slaves, who carried smallpox into virgin-soil territory. As Europeans looked on while whole native tribes expired, Catholic dogma accounted for the astounding death rate as the will of God that Spain should rule the New World.

1508 On royal appointment, Captain Diego de Nicueza (or Nicuesa), the royal meat carver for King Ferdinand and Queen Isabella and a popular courtier-turned-adventurer, led a five-ship convoy from Spain to the Caribbean. His followers intended to colonize a strip of Mesoamerica on the Caribbean side from Darien, Panama, to Cape Gracias a Dios, called Castilla de Oro. Traveling from Santo Domingo en route to the Isthmus of Panama, his ship ran aground during a storm. He and 700 men were marooned on an unknown islet apart from the fleet, where they suffered thirst and hunger. Epidemic yellow fever afflicted 86 percent of his crew. Illness, starvation, and hostile Indian attack reduced his followers to 100, forcing him to abandon his mission.

1512 After the Portuguese carried venereal disease to Asia's mainland, the arrival of syphilitic Chinese traders or pirates to Nagasaki, Japan, introduced the disease to a virgin-soil population. The Japanese dubbed the resulting chancre the "Chinese ulcer" or the "T'ang sore," a snide reference to the Chinese ruling house.

1515 The native population of Puerto Rico suffered exhaustion and depletion when enslaved by the Spanish in 1509. As a result of fatigue, despair, and lack of immunity to new diseases, only one-third of the Arawak and Calusa survived epidemic smallpox, which African slaves introduced on the island. In 1516, Bartolomé de Las Casas, a Dominican friar and author of *Historia de las Indias* (*History of the Indies*) (1532), obtained a royal order to free the natives. From 50,000 in 1507, within a quarter century, their number shrank to 600.

1517 A lethal outbreak of diphtheria in Basel, Switzerland, and the Netherlands killed within 24 hours of the appearance of symptoms. Cause of death was suffocation after the formation of a membrane in the throat.

1518 An outbreak of alternating depression and spontaneous dance frenzy or tarantism seized people in Strasbourg, France, after a single crazed woman began the hysteric motion. For four weeks, 400 participants joined a manic dance, which historians connect with the emotional trauma of famine and epidemic disease. Some who carried their frolic to a grain market fell down dead with exertion. A rescue mission to Saint Vitus on the Rock at Zabern applied holy water and oil and the sign of the cross to alleviate the ecstasy.

December 1518 The colonial importation of thousands of West African slaves to the Caribbean isle of Hispaniola introduced a second wave of smallpox that quickly engulfed slave and native populations. The pox was an unknown disease among the Taíno, who dubbed it the "great leprosy." Contagion killed one-third of their number. The odor of death seeped through their compounds; contagion made them extinct within a generation. Worsening the effects of the scourge was the Indian healing regimen of communal bathing, which spread the pathogen. From an outbreak at Santo Domingo, the disease invaded Colombia, Cuba, Mexico, Peru, Puerto Rico, Venezuela, and Yucatan. Augustinian priests petitioned King Charles V of Spain for aid.

1519 A German physician, Ulrich von Hutten of Almayn, author of *De Morbo Gallico* (*On the French Disease*) (1519), commented on the prevalence of syphilis:

> It is thought that this Disease in our days ariseth not, unless by infection from carnal Contact, as in copulating with a diseased Person, since it appears now that young Children, old Men and others, not given to fornication or bodily lust, are very rarely diseased: Also the more a Man is addicted to these Pleasures, the sooner he catcheth it.... In Women the Disease resteth in their secret Places, wherein are little pretty Sores, full of venomous Poison, being very dangerous to such as unknowingly meddle with them [Major 1932, p. 34].

fall 1519 The re-emergence of bubonic plague at Dublin lasted into the cold months, subsided, then returned in spring, when it spread down Ireland's east coast, which was held by the English. When typhus struck in summer, people fled in terror and died in glades and plowed fields. The major Irish cities continued to suffer contagion for the next five years as the pathogens of flu, plague, smallpox, and typhus stalked the island.

April 23, 1520 The introduction of European diseases among indigenous Mesoamericans began with the dispatch of Spanish *conquistadores* to Texas. Chronicler Bartolomé de Las Casas, a Dominican friar and author of *Historia de las Indias* (*History of the Indies*) (1532), disclosed that the Spanish expeditions precipitated three centuries of epidemics in Mexico, reducing the population from around 30 million to around three million within a century. The bearer of contagion was Hernán Cortéz, the conqueror of Mexico, who arrived at Yucatan on Easter 1519 with 550 troops and suffered a defeat by Montezuma's Aztec forces. Cortéz prepared for an invasion of the capital city of Tenochtitlán from July 2 to December 20, 1519. Unknown to him, the soldiers whom Diego Velázquez, Hispaniola's governor, had sent from Cuba included a slave ill with smallpox.

In late spring 1520, the first outbreak of *hueyzahuatl* (smallpox) began in Vera Cruz, a city founded by Cortéz, and surged into Tenochtitlán by late summer. The Spanish forced their way into Tenochtitlán with cannon fire, fought their way to the administrative center in block-by-block combat, and triumphed on August 13, 1520. On entry to the fallen city, Cortéz discovered that a single exposure to smallpox infected the entire virgin-soil population, including the Indian mercenaries enlisted by the Spanish. After the introduction of contagion at Cempoalla, the general mourned his old friend and ally Maxixca, the old lord of Tlazcala, felled by the pox that swept the land, striking aristocrat and peasant. The natives immersed themselves in cold water, which hastened their deaths. From Cempoalla, the disease moved to Tlazcala, where Cuitlahua, Montezuma's successor and nephew, fell dead.

After Cortéz destroyed the city's feeder spring, the Aztec had little choice but to drink from a polluted lake. Cadavers choked canals, raising a repulsive stench. A third of the people died of hunger and of horrible eruptions they called the "great fire"; survivors were hideously pockmarked. Only nine of the 100 Spanish who died during the invasion succumbed to the disease. When Álvar Nuñez Cabeza de Vaca arrived in West Texas in 1535, he found the results of smallpox scarring in the blind eyes of a disproportionate number of natives. In 1802, Spanish historian Joaquin Villalba published *Epidemiologia Hispañola*, an account of the Mesoamerican smallpox epidemic of 1520, which he based on Francisco Bravo's *Opera Medicinalia, in Quibus Quam Plurima Extant Scitu Medico Necessaria* (*A Medical Encyclopedia Containing Everything Known to Science*) (1570).

April 14, 1521 Among the Maya, *hueyzahuatl* (smallpox) and other European diseases ravaged an unsuspecting population. On the 40th day of the scourge, the pox killed King Huanya Capac and his heir. As contagion felled peasants, the population of Mexico dropped 90 percent. One estimate of the death rate among natives in

Mesoamerica and South America is 98.8 percent, from 130 million to 1,600,000.

Various writings of the period concur on the total tragedy, but differ in their interpretation. Natives, who compiled a pictorial record of the progress of smallpox described in Bernardino de Sahagún's *Historia General de las Cosas de Nueva España* (*General History of the Things of New Spain*) (ca. 1575), known as the *Codex Florentino*, assumed the superiority and divine favor of their European conquerors. Castilian historian Bernal Díaz del Castillo of Medina del Campo, Spain, a common foot soldier, retired to Guatemala and compiled *Verdadera Historia de la Conquista de Nueva España* (*True History of the Conquest of New Spain*) (1568). Unlike scholarly chroniclers, he gave an anecdotal account.

Another who regretted the terrible loss to disease was Franciscan chronicler Toribio de Benevente Motolinía, who summarized in *Historia de los Indios de Nueva España* (*History of the Indians of New Spain*) (1542) the devastation from smallpox and starvation. He remarked on the disease the Indians called the "great leprosy":

> [The Indians] died in heaps, like bedbugs. Many others died of starvation, because, as they were all taken sick at once, they could not care for each other, nor was there anyone to give them bread or anything else. In many places it happened that everyone in a house died, and, as it was impossible to bury the great number of dead, they pulled down the houses over them in order to check the stench that rose from the dead bodies so that their homes became their tombs [Motolinía].

He commented that those who survived bore pockmarks all their lives.

Additional voices corroborated the terrible loss. In *Relación de Texcoco* (*Concerning Texcoco*) (1580), author Juan Bautista de Pomar, the mestizo historian of the city of Texcoco, called the plague of 1520 the worst of the era. When he compiled his text over five decades after the great plague epidemic, he noted the decline of citizens from 15,000 to 600. In 1565, Francisco Ceynos, a member of the Royal Council of the Indies, summarized in a letter the *grandes muertes* (great death) as a demographic catastrophe.

A treatise on the outbreak of smallpox in Mexico in 1520 appeared under the title *De Febris Epidemicae, et Novae Quae Latine Puncticularis* (*On Epidemic Fevers and New One Particularized in Latin*) (1574). It appears to have been the compilation of Alonso de Torres, whom the Marques Don Luis de Astui'iga y Avila employed to compose a history of the emergence of the disease in the Western Hemisphere. De Torres may have based his work on *Opera Medicinalia, in Quibus Quam Plurima Extant Scitu Medico Necessaria* (*A Medical Encyclopedia Containing Everything Known to Science*) (1570), compiled by Francisco Bravo of Ossuna, Spain. His treatise, the first medical monograph printed in America, introduced to European medicine Mesoamerica's native plant remedies and treatments. *See also* **1524.**

1522 When bubonic plague resurged in Florence, Italy, mannerist religious painter Jacopo Carucci retreated into his work. A student of Andrea del Sarto called Pontormo for his birth in the town of Pontormo outside Florence, he painted under the patronage of the Medicis and was decking their villa at Poggio a Caiano with mythological scenes in 1521. Because of his terror of illness, at the outbreak of plague, he took refuge from contagion for three years at the Carthusian monastery at Certosa del Galluzzo. In the cloisters, he painted frescoes of Christ's passion, Christ before Pilate, and the supper at Emmaus. Carucci's melancholia forced him into more neurotic scenes, including heaps of corpses in the frescos for the choir of San Lorenzo.

Lent 1522 At a session of county courts at Cambridge, England, called the "black assizes," lawyers, judges, and staff fell ill with a lethal outbreak of typhus. Medical historians speculated that contact with prisoners from foul, overcrowded jails spread disease-bearing body lice through courtrooms.

1524 Hernán Cortéz, the conqueror of Mexico, atoned for the spread of cholera, dysentery, measles, pertussis, plague, scarlet fever, smallpox, tuberculosis, typhoid, and typhus among Mesoamerican natives by building in Mexico City the Hospital of the Immaculate Conception, the New World's first European hospital. As smallpox spread northeast around the Gulf Coast, it targeted the Oneota, an agrarian culture along the Mississippi and Fox rivers in Missouri and Iowa.

1525 Before Europeans arrived in South America, smallpox, influenza, plague, measles, typhus, and other new diseases swept south through the Andes from an unknown source, slaughtering natives who had no immunity. The contact between the Peruvian Inca and a Spaniard ill with smallpox introduced contagion that reduced the indigenous population by half. In the words of Franciscan friar Toribio de Benavente Motolinía, a scholar, translator, and author of *Historia de los Indios de Nueva España* (*History of the Indians of New Spain*) (1542), natives died like insects.

1526 Gonzalo Fernández de Oviedo y Valdés, a courtier of Spain's Ferdinand and Isabella and author of *De la Natura Historia de las Indias* (*On the Natural History of the Indies*) (1526), interviewed Christopher Columbus's newly arrived crew and captive Indians following the first voyage to the New World. Based on clinical findings, Oviedo later claimed to King Charles I of Spain that syphilis in Europe derived from expeditioners' contacts with Caribbean Indians. The following year in Switzerland, Johannes Senff (also Sinapius) summarized costs of treating the disease with guaiacum, the resin of a tree found in the American tropics, and of feeding patients a special diet based on chicken, eggs, bread, and purgative herbs. He concluded that a laborer would spend a year's salary under the regimen.

late fall 1526 After explorer Lucas Vázquez de Ayllón, a Spanish attorney, and his Shakori slave Franciscano Chicorano sailed from Hispaniola to continental North America, Ayllón founded San Miguel de Gualdape, the first European town in the United States, near the mouth of the Peedee River at Winyah Bay, South Carolina. Although supplied amply with horses, pigs, and farm implements, his 600 colonists, three Dominican friars, and some African slaves suffered cold and epidemic swamp fever. The settlement lasted less than two months. Following Chicorano's escape and Ayllón's death on October 18, when slaves mutinied and Indians attacked, the remaining settlers fled the area. Only 150 survived to reach Santo Domingo.

1527 In Puerto Rico, epidemic smallpox killed one-third of the remaining Taíno, a fragile population of Native Americans devastated by Spanish colonizers. That same year, native slaves rebelled against their European overlords.

1527 When plague re-emerged in Austria, Germany, and Switzerland in the 1500s, the death count mounted into the ten thousands. To spare their students, the universities of Tübingen and Wittenburg relocated departments away from contagion. Martin Luther, father of the Protestant Reformation, scolded his followers for fleeing the sick. To set a worthy example of Christian love for his neighbor, Luther received plague victims at his residence.

1528 Epidemic typhoid fever killed half of the Gulf Coast Islands Indians who occupied lands from Louisiana west and south over Texas and Mexico. *See also* **November 1528.**

1528 An outbreak of *sudor Anglicus* or English Sweate sickness (also called bloody fever and day sweat) felled Margaret Roper, the daughter of Saint Thomas More. The disease killed Prince Arthur Tudor, the heir who passed along succession to Henry VIII. The king developed such a morbid fear of disease that he continually shifted his residence. When women in Anne Boleyn's court fell ill, the king departed in terror to Essex and from there to Hunston, Hartfordshire. Anne fell ill and recovered. On July 25, 1529, the disease swept Europe, striking in Hamburg, Lubeck, Zwickau, Stettin, Danzig, Strasbourg, Cologne, Augsburg, Vienna, Amsterdam, Antwerp, Lithuania, Russia, and Scandinavia. German Catholics linked the disease to Martin Luther's Protestantism.

July 1528 During a French siege on the Po Valley, typhus killed 19,000 of the 25,000 imperial troops of Charles V, including General Charles Lannoy, the Viceroy of Naples. Survivors barricaded themselves inside Naples. At the command of Francis I, the 28,000 attackers led by General Odet de Lautrec, Maréchal de France, surged to a sure victory when typhus began undermining the formerly healthy French, reducing them by 85 percent and killing Lautrec. As the remaining 4,000 mounted a disorderly retreat, forces of the Holy Roman Empire, led by the Prince of Orange, cut them down. As a result, in August 16, Francis I ceded his claims on Italy.

November 1528 Álvar Núñez Cabeza de Vaca, author of *The Castaways* (1555), struggled to join Cortéz in East Texas. Accompanied by an Arab slave and two adventurers, Cabeza de Vaca lived with Gulf Coast Indians at Galveston Island and kept records of the spread of smallpox and leprosy. He noted that, after half the natives died from a bowel disease, they blamed the Spanish and no longer welcomed European explorers. As

Cabeza de Vaca continued to explore Texas, he developed into a healer.

1529 An outbreak of bubonic plague north of the Firth of Forth required the restoration of stringent laws in Edinburgh, Scotland. The city imposed a one-time head tax to pay the salary of a cleaner and orderly, John Barbour, who aided the sick and shepherded those exposed to disease to a cluster of wood huts on open ground at Burgh Muir. After washing and smoking infected clothes, he stored them at the Chapel of Saint Roche. Those who violated regulations risked branding, banishment, and the confiscation of goods. *See also* **May 1530.**

1530 A Veronese physician and medical writer, Girolamo Fracastorius (also Jerome or Hieronymus Fracastor or Fracastoro), a graduate of the University of Padua, observed a syphilis epidemic. In a three-book poem, *Syphilis sive Morbus Gallicus* (*Syphilis or French Disease*) (1530), he assigned the disease the name "syphilis" and recognized as the source Christopher Columbus's first contact with Hispaniola in 1492. Fracastorius hypothesized a rudimentary germ theory about the way in which microbes pass from humans, animals, and objects to human victims. Alternate explanations of syphilis included scapegoating Hindus from India or Jews, who allegedly imported the *pestis Marrana*.

ca. 1530 During the Ming dynasty, when venereal disease spread over Canton's southern coast, a Chinese syphilis specialist, Wang Ji, treated the illness internally and externally with ginseng, atractylodes root, Indian bread, and licorice root and with applications of honeysuckle ointment and garlic cautery to chancres.

May 1530 From ships in port at Edinburgh, Scotland, bubonic plague spread to Aberdeen and Leith and into the north country. At Leith, contagion reduced the soldier-monks of the Preceptory of Saint Anthony to two survivors. Amply forewarned, Edinburgh city authorities blamed Katryne Heriot with infecting local people by importing two rolls of stolen buckram that bore contagion. Health monitors compiled a registry of the sick and provided food for the poor, both in town and on the surrounding moors. A shelter at Saint Mary's Wynd received homeless children.

Stringent town rules dating back to 1498 required the execution of wandering cats, dogs, and swine and forbade vagrancy and the concealment of disease or the attending of public functions during the contagion. Servants washing infected clothing and bedding required city supervision to prevent the mixing of sanitary with contaminated goods. Punishments for violations ranged from exile, confiscation of belongings, and jailing on bread and water for 15 days to branding, drowning, and hanging.

1531 A measles epidemic that killed 67 percent of residents of Cuba moved south through Central America, striking Nicaragua and Panama, slaying about the same proportion. The infection advanced north into the Great Basin and killed Pueblo Indians.

1532 Because a European disease killed their emperor, Huayna Capac, the Inca were so weakened by loss and civil war that a standing army of 40,000 lacked the might to repel Francisco Pizarro, an illiterate expeditioner from Castile whom Charles I, the king of Spain, had elevated to governor of Peru. Pizarro and his 168 soldiers had plundered south from Panama to Cuzco over a nine-year period from 1524 to 1533. After the Inca confronted typhus in 1546, flu and smallpox in 1558, smallpox in 1589, diphtheria in 1614, and measles four years later, their moribund culture was beyond salvation.

That same year, chronicler Garcilazo de la Vega, historian of the Peruvian Inca, attested to an epidemic of skin disease in Coaque, Ecuador. The infection targeted armies and public works projects. Francisco Pizarro's forces suffered an outbreak of bleeding warts that covered the body. The potentially fatal disease, called *Bartonellosis bacilliformis* or Peruvian wart, survives in sandflies. It is endemic in Colombia, Ecuador, and Peru and causes acute hemolytic anemia and pericarditis, killing half its victims.

1535 Expeditioner Jacques Cartier, founder of Sainte Croix, Quebec, spread infection that threatened the existence of the Laurentian Iroquois. According to Richard Hakluyt's *The Principall Navigations* (1600), Cartier also had problems within his crew with scurvy, which killed 25. He complained:

> Some did lose all their strength, and could not stand on their feete, then did their legges swel.... Others also had all their skins spotted with spots of blood of a purple color.... Their mouth

became stincking, their gummes so rotten, that all the flesh did fall off, even to the rootes of the teeth, which did also almost all fall out [Major 1932, pp. 587–589].

When Domagaia, a local Indian living up the Saint Lawrence River at Stadacona, Quebec, recovered after a severe bout of scurvy, he told Cartier that he had healed himself by drinking the juice and sap of the white Canada balsam. As described in Pierre François Charlevoix's *History and General Description of New France* (1744), following the directions of the Indians, Cartier poured boiling water on white evergreen needles to extract a juice that cured the vitamin deficiency. Charlevoix added, "Some even, it is said, who had had the venereal disease, and had not been perfectly cured, in a short time recovered perfect health" [1900, p. 121].

1535 Under the dominion of the English, Dublin encountered the same ill fortune it had suffered in the previous decade with the emergence of multiple scourges, probable dysentery, plague, typhus, and *galar breac* (smallpox). So many soldiers died that Henry VIII could expect from Ireland no strong contingent of reinforcements for the royal army.

May 12, 1535 Governor Georg Hohermuth von Speyer (also Jorge de Espira or George of Spires) set out with 400 men from Coro to the Barquisimeto valley of Venezuela, raiding villages for slaves and searching for treasure. As the expedition pressed on into the summer, the heat, swarms of mosquitoes, and an unidentified sickness (endemic malaria) felled all of the Indians and 310 Germans. On November 3, as the party crossed the *llanos*, Philip von Hutten reported in *Zeitung aus India Junkher* (*Chronicle of the Indies Expedition*) (1785), "We set off with over eighty Christians ill. It was necessary to carry thirty of them across the saddles of the horses, tied on like sacks, which was lamentable to see. Four died before reaching the next camp" (Hemming 1978, p. 57). *See also* **1543.**

1538 A former Portuguese soldier of fortune, Saint John of God, ministered to the poor when plague spread in Granada, Spain. As a means of containing the disease, he treated patients individually in isolation wards. The city became the headquarters of the Brothers Hospitallers, who influenced medical workers in Cordoba, Madrid, Naples, Paris, and Toledo. By 1602, the brothers had introduced their healing mission to the poor of Central and South America.

May 30, 1539 Hernando de Soto landed at Tampa Bay with 600 soldiers and encountered vigorous attacks by indigenous people as he made his way north through Florida, Georgia, and the Carolinas and west and south through Tennessee, Alabama, Mississippi, Arkansas, and Texas. No European sighted the territory again until 1682, when Réné-Robert de la Salle canoed up the Mississippi River. The absence of thriving Indian villages that had opposed de Soto leaves scholars to surmise that a massive epidemic or series of contagions may have wiped out a huge native civilization. Whatever the number, smallpox chronicler Elizabeth A. Fenn, an assistant professor of history at George Washington University and author of *Pox Americana: The Great Smallpox Epidemic of 1775–82* (2001), noted the all-encompassing sweep over civilization. Speaking for her, Charles C. Mann wrote in "1491," *Atlantic Monthly*: "Whether one million or 10 million or 100 million died ... the pall of sorrow that engulfed the hemisphere was immeasurable. Languages, prayers, hopes, habits, and dreams—entire ways of life hissed away like steam" (March 2002, n. p.).

1540 Pedro de Castaneda, historian and author of *The Journey of Francisco Vasquez de Coronado, 1540–1542*, recorded healing techniques of the Kansa Indians. The Wichita guide Ysopete demonstrated the effectiveness of milkweed and dock root for dysentery and foxglove to assuage chills and fever. When contagion swept up the Atlantic coast to Canada and south of the Amazon basin, diseases bludgeoned some 100 million Amerindians, devastating the Iroquois of the Ohio Valley, the Yanomami (or Yanonami) and Tupinambá of Brazil, the Creek in Georgia and the Catawba, Cherokee, and Etiwan in the Carolinas. Made virtually extinct by the diseases were the Assiniboin and Mandan of the upper Missouri Valley, Arawak in the Caribbean, and the Calusa of Cuba and Puerto Rico.

Central and South America were no better off after contact between indigenous people and Europeans. In Guatemala, a third of the population died from smallpox; the proportions were twice as high among the Maya of Yucatán and even more devastating in Panama, Nicaragua, and Costa Rica. Malaria so devastated Spaniards settling at Limón, Costa Rica, that they were forced to build their enclaves away from mosquito-infested areas while their African slaves, who carried some natural immunity to infection, were able to remain in the province. By the 1800s, every tribe from the Aleutian Islands to Tierra del Fuego, Argentina, had suffered some form of catastrophic disease.

1542 During a power struggle for control of Hungary, an epidemic of typhus killed 30,000 of the 55,000 German imperial troops of Ferdinand I, led by Count Joachim of Brandenburg. At Buda, the epidemic hindered a two-pronged attack against Turkish invaders, killing more Germans than Hungarians, who had developed immunity to endemic typhus. The severity of infection earned Hungary the reputation as the graveyard of the Germans. As the surviving imperial forces marched home, they inflicted the "Hungarian disease" on districts in their path.

1543 In the Venezuelan interior, Philip von Hutten's German expeditioners encountered rains and a mange-like pestilence that turned their skin a dim orange. He reported that they lost their hair as scabies crept over them. He took survivors to San Juan for treatment.

1545 In Geneva, Switzerland, during an outbreak of plague, the judicial system arrested people connected with hospitals. Judges exiled or tortured 62 persons, 49 of them female, on charges of being *engraisseurs* (plague spreaders). Executioners burned 32 at the stake in a series of persecutions more lethal than witch trials.

1545 After a four-year epidemic of *cocoliztli* (hemorrhagic fever) struck Morelos, Mexico, victims complained of jaundice, torso pain, headache, fever, vertigo, and bleeding from mouth, nose, and ears. The scourge killed 800,000 and eradicated 80 percent of indigenous peoples. The Spanish colonial government settled peasants at planned villages and converted unoccupied farmland into pasturage for herds or fields of sugarcane.

1546 In *De Contagione et Contagiosis Morbis* (*On Contagion and Contagious Diseases*) (1546), Veronese physician and medical writer Girolamo Fracastorius (also Jerome or Hieronymus Fracastor or Fracastoro), the father of epidemiology, recognized the endemic nature of tuberculosis in Italian families. Some had passed the fatal disease over six generations. He commented on the syphilis epidemic in Europe, spread by rampant prostitution, and proposed treatment with the bark of the guaiac tree.

March 1546 When expeditioner Jerónimo Dortal set out over Venezuela toward the Orinoco River, his forces mutinied, banished their leader, and pressed on. They suffered an unidentified fever that caused them to crave salt. In their frenzy, they rejected corn and tried gnawing the salty sweat from their clothes and saddles. The fever plunged the men into a coma-like stupor. One report described their emotional deterioration:

> They were very inhuman to one another. When a man was ill, if he was a footsoldier he was given to a horseman to carry on his horse so that he would not be left behind. But the sick man would keep falling, unable to stay in the saddle because of his illness. The horse's owner would put him across the saddle like someone throwing a sheep, tying his hands and feet to the girths on either side with cords [Hemming 1978, p. 65].

In addition to disease, the expeditioners fought the alligators and snakes of the swamps and battled famine by eating snakeskins and chewing their swordbelts.

1551 The unidentified *sudor Anglicus* or English Sweate sickness moved once more over the land, from Shrewsbury, England, to Chester, Coventry, London, and the eastern shore. It died out in November, then disappeared, leading modern researchers to assume it was an extinct Hantavirus. John Caius, author of *A Boke or Counseill against the Disease Commonly Called the Sweate or Sweatyng Sicknesse* (1552), recorded the outbreak at Shrewsbury and many fatalities at Ludlow, Presten, and Wales. He surmised that the disease spread through filth.

1552 *Brevissima Relacion* (*A Short History*) by Bartolomé de las Casas, a conscience-stricken Dominican convert and author of *Historia de las Indias* (*History of the Indies*) (1532), denounced the Spanish for depopulating the Caribbean and Central Mexico since his arrival in 1502. He blamed the fearful decline of native peoples on the conquerors' enslavement of workers and on exposure of virgin-soil populations to European-borne measles, pneumonia, and smallpox, however unintentional the infecting of vulnerable people.

November 1552 As the Emperor Charles V's forces triumphed at Toul and Verdun and advanced on Metz, France, his 14 regiments of *landsknechtes* (cavalry) and 120,000 hired infantrymen suffered hunger because their number outstripped provisions. In sight of their quarry, the men had to fall back after encountering the double blow of dysentery and typhus, the standard scourges of the military. Of the 75,000 men camped in the open, especially the warm-climate mercenaries from Iberia and Italy, over 13 percent died of the epidemic, cold, and starvation.

Across no-man's land, Charles faced François de Lorraine, second Duc de Guise, a stickler for health measures to protect his 4,600 soldiers and 920 local militia. He demanded daily street cleaning, supervision of clean water and provisions, and the quarantining of the sick, both military and civilian. Metz held out for 65 days while the emperor's men sickened in great number and around 30,000 died. By January 1553, Charles was ready to capitulate and to accept the loss of western Europe. Sick and disappointed, he marched home and remained incommunicado in his castle at Luxembourg for a month to plan a subsequent campaign in the Netherlands.

1555 At Rio de Janeiro, smallpox imported with cargos of African bondsmen sickened French Huguenot settlers and fanned out to engulf half of Brazil's indigenous people. A small staff of Jesuit fathers, the original carriers of the contagion upon their arrival in 1550, cleansed and treated patients with leeches. Worsening the misery of Indians was a lack of medical care, overcrowded huts, and hunger as the stock of manioc flour diminished.

To keep from starving, some had little choice but to hire themselves out as agricultural slaves on *fazendas* (plantations). To men who disdained farming as woman's work, the decision to plant and harvest heaped them with shame and despair, leading some to commit suicide. Settlers of São Paulo supplied themselves with replacement workers by leading slave raids deep into the interior. After a seven-year onslaught, when the disease hit the Bahía and Ihléos, the coastal population of Amerindians declined by 95 percent.

1556 At Palermo, Sicily, anatomist Giovanni Filippo Ingrassias, the city's *protomedicus* (health officer), battled endemic malaria. He succeeded in controlling outbreaks by draining wetlands, instituting the city's first sanitary code, and building isolation hospitals. For his brilliance, he earned the nickname "the Sicilian Hippocrates."

1557 Pandemic malaria invaded continental Europe, killing Charles V, Europe's last Holy Roman Emperor, who had retired to a monastery at Yuste, Spain, to meditate in private quarters overlooking a mosquito-infested pond. In the fens of Italy, the disease targeted farm laborers and soldiers. Residents of Maremma Grossetana avoided malaria by resettling in hill towns for the summer. In Britain, high death rates in the marshy coasts of southern and eastern England inhibited agriculture. The survival rate of farm families from outbreaks of "marsh fever" did not improve until the advent of better drainage of mosquito-breeding areas and upgrades to hygiene and immunity.

1559 The first of eight epidemics of smallpox in Paraguay killed 10,000 Indians. Thousands more died over the next 150 years, obliterating the Guarani and other native tribes in Altos, Ita, and Yaguaron. The loss of Indian labor to Hispanic colonizers required the importation of black African slaves.

1560 After a year of epidemic influenza throughout the Southeastern Indian tribes, Gulf Coast peoples, and natives of central New Spain, outbreaks of the disease along with hemorrhagic dysentery and an epidemic of smallpox weakened Portugal's Brazilian colony. As Marshall C. Eakin summarized in *Brazil, the Once and Future Country* (1997):

> Starving and sickly Indians staggered into the colonial settlements begging for food and work. With the breakdown of barter, the difficulties of fighting both Indians and Jesuits, and the rising demand for labor on the sugar plantations, the Portuguese began to look for an alternative to native labor [p. 18].

European colonists imported black Africans to harvest sugar cane. The buildup of a slave-based economy by nearly four million malaria-resistant African laborers developed the coastal areas into a thriving colony.

1562 As described by Portuguese Jesuit historian Antonio Serafim Leite in *Monumenta Brasilae* (*Monuments of Brazil*) (ca. 1560), Portuguese colonists introduced the scourge of smallpox in Brazilian villages. Multiple deaths produced a revolting odor among its victims as whole tribes died out. Survivors fled to the interior and refused to return to the *aldeias* (mission compounds) until contagion had passed.

After plague and smallpox reached Itabaracá, Brazil, from Portugal, Jesuit missionaries fought disease by a crude method: "[peeling] part of their legs and almost all their feet, cutting off the corrupt skin with scissors and exposing the live flesh ... and washing that corruption with warm water" (Bakewell 1997, p. 305). The populations at Bahía and Olinda grew too sick to grow manioc and died out, leaving ghost towns.

October 1562 Plague in England sickened aristocrat and peasant alike, generating a backlash against doctors and pharmacists. The government prohibited fairs, closed theaters, and killed stray dogs. At the first sign of disease, health officials marked homes with the blue X to indicate infection. Inside, they collected disease-ridden bedding and clothing in the fields; late-evening bonfires cleansed disease from the atmosphere. Churches removed windows to create cross-currents of ventilation to rid worshippers of infection. At Chelmsford, Essex, a law forced citizens to scour their gutters once a month and forbade butchers or other residents from tossing animal offal into streets or rivers. A watchman stood at every infected household to catch lawbreakers, who were seared on the earlobe with a hot iron for infractions.

When Elizabeth I realized the danger, she forbade importation of trade goods from London and erected a scaffold at Windsor on which to execute anyone arriving from the city. After she sickened with smallpox, court physicians wrapped her in blankets near the hearth. She survived, but her face was heavily pockmarked and her head left bald. Her makeup and wig influenced styles among courtiers. By autumn, the weekly death toll rose to 18,000. The next year, the privileged abandoned townhouses, parliament, and law courts. The queen's privy council fled plague in London by reconvening 20 miles north at Hertford Castle. *See also* **summer 1578; April 1593.**

1564 At Zaragoza, Spain, a lengthy scourge of plague slew 10,000. One of the first men to comprehend the nature of the disease was Juan Tomás Porcell, who treated the sick and studied corpses, the source of his text *Informacion y Curacion dela Peste* (*Information and Treatment of Plague*)

(1565). Inhibiting treatment of the sick at the height of the Spanish Inquisition was King Philip II's complicity with anti–Semite church leaders in refusing Jewish apothecaries the right to compound medicines for Gentile patients.

1564 In reference to an outbreak of plague at Bremen, Germany, town physician Johann von Ewich, a learned man educated at Padua and Venice, issued *De Officio Fidelis* (*On the Duty of the Faithful*) (1582), a monograph on the health community's responsibility to the sick. In addition to clinical observation, he commented on the need for sanitation to prevent scourges.

1564 When the sweating disease migrated to London from Le Havre, France, William Bullein issued *A Dialogue against the Fever Pestilence* (1564), a popular work that was reissued in 1573. The text offered remedies, cheery anecdotes, and encouragement to people terrified by the epidemic.

spring 1564 At the industrial center of Lyon, France, plague assailed the populace, perhaps from foreign traders visiting the area. Officials housed victims in tents outside the city limits. Because the outbreak coincided with a royal progress made by King Charles IX and his mother, Catherine de Medici, through Provence, the touring court remained on the move for several months to avoid contact with emerging pockets of disease. The party failed to avoid contagion, which killed some of its members.

summer 1566 While preparing a siege against Hungary, Maximilian II of Germany and his 80,000 troops suffered a devastating outbreak of typhus in camp on the island of Komorn on the Danube River. According to the regimental physician, Viennese physician Thomas Jordanus, the imperial Hapsburg forces were too sapped by typhus, malaria, scurvy, hunger, and losses to fight Sultan Suleiman I and his Turks. Instead, the Germans abandoned the expedition before firing a shot. On Maximilian's withdrawal, his men infected the civilian population of Hungary as well as the countries of central and north-central Europe.

1567 According to Yu Maokun's text *Douke Jinjing Fujijie* (1727), variolation had protected the Chinese from smallpox since 1567. Yu T'ien-Chih, author of *Miscellaneous Ideas in Medicine* (1643), described the insufflation of dry scabs or smallpox matter, which doctors ground fine for inhaling into the nostrils through an ivory or silver tube.

September 1568 Another bout of bubonic plague at Edinburgh, Scotland, gripped the city until fall 1569, killing 2,500 or 10 percent of the citizenry. Authorities surmised that the carrier was a merchant and banned Danish ships from the harbor. Additions to earlier strict regulations required fumigation of residences, boiling of infected clothing and bedding, and burial of plague victims under seven feet of soil.

Gilbert Skeyne, royal physician and professor of medicine at King's College, Aberdeen, wrote a 46-page monograph, *Brief Description of the Pest* (1568), for which he abandoned scholarly Latin to address ordinary readers in the common tongue. He described how the fever cut down victims and observed:

> Every ane is become sae detestable to other (whilk is to be lamentit), and specially the puir in the sight of the rich, as gif they were not equal with them touching their creation, but rather without saul or spirit, as beasts degenerate fra mankind ["Domestic Annals"].

He shuddered at the shunning of the poor, who died ignobly without treatment or support.

Skeyne's booklet, the first medical treatise published in Scotland, blamed corrupt air and human wickedness, which God punished with the scourge. Of inferior causes, he listed stagnant water, autumnal humidity, rotting animal waste, filth, unwholesome food, and polluted drinking water. Because of famine, the poor were forced to take what nourishment they could find. Worsening their ill fortune were comets, portents of calamity.

summer 1572 At Haarlem in the Netherlands, an uprising against Spanish rule produced an eight-month siege of the urban center. The soldiers of Philip II of Spain, on orders to force obedience through terrorism, sacked the capital, desecrated the cathedral, and slew the garrison. Rebels held firm until starvation and epidemic disease overcame them, forcing capitulation. Four years later, William I of Orange freed Haarlem from Spain and incorporated the city into a united Holland.

1574 The establishment at Valletta of the Holy Infirmary Hospital of the Knights of Malta, Europe's most famous medical center, began 224 years of patient care that offered free treatment in pleasant surroundings. Known for a smoke cure for rampant syphilis, the 130-bed facility advanced the use of anesthesia, pharmaceuticals, and therapeutic diet that offered daily wine rations.

summer 1574 When plague besieged the port of Dublin, Ireland, for two years, the city was bereft of administrators, most of whom had fled to the countryside. The remaining enforcers of health regulations took charge, established isolation wards for the contagious, and appointed a public physician to treat the sick.

1576 An account of widespread diphtheria in Paris derives from *Epidemiorum et Ephemeridium* (*Epidemics and Journals*) (1640), compiled by Guillaume de Baillou (also Gulielmus Ballonius), court physician to Henry IV of France and renowned epidemiologist. He noted: "The greatest difficulty was in ... difficult breathing...; sluggish resisting phlegm was found which covered the trachea like a membrane and the entry & exit of air to the exterior was not free: thus sudden suffocation" (Major 1932, pp. 154, 156). The contagion spread widely among an international population comprised in part by visiting scholars, university students, soldiers, diplomats, prelates, and traders. *See also* **1578.**

April 1576 The onslaught of *cocoliztli* (also *hueyzanuatl* or *matlazáhuatl*), the Nahuatl term for "pest," spread from the Gulf of Mexico west from the lowlands of the Yucatán peninsula into the highlands to Chichimecas. By the end of the calendar year, nearly 400,000 perished from it. In June at Mexico City's Royal Hospital, Francisco Hernández, a court physician to King Philip II of Spain and health official in the Indies, battled the unidentified plague that ravaged the city from mid–1576 to January 1577. He described its symptoms to the viceroy. The contagion did not die out completely until December 1578, but it returned the following August.

A Jesuit associate, Alfonso Lopez de Hinojosos, the king's personal physician, compiled *Summa y Recopilación de Chirugia* (*Treatise and Recompilation of Surgery*) (1578), which contained anatomical studies of over 200 patients in the wards and homes and of autopsies of victims. Written in Latin, the monograph did not identify the plague, which may have been either smallpox or typhus. Paleopathologists identify the scourge as typhus.

The puzzling term *huey cocoliztli* (also *matlazáhuatl*) suggests that the pathogen causing the great pestilence was actually an indigenous rodent-borne hemorrhagic fever virus causing tumors, delirium, congested lungs, convulsions, dysentery, infected mastoid, netlike rash, and bleeding from the nose and ears. Other diagnoses of the outbreak include measles, dysentery, pneumonia, pleurisy, influenza, measles, spotted fever, pneumonic plague, yellow fever, and alastrim or *variola minor.*

August 1576 During an outbreak of plague at Milan, Saint Carlo (or Charles) Borromeo of Arona, Italy, the Archbishop of Milan and father of the clergy, earned sainthood for his devotion to patient care. Unlike self-serving prelates who fled the city, he refused to abandon the sick, whom he tended for four months in homes and hospital wards. Another beloved citizen, Lodovico Settala, faithfully treated plague victims at great personal risk.

The Milanese overreacted by tormenting and executing suspects of spreading contagion. When they killed all the most despised suspects—Jews, beggars, and Gypsies—they began torturing health workers. At the end of the contagion, surviving citizens pledged to erect a church dedicated to Saint Sebastian, whose body, pierced by arrows, had come to symbolize the onset of plague.

fall 1576 The outbreak of plague at Venice squelched an exuberant Italian city. Strollers deserted the Mercia, Piazza, and Rialto. Shops were shuttered and cafes, inns, and courts closed. Prison populations soon dwindled from multiple fatalities. Officials imposed a week-long quarantine for the whole city, but made no inroads against the disease. As the contagion killed 180,000 citizens, whole families died.

As patients overran the lazaretto and overworked staff, the *Provveditori di Sanità* (*Superintendent of Health*) hired two doctors from Padua. City authorities towed defunct galleons to the center of the bay and transformed them into isolation wards. While senators and prominent families decamped to the Italian hills, the heroic Doge Alvise Mocenigo I and his lady remained on duty. The onslaught that killed the Renaissance painter Titian continued until the official proclamation of health on July 21, 1577.

In gratitude, Doge Sebastiano Venier and the Venetian senate hired architect Andrea Palladio to honor the Virgin Mary by building the basilica of Il Redentore (The Redeemer) across the Canale della Giudecca on the island of the Giudecca. At secular festivities, Venetians gave thanks for deliverance from the epidemic. In subsequent years, the Feast of the Redeemer, scheduled in the third Sunday in July, honored the city's courage.

July 5, 1577 The grip of typhus in an Oxford courtroom proved fatal for some 500 English citizens, 40 percent of whom were employees of the university. The freakish epidemic started among court officials, jury, and viewers during a two-day assize. At the trial of the miscreant Rowland Jencks, whom the crown accused of treason and profanation of the Church, a throng packing the room spread lice-borne illness that killed within two or three weeks. Among the dead were the chief baron of the exchequer, sergeant-at-law, five justices of the peace, two sheriffs, a knight, and most of the jurymen. Jencks survived and was found guilty and sentenced to have his ears lopped off. *See also* **March 1586.**

1578 Guillaume de Baillou, court physician to Henry IV of France and noted epidemiologist, correctly identified an outbreak of pertussis that threatened the infants and small children of Paris. He named the disease *Tussis quintina.* From diagnosis and treatment of pertussis in Venice, he spoke of male infants killed in great numbers from quintana (cough) in paroxysms strong enough to expel blood from mouth and nose.

1578 After severe, often fatal syphilis spread widely among Swedes previously unexposed to the disease, medical writer Bengt Olsson (also Benedictus Olai), the court physician of Eric XIV and John III of Sweden, compiled *Handbook of the Healing Art* (1578). Olsson described syphilis as an unprecedented pestilence and declared that it entered Europe in 1493 at Naples.

summer 1578 Epidemic plague in London killed 8,000. Queen Elizabeth I ordered doctors to offer nostrums to the populace, distributed relief funds to peasants, and dispatched visiting nurses to each parish to tend the sick. Hawkers sold nosegays of herbs, bags of arsenic to be worn close to the chest, sponges of rose water and vinegar or wormwood, syrups of wine and gunpowder to sweat disease out of patients, and amulets of hyacinth stone for energizing the vitals. Broadsides suggested home remedies, for example, the list printed by Edward Allde, which named "A remedie against the Plague, sent to the Lord Maior of London from King Henry the eight," "A Plaister to draw the Sore to a head, and to breake it," and "A drinke to be taken every Morning for a preservative against the Plague, and for voyding of infection" (Wilson 1927, p. 9). To assure quality care of the sick, the city of London appointed nurse-keepers, who visited plague victims at home. Playwright Thomas Dekker smirkingly referred to the women as "women-sleepers" for their tendency to make themselves comfortable while they waited for the dying to expire (ibid., p. 67).

The city grew somber with the closing of theaters and the marking of each infected household with a blue X. As with previous outbreaks, city officials killed stray cats and dogs, cleansed streets, and lit bonfires to burn contagion from the night air. London's Court of Aldermen warned of assemblies at fencing, bearbaiting, and spectacles where "the basest sort of people and many infected with sores running on them ... which be otherwise perilous for contagion" (Wilson 1927, p. 53). The Queen's Players reached a compromise by closing theaters only when the week's death toll rose to 50. *See also* **April 1593.**

1580 According to Jean Baptiste Bertrand, a city physician and inspector of corpses and author of *A Historical Relation of the Plague at Marseilles in the Year 1720*, Aix-en-Provence, France, suffered a severe plague epidemic that afflicted nearly every residence, wiping out whole households, including servants. An itinerant Franciscan hermit and survivor of the plague treated the sick and encouraged all to repent to ward off illness. The devout raised altars to him, extolled his sanctity, and purchased pictures of him to hang in their homes to protect their families from contagion.

When an upsurge of plague struck southern France seven years later, people grew suspicious of the holy hermit because of the connection between his visits and deaths from plague. The courts tried him on multiple charges of fakery, desertion from the army, murder, womanizing, and spreading contagion as an agent of King Philip II of Spain. Angry citizens burned him at the stake and whipped his mistress, Joan Arnaud.

1580 Among the indigenous Caracas and Carib of Venezuela, smallpox reduced the populace by two-thirds. The disease entered the Caracas Valley during the exploration of Diego de Losada, who founded the city of Santiago de León de Caracas in 1567 for the Hapsburg King Maximilian II. Native resistance to Losada's 136 Spanish insurgents halted after infection carried off some 20,000 Indians.

1580 At Istanbul, Islamic fundamentalists sought a scapegoat to blame for the advance of the plague. They focused on the prestigious Arab astronomer Takiyuddin Mehmed and demolished his observatory.

1580s In Porto Seguro in southern Bahía, Brazil, European diseases, combined with the near enslavement of the Tupinikin, strained relationships between Portuguese colonists and indigenous people. The Indian population declined rapidly from epidemics and from the flight of the surviving Tupinikin. In their absence, the colony encountered attacks and raids by non–Tupi Indians of the inland forested hill country on cane fields and sugar mills.

August 1580 Flu ravaged Turkey, Malta, and North Africa and infected Italy and Sicily. From Islamic raiders and legitimate traders in the port of Naples, the disease migrated to major cities in central and northern Italy.

1583 As reported by physician Ildefonso Nuñez de Llerena of Seville and by Alfonso Ruizibus de Fontecha, author of *Disputationes Medicae* (*Medical Controversies*) (1611), when diphtheria struck Spain, peasants named the disease *el garrotillo* (the little strangler). For the next 35 years, the disease moved from region to region, producing the highest death rate in 1613. Francisco de Goya preserved the tensions of the outbreak with a painting, "El Garrotillo" (ca. 1812), in which he depicted a doctor reaching into a child's throat to rip free the coarse membrane that chokes him.

1584 Renaissance medicine began to divest itself of medieval superstition and astrology and introduce scientific causality and treatment of disease based on the rediscovery of Greek medical texts. At the University of Paris, Jean Riolan le Père (also Johannes Riolanus or John Riolan the Elder), a lecturer and anatomist on the medical faculty, taught a course to prepare clinicians to diagnose and treat endemic syphilis. Dubbed the "Indian disease," it responded to fumigation, ointment, and pills compounded of mercury, but not to treatments of guaiac bark.

July 1584 The outbreak of bubonic plague at Flanders was supposedly initiated through contact with an unnamed woman. It advanced to Wester Wemys, Scotland, and on to Dundee, Edinburgh, Falkland, Perth, Saint Andrews, and Stirling. To stem contagion, the city of Edinburgh closed public markets. Despite hygienic measures, disease killed 10 percent of the Scots and forced King James VI to shelter at his castle at Ruthven away from urban disease. By December, so many people had fled contagion that the church closed for lack of deacons and elders.

1585 When Sir Francis Drake's 25-ship flotilla advanced from Plymouth, England, to Santiago in the Cape Verde Islands, around 600 of his 2,300 privateers contracted typhus. Their loss forced an abandonment of the expedition and Drake's return to home port.

1585 In northern Peru, smallpox swept through peasant villages, turning them into ghost towns. Starvation resulted from the lack of workers to harvest grain or earn money from trade or mining.

winter 1585–1586 Captain John Smith's *A True Relation of Such Occurrences and Accidents of Noate as Hath Hapned in Virginia since the first Planting of That Collony* (1601) and *Generall Historie of Virginia, New England, and the Summer Isles* (1624) and geographer and astronomer Thomas Harriot's *A Briefe and True Report of the New Found Land of Virginia* (1588) characterize the arrival of epidemic disease among the English colonists of Roanoke. Harriot, the more scholarly reporter, detailed symptoms and courses of sickness. After making landfall in the Chesapeake Bay region, Smith collected accounts of epidemic influenza, which was deadly to the Accomac, who lived near the Roanoke colony on Virginia's eastern shore.

1586 Epidemic typhus struck Indians of the Atlantic shore. Beginning at the tidewater Carolinas, it moved south to the Apalachee, Creek, and Timucua of Florida. To the north, the scourge infected southern New England tribes.

March 18, 1586 In Devonshire, England, the onset of typhus targeted an Exeter courtroom, felling Judge Sir B. Drake, several magistrates, and all but one juryman as well as numerous prisoners awaiting trial. Medical historians link the epidemic in Exeter's Black Assizes to Portuguese prisoners in the dock, who transported "gaol fever" from their ship to a stifling, unsanitary lockup.

September 1588 Sir Francis Drake, aboard the *Revenge*, joined captains John Hawkins and Martin Frobisher in trouncing the Spanish Armada, which Philip II of Spain had boastfully christened his "Felicissima Armada" (luckiest fleet). As a result, the Spanish lost all claim to the rule of the seas. The ten surviving vessels sailed back to the port of San Sebastián at Guipúzcoa, Spain, bearing 2,475 Spanish soldiers, 327 suffering wounds, typhus, or both. Local people rallied to the need with food for malnourished sailors as well as bandages and medicines. Typhus spread through the harbors, killing Miguel de Oquendo, captain of the 30-gun flagship *Santa Ana* and commander of the ten-galleon squadron of Guipuzcoa.

1589 At Quito, Ecuador, a smallpox epidemic spread from Colombia and moved southwest to Trujillo, Peru. Don Ferdinando (or Fernando) de Torres y Portugal, viceroy of Lima, halted the advance of the disease by establishing civic prevention measures. According to his letter to Diego García de Paredes y Ulloa, both measles and smallpox struck the Spanish and Araucana Indians of Chile. In 1591, smallpox began in Mixteca in west central Mexico and swept up the west coast.

1591 A Jesuit healer, Saint Aloysius (Luigi de Gonzaga or Louis de Gonzague) of Castiglione, Italy, fought plague at a hospital in Rome, where he treated the sick with cool baths, meals, and consolation. He contracted the disease and died on March 3 at age 22. For his courage and kindness to the sick, he was canonized in 1626 and became the patron of AIDS patients.

1591 When a Spanish ship brought smallpox from Mexico, the disease struck the Philippines for the first time, generating high mortality rates. The epidemic was witnessed by a Jesuit priest-historian, Pedro Chirino, author of *Relacion de las Islas Pilipinas* (*History of the Philippines Islands*) (1604) and co-founder of San Carlos University in Cebu City. The scourge was especially virulent at Batangas. Over the next two years, infection spread to Central Mexico as far as Sinaloa on the Gulf of California.

1592 Over a four-year span, outbreaks of measles among the Seneca of central New York State may have caused thousands of fatalities, particularly among the very young. Survivors incurred collateral damage to sight and hearing. Medical historians surmise that the transference of the pathogen occurred from trade with southeastern tribes, who had contracted infection from Spanish *conquistadores*.

1592 Simon Forman, a noted folk healer, magician, alchemist, and diarist from Quidhampton, near Salisbury, England, opened a treatment center and surgery in London at Billingsgate. During the 1592 outbreak of plague, when other doctors fled town, he saved lives, including his own, by applying unorthodox methods of treatment. After the second epidemic in 1594, he compiled *Discourses on the Plague* (1595), which offended the staff of the Royal College of Physicians in London. Imprisoned and fined for practicing alternative medicine, Forman wrangled with his detractors for seven years before acquiring certification from Cambridge University in 1603.

1592 The Western Sudanese expedition that the Sa'did Sultan Mulay al-Mansur's mercenaries mounted against the Songhay state in the middle Niger River valley resulted in catastrophe. Heat, foul water, and tropical fevers decimated the troops and horses of Judar Pasha, who marched south from Marrakesh, Morocco, on October 16, 1590. Encumbered by heavy armaments and confused by unfamiliarity with the locale and its people, the invaders encountered resistance at Dendi led by Askiya Nubu. According to chronicler Es-Sa'di's *Ta'rikh es-Sudan* (*History of the Sudan*) (1591):

> The Moroccan troops were badly affected by the exhaustion which they endured, the lack of food, the destitution in which they lived, and the disease which the insalubrity of the land caused. Water killed many beyond those who perished in the battles [Kaba, 1981, p. 469].

When reinforcements arrived to Timbuktu, Mali, they also succumbed in huge numbers to Songhay attack and swampland pathogens borne by mosquitoes and tsetse flies.

April 1593 Over 11 percent of greater London's citizenry died of plague. During the six-month epidemic, Queen Elizabeth I withdrew into self-exile. The poor suffered the highest fatality rate because they could not vacate the city like the rich to enjoy clean country air. To contain the disease, port authorities took action against the importation of suspect cargo, but the area lacked methods of isolating the sick. Playwright Thomas Nash charged the city of London for its lack of pesthouses and claimed that citizens were "mixing hand over heade the sicke with the whole" (Wilson 1927, p. 77).

Like other cities, Cambridge erected a pesthouse, which did little toward halting infection. London's city council appointed warders, street observers armed with halberds who earned sixpence per day for enforcing regulations. Officials posted these civil guards outside infected houses from 5:00 a.m. until 9:00 p.m. The job of warders was to oversee quarantine, secure the locks on marked residences, and arrest people who violated the law requiring isolation of the sick and those exposed to contagion. One warder who used quarantine as an opportunity to enrich himself on valuables of the sick received a sentence of three weeks at hard labor.

1594 After the crowning of Henry IV of France at Chartres, he received 1,500 victims of scrofula, a tubercular condition in the glands of the neck and throat called the king's evil in England and Saint Marcoul's evil in France. The royal physician Laurentius identified 750 patients whom the king's royal touch cured. As described in Peter Lowe's *Discourse of the Whole Art of Chyrurgie* (1612), the king knelt at the hospital of Saint-Marcoul in Rheims and sketched the sign of the cross on each seeker's forehead. He spoke a comforting phrase: "Le roi te touche, Dieu te guerisse" ("The king touches thee, may God heal thee") (Brewer).

1595 During the administration of the Florida Colony by Governor Pedro Menendez de Aviles, Franciscan missionary friars evangelized the 13 Timucua chiefdoms of north central and northeastern Florida. A headcount of 50,000 natives indicated that only 25 percent had survived epidemics of European diseases. In the same period, at Toluca, Mexico, an outbreak of contagion, possibly a mix of diseases, moved into Oaxaca and south to Guatemala over the next two years.

1596 When the *Rodamundo* docked at Santander on Spain's west coast, stevedores offloaded cloth from Dunkirk, France, that concealed plague-infested rats. A national catastrophe emerged as the scourge raced across Asturias, Galicia, Logrono, Segovia, and Vizcaya, killing 600,000. In Old Castile, Spain, in 1597, municipal authorities recognized a connection between epidemic plague and malnutrition. To soothe the panicked citizenry, agents distributed wheat loaves, which stemmed the need to forage for food. Indirectly, the feeding of the hungry halted random movement of the peasantry, thereby limiting the spread of infection.

June 1597 The emergence of bubonic plague at Inveresk, Scotland, produced the fourth of Edinburgh's great outbreaks. By this time, civic readiness was in place. Enforcers of health regulations activated stringent emergency measures involving isolating crew from ships in the harbor, destroying stray animals, and banning travelers, vendors, and beggars from entering the city.

1600 During an emergence of plague in Lima, Peru, Dominican friar Saint Martin de Porres learned healing from a barber-surgeon. He treated the city's outcast blacks and mulattos and established a home to receive children orphaned by the epidemic. In Bordeaux, France, his contemporary, Saint Jeanne (or Joan) de Lestonnac, dispatched visiting nurses to battle plague from house to house. From her effort grew the Sisters of Notre Dame.

ca. 1600 The Manchus of China developed natural immunities to smallpox in areas where the disease was endemic. To control contagion, Chinese medical experts set up European-style public health methods of investigating outbreaks and quarantining the sick.

1600s After contact with European expeditioners, the Winnebago, Menominee, and other Great Lakes tribes suffered the first of waves of lethal epidemics among virgin-soil populations. One-quarter of the Winnebago died from smallpox. By 1665, around 500 or 6.3 percent survived from a previous headcount of 8,000.

1601 Historians posted varied reasons for the fall of the Asirgarh fort, a stronghold of Faruqi leaders at Khandesh in Southern India between the Tapti and Narmada rivers. Jerome, a Jesuit missionary, blamed bribery. Bijanpuri chronicler Hashim Beg Fuzuni claimed that the victor, Akbar, the Moghul emperor of Hindustan who was planning an all–India empire, paid his agents to poison the fort's cistern. Other sources stated that, after an 11-month siege, the soldiers within were weakened by an epidemic. Akbar himself died of dysentery four years after his victory.

1601 As proven by archival records, epidemic *modorra* (plague) transported from Seville, Spain, struck the Canary Islands during a military invasion by the Dutch. Authorities set up infirmaries, quarantined the sick and burned their belongings, established *cordons sanitaires*, and restricted ports. To restore the people's spirits, the clergy launched religious processions featuring holy relics and icons of St. Roch. During five years of intermittent outbreaks, islanders suffered famine that lessened their chances of survival.

August 1, 1601 Four months into a three-year voyage from England to India by the first East India Company convoy, British Navy Captain Sir James Lancaster began treating rampant scurvy on his own vessel, the flagship *Red Dragon*, with three teaspoons of lemon juice per man each morning before breakfast. Samuel Purchas's five-book *Hakluytus Posthumus, or Purchas His Pilgrimes, Contayning a History of the World in Sea Voyages and Lande Travells by Englishmen and Others* (1625) described how Lancaster's crew outpaced the other sailors of the ships *Ascencion*, *Hector*, and *Susan*, corroborating Lancaster's faith in citrus juice to prevent a deficiency of vitamin C. The situation was so precarious on the three traders that merchants on board spelled the weakened sailors at the helm and manned the sails until the convoy approached the Cape of Good Hope. Of a total of 278 men, 110 died of scurvy.

The positive results of lemon juice against scurvy influenced James Lind, who eventually proved that the disease was the result of vitamin C deficiency. Unfortunately for Lancaster, at a stop at Antongil Bay, Madagascar, where his crew took on more oranges and lemons, some contracted dysentery and died before the ship could reach harbor in Sumatra. By 1607, the East India Company began provisioning all its ships with lemon juice.

1602 During epidemic plague in Latvia, Lithuania, Poland, and Prussia, a fierce onslaught of contagion at Gdansk killed 18,700 Poles, two-thirds of whom died within a week. Tended by Joachim Oelhaf, who trained in Padua, Italy, the city still lost 30 percent of its residents.

1604 In Mesoamerica over a three-year period, the indigenous and peace-loving Otomie, a virgin-soil population, outpaced the rest of the Tlascala Valley of Mexico in fatalities from epidemics of unidentified origin and type. Sapped—probably by smallpox, measles, and influenza—the population dwindled by some 95 percent.

1604 When bubonic plague struck York, England, contagion thrived among 11,500 residents living in crowded environs. To allay the stench of some 3,512 rotting corpses, pub owners sprinkled lavender through their taprooms. The country vendors brought produce to an ancient stone cross for sale and ordered buyers to dip the purchase-money into water before handing it over.

early spring 1604 Dubliners encountered a new outbreak of plague, no doubt imported by ships from England to Ireland. City appointees superintended a lazaretto outside of town and set a watch on frenzied victims running through the streets to escape death. To protect the healthy, town dwellers cleansed the air outside their doors at night by burning sulfur and aromatics. Only those visitors certified free of disease were allowed entry.

April 1604 As disease swept over England, violations of plague statutes were common, notably at Gloucester, where 350 or one-twelfth of the population died after contagion advanced from Bristol. City alderman John Taylor deliberately concealed from Gloucester health officials his servant's death from plague. When a second staff member fell ill, Taylor again failed to report contagion in his household and allowed the man to serve at a gathering that included the mayor and other city dignitaries. Another maidservant died alone in the cellar. Disclosure of Taylor's crime cost him £100 and the loss of his title. When enforcers of health regulations sealed and quarantined the Taylor residence, his son ripped open the door and menaced anyone trying to nail up the entrance. For his rampage, the courts fined and pilloried him.

1605 During a famine in Santiago, Cape Verde, smallpox spread among islanders from Marrakech, Morocco. Worsening their misery were swarms of flies that sucked the blood of humans and cattle.

June 1605 As reported in a letter from Jesuit priest Pierre Biard to the Reverend Father Provincial in Paris, the exploration of Maine by the French resulted in serious loss to *mal de la terre* (scurvy). While Jean de Biencourt, Seigneur de Poutrincourt, returned to France for supplies and colonists, Pierre Dugua Sieur de Mons, explorer Samuel de Champlain, and 77 expeditioners spent the winter at a camp on Saint Croix island in the Saint John River, one of the earliest European settlements in North America. By June, 35 of the men had died of inadequate diet, including a priest and a minister; another 20 lay gravely ill. Champlain supplied a detailed description of symptoms. Marc Lescarbot, Nova Scotia's first historian and author of *Histoire de la Nouvelle-France* (*History of New France*) (1609), surmised that the disease was the result of bad food, malignant air, or some physical indisposition. Because of the dismal experience, Champlain chose to decamp and try his luck at Port Royal, Nova Scotia.

1606 During rampaging plague in Europe, in Paris, health officials constructed the Hospital of Saint Louis, the city's oldest medical center, to accommodate victims. In London, King James I wrote to Lord Mayor John Watts requesting a test of a plague remedy concocted by the German healer d'Ashenbroke and Henry de Ommeren.

summer 1607 The "Great Mortality" that struck the Jamestown colony in Virginia took over half of the original 105 settlers of marshy, undrained ground as well as local Powhatans, who had no immunity to imported bacteria. Refuting claims that malaria killed them, medical historians surmise that the disease was typhoid fever, complicated by malnutrition, which precipitated beriberi and scurvy.

September 12, 1609 The arrival of expeditioner Henry Hudson and 20 Dutch and English sailors aboard the *Half Moon* up river to Albany, New York, was devastating to native Mahican. Among the Iroquois, the newcomers spread smallpox, measles, diphtheria, and scarlet fever, European pathogens that reduced the natives' number significantly. By 1662, the Iroquois League lost half its tribe to contagion and fell in number to 10,000. By 1720, the Mahican no longer survived as a Hudson Valley tribe and resettled in Massachusetts as the Stockbridge Indians.

1610 Basel, Switzerland, experienced its first outbreak of bubonic plague, which struck 15,000 or 40 percent of the populace, killing 3,968. After his experience in practice at Basel, Danish anatomist Caspar Bartholin the Elder (born in Malmo, Sweden) treated Christian IV and his nobles at Copenhagen. On the king's order, Bartholin published a handbook, *A Short Instruction* (1619), on how to avoid the plague.

1610 After observing epidemic scarlet fever and the common cold, Korean physician and medical writer Ho Jun of Seoul, an army doctor, became the nation's first contributor to epidemiology. During difficult times when war raged with Japan, the ruler Sun Jo commanded Ho Jun to collect medical precepts and pharmacopia in the 25-volume *Tongui Pogam* (also *Dong-ui Pogam, Dong Ui Bo Gam,* or *Tongipogam*) (1610). In his compilation of empirical data, he made the initial clinical record of pulses, fever, and other symptoms devoid of supernatural content or Chinese herbalism. The text was a cornerstone of Korea's scientific renaissance.

1613 The Timucua, a nation of 50,000 natives of north central and northeastern Florida, were among the earliest Southeast tribes to experience sustained interaction with Spanish expeditioners and missionaries and, for four years, suffered the consequences. Along with warfare and enslavement by Spaniards, a four-year epidemic of smallpox and other European diseases halved the population and reduced the number of chiefdoms. Survivors appear to have joined the Seminoles. By 1700, only 1,000 Timucua survived. Preserving their pre-conquest way of life were sketches of Frenchman Le Moyne de Morgues.

1614 Following incursions by slave traders to the Plymouth Company of England in 1614, the Massachusett, Micmac, Pawtucket, and Wampanoag incurred a four-year onslaught of smallpox. The scourge reduced the Wampanoag by 70 percent and the Massachusett by 90 percent, leaving largely unoccupied land for the Pilgrims to settle after their voyage to the New World in 1620. When the Micmac raided Massachusetts, they spread contagion to their enemies, the Pennacook, who lost over 75 percent of their number. To the south, the Powhatan Confederacy of Virginia declined from combined effects of disease and war.

1615 Among the Otomie, the indigenous people of the Tlascala Valley of Mexico, drought and famine worsened their sufferings following a second great wave of pestilence following the unidentified scourge of 1604–1607.

1616 Disease broke out among the Eastern Abenaki of Maine after the wreck of a French vessel off Cape Cod. Infection spread to the native towns of Namasket and Massachusett. As epidemics of European diseases invaded Northeast American tribes, natives suffered the "Great Dying," which killed up to 90 percent of indigenous peoples. Epidemiologists have pondered identification of the disease, alternately considering cerebrospinal meningitis, chicken pox, measles, plague, smallpox, typhoid fever, typhus, and yellow fever. Consensus leans toward a diagnosis of smallpox with some holdouts insisting on typhus, plague, or meningitis as the source of contagion.

1617 A smallpox epidemic that fishermen spread among Indians in Massachusetts killed 90 percent of Algonquian natives over a two-year period. Among the most noted of New England's victims slain by European diseases was Pocahontas, who died of smallpox on her return voyage from Gravesend, England. At Patuxit, the corpses of numerous natives attested to the virulence of the infection.

John Winthrop, a physician and governor of the Massachusetts Bay Colony, wrote in his journal that the disease was God's divine plan to reduce the native population so Puritans could settle on Indian homelands in 1620. Speaking the opinions of numerous English settlers, New York justice of the peace and historian Daniel Denton observed in *A Brief Description of New-York: Formerly Called New-Netherlands* (1670):

> It is to be admired, how strangely they have decrease by the Hand of God, since the English first settling of those parts; for since my time, where there were six towns, they are reduced to two small Villages, and it hath been generally observed, that where the English come to settle, a Divine Hand makes way for them, by removing or cutting off the Indians either by Wars one with the other, or by some raging mortal Disease ["Daniel Denton"].

1618 During the Thirty Years' War, Europe incurred serious pestilence. In Germany, misery resulted from repeated onslaughts of dysentery, plague, and typhus, called the "Hungarian sickness." Frankfurt lost 1,785 people; Nuremberg suffered 2,487 deaths. When the pestilence advanced on France, it struck Lorraine and cost Metz 3,000 victims. Farther south, *male in cana* (diphtheria) outbreaks beset Italy, killing 8,000. Beginning among the young in Chiaia on the west coast, the scourge swept through Naples, Italy, and west to Messina and Palermo, Sicily. *See also* **1629, 1634.**

1619 In Denmark, health officials established a century-long regulation concerning the responsibilities of physicians during epidemics. Although not required to make home visits to the sick, health professionals were obligated to publish lists of curatives that the public could access during epidemics of dysentery, fever, plague, and scurvy. A century of these printed instructions were issued from the teaching staff at the University of Copenhagen and were housed in the Royal Library and in Copenhagen's Danish National Library of Science and Medicine.

1619 Squanto, a Patuxit Wampanoag from Cape Cod whom Captain George Weymouth captured in 1604 and enslaved in Malaga, Spain, returned to New England fifteen years later to interpret for the English. He found his whole tribe dead from European diseases, primarily smallpox. On their uninhabited homeland, the Puritans built a colony. In fall 1622, while bartering as agent for the Puritans at Chatham Harbor, Massachusetts, Squanto himself died of "Indian fever," a folk term for smallpox.

late August 1619 As explained in a letter written at Jamestown, Virginia, by John Rolfe, colonists witnessed the first merchandising of black "Negar" slaves in the New World. Dutch privateers captured a Spanish slaver carrying 200 African captives and ferried about 20 aboard their man of war to Old Point Comfort to trade for supplies. The Africans, most of whom were bought by Governor Sir George Yeardley and Abraham Peirsey, a wealthy merchant, introduced yellow fever, malaria, and uncinarisis or hookworm disease in North America. Studies of slave mortality suggest parasitic disease and malnutrition were causes of low birth weight and poor health among slave children. The English added the diseases to their own list of lethal maladies that threatened their lives and the colony's survival.

1623 England suffered a series of floods, famine, and epidemic dysentery, plague, smallpox, and typhus. Plague alone killed 20 percent of the people of London, causing Sir William Paddy, physician to King James I, to create elaborate prevention and treatment for royal ills lest they be the onset of plague. In 1625, the London city council voted to pay over £290 to pharmacists, doctors, and surgeons to keep the populace healthy. After typhus felled poet John Donne, he came under the care of London court physician Sir Theodore Turquet de Mayerne, an advocate of clinical medicine. In reference to the experience, Donne wrote *Devotions upon Emergent Occasions and Severall Steps in My Sicknes* (1624).

spring 1623 When European settlers encountered the Nauset or Cape Indians and the Wampanoag or Massasoit of Cape Cod, Massachusetts, natives died of unidentified diseases introduced in 1614 by sea captain Thomas Hunt and his sailors and spread to the community of Wessgussett in 1622. Among the dead were Aspinet, sagamore of the Wampanoag Confederacy at Truro, and his ally Iyanough, sachem of the local Cummaquid at Barnstable, Massachusetts. Another victim, Chief Massasoit, became seriously ill during the winter, when the English took him in and successfully treated his malady.

1626 A three-year siege of plague assailed France. At Angers, officials spent 100,000 *livres* for the treatment of 8,000 plague victims. They used the money to hire doctors and watchmen, gravediggers, provisioners to provide food and medicines, and carpenters to build plague huts.

In the summer of the third year, Lyon suffered the largest number of fatalities. Burdened by high prices, heavy taxation, and unemployment, the city reeled from contagion begun by soldiers marching through the area. City administrators adapted Saint Laurent Hospital as a contagion ward for 4,000 and added huts outside the facility to shelter additional cases. A recrudescence in March 1629 restored the disease to its former virulence. In all, 35,000 died.

The lethal contagion influenced the expansion of earth science. Nicolas-Claude Fabri de Peiresc, leader of a consortium of scientists, had to postpone measurements of longitude, which he intended as a contribution to scientific observation and inquiry. He also experienced delays and work stoppages as postal agents quarantined and decontaminated letters and books with herbs and vinegar, and as he abandoned quadrants and five telescopes at Aix while sheltered at a country retreat at Belgentier for three years. During this hiatus, Galileo Galilei came into conflict with the Inquisition for persevering in his study of the heavens. Galileo's example was a warning to Peiresc, who made obvious efforts to placate the anti-science contingency of the Catholic hierarchy.

1629 Scots explorers stopped in St. Johns, Newfoundland, and found evidence of endemic scurvy. During the first months of the settlement of Port Royal, Acadia, founded by Sir William Alexander, disease threatened the 70 colonists. Nova Scotian medics and Jesuit nurses, dispatched by the Holy See, tended victims of the plague. Their efforts constituted the first Christian nursing program in the New World.

1629 In the first of the Dutch wars at Batavia (modern Jakarta) in the Dutch East Indies, the Javanese lost because of hunger and pestilence. The first practitioner of tropical medicine, Dutch naturalist Jacobus Bontius of Leiden, author of *Historia Naturalis Indiae Orientalis* (*Biology of the Indies*) (1629), also compiled *De Medicina Indorum* (*On Medicine in the Indies*) (1629) containing his clinical observations of cholera and beriberi. From professional and personal experience with beriberi, he reported on the nature of disease, which was named for the Sinhalese word for "sheep" because of the tottering gait of victims. *See also* **1652**.

1629 In the valley of Mexico, indigenous people incurred the third in a series of epidemics of an unknown disease. Natives called it *cocoliztli*, the Nahuatl for "pest," a catastrophic malady that raged for two years, severely lowering their headcount. Medical historians have linked the symptoms to indigenous rodent-borne hemorrhagic fever.

1629 In the midst of the Thirty Years' War, Mantuans suffered a devastating plague that killed 280,000. Carried by the French and German forces, the disease advanced along troop routes throughout sub–Alpine Italy. The outbreak moved into Milan and exploded into a virulent attack on pre–Lenten carnival celebrants, killing 3,500 daily. Extending into mid–1631, the epidemic resulted in mass human loss. Following

the scourge, Venice, the great Renaissance sea power and banking center, never regained its former prominence.

Pestilence made huge changes in the social fabric, especially for minorities and women. After plague curtailed Italy's trade and industry, princes and mercantile magnates encouraged Jewish immigration to replace the invaluable Jewish pawnbrokers and moneylenders who fled or were harried or murdered as alleged plague-spreaders. Filling the void were merchants and bankers arriving from Brazil, Holland, North Africa, and Spain. At Bologna, Italian women were accepted as testators to wills. Widows with children were more likely to receive most or all of a legacy; dying patriarchs named them as heirs to protect family honor or social standing.

1630 Smallpox overran the Indians of New England, New York, and Canada and spread across the North American frontier to tribes who hadn't encountered white Europeans. In the Saint Lawrence Valley, a series of epidemics into the decade reduced a native population of 500 by 60 percent. Adding new vigor and hope to indigenous tribes was an influx of Huron, Iroquois, and Abenaki, who restored the headcount to the original figure by 1675. *See also* **fall 1633; fall 1636; 1637.**

1630 When plague dogged England, compilers of the *Books of Orders of 1630* summarized problems at Cambridge with high prices and grain hoarding, health and sanitation management, Irish immigration and poor relief, and the lack of urban control. The Privy Council heard from the College of Physicians on April 20 concerning the overbuilding of cities, ponds in innyards, unclean streets and sewers, stalls selling unwholesome corn and fish, foul city slaughterhouses where butchers killed diseased livestock and discarded their entrails on dunghills, reeking burial vaults, and churches and churchyards "overlaid" with burials of plague victims (Wilson 1927, p. 24). An essential to city cleanliness was the rat-catcher, who walked the streets with his staff and bag crying:

> Ha' ye any rats, mice, polecats or weasels?
> Or ha' ye any old sows sick of the measles?
> I can kill them, and I can kill moles,
> and I can kill vermin that creepeth up and
> creepeth down and peepeth into holes [ibid., p. 37].

Crucial to England's social engineering were elements of transportation, provisioning, quarantining, warehousing the sick, and data collection. The laws, which the Royal College of Physicians revised, set up bureaus to study enforcement of regulations as deterrents to plague. Another record book, *Orders and Directions* (1630), restricted the movement of goods and people to prevent contagion from spreading and set up the machinery by which civic authorities—comprised of a commission and its deputies—enforced the law to maintain a degree of normal function until the epidemic abated.

In London, a Swiss Huguenot diagnostician and pharmacologist, Sir Theodore Turquet de Mayerne, senior court physician to Henry IV of France and the English kings James I and Charles I, supervised public health during an outbreak of plague. Acting in the capacity of clinician and epidemiologist, he set up a plague policy based on similar urban planning at Paris. He immediately advised on the need for adequate food to avert famine. He also recorded case histories in notebooks and Latin casebooks that comprise the 20-volume *Ephemerides Morborum* (*Diaries of Diseases*), which remained unpublished at his death in 1655.

As chief sanitarian during the pestilence, de Mayerne took recommendations from the Royal College of Physicians concerning unhealthful conditions, notably widespread beggary, foul meat and fish stalls and abattoirs, unsanitary sewers and streets, standard ponds, overcrowded residences and cemeteries, and burial of the dead in church floors. Acting on these advisos, he advocated the collection of dung from streets and alleys, the removal of slaughtering and tanning from the city limits, the killing of stray animals, the quarantining of ships, and the erection of five lazarettos and a plague hospital designed in the style of L'Hôpital Saint Louis at Paris. He also proposed the razing of slums and grog shops to rid cities of idlers and vagrants and the ejection of an "unruly base sorte of people," a euphemism for Irish immigrants and other social pariahs (Gallagher 1994). Among the outcast Waldensians, the plague devastated their followers. Of their 13 leaders, 11 died. To provide enough leadership, the survivors petitioned Geneva for more pastors.

winter 1630 The network of health magistracies, an intercity consortium of Italian plague doctors, oversaw protective measures at Bologna, Florence, Genoa, Lucca, Mantua, Milan, Modena, Parma, Venice, and Verona. They standardized urban sanitation, reports of disease, quarantine, and treatment and issued official death certificates. Sextons called *manotti* or *baccamorti* wore bells on their ankles to warn of their approach as they collected corpses to heap in burial ditches. The greedy scavenged infected garments and bedding to spread around the area to perpetuate the plague, the source of their income.

At Florence, an influx of hireling soldiers marching back to Milan from a turf war in Mantua may have introduced contagion. After civil sanitarians enacted emergency statutes, authorities pressed 330 criminal actions against citizens and outsiders who violated health codes. Those seeking wellness in a time of scourge applied bizarre home remedies—extreme purging, dieting, soaking towels and handkerchiefs in vinegar, and dispersing acid into the air. Despite these efforts, the disease cost the city many of its painters, marble workers, and Capuchin monks as well as nurses and gravediggers. In terror, survivors turned to God with exhibitions of piety and charity to the poor that continued for several years. The canonization procedure of Domenica da Paradiso, a late medieval mystic, resulted from beliefs that the healing powers she displayed a century before would end the epidemic.

In Milan, authorities ushered all mendicants to the moated lazaretto, where the city fed and cared for 3,000. During a revolt against sanitary measures, police officers chained protesters and forced them into the pesthouse, where occupancy rose to 10,000. Inmates roomed together on straw at the rate of 30 per cell and had no choice but to draw drinking water from the foul moat, which was clogged with refuse and filth.

At Pistoia, plague victims had access to a crowded lazaretto that admitted 1,198 patients in a ten-month period. The sick slept five to a bed. Overcrowding delayed the healing of survivors, who typically suffered relapses and died at the rate of 51 percent of all in-house patients. Those immured at home tended to be from the underclass:

Number of Homes	Class
99	low-level artisans
15	middle-class artisans
11	ruling class

To pay for the heavy burden on city coffers from a death rate of 12 percent overall, health officials relied primarily on loans and donations to the poor box.

At Venice, where 33 percent of 150,000 citizens died of plague, Doge Nicolo Contarini propitiated the Virgin Mary with a sacred vow to build a church if she would intervene on behalf of the citizens. When the arrival of cooler temperatures reduced problems with rat-borne fleas, the epidemic ended. Officials, as they had promised, chose land at Dorsoduro at the juncture of the Grand Canal with Saint Mark's Basin and held a contest among architects to choose a builder and design. The winner, Baldassare Longhena, planned Santa Maria della Salute, a huge octagonal basilica with white Istrian stone steps, completed in 1682. In memory of the Virgin's rescue of the city, each November 21, Venetians erect a pontoon connector between the Grand Canal and Longhena's church and humbly process over the bridge. At the portal, a priest recites prayers and blesses the oars of gondoliers, who still paint their boats black to commemorate the terrible plague.

1631 Along 200 miles of the southern Massachusetts coast, a new outbreak of smallpox or leptospirosis among the Massachusett Indians reduced some 3,000 indigenous people to about 500. Twenty-first century research targeted black rats (*Rattus rattus*) from European ships as sources of resultant fever, headache, jaundice, and skin lesions. Increasing exposure to pathogens were walking barefoot and burying corn in ground caches, two suspect customs that brought about the extinction of the Patuxet.

1632 During a century of serious endemic syphilis in Canton, China, Chen Sicheng developed a reputation as a skilled syphilologist. To halt the endemic disease, he compiled a revolutionary text, *Meichuang Mi Lu* (*Secret Records on Syphilis*) (1632), which contained his conclusions about the transmission of infection and about his formulation of an inorganic compound of diluted arsenic that cured some of his patients. His applications of both arsenic and mercury preceded their use in Western pharmacology by three centuries.

1633 An Augustinian friar, Antonio de Calancha of Peru, recorded in *Chronicle of Saint Augustine* (1633) how bitter cinchona bark relieved the pain and fever of malaria, a pandemic disease in equatorial countries. He exulted:

> A tree grows which they call "the fever tree" in the country of Loxa, whose bark, of the color of cinnamon, made into powder amounting to the weight of two small silver coins and given as a beverage, cures the fevers and tertiana; it has produced miraculous results in Lima [Burba].

Seven years later, missionary priests began processing the alkaloid bark and shipping it to Europe, where it was called quinine.

1633 During a year-long cycle of sickness and starvation, plague struck the people of Mexico City, killing large numbers. Stagnant water and garbage increased the population of rats, vectors of the pathogen. Of the 60 Dominicans felled at the convent of Santa Catarina de Sena, 40 percent died. The people were still weak five years later when measles overran the city, killing even more.

fall 1633 When Dutch traders sailed from Amsterdam for Massachusetts, 15 children died en route of smallpox. Upon the vessel's arrival, the disease killed 20 colonists living near the colonial port. While conspiring with the Pequot against the English, the Dutch spread smallpox the next year to 95 percent of the indigenous people living along the Connecticut River. William Bradford, the Puritan governor of Plymouth Colony and author of *History of Plymouth Plantation, 1606–1646*, reported that over 900 died. Of the sufferings on Long Island, he produced a graphic description:

> [They] died most miserably. [They] fall into a lamentable condition as they lye on their hard mats: ye pox breaking the mattering, and running into one another, their skin cleaving ... to the mats ... when they turn them a whole side will flea off at once and they will be all of a gore of blood, most fearful to behold; and they being very sore ... they die like rotten sheep [Underhill, 1998].

Some expired rapidly at the very onset of fever. The people were so depleted that they were unable to bury the dead, who lay decaying. Those who fled carried contagion into the New England interior. The catastrophe freed the Indians' homeland for European colonization.

That summer, the disease migrated north. Carried along the Saint Lawrence Seaway by French ships, smallpox struck first the Montagnais, then the Algonquian, and traveled with traders to the Huron of Trois-Rivières (Three Rivers) in Quebec. Into the fall months, pestilence more than halved the Huron headcount. From Quebec, the disease spread to the Neutral and Petun Indians of southern Ontario. *See also* **1670.**

In the words of Nebraska historian Mari Sandoz in *The Beaver Men: Spearheads of Empire* (1964), after half the populace lay on the ground, "Those still able to rise fired their long houses and fled. As they ran and paddled away with their pathetic bundles, trying to carry the sick children along, some thought that a woman's shape appeared in the smoke of the sky" (Sandoz 1964, p. 44).

In *Relation de Ce Qui S'est Passé en La Nouvelle France en l'Année 1638* (*Account of Events in New France in 1638*), a journal of life among the Huron of Quebec, Father Paul Le Jeune of Normandy, superior of the Jesuits, described the natives' rush to abandon a village stricken with disease that traders imported by canoe. He stated:

> On the fifth day of April, a Montaignais Savage came to report to Father Buteux that our Fathers and our Frenchmen who accompanied them had been abandoned in the woods and tied to trees, by the Hurons who ... believed that this malady was caused by the French, and it was this which made them treat the French in this way; this savage declared that he had heard the news from the lips of some Bissiriniens, neighbors of the Hurons [Thwaites 1901, p. 319].

Another charge derived from Tonneraouanoné, a nostrum salesman at Ossossané, who accused the Jesuit black robes of spreading disease and of dosing Indians with medicine that sped their deaths. He later denied the charge.

Le Jeune surmised that no one escaped contagion, especially in autumn during the fishing and harvesting seasons. At a height of agricultural productivity, the people sickened and collapsed, leaving their crops unreaped. Those who survived measles and smallpox experienced blindness. Others were carried off by intestinal pathogens.

1634 At Bogota, Colombia, the Hospital of Saint John of God of the Holy Faith was a source of treatment for victims of pandemic typhus, which local people called the "tabardillo" (little vest) or "plague of Saint Giles." Symptoms began with fever and headache and advanced to vomiting, paralysis, torso aches, chills, throat ulceration, and delirium. Hispanics died at the rate of 20 percent, Muisca Indians at 80 percent. The scourge, contemporaneous with locust infestation and starvation, started in slaves from Cartagena and encompassed Guatemala, Ecuador, Paraguay, and Mexico, where locals were recovering from epidemic typhus. Survivors became deaf, crippled invalids and remained debilitated for life.

1634 Following a plague of locusts, Tigray, Ethiopia, endured epidemic cholera, which originated in Dembiya and spread north.

1634 The citizens of Munich, Germany, erected the Mariensäule to honor a pledge the Elector Maximilian I made after Gustav Adolf led a Swedish invasion force into the area during the Thirty Years' War. The immense Corinthian column topped by the Virgin Mary, Queen of Heaven, is the design of Hubert Gerhard. It commemorates 7,000 citizens who died in an epidemic of plague, leaving only two-thirds of the populace alive. The symbols on the base—the lions of war, dragons of hunger, snakes of disbelief, and a basilisk of pestilence—face triumphant cherubs.

1635 In the months after the Winnebago met Jean Nicolet at Doty Island, Lake Winnebago, Wisconsin, smallpox reduced a population of 25,000 to 16,000 people. A second outbreak in 1636 killed 6,000 more, leaving them at 40 percent of their original headcount. At the end of the 17th century, Cotton Mather, a Puritan minister in the Massachusetts Bay Colony and New England's most famous sermonizer, declared that Indians deserved God's curse. Mather used as proof of their inherent evil the fact that smallpox had killed them at a rapid rate in the epidemic of 1633–1635. The scourges continued into 1640, the worst occurring in 1639.

1635 A severe outbreak of plague in France that lasted until 1636 created 3,000 orphans. Capuchin brothers collected the children and sheltered them at Amiens. Jesuit mystic poet Friedrich Spree von Langenfeld succumbed to plague at Trier while attending sick and dying soldiers. Another Alsatian-German Jesuit poet, Jacob Balde at the court of Munich, grieved for the dying in original odes and elegies.

May 1635 Jean de Brébeuf, a French Jesuit missionary and explorer, described in *Relation of What Occurred among the Hurons in the Year 1635* (1901) his experiences with an epidemic among the Montaignais in Ihonatiria, Ontario. He described the "sickness among our Savages" at Three Rivers the previous autumn as fatal to many and infectious to almost everyone who traded in the area (p. 87). He remarked,

> A large number of persons are dead; there are still some who have not recovered. This sickness began with violent fever, which was followed by a sort of measles or smallpox, different, however, from that common in France, accompanied in several cases by blindness for some days, or by dimness of sight, and terminated at length by diarrhea which has carried off many and is still bringing some to the grave [p. 89].

The unusually virulent disease limited hunting, fishing, and harvesting, thus leaving the people hungry and feeble, despite the intervention and potions of necromancers.

spring 1636 Plague at the harbors of Hull and Yarmouth, England, derived from trading vessels from the Netherlands. Still virulent in late winter, the infection caused prose writer Thomas Brewer to print a pious, hand-wringing pamphlet, *Lord Have Mercy Upon Us* (1636). The disease resurged in spring 1637 in London, killing 3,000. When provisioners ceased to supply towns, people scavenged for food. Demographer John Graunt of Cornhill, compiler of *Natural and Political Observations Made Upon the Bills of Mortality* (1662), declared that the death toll reached 10,400 for the year as a result of fetid air rather than from the "effluvia" of human bodies. Citizens bought copies of bills of mortality, an official census listed by disease. They scanned each issue for data on contagion and commented on the predations of plague in their letters and journals.

fall 1636 In southern Quebec, an unidentified outbreak, possibly influenza, struck the beleaguered Huron at Ossossane and infected four Jesuit missionaries at Ihonatiria, Ontario. In this same era, the Huron suffered a scourge of measles or smallpox that reduced them to around 10,000 and deprived them of influential leaders. As a result of sickness, more Indians turned to Catholicism as a protection from contagion.

1637 As described by Franciscan friar Juan de Prada and others, smallpox raged among North American Indians for four years from 1637 to 1641. Among the Hopi Pueblo Indians, 14,000 newly converted Christians died of the pox. Baron Louis-Armand Lahontan's *New Voyages to North America* (ca. 1700) declared that the Attikamek, a Montaignais tribe of the upper Saint Maurice River, Quebec, were annihilated. Sickness and famine felled villagers at Andiatae and Onnentsati. Visitors to Nipissing were decimated. At Trois-Rivières (Three Rivers), Father Simon Baron gained respect for medical treatment of the sick. The scourge further reduced the population of Huron in southern Ontario to 9,000. The surviving Huron returned to northern Ontario, spreading infection along their route.

1639 At New France (Quebec), terror erupted with an epidemic of smallpox, introduced perhaps by 30 Huron traders bearing a load of furs. In the words of Pierre François Charlevoix, author of *History and General Description of New France* (1744):

> The servants of the Lord took all the inconvenience for themselves, the sick and their pupils experiencing none of it. Yet God wished to subject both to the severest trials. The Ursuline seminary was first attacked by small-pox, and an epidemic brought to the hospital more sick than there were beds or even rooms to put them in. These afflictions did not disconcert the nuns. They provided for all in a way not easily conceived, and never was seen more clearly what the power of charity can effect [1900, p. 287].

As a result of contagion in close, foul-smelling quarters, all the nuns contracted the pox. The Jesuits of Trois-Rivières (Three Rivers) in Quebec remained healthy during the epidemic, but later developed *mal du terre* (scurvy), which killed some of them.

The Huron population of Quebec, which had declined in previous epidemics, suffered additional outbreaks of smallpox introduced by the Kichesipirini via the Abenaki of Maine. Jerome Lallemont, a Catholic missionary who filled in graphic details of the suffering, explained how lack of quarantine worsened the pestilence:

> The Hurons—no matter what plague or contagion they may have—live in the midst of their sick, in the same indifference, and community of all things, as if they were in perfect health. In fact, in a few days, almost all those in the cabin of the deceased found themselves infected; then the evil spread from house to house, from village to village, and finally became scattered throughout the country [Lallemont 1901, p. 89].

Into the winter months, the outbreak caused thousands of fatalities, unintentionally spread by ministering missionaries. In *Relation de Ce Qui*

S'est Passé en La Nouvelle France en l'Année 1639 (*Account of Events in New France in 1639*), Father Paul Le Jeune of Normandy, superior of the Jesuits of Quebec, recorded the gift of a native garment from the Indians to the dauphin of France. He hesitated to pass along a gift that might endanger the life of French royalty.

Le Jeune was puzzled by a twofold response to his mission. He remarked that Indians accused the clergy of spreading smallpox. They believed that God favored white priests and spared them because they knew the right words for prayers. Despite their fears, the Huron turned to Jesuit missionaries for baptism to save their souls from perdition. He also noted the cruel actions of a woman who abandoned her brother and his children to avoid contagion. The woman fled with her own children, but died before she could reach a hospital. Rescuers brought her son to white healers for medical care, but he, too, succumbed to the pox.

One hero, Saint Anthony Daniel, a Jesuit priest from Dieppe, taught catechism, directed a Huron boys' school, and founded fourteen mission stations. At Ossossane west of Lake Simcoe, he traveled from lodge to lodge treating victims of smallpox. The Indians charged him with spreading disease through baptism, forcing him to plead his innocence in a council session. He explained that the illness was already taking hold when he arrived and reasoned that he would not nurse suffering people if he intended to murder them.

Because of Saint Anthony Daniel's worthy example, he converted several hundred Huron by spring, but others rejected the Jesuit concept of heaven. A decade later, an attack by Iroquois during mass left him pierced through the vestments with arrows. His killers dipped their fingers in his blood, then tossed his remains into his burning church.

1639 An army surgeon and surgeon-general of the East India Company, John Woodall, father of naval hygiene, composed a classic text, *The Surgeon's Mate, or Military & Domestique Surgery* (1617). In 1639, during his tenure as surgeon to Saint Bartholomew's Hospital in London, he issued the third edition along with a list of items and medicines belonging in the surgeon's chest, an essay on plague and gangrene, and a commentary on scurvy. To cure the rampant debilitating disease that struck crews on long sea voyages, he recommended scurvy grass, horseradish, nasturtia aquatica, wormwood, and sorrel. For galley cookery, he advised cooks to provision the ship with lemons, limes, oranges, and tamarinds.

summer 1639 Scarlet fever invaded the Tobacco, Neutral, and Erie Indians of the Great Lakes region of North America. The Huron appear to have contracted scarlet fever from the Susquehannock. The sick living along the Ottawa and Saint Lawrence rivers often perished within 48 hours of their first symptoms. Because shamans treated fever with sweat lodges, French missionaries used their failure to cure disease as proof that Christianity was a stronger faith than animism. The Iroquois also exploited the weakened Huron with well-timed attacks, from which the Huron never recovered.

November 1639 During the Eighty Years' War, a face-off between the Dutch and Iberian fleets resulted in mass death from epidemic disease rather than combat. At Cape Verde, 3,000 Spaniard and Portuguese crewmen of 77 ships died in the harbor, forcing leaders to regroup before departing in January 1640 with only 1,200 marines capable of fighting.

1640 After Dutch fur traders encountered Huron of the Hudson Valley, in New York, European diseases, particularly measles, smallpox, and typhus, reduced natives to only 10 percent of their original number.

August 1640 In *Relation de Ce Qui S'est Passé en La Nouvelle France en l'Année 1638* (*Account of Events in New France in 1638*), Father Paul Le Jeune noted the work of Ursuline nuns at the hospital in Quebec during a period that greatly diminished the population of Huron-Wendat (or Wyandot). In the wards, more than 100 white patients and twice that many Huron were infected with smallpox. The sickness produced a terrible stench. The nuns remarked on the courage, cooperation, and patience of the Indians. Father Claude Pijart evangelized them all and baptized twenty.

The loss of native workers made vast changes in the French trading system. The French expanded their territory and settled in the wilderness to begin trapping, collecting, skinning, drying, and shipping their own furs. From this era of hands-on, unlicensed labor came the legendary "Coureurs de Bois" (woods runners), unscrupulous traders who further debauched the weakened Indians by selling them whisky.

1641 In Mexico City, a drought worsened an outbreak of an unknown pestilence called *cocoliztli*, the Nahuatl term for "pest." Infection reached its height among indigenous Indians a year later. Subsequent medical investigation connected the scourge with a form of hemorrhagic fever.

1641 Epidemic plague during China's Ming period killed large numbers of people. Epidemiologist Wu Youxing (or Wu Yu-hsing) issued *Wenyilun* (*On Pestilence*) (1642), which characterized the disease as the result of excess. He advised herbal cures. His proposal of a transmissible pathogen invigorated medical debate over how the disease traveled from person to person.

December 1642 Abel Janszoon Tasman, an employee of the Dutch East India Company, sailed from Batavia (modern Jakarta), Indonesia, to New Zealand in the *Heemskerck* and the *Zeehaen*. After Maori canoeists of the southern island of New Zealand rammed his cockboat, they killed three of his Dutch sailors. Consequently, Tasman left the island group without disembarking. Medical historians surmised that the native cannibals may have eaten the Dutch crew members' remains, thus incurring European diseases.

spring 1643 One of the most important medical writers of his day, English clinical neurologist Thomas Willis North Hinksey of Berkshire was a beginning practitioner of medicine when he encountered epidemic typhoid. Living a mile and a half from Oxford, where he had graduated from Christ Church the previous year, he was employed as servitor to a cathedral priest. A member of the Royal Society and author of *Diatribae Duae: De Febribus* (*Two Texts: On Fevers*) (1659), he became the first to describe typhoid fever, which he observed among the troops of King Charles I at Oxford. His Latin treatise carefully delineated the disease from other look-alike fevers.

In April, typhus overwhelmed the people of Reading during an 11-day siege. Infection spread rapidly and felled some children and youth, but mostly elderly males, effectively halting hostilities until both sides recovered. On the march into the summer, both the crown's troops and the Puritan army carried typhus and typhoid into the countryside. At garrison headquarters at Tiverton in early July, the troops of the Earl of Essex spread disease to civilians, causing the deaths of 105.

fall 1644 As the English Revolution continued, plague struck Newcastle, Scotland, and advanced to Edinburgh. City officials closed the university and appointed Paulitius as town physician to diagnose cases. Liveried cleaners patrolled the streets, burning thatch and refuse and purifying infected houses. Courtrooms closed; starving debtors were released from the tollbooth. In the absence of a standing militia, authorities posted a watch against disease-bearing outsiders entering the city.

April 3, 1645 As chronicled by David Aldinstone, session clerk of South Leith Church, the citizens of Leith, Scotland, suffered severe bubonic plague spread by Alexander Leslie marching his covenanters from Newcastle and extending north to Perth. When three residents of Yardhead fell ill, health officials isolated them indoors and suspended a white sheet from their casements in token of quarantine. Also confined to his home, Aldinstone paid the penalty for praying with another plague victim, Margaret Gilmuir, in her last hours. Church authorities immediately began collecting money to aid the poor, garnering supplies, closing middens, and hiring cleaners and sledders to collect garbage. City officials appointed Aldinstone's friend, Alexander Abercrombie, a maltman from Yardhead, to gather heather to burn in infected residences as a fumigant.

The disease moved to the King James VI Hospital and into other residential areas. Local Scots built temporary pest huts called booths or lodges on Leith Links. To stem contagion, survivors burned bedding and boiled infected clothes nearby. Gravediggers quickly exceeded space in churchyards and began filling plague pits with corpses. The Scottish parliament noted:

> The number of dead exceeds the number of the living, and amongst them it cannot be decernit quha [who] are clean and quha are foulle: and to make the calamatie greater, they are visit with ane lamentable famine, both for penurie and also for laicke of means ["History of Leith"].

The collection of unclaimed goods from plague victims produced surplus cash amounting to £500, which the city spent on a silver baptismal font for South Leith Church. By November 1645, when heavy rains rinsed the streets clean, the official death toll for Leith stood at 2,736. As a result of the experience in their town, citizens instituted rules of hygiene that required bathing before church attendance.

April 4, 1646 When plague moved on from Leith to Menmuir and Brechin, Scotland, churches closed for seven months and refused to collect money because it might be contaminated. Commerce continued via stone bowls in which vendors placed dairy food and meal. Buyers removed what they wanted and left coins covered with water to cleanse them of disease.

Officials of Brechin hired cleaners from Edinburgh to carry out corpses and to fumigate streets by pushing flaming tar barrels down thoroughfares. City workers set up straw and branch pest huts at Murlingden to receive the dying. After their deaths, examiners prevented recontamination from their corpses by setting the huts aflame. In all, the disease killed 600 at Brechin.

1647 In the environs of Rome, Italians suffered a two-pronged scourge of malnutrition from an inadequate harvest of wheat and from the annual snowmelt flooding of the Tiber River, which increased mosquito breeding grounds. As a result, hungry people succumbed in large numbers to malaria.

1647 In Bridgetown, Barbados, yellow fever killed over 5,000. As attested to by chronicler and cartographer Richard Ligon, author of *A True and Exact History of the Island of Barbadoes* (1657), slaves imported from West Africa carried the disease to the Caribbean. As the breeding ground for the *Aedes aegypti* mosquito proved stable, an insect-borne fever, called "nova pestis" (the new pestilence), "bleeding fever," or "Barbados distemper," attacked primarily European settlers and visitors, indentured laborers in the sugar industry, traders, sailors, and the few remaining Carib.

The disease rapidly invaded Cuba, Guadeloupe, Jamaica, Saint Kitts, and the Yucatán peninsula of Central America. White overlords in Jamaica died at a high rate, mainly because of their risky lifestyle and their lack of immunity to yellow fever. The rise of disease throughout the Caribbean inhibited shipping to the New England colonies, where port authorities, fearing the spread of contagion, refused their vessels docking privileges. The Puritans who settled Massachusetts Bay quarantined incoming ships from Barbados to prevent infection. They added to their deterrents a red flag to be raised outside residences harboring the incoming sick.

1648 In the New World, a crewman aboard a slaving vessel from West Africa carried the yellow fever virus to Havana, Cuba, the first identifiable patient zero in the Western Hemisphere. Mosquitoes breeding in stagnant water spread contagion, which proliferated in human salivary glands. The brief fever abated after four days. When it returned, the disease ran its fatal course in eight days, producing jaundice and liver malfunction that caused bleeding from mouth and nose and vomiting of black blood.

When the fever migrated to the Yucatán peninsula, it invaded virgin-soil populations. One horrified Indian mourned:

> Great was the stench of death. After our fathers and grandfathers succumbed, half the people fled to the fields. The dogs and vultures devoured the bodies. The mortality was terrible. Your grandfathers died, and with them died the son of the king and his brothers and kinsmen. So it was that we became orphans, oh, my sons! So we became when we were young. All of us were thus. We were born to die! ["Epidemic Disease"].

early 1648 A wave of deadly black smallpox enveloped the Massachusetts Bay Colony, striking the children born in the colony who lacked immunity from earlier outbreaks. The disease earned its name for subcutaneous hemorrhaging that turned victims black. On March 9, Angus Montrose wrote that the pestilence enlarged daily, forcing session members to flee contagion. To assure no further infection, residents banned the sale of food and goods in towns on pain of a fine of £4. Worsening the scenario was an increase in pertussis, which raised the death toll. Hardest hit were residents of Barnstable, Boston, Cape Cod, Roxbury, and Scituate. Survivors bore evidence of the pox in blindness, disfigurement, or sterility.

Devout victims and their families tended to side with moral environmentalists, who interpreted pain and death as a punishment from God for personal failings and lack of piety. A year after the scourge in March 1649, church-goers fumigated their sanctuary and established the second Thursday of the month as a day of humiliation or fast day for the humbling of participating sinners. The ceremony climaxed with individual prayer and contemplation of sins.

1650 A pioneer of field rocketry under kings Wladyslaw IV and Jan II Kazimierz of Poland, General Kazimierz Siemienowicz of Lithuania compiled Book I of a classic ordnance manual, *Artis Magnae Artilleriae* (*The Great Art of Artillery*) (1650), denouncing biological warfare. He castigated poisoned spheres and other forms of poisoned weapons as unworthy acts of a soldier. Because of his respect for honorable warfare, his writings remained influential for two centuries. Despite his noble words, he increased the killing power of Polish guns by placing saliva from rabid dogs into hollow bullets to shower on the enemy in an effort to start an epidemic.

1650 When a Spanish crew brought plague from Iberia to Galway, Ireland, Dubliners revived previous plans for disease containment and eradication. Taxes paid for the hiring of a city surgeon and the treatment of the sick at a municipal lazaretto. By June, the city was nearly barren of life as the living fled the scene, leaving behind the dead and dying. In the wake of pestilence, the area suffered several years of hunger and outbreaks of dysentery.

1650 Until the arrival of Europeans, the Wendat (or Wyandot), who lived along Lake Huron, consisted of some 30,000 members who occupied longhouses and grew vegetable gardens. They disappeared from history after the establishment of the Sainte Marie mission in Midland, Ontario, in 1640. Within a decade, along with starvation and raids by Mohawk and Seneca, the diseases carried by the 23 Jesuit priests (called black robes) reduced the settlement to under 300.

September 7, 1650 When the *Saint André* sailed for Canada, bringing priests and hospital nuns and 109 settlers, a shipboard pestilence killed several before the docking. At the Tadoussac mission in Quebec, Jesuit priest and explorer Jean de Quen treated the sick and himself contracted the disease in early October. The infection spread to Montagnais throughout the region.

ca. 1650 At Genoa, a sizeable pesthouse and quarantine system protected the city from contagion. Guarded by German mercenaries, the building contained a ward for 55 patients diagnosed with plague and a separate isolation center for 238 people exposed to plague or for those entering the city from areas known to be reporting an outbreak. The city government attempted to develop inter-city rules consistent with those in Naples and Rome, but failed to coordinate efforts.

1652 A Dutch anatomist, Nicholas Tulp, author of *Observationes Medicae* (*Clinical Observations*) (1652), examined a young man returning from the Dutch East Indies with beriberi or polyneuritis, an endemic scourge that killed many who ventured to the South Pacific. Tulp commented that beriberi, called "the lameness disease," caused unusual cold and paralysis that was difficult to treat.

1652 Plague had devastating effects on France. A corps of lay nurses, the Sisters of Saint Charles of Nancy, France, organized their order to feed and alleviate the suffering of plague victims. The favorite plants for cleansing—cloves, garlic, juniper, lavender, melissa, rose, and rue—protected worshippers and mourners at churches. Chapter IX of English journalist Daniel Defoe's *Journal of the Plague Year* (1721) describes the aromatics, balsamics, drugs, and herbs that made one church smell like bottled fragrance in an apothecary shop.

Valerie Ann Worwood's *Complete Book of Essential Oils and Aromatherapy* (1991) described how, at Toulouse, Father Arnaud Baric became an expert at disinfecting cities through fumigation and a form of aromatherapy dispensed by a health captain, who burned the plants on hot coals. Under Baric's direction, city employees visited infected residences and wards to expel fetid air. When the fumigators completed one round, they boiled fragrant plants in cauldrons of water in their cloth tents, which served as steam rooms. For Baric's sympathy with the poor, he aroused the ire of Jesuits, who considered him a Jansenist. A street in Toulouse preserves Baric's name.

1655 When plague journeyed from Algiers to Amsterdam, Holland, it yielded 16,727 fatalities. Simultaneously, Bremen and Hamburg, Germany, lost 60 percent of their populace to the scourge. Surgeon-anatomist and medical author Paul Barbette, who trained at the University of Leiden, issued an immediate clinical account of the disease. His work, *De Peste Tractatus* (*Treatise on Pestilence*) (1678), was widely translated and remained in circulation into the 18th century.

1655 A second smallpox epidemic struck the Timucua of north central and northeastern

Florida, halving the population a second time and producing a shortage of press gangs at St. Augustine from the pool of exhausted, malnourished Indian laborers. Colonial governor Diego de Rebolledo responded by launching a raid at Ybiniyutti on the upper St. John's River to muster men to grow provisions for his military installations. During the brutal Spanish reprisal of 1656, Father Gomez de Palma lamented, "I have been here in Florida for 46 years seeing the labors and persecutions of the poor Indians, but never have I seen them so unfortunate as under this governor" (Wasserman, 75). As Timucuan society diminished, individual survivors appear to have merged with the Seminole, who moved into the vacated area around 1700.

May 1655 Jamaicans suffered their first yellow fever epidemic, which arrived from West Africa aboard a slaver or from a Cuban vessel. Contagion killed 1,000 islanders plus hundreds of European occupation troops. The soldiers of Sir William Penn, commander of the expedition fleet that Oliver Cromwell dispatched to the West Indies, died at the rate of 560 per month. Because black slaves were naturally immune to the disease, they quickly outnumbered indigenous people.

1656 When plague migrated from Algiers to Genoa, Naples, and Rome and east to Barletta, Italy, the death toll reached 218,000, nearly 44 percent. Individual mortality varied:

City	Deaths from Plague	Fatality Rate
Barletta	12,000	60%
Genoa	60,000	60%
Naples	150,000	50%
Rome	23,000	19%

At Genoa, as recorded in *Li Lazaretti della Città* (*City Pesthouses*) (1658) by a priest, Antero Maria Micone da San Bonaventura, the poor had less chance of avoiding an epidemic of pneumonic and septicemic plague. The underclass had the advantage of lean bodies and hardiness compared to the gouty, overfed rich and the services of the Order of St. Camillus and of a Bavarian physician, Martinus Ludheim.

A report on 14,000 deaths, health precautions, quarantining, and sanitary measures in the slums at Naples and Rome appeared in Cardinal Hieronymi Gastaldi's *Tractatus de Avertenda et Profliganda Peste* (*Treatise on Averting and Avoid-*

ing Plague) (1684), one of the valuable texts on quarantine. As the disease spread, Tuscany reported no disease because of its stringent regulation of trade, but the terrible era devastated Minorca to the west.

1656 East of the Baltic Sea, Sweden's occupation forces in Finnic Livonia (modern Latvia and Estonia) suffered famine and an unspecified plague. Sickness left them at 60 percent of their original strength and reduced Krakow and Warsaw, Poland, by two-thirds and one-half. The virulence of the disease kept Livonia secure from attack by scaring off the Russians.

1657 Archaeological examination of the remains from two ancient cemeteries begun at Notre-Dame-de-Québec Basilica in Quebec City produced skeletal proof of an epidemic striking children under age ten. Bone lesions point to rickets from insufficient vitamin D in the diet as the source of high mortality rates among the offspring of European settlers.

1658 At Montreal, nurse-missionary Jeanne Mance arrived from France with the first corps of Hospitallers of Saint Joseph. On the 67-day sea voyage, typhus threatened everyone on board. During the Iroquois War, when the French survivors reached the city, they received the support of Louis XIV. The staff replaced the original stone hospital with the Hôtel Dieu, a two-story facility containing separate wards for men and women and a dormitory for the Hospitallers on the second floor.

1658 After the Dutch East India Company took control of Ceylon from the Portuguese, officials began combatting widespread leprosy. They institutionalized diseased islanders at the leper asylum outside Colombo or banished them to the Tuticorin leper colony on India's southeastern tip. The Dutch themselves remained free of the disease until 1694.

1659 Puritan cleric Cotton Mather confirmed the emergence of respiratory diphtheria at Roxbury in Massachusetts Bay Colony. Within 14 days in December, contagion killed five-year-old Mary, three-year-old Elizabeth, and one-year-old Sarah, the children of Puritan preacher and missionary Samuel and Mary Wilson Danforth. On its first appearance in the Western Hemisphere, the scourge aroused fears of throat distemper" or "putrid throat," the creator of "windpipe bladders." The pestilence thrived in young children and among the Abenaki and Pennacook.

1659 In central Mexico, measles killed Maya children at a devastating rate. Simultaneously, a smallpox epidemic filtering eastward from the Great Lakes to Canada and New England invaded Long Island, killing the Montauk (or Montauk-ett) sachem Wyandanch (also Wyandance or Waiandance) of Paumanack, a respected peacemaker. His friend Captain Lion Gardiner, author of *So Must We Be One, Otherwise We Shall Be All Gone Shortly* (1660), suspected poisoning, but it is more likely that the elder statesmen died of contagion like thousands of other Algonquian. To be near Gardiner and the Rev. Thomas James, who sheltered them during hostilities with the Narragansett, the Montauk moved to East Hampton, leaving native village fires cold and blackened. Ethnographer William Wallace Tooker, author of *The Indian Village of Wegwagonock* (1896), noted that, gradually, wigwams disappeared as bark coverings decomposed. Within a year, the Montauk gave up their ancestral lands to whites.

August 1659 When an epidemic assailed victims at New Haven, Connecticut, with headache, fever, and delirium, the Rev. John Davenport attended the sick. Assisting the main caregiver, his wife, herbalist Elizabeth Wooley Davenport analyzed urine and prescribed homemade nostrums. She consulted Governor John Winthrop's manuals and medical texts and organized a network of nurse-practitioners to reinforce patient care. Another source of practical healing came from the letters of Sarah Hawkredd Cotton of Boston.

1660 A famous survivor of widespread smallpox, the Blessed Kateri Tekakwitha, born to Mohawk-Algonquian parents at Auriesville, New York, in 1656, contracted the virus at age four during a lethal three-year epidemic that killed 1,000 members of the Iroquois Confederacy. After her parents and younger brother succumbed to the pox, the orphan, horribly pitted and scarred, lived with aunts and an uncle. At age 20, she was baptized by Catholic missionaries and lived in Canada in a Christian settlement south of Montreal, where she treated the sick and elderly. Called the "Lily of the Mohawks," she died on April 7, 1680, and was beatified three centuries later, becoming the first Native American to be declared blessed.

1660 Following bouts with influenza or tuberculosis, an outbreak of severely infectious and lethal hemorrhagic smallpox in Maranhão and Belem, Brazil, arrived from Angola. A missionary from Luxembourg blamed infection on stubborn Indians who rejected Christianity. The scourge spread up the Amazon, killing off much of the native labor on which European colonizers depended. To rescue their animistic congregation from a pestilence that blackened their skin, Jesuits worked harder to convert the Indians, worsening the contagion by gathering the people into sessions for prayer and exhortations.

Terrified, the indigenous Tupinambá were felled by a disease they called the "great fire" which left them unsightly and reeking of putrid excrescence. Some abandoned their homes and ran into the wilderness, carrying pestilence with them. Only 10 percent survived. Jesuit father João Felipe Betendorf's *Cronica* reported that priests dug graves with bare hands because no Indians survived to perform the service for their kin.

The muddle of affairs that followed widespread death from disease affected the social, ethnic, religious, and economic order. In retaliation against meddling priests, in January 1661, Spanish overlords complained to King Alfonso VI, who sanctioned replacement of an Indian work force with African abductees. In May, the settlers of Maranhão rebelled against Jesuit missionaries and expelled their Portuguese leader, Father Antonio Vieira. The Portuguese Inquisition found him guilty, forbade him to preach, and imprisoned him for five years.

1661 Typhus, called "spotted fever," struck London and remained virulent for five years. The disease often emerged among prisoners at Newgate Gaol and spread about the city center, killing 15,700. The end of the outbreak left citizens ill-prepared for the Great Death of 1665. Epidemiologist and medical writer Thomas Sydenham, called the "English Hippocrates," referred to this new and intriguing disease in his writings, *Praxis Medica* (*The Practice of Medicine*) (1707) and the encyclopedic *Opera Medica* (*Medical Works*) (1715). His commentary on acute infection earned him the title of "father of English medicine."

1661 The Susquehannock, rivals of the Iroquois, gradually wore down from persistent battles. When smallpox overran their remaining village, Indians living in close quarters contracted the disease. When the western Iroquois assaulted them once more in 1668, only 300 Susquehannock survived to fight for their people. Survivors ceded to the Iroquois in 1675 and fled west to merge with the Shawnee.

1662 In New York State, smallpox wiped out over 1,000 members of the Iroquois League, comprised of Cayuga, Mohawk, Oneida, Onondaga, and Seneca. Already fallen to around 10,000 natives, the league may have contracted the pathogen from the Delaware, the Great Lake Cree, or Susquehannock traders. Worsening the chances of survival was the arrangement and inseparable lifestyle of the Iroquois longhouse, the traditional multi-family residence. The disease blinded and marred the faces of survivors and carried off infants and toddlers as well as elderly kin, the repositories of tribal wisdom and culture. The Susquehannock suffered a major blow that reduced their number to 300.

1663 After an eruption of plague in Turkey in 1661 and impeded enemy Venetians at Zadar, Croatia, during the Cretan War, the disease reached Holland two years later, killing 35,000 at Amsterdam by 1664. In the words of English journalist Daniel Defoe, author of *Journal of the Plague Year* (1721), "It mattered not from whence it came; but all agreed it was come into Holland again" (1960, p. 11). Fortunately for the city, it was the final great siege of smallpox. Among the dead was Hendrickje Stoffels, long-term companion of painter Rembrandt von Rijn and mother of his daughter Cornelia. Hendrickje was buried on July 24 at Westerkerk. Because of the city's primacy to European trade, contagion spread to England and Germany. Baron Frey von Dheren, a civil servant at Bingen in southwestern Germany, honored the town's 1,300 dead with a chapel named for Saint Roch.

1663 In Mexico, waves of plague amid drought elevated malnutrition and death rates among indigenous people and their Spanish masters.

1664 As reported by gynecologist Philippe Peu, author of *La Pratique des Accouchements* (*The Practice of Obstetrics*) (1694), an epidemic of puerperal fever or septicemia killed many patients in the maternity ward of the Hôtel de Dieu in Paris, a repeat of an outbreak two years before. Originally called childbed fever or lying-in fever,

the disease acquired its name from Thomas Willis, an Oxford-educated physician and prolific 17th-century medical author. He formulated the exacting medical term *puerperaum febris* (puerperal fever) from the Latin *puer* (child).

1665 A recurrence of severely infectious hemorrhagic smallpox struck coastal Brazil, invading the sugar export center of Pernambuco (modern Recife) and advancing on Rio de Janeiro two years after an upsurge in yellow fever. Victims complained of fever, torso pain, and rash as well as bleeding pustules. The death toll reached half the populace, targeting African slaves with high mortality. The loss of blacks had disastrous effects on the Brazilian economy. Without workers, European entrepreneurs lost their investments and abandoned huge plantations, a reduction in planting that resulted in starvation.

June 1665 During the Great Plague of London, epidemic numbers climbed in June, when 100 died in a week. Theaters closed, as did universities and the Inns of Court. On June 7, diarist Samuel Pepys, England's first secretary of the admiralty, noted: "This day I did in Drury Lane see two or three houses marked with a red cross upon their doors and 'Lord have mercy upon us' writ there, which was a sad sight to me" [1960, p. 256]. He added details about death knells, fear and sadness, and the crown's desertion of Whitehall, the main royal palace. By June 30, the month's death count had climbed to 300 per day or 9,000 total.

Within one month, London's mortalities from epidemic plague rose to 5,667 for the month. Pepys reported empty streets and many touching stories, including the plight of starving sailors quarantined in port. He complained that "plague [made] us cruel as doggs one to another" (ibid., p. 274). On July 12, a common prayer and fast held in London and Westminster initiated a series of worship services on subsequent Wednesdays. A reprint of John Featly's *Tears against the Plague* (1646) declared:

> Pestilence consumes and hunger cries; thus the visited die they know not of what; for sickness calls, and hunger calls,
> And want calls, and sorrow calls; all of them join in their hideous concord, in their horrid discord, and call for our
> Ruin and yell for our destruction (Byrne, 135).

By mid–July, the privileged began a mass emigration from the city; the royal family of Charles II followed the next week and took up residence at Hampton Court, where he remained until February 1, 1666. His mother, Henrietta Maria, fled to France. On June 21, Pepys went to his office at the Naval accounting bureau, leaving his wife Elisabeth and her household to move to the country. He noted, "I pray God bless her, but it seems she was to the last unwilling to go" (ibid., p. 259).

By August 1665, the death count from epidemic plague in London mounted to 238,700 for the month, forcing thousands to stay indoors until the end of their quarantine. City provisioning lagged, leaving some citizens to starve. So many died that families could no longer bury all their dead at night, when a curfew allowed the ailing to walk outdoors apart from the healthy. To accommodate so many plague-infected corpses, the city opened communal pits north of the tower of London and two more outside Blackfriars. See also **August 1665, September 1665**.

August 1665 The town of Colchester, which had suffered severe outbreaks of plague in 1348, again battled the disease, which spread from London. The epidemic killed 4,731, half the citizenry. When the disease reached Braintree, it slew one-third of the people, leaving only 22 of 687 victims alive. After business and traffic ceased, grass grew in the rutted streets. Because disease left both Colchester and Braintree with a high proportion of indigent people, the government exempted them from paying hearth taxes.

September 1665 As the London plague epidemic reached its height in the hot weeks of September, enforcers of health regulations began fumigating streets, disinfecting ditches and public latrines, and killing stray animals. Parishes were sealed and ships' crews confined on board in the harbor. On the third of the month, diarist Samuel Pepys reported the removal of a healthy child from the home of his parents. Friends received him naked and carried him to Greenwich along with a new, uninfected suit of clothes.

Philanthropist Lady Anne Clifford, donor of the Hospital of Saint Anne at Appleby in 1651, explained in her diary that, for several weeks, the death toll in London was 8,000. Attendants at the city's metropolitan hospitals refused to receive the sick and dying lest the staff become infected and die. Officials required burial of the dead under six feet of earth and enforced sequestration of their homes by stationing armed guards and halting the sale of furnishings, clothes, and belongings of the dead. Those still alive inside the houses fired sickrooms with sulfur or a blast of gunpowder and freshened bedclothes and drapes with fragrant candles, benjamin or spice-bush, rosemary, and rosin or purified them with sulfur. A hero of the London plague was its lord mayor, Sir John Lawrence, who underwrote the livelihood of 40,000 out-of-work servants and supplemented cash with donated goods.

September 6, 1665 At Eyam, England's famous "plague village" ten miles south of Sheffield, Derbyshire, a package of damp fabric carried fleas that killed the tailor, George Viccars, on September 6, 1665. The Rev. William Mompesson, newly married and chaplain to Sir George Saville, convinced villagers not to venture out lest they infect all of northern England. At his urgings and those of retired rector Thomas Stanley, residents built huts for the sick, buried their family members at home, and quarantined themselves, surviving on supplies that the emissaries of the Earl of Devonshire left at the parish stone boundary markers. In payment, villagers left vinegar-soaked coins in a slab-covered spring called Mompesson's Well.

Mompesson's heroism saved northern England's citizens from plague, but cost Eyam 267 of its 350 residents. To limit contagion, the vicar held village funerals outdoors at a church in a limestone cavern in Cucklet Dell. He buried his own wife, Catherine, in the churchyard on August 25. Gravedigger Marshall Howe assisted with interments in exchange for the rights of plunder in the homes of the deceased.

The era produced diaries, journals, sermons, verse, and reflective and clinical essays on bubonic plague, including Nathaniel Hodges's *Historical Account of the Plague in London in 1665*. In *The History of the Royal Society of London* (1667), poet and historian Thomas Sprat immortalized plague's random menace as an arrow in the dark. From a spate of heavy reading came a seemingly harmless children's rhyme:

> Ring-a-ring of roses,
> Pocketful of posies,
> Attischo, Attischo,
> We all fall down.

The lines denote the appearance of tell-tale red blotches on the skin and the use of pocketfuls of herbs to ward off contagion. The third line mimics the sounds of a sneeze; the conclusion notes that no one is exempt from the fatal disease. For good reason, on November 15, Pepys exulted, "The plague, blessed be God! is decreased" (Pepys 1960, p. 282).

December 19, 1665 A tabulation of christenings and burials in London by parish summarized as well the number who had died of plague, ranging from a low of zero at Saint John Evangelist to a high of 4,838 at Saint Giles Cripplegate. Compared to a total of 9,967 christenings, there were 97,306 burials—48,569 males and 48,737 females. Of the lot, 68,596 or 70.5 percent were plague victims. A tabulation of causes of death was revealing (Naphy and Spicer 2000, p. 84):

Cause	Number
Abortive and Stilborn	617
Aged	1,845
Ague and Feaver	5,257
Bedrid	10
Blasted [infected]	5
Bleeding	16
Bloody Flux, Scowring & Flux [dysentery]	185
Burnt and Scalded	8
Calenture [delirium]	3
Cancer, Gangrene and Fistula	56
Canker, and Thrush [candidiasis]	111
Childbed	625
Chrisomes [month-old] and Infants	1,258
Cold and Cough	68
Collick and Winde	134
Consumption and Tissick [tuberculosis]	4,808
Convulsion and Mother [hysteria]	2,036
Distracted [deranged]	5
Dropsie [edema]; Timpany [bloat]	1,478
Drowned	50
Executed	21
Flox [hemorrhagic smallpox] and Smal Pox	655
Found dead	20
French Pox [syphilis]	86
Frighted	23
Gour [gout?] and Sciatica	27
Grief	46
Griping Guts [colitis]	1,288
Hangd & made away themselves	7
Headmouldshot & Mouldfallen [skull defect at birth]	14
Impostume [abscess]	227
Jaundies	110

Cause	Number
Kild by several accidents	46
Kings Evill [scrofula]	86
Leprosie	2
Lethargy [depression]	14
Livergrowne [liver enlargement]	20
Meagrom [migraine] and Headach	12
Measles	7
Murthered, and Shot	9
Overlaid [oppressed] and Starved	45
Palsie	30
Plague	68,596
Plannet struck [evil horoscope]	6
Plurisie	15
Poysoned	1
Quinsie [tonsilitis]	35
Rickets	557
Rising of the Lights [lung abscess]	397
Rupture	34
Scurvy	105
Shingles and Swine Pox	2
Sores, Ulcers, broken and bruised Limbes	82
Spleen	14
Spotted Feaver [typhus] and Purples [skin hemorrhage]	1,929
Stopping of the Stomach	332
Stone and Strangury [urinary disease]	98
Surfet [overindulgence]	1,251
Teeth and Worms	2,614
Vomiting	51
Wenn [tumor]	1

1666 After the arrival of settlers from England to the Massachusetts Bay Colony, Boston experienced its first smallpox epidemic, which produced head and torso pain, fever, pustules, convulsion, and delirium. Those susceptible to complications suffered death, disfigurement, infertility, and blindness. As shown by the journals of statesman Simon Bradstreet and mintmaster John Hull, between 40 and 50 died.

1666 During a five-year scourge of plague in the northwest Parisian Basin, northern Normandy, and Picardy, health officials battled France's last major outbreak of bubonic plague. Their weapons were limited to fumigation, isolation, and quarantine.

1667 A sailor disembarking at Accomac on the northern shore of Virginia slipped through harbor quarantine and spread smallpox. The disease killed hundreds of colonists and infiltrated the Accomac and Powhatan tribes.

1667 John Locke, a physician and philosopher, affirmed that tuberculosis was a major killer in London. The disease carried off one in five Londoners, generating close to 1,250 new infections per year. Among the victims of Europe's white plague was philosopher Baruch Spinoza of Amsterdam, whose mother and two older siblings died of the disease. Locke himself was infected, but he managed to survive consumption for 38 years.

1667 A recrudescence of smallpox in London cost 3,200 lives. It lasted into the late winter of the next year, forcing self-immurement among city dwellers. Samuel Pepys wrote in his diary for February 9, 1668, "Hardly ever was remembered such a season for the smallpox as these last two months have been, people being seen all up and down the streets, newly come out after the smallpox" (Pepys 1960, p. 456).

spring 1667 For two successive springs, plague and European diseases struck Mexico with a high fatality rate among the Tarahumana and the Tepehuan of Chihuahua. A Jesuit chronicle recorded cycles of ill fortune:

As its inseparable companion, disease followed hunger, attacking the villages oppressively and without remedy. Wretched

> and starving, people went from village to village or into the countryside seeking help, finally surrendering their lives to hunger and disease along the royal roads [Deeds, 78].

late summer 1668 One of the first epidemics of yellow fever recorded outside the tropics engulfed New York. Imported from Barbados, the scourge raged into early fall, killing 370. Chronicler Noah Webster, famed lexicographer and author of *A Brief History of Epidemic and Pestilential Diseases* (1799), called the scourge "autumnal bilious fever," a reference to the yellowing of victims' skin. In September, New York's second governor, Francis Lovelace, an English Puritan, called for a day of humiliation, a time of fasting and prayer to beg divine grace for sinners who brought on disease as a punishment from God.

1670 The Montaignais of Quebec suffered attacks of the Iroquois that pushed them from the St. Lawrence. A smallpox epidemic diminished their number, forcing the poor wandering Montaignais to seek aid from the Jesuit fathers at the Maison des Jésuites de Sillery outside Quebec. Survivors determined to hold fast to the Christian faith to protect them from future sufferings.

winter 1670 When smallpox hit Sault Sainte Marie in eastern Michigan, at Bawating, the population of the Saulteur Ojibwa Indians fell below 200. In contrast to the losses of other Great Lakes tribes, the Ojibwa suffered far fewer deaths. To preserve the culture of decimated native societies, the Amikwa, Nikikouek, and Marameg merged with the Saulteur of upper Michigan.

September 1671 As reported in *Relation of What Occurred in New France in the Year 1672*, a French trader, Charles Albanel, passed through Quebec on the way to Hudson's Bay, and reported that local Indians had diminished greatly in number because of war with the Iroquois and smallpox. To replenish their number, they received into the tribe people from other tribes.

1674 The founder of Coahuila Mission in Mexico, Franciscan Friar Juan Larios, who explored the Southwest with expeditioner Fernando del Bosque, reported an epidemic at San Ildefonso de la Paz in northeastern Coahuila south of the Rio Grande River. The priest took pride in the baptism of 300 dying Indians.

December 25, 1675 The death of a Maltese girl named Anna preceded the island's worst onslaught of plague. During the previous year, harbor masters had been on the lookout for diseased crews on slavers, warships, and merchant ships approaching the harbor, especially those that embarked from Constantinople or Alexandria, Egypt. In June 1674, the Grand Master traced Giovanni Vella, an escapee from a quarantined ship, to the Carmelite Church at Vittoriosa, where he sheltered under papal protection. The Grand Master petitioned church authorities to end ecclesiastical sanctuary for a decade until the plague died out.

Over an eight-month period, 11,300 Maltese victims died. Chroniclers Dal Pozzo and Lorenzo Hasciac blamed sin as the reason for God's visitation in the form of pestilence. On January 29, 1675, authorities established a lazaretto on Marsamxett island and opened a contagion ward in a former ammunition magazine. Knights conducted an island census, divided areas into districts, and enforced health regulations. They transported victims to hospitals, supervised burials, and disposed of infected belongings. On April 11, to supply enough graves, a parish priest blessed a cemetery for use as a plague burial ground. Where space was limited, villagers reopened abandoned chapels or designated burial fields to receive the dead.

Health workers who bore the titles *medico infetto* (contagion doctor) and *chirurgo infetto* (contagion surgeon) received hardship pay; for spiritual matters, the *cappellano infetto* (contagion chaplain) or *sacerdote infetto* (contagion priest) comforted the dying and conferred last rites. Many staff members and oarsmen rowing victims to the lazaretto were infected. Two physicians, Silvano Mancarella and Pietro Turrensi, along with their *buonavoglia* (goodman or medical aide) attended Chaplain Giovanni Garzin, who died of plague at Porta Reale under the care of Dominican chaplain Clemente Maldonato. One holdout physician, Gian Francesco Bonamico, issued a publication, *Lettera Missiva* (1668), describing clinical symptoms and therapies and explaining why he chose to barricade himself in his home rather than to treat patients.

By June 24, the plague tally began dropping. Islanders credited the waning of the scourge to Saint Rosalia of Palermo, whose relics protected Sicilians from the Black Death in 1624. Over a six-week period, to ready 26,000 Maltan residences for reoccupation, workers fumigated buildings, collected garbage, and incinerated infected bedding and clothes. Cleansing ended with boiling washable fabrics, scrubbing and whitewashing walls, and airing rooms. Authorities tamped down earth over graves and walled off plague cemeteries. For personal effects, villagers accessed public washing stations along the seashore.

To end famine, harbormasters began admitting provisioners and grain ships. Necessary duties on wine, cheese, meat, seed, firelogs, and coal recovered the 150,000 *scudi* the island government incurred for plague expenses. Quarantine ended officially on September 9. On September 26, a public thanksgiving processional and discharge of muskets celebrated the return of health to Malta. The pious dedicated new altars to saints Anne, Nicholas, Roch, Rosalia, and Sebastian, the patron of plague victims.

1676 An Oxford-trained physician and witness of epidemic typhus, Thomas Sydenham, the English Hippocrates, classified fevers in *Observationes Medicae* (*Clinical Observations*) (1676), including data on plague, cholera, measles, typhus, smallpox, dysentery, and flu. He spoke of *febris scarlatina* (scarlet fever) as a pandemic striking 2,112 infants late in summer with chills and red spots different from those caused by measles. He described the end of the second or third day, when the disease left a delicate squamous covering that resembled a sprinkling of flour.

summer 1677 English trading vessels carried smallpox to Charleston, South Carolina. From there, it spread up the Atlantic coast to Boston, where it killed several local dignitaries before peaking on September 30. The Rev. Thomas Thacher, minister of Old South Church and licenser of the first Boston press, compiled *A Brief Rule to Guide the Common-People of New England How to Order Themselves & Theirs in the Small-Pocks or Measels* (1678), the nation's first health directive on diagnosis and treatment. He advised, "Let the sick abstein from Flesh and Wine, and open Air, let him use small Bear warmed with a Tost for his ordinary drink, and moderately when he desires it" ("Smallpox, Scourge of the Colonies"). When another outbreak occurred in 1702, printers re-issued the informative broadside.

1678 The onset of malaria in Europe arrived with soldiers and ships' crews from service in the tropics of Africa, India, and the West Indies. Sir Robert Talbor, author of *Pyretologia: A Rational Account of the Cause and Cures of Agues* (1672) and *The English Remedy, or Talbor's Wonderful Secret for Curing of Agues and Feavers* (1682), rid Charles II of England of "tertian fever" and received a referral to the court of France. On his next professional call, Talbor earned 2,000 guineas in gold *louis d'or* from Louis XIV for treating his son and heir for malaria. A self-trained apothecary and "feverologist," Talbor had studied malaria at the port of Essex, a major nexus for Caribbean drug smuggling, and evolved his own nostrum—cinchona powder, called "Peruvian bark" or "Jesuit's powder," dissolved along with opium and wormwood in white wine. (Other sources list rose leaves and lemon juice as ingredients in Talbor's formula.)

The powerful alkaloid was a volatile subject among the English. In an era of anti–Catholicism, Oliver Cromwell had cursed cinchona as an element of popery and chose to die of malaria rather than submit to treatment with the drug. To avoid persecution, Talbor was careful to conceal the source of his drug. Advanced to the rank of knight, he used court connections to become personal physician of Louis XIV, Louisa Maria of Spain, and a host of nobles. Talbor's secret remained secure until his death in 1681.

late May 1678 Plague migrated on a trading vessel from Orán, Algeria, to Málaga, Spain. The next year, the pestilence spread to north and west from Turkey to central Europe, killing 83,000 in Prague, Czechoslovakia, and 4,397 or 44 percent of the citizenry of Halle, Germany. Thousands of Austrians perished, especially in Vienna, where 150,000 died. Other Europeans sneered at the unhygienic city as suffering the "Viennese death" (Cohen 2001, p. 375). Health officials set up plague hospitals, where the Brotherhood of the Holy Trinity nursed patients.

An Augustinian friar, Abraham of Sancta Clara, minister to Leopold I, treated the sick in Vienna. He recorded details of the epidemic in *Merk's, Wien!* (*Pay Attention, Vienna*) (1680), *Lösch Wien* (*Vienna Blotted Out*) (1680), and *Die Grosse Totenbruderschaft* (*The Great Brotherhood of the Dead*) (1681), which lists prominent citizens who succumbed to plague. As a means of leveling the proud, he remarked that "after death the prince royal is as frightfully noisome as the newborn child of the peasant" (*New Catholic Encyclopedia*, 1967). In thanks for the end to the scourge, citizens donated the Pestsäule, a baroque plague column commissioned by Emperor Leopold I, and the richly domed and decorated Karlskirche, erected by Johan Bernhard Fischer von Erlach on the order of the Emperor Charles IV.

1680 From October to May 1681, a smallpox outbreak in the Upper Huallaga Valley of Ecuador killed over 66,000 and, within a year, advanced from Quito across the Andes into the Amazon basin. At Bahía, Salvador, the loss preceded three years of drought. On large plantations, contagion felled the African slaves who harvested sugar cane, leaving inadequate work forces to sustain the economy. After priests recruited 40,000 more natives, some 600 Indians embraced Catholicism as a respite from the scourge. Others fleeing by canoe imported the infection to the Amazonian Omagua.

October 27, 1682 When William Penn arrived at New Castle in northern Delaware aboard the crowded ship *Welcome*, many of the 100 passengers were suffering from smallpox. During the Atlantic crossing, the disease had already killed 30 passengers. Because Penn was immune to the pox following his own infection at age 3, he tended to the sick and comforted the dying during the miserable sea voyage.

April 1684 A virulent outbreak of rubella invaded Japan, killing 8,000 in Nagasaki and Sakai alone. At Kyoto, peasants held late-evening drum rituals and made fetish dolls to drive away contagion. By mid-summer, the disease had weakened. It continued to infect, but not kill, as it advanced on Tokyo.

Easter 1686 Louis XIV of France, who touched 2,600 scrofula victims after his crowning in 1654, treated 1,600 more on Easter 1686 at Corbeny near Saint Marcoul's relics. In addition to distributing 15 *sous* in coin and 30 *sous* to foreigners, he recited a comforting phrase. His successor, Louis XV, simplified the annual pilgrimage to Corbeny by moving the holy relics to the Abbey of Saint-Remi in Rheims, where the new king touched over 2,000. The ceremony died out with Louis XVIII and returned to use in 1824 under Charles X, who added a personal touch to the comforting of the sick.

August 27, 1687 Chronicler Bishop Cartwright commented on how James II of England reinstituted the holy office and the sign of the cross for use during public and private healing ceremonies for the king's evil or scrofula, a tubercular condition causing swollen glands of the throat and neck. Of the hordes of sufferers arriving at 9:00 a.m., he granted audiences to only 350. To aid the poor in journeying to the ritual, private donors and guilds paid for their travel. The modified holy liturgy the king's staff recited appeared in Hamon L'Estrange's *Alliance of Divine Offices Exhibiting All the Liturgies of the Church of England since the Reformation* (1699). The last official enactment of touching for the king's evil occurred in 1712 under Queen Anne.

late summer 1687 Pierre François Charlevoix, an explorer and missionary and author of *History and General Description of New France* (1744), observed widespread illness among the Indians of Quebec. After coming in contact with

European ships, natives contracted measles and spotted fever or typhus. One historian estimates that 1,700 died of European diseases.

spring 1689 During an English siege of Ireland that began in April, when competing Catholics and Protestants laid claim to the throne, epidemic dysentery, syphilis, and typhus assaulted an army camp outside Dundalk, killing half the 12,000 men. Typhus, in the most lethal outbreak in British military history, spread to civilians as well. A recrudescence in late summer and early autumn derived from tenting on boggy land. On the retreat from Ireland, the English transported patients in wagons and loaded them on evacuation vessels at Belfast. Some of the patients and crew died before they could depart for England. Some of the men who marched south collapsed from sickness along the way.

1690 The threat of an epidemic of plague, which had killed Count Piccolomini, and the unrest during the Austro-Ottoman wars precipitated the exodus of Serbs from Kosovo, Serbia, over the Danube River. The horde of peasants traveled under the leadership of Janjeve Toma Raspasari, an Albanian priest, and Anton Znoriq, a Serb. The depopulation of towns and villages, known as the Great Migration, resulted in the loss of young, vigorous families and their spiritual leaders to Austria and Hungary. In their wake, Turks looted their homes.

March 1690 A new outbreak of measles felled much of Japan, including people of both genders and all ages. By April, it advanced on Tokyo. The following April 1691, Hida to the west and Sendai to the north bore the brunt of the epidemic.

October 1690 During King William's War, the first of the French and Indian Wars, Sir William Phips led a convoy of 14 vessels and 750 English troops of King William III from Boston to seize Port Royal, Acadia. Rebuffed by Count Louis de Frontenac near Quebec on the Saint Lawrence River, Phips fled. The episode cost lives to combat and the scuttling of five vessels. In the turmoil, 200 English colonists contracted smallpox and died.

summer 1691 Yellow fever made a second pass at the British colony of Barbados, felling those too young to have built immunity during the outbreak of 1647. Sickness devastated the 3,100 crewmen aboard the 18 ships of the British fleet that anchored at Bridgetown, causing the high command to jettison plans for an attack on the French at Martinique. The return to English ports was a hardship on those few healthy enough to man the fleet.

1692 For five years, a parallel outbreak of diseases overran the valley of Mexico and migrated northwest into Baja California by 1697. Worsening the situation, hunger, inflated food prices, and grain riots involved Indians and creoles, who suffered a ban on pulque, exile to barrios, lashing, and imprisonment at hard labor.

1694 Smallpox epidemics in Europe killed Queen Mary of England, Joseph I of Germany, Peter II of Russia, and William II of Orange and Louis XV of France. Queen Mary produced symptoms of hemorrhagic smallpox, which caused her skin to blacken. Before her death in late December at age 32, she lay expiring in full view of courtiers. A Dutch engraving pictured her corpse displayed in state under a curtained pavilion marked at each corner with an eight-foot candle. Composer Henry Purcell issued *The Funeral Music for Queen Mary* (1695) at Westminster Abbey, comprised of an anthem, four pieces for trumpets and trombones, and two elegies.

1694 Syphilis which troops spread in the Scottish highlands was a remnant of the disease left behind by the invasion force of Oliver Cromwell in summer 1644.

1695 When epidemics felled Spaniards, a Dominican missionary to Mexico, Antonio de Monroy became archbishop of Santiago de Compostela, a famous pilgrim destination in northwestern Spain. He welcomed the sick to a church receiving center and distributed aid and food to the hungry locally and throughout Galicia. When dire circumstances recurred in 1710, he repeated his generosity.

1695 Among a virgin population in the Jesuit missions of Paraguay, 16,000 Indians died of a measles epidemic.

1698 Contagion swept the eastern seaboard of North America. An outbreak of smallpox at Hopkins, South Carolina, reduced the Congaree Indians significantly because of their lack of immunity. In one band, the handful of survivors were so traumatized that they fled their swampland village, leaving corpses unburied.

Among the Tonica (or Tunica) of the lower Yazoo River in Mississippi, smallpox devastated the tribe. That summer, fathers Davion, La Source, and Montigny, missionary priests from the Quebec Seminary of Foreign Missions, found moribund tribe members, including the chief. The priests baptized the dying and made plans to learn the Tonica language and set up a mission.

1699 A serious outbreak of yellow fever struck Philadelphia, killing 220, including around 90 Quakers, and forcing other citizens to desert the town. The journal of English Quaker Thomas Story records that "Great was the fear that fell upon all flesh! I saw no lofty or airy countenances, nor heard any vain jesting; but every face gathered paleness, and many hearts were humbled" ("Watson's Annals").

late August 1699 In addition to an earthquake, hurricane, fire, and epidemic of smallpox, the first outbreak of yellow fever tested the city of Charleston, South Carolina. The savage three-month onslaught killed around 300 people. Into the fall, the deaths robbed the city of its vitality by taking half of the assemblymen and other officials, leaving survivors to pull together a civic plan to allay panic. Fear ruled the citizens until the abeyance of the disease near Thanksgiving.

1700 As a barrier against imported yellow fever, called the "black vomit" or "Barbados fever," which had ravaged Charleston, South Carolina in 1699, pesthouses at Sullivan's Island across from the port confined 200,000 blacks, over one-third of incoming African bondsmen from Sierra Leone, Ghana, and Benin. For over a century, before newcomers could pass through the customs house for sale, those with obvious disease were immured on slave vessels. Those who passed inspection remained for two weeks to 40 days at a cramped wood quarantine station, where they were groomed with oil and pitch to hide skin imperfections before buyers' inspection.

Duke University Professor Peter H. Wood ennobled the offshore inspection system as black America's Ellis Island. Those who died on the slave ships during the middle passage were buried at sea; those expiring at the stations were unceremoniously dumped into mass graves dug in sand dunes. In 1752, a hurricane swept the pesthouse away along with 15 people, of whom nine drowned. The other six rode the roof up the Cooper River to safety.

1700 Along the lower Mississippi, the Houma, who had numbered 3,000 in 1650, shrank to 1,800 in 1699 as European diseases took their toll. The next year, dysentery halved the remaining population. Over the next three decades, war, alcohol, and a dozen scourges cut them down to about 10 percent of their original number. By 1718, the 400 who survived allied with the Acolapissa and Bayougoula, who had also lost about 90 percent of their former tribe members. Additional smallpox infection in 1721 whittled the headcount to a few hundred.

1700 An advocate of preventive medicine, Bernardino Ramazzini of Capri, Italy, earned the title "father of occupational medicine" for publicizing the rampant disease suffered by laborers. While chairing medical theory at the University of Modena, he began a controlled study of workplace disease and published his findings in *De Morbis Artificum Diatriba* (*A Treatise on the Diseases of Workers*) (1700). Chief among the problems were skeletal and nerve disorders from the handling of toxic chemicals by fullers, metalworkers, and apothecaries; sitting at inappropriate chairs and tables in low or flickery light; stooping, the cause of bowed legs in bakers and bent backs in porters; squinting, causing dim eyesight in artisans and lathe operators; and constant lifting that strained the body, lowering the life span of workers on the docks of port cities and in the grain, baking, weaving, and woodworking industries. He identified grain itch from contact with arum lilies and farmer's lung and respiratory lesions in chalk workers and stonecutters from inhaling irritants, a forerunner of silicosis. For controlling malaria, he recommended quinine. For his interest in the daily lives of nurses, midwives, soldiers, potters, miners, masons, notaries, sawyers, blacksmiths, wrestlers, female weavers, and nuns, Ramazzini became the patron saint of industrial medicine and ergonomics.

early 1700s As confirmed by the Archive of the Indies and data from the Ocopa monastery and the National Library of Lima, the Franciscans who established the Amuesha missions of the Cerro de la Sal, Peru, unintentionally created disaster by spreading European diseases among Campa Indians of the upper Ucayali River. Overcrowded in mission compounds, the Indians suffered such terrifying outbreaks that they revolted against the outsiders and their religion by fleeing to the traditional rainforest.

1702 In Holland, epidemic dysentery at Venlo in Limburg was worsened by famine, hot weather, and other privations generated by the War of Spanish Succession. Troop movements spread infection that killed 511 or 10 percent of the inhabitants before following the army to Germany. Venlo suffered a second outbreak when the town fell under siege.

summer 1702 In a "malignant distemper" of yellow fever in New York, over 570 people perished at the rate of 20 per day, totaling about 10 percent of the citizenry. Called the American plague, the disease spread north from South Carolina to Philadelphia and New York, possibly carried by slave coffles. Municipal authorities distributed coal dust and lime as antidotes to infection and ignited bonfires to detoxify the air.

late 1702 A smallpox epidemic killing 300 colonists of Boston coincided with the spread of scarlet fever. The assembly enacted quarantines of the sick, including the children of the Rev. Cotton Mather, Massachusetts chronicler and spiritual leader. From his pulpit at the Congregational Church, he thundered, "The Pale, the Swollen, the Wasted, and perhaps the Spotted Faces of the Sick in the Family, are such as our Heavenly Father has been spitting upon" (Poten 1999). A decade later, he recanted his ignorance of contagion and supported inoculation.

1703 At Quebec's Hôtel de Dieu, nuns inspired by Jesuits nursed victims of a smallpox outbreak. The stone mission, founded on August 1, 1639, employed three Ursuline sisters led by Marie de l'Incarnation of Tours, France, founder of the Ursuline Order of Canada. Funded by Madame de La Peltrie, the outreach ministered to the Huron and educated the daughters of French whites.

1703 Amid a widespread crisis from earthquake, flood, fire, and an eruption of Mount Fuji, Japan suffered an outbreak of measles. A recrudescence five years later covered the entire island chain except for Hida on Honshu Island.

August 28, 1706 During the uncertain times of Queen Anne's War, the return of yellow fever to Charleston, South Carolina, killed 5 percent of the colonial population. The epidemic encouraged the French and Spanish to take advantage of weakened coastal forces. After Governor Sir Nathaniel Johnson refused to cede the port city, he surprised his attackers by alerting a militia that he had posted beyond the city, well away from contagion. For his military acumen, he received a grant of 1,200 acres. Officials named a fort for him. *See also* **May 1732.**

1707 An epidemic of smallpox in Iceland reduced the population of 50,000 to one-third its previous headcount. Lessening chances of survival were lack of immunity because of isolation from the rest of Europe, overcrowded homes with smoky interiors, and a daily diet and lifestyle that encouraged scurvy and tuberculosis.

1708 After Sweden's King Charles XII led his army of 40,000 Caroleans toward Moscow, Swedish forces failed to subdue southern Russia because of an epidemic of plague. Worsened by cold and exhaustion, it originated among the Polish military and took one-half of the population of Danzig, Poland. On July 27, 1709, the Swedes capitulated to Tsar Peter I the Great at the Battle of Poltava. Of the 10,000 residents of Tallinn, Estonia, only 2,000 survived disease and starvation to surrender to the Russians without a fight. The loss from four years of pestilence boosted Russia to mastery of the Baltic Sea. *See also* **June 1710; 1711.**

1708 For three years, scarlet fever swept through Native American communities. It invaded the Saint Lawrence River Valley and Great Lakes area and may have reached the Gulf Coast, Lower California, and southwestern Pueblo Indians as well.

1709 In Cologne, Germany, an Italian, Jean Marie Farina, marketed eau de cologne. The toilet water formula, called *aqua mirabilis* (wondrous water), was originally concocted by Jean Paul Feminis out of bergamot and other citric oil and local water. The superstitious believed the fragrant, diluted perfume countered bubonic plague.

June 1709 A plague epidemic at Gdansk (Danzig) and Königsburg, Poland, killed over 32,600 people within six months. The city council hired two special priests to administer last rites and to distribute loaves to the sick. A tract from the period describes how guards warded off infected visitors as well as unwelcome Jews trading in cloth, wool, and incense. The gate to Brandenburg, Prussia, remained locked. City agents killed stray cats and dogs, lit sulfur fires to purify the air, and removed corpses for covering with quicklime to hasten decay.

Individuals did what they could to stay well. Citizens sprinkled vinegar, drank plague elixirs, smoked tobacco with alum, and inhaled plague incense powders before setting out on public streets. A local doctor, Manasse Stöckel, who visited 200 sick each day, issued a brochure warning of spurious cures. Another plague essayist, Johann Christoff Gottwald, summarized the course of the disease, which he linked to a clash of Swedes with Saxons at Pinczow.

Despite concerted efforts, burials rose to over 100 per day. Harried gravediggers were forced to widen pits to hold as many as 50 corpses each. A failed harvest extended the loss with deaths from malnutrition. In 1710, Frederick I of Prussia, fearing for the citizens of Berlin, erected a 400-bed pesthouse at the Spandauer Gate near the river Spree on the Czech border.

fall 1709 During the Great Winter of 1709, a puzzling weather phenomenon perhaps brought on by multiple volcanic eruptions and ash blotting out sunlight, a virulent respiratory disease struck Europe. The pestilence afflicted the British Isles, Scandinavia, Russia, and the Balkans before advancing south to Egypt and Ethiopia. Identified by Roman epidemiologist Giovanni Maria Lancisi, medical officer to the papal state, as *epidemia rheumatica*, the pestilence was probably flu. In *Dissertatio de Nativis* (*Dissertation on Natives*) (1711), a major work on public hygiene and epidemics, he described the upsurge of influenza in Rome and attributed the outbreak to unseasonably warm fall weather. The text advised the government to halt deforestation because trees filtered the air of pathogens.

fall 1709 In London, where people huddled together to defeat intense cold, epidemic typhus inflicted high death rates for over two decades. The figures in the annual Bill of Mortality reached 4,740 in 1710 and remained alarmingly grim until 1720:

Year	Deaths from typhus
1714	4,781
1715	3,750
1716	3,100

Year	Deaths from typhus
1717	3,100
1718	3,607
1719	3,927
1720	3,976

1710 During the Great Northern War, residents of Skåne and of southern Sweden suffered a three-year outbreak of camp fever that may have been petechial fever, pneumonic plague, or typhus. Precipitated by Danish military insurgents, the disease caused widespread suffering and death. After the disease abated, at Blekinge, Sweden, families opened a common grave of scourge victims and retrieved the remains of family members for reburial according to parish custom and Christian ritual. See also **1711, spring 1714**.

summer 1710 The nursing Ursuline sisters of the Hôtel de Dieu in Quebec accepted patients during a fearful *mal de siam* (yellow fever) epidemic. The nuns' convent notes explained how ailing passengers and crew of the vessel *La Belle Brune* from the Caribbean introduced the infection. After felling some of them before they could receive aid, the disease raced through the harbor, slaying all the visiting priests and one-quarter of the nuns who contracted the fever.

June 1710 During combat with Sweden along the Baltic coastline during the Great Northern War, an epidemic of plague assailed the Russian forces of Tsar Peter I the Great. In an act of biological warfare, his artillery, commanded by Field Marshal Sheremetev, bombarded Reval, Estonia, with plague-infected corpses. When Estonians contracted the disease, they surrendered. *See also* **1711**.

After the disease inundated Latvia that summer, local people realized that the Russian infantry had introduced contagion from Prussia. When whole villages expired, the juggernaut pestilence became a part of national folklore. Plague stories permeated incantations, folk music and hymns, prayers, sermons, and fiction.

1711 Epidemic typhoid fever killed 7 percent of the Choctaw and Chitimacha of Louisiana along the Mississippi River Delta. Few records remain of the devastation to native people and settlers.

1711 The Great Northern War of 1709–1720 ground to a halt after Swedish ships' crews contracted plague. Both Copenhagen, Denmark, and Karlskrona, Sweden, suffered epidemics, as did the Baltic States. Stockholm lost 15,000 or one-quarter of its 60,000 citizens to the disease. At Karlshamm, Sweden, half of the 1,800 citizens died of plague. The city established a graveyard for pestilence near Surbrunnsparken.

In this same period, like many Danes, King Frederick IV of Denmark fled Copenhagen, which was beset with illness and death, and immured himself at Koldinghus Castle. While dancing at a masked ball in the Great Hall, he met Countess Anna Sophie Reventlow, daughter of a Danish chancellor, who became his morganatic wife. After Louise, the official Swedish queen, died in 1721, Frederik remarried Anna Sophie in a formal state ceremony. He crowned her Denmark's new queen and settled at Fredensborg Castle.

1712 When a ship from Barbados brought small-pox to Boston, the disease advanced from one case to an epidemic striking half the populace and killing one-twelfth.

February 13, 1713 A fleet of Dutch ships from India bearing smallpox among its passengers reached Cape Colony and Cape Town, South Africa. Among the sick at Table Bay were crew and passengers, including the children of Commissioner Joannes van Steeland. After porters carried soiled bed linen and garments to a slave-run laundry operated by the Dutch East India Company, contagion spread in the area for the first time. By April 19, blacks were dying at the rate of eight per day. Six weeks later, 110 whites were dead. To maintain nurse care, health officials paid each slave one rix dollar per day. By the end of June, only 20 whites in Drakenstein district remained uninfected. Without field workers, famine threatened until authorities imported food and supplies from Batavia (modern Jakarta), Indonesia. *See also* **late April 1713**.

late April 1713 Smallpox cruelly targeted the South African Khoikhoi or Khoekhoen, peaceful shepherds and hunter-gatherers known in slang as the Hottentot, with whom Europeans had traded since 1500. Native healers lacked any knowledge of a disease that slew the unsuspecting blacks in their *matieshuise* (round huts). In May, even with aid from the Dutch Reformed Church, the outbreak among the Khoikhoi caused survivors of the pox to flee the Drakenstein area, abandoning their tribal structure to join the San people, called Bushmen.

Despite the condescension of whites toward natives, Europeans pitied the suffering of the Khoikhoi. In *Description of the Cape of Good Hope; with Matters Concerning It* (1726), Dutch merchant, chronicler, and cartographer François Valentyn stated:

> The Hottentots died in their hundreds. They lay everywhere on the roads. Cursing at the Dutchmen, who they said had bewitched them, they fled inland with their kraals, huts, and cattle in hopes there to be freed from the malign disease [*Reader's Digest Illustrated History,* 1994, p. 45].

Natives of the northern interior were so terrified by the pocked refugees that they slew them on sight.

Reduced to less than one-tenth their original number, the Khoikhoi were so dispirited by deaths, alcoholism, drought, and livestock disease that they ceded to the Dutch all claims to ancestral pastures and lands. As a result of catastrophic loss, the dispossessed Khoikhoi lost their language, chiefdoms, clan names, and cultural distinction by intermarrying with the slaves imported to beef up the work force. Thus, the Khoikhoi, though free, entered a virtual serfdom as the Cape Colored farm laborers and herders. Westernized and Christianized by Huguenot minister Pierre Simond and Moravian missionary George Schmidt, the Khoikhoi abandoned rigorous farm work superintended by greedy *trekboers* (Dutch settlers) and sought residence at missions, where they adopted European clothing and spoke Dutch, the basis of modern Afrikaans. *See also* **May 1755.**

1714 During widespread outbreaks of smallpox, the personal physician to the British ambassador in Constantinople, Emanuel Timoni (or Timonious), a Greek-Italian physician educated at Oxford and Padua, informed London doctors of the methodology of vaccination by issuing an eight-page article, "An Account or History of the Procuring of the Small Pox by Incision or Inoculation" (1714) in the *Philosophical Transactions of the Royal Society*. His crude method involved extracting fresh smallpox matter and rubbing it into a scratch on the healthy patient's arm.

1714 Throughout Mexico, an unidentified fever felled locals, killing 14,000. At the port city of Veracruz, officials modeled public health policies and containment procedures on European standards. The city surpassed other areas of New Spain in controlling yellow fever and other epidemic diseases that stymied the viceroy, Don Fernando de Silva, in his efforts to raise profits from the Western Hemisphere. Because of the city's success in improving public health, its program set the standard for Spain's other New World colonies.

spring 1714 The sinking of the 56-gun man-of-war *Fredericus Tertius* off Bergen, Norway, during the Great Northern War, derived in part from shipboard conditions and inadequate diet of the 400-man crew, who ate salted provisions. They suffered a high incidence of illness and death from scurvy and epidemic diseases. More Danish-Norwegian crewmen succumbed to scurvy than to combat injuries.

February 1717 Portuguese colonial records detailed how war, drought, famine, and the slave trade severely threatened residents along Africa's Atlantic shore. Survivors faced an epidemic of smallpox and other water-borne diseases that emerged during the rainy season. Historian John Reader's *Africa, a Biography of the Continent* (1998) summarized the irony of relief from drought in an African proverb: "Hunger does not kill, it is sickness that kills" (p. 439).

1718 At Abbeville, Amiens, Saint Quentin, and Saint Valéry, France, *sudor Anglicus* or the English sweating disease emerged under the names *suette des Picards* or the Picardy sweating disease (also Schweiss-friesel or miliary fever). The initial outbreak prefaced a spread across Burgundy, Flanders, Ile de France, Normandy, and Poitou.

March 1718 In the war on smallpox, Lady Mary Wortley Montagu, wife of the British ambassador to Turkey, made a valuable contribution. Having suffered physical disfigurement in 1715 from the disease and having lost a brother to it as well, in 1717, she wrote about "ingrafting" or variation, which she observed at the Ottoman Court in Constantinople, where her husband was the English ambassador. She explained in a letter to Lady Criswell how elderly female inoculators carried smallpox matter in a shell and tipped a tiny drop into the open veins of their patients. To protect her five-year-old son Edward, Montagu had a physician, Charles Maitland, vaccinate him.

In April 1721, Montagu returned to England and had her four-year-old daughter Mary inoculated, a procedure witnessed by the royal doctor, Sir Hans Sloane. The treatment was the first professional inoculation in medical history. On August 9, Maitland received royal permission to perform the Royal Experiment, a trial of smallpox vaccination on six condemned inmates of Newgate Prison. After subsequent tests on poor Londoners, on April 17, 1722, Maitland successfully immunized two royal princesses, four-year-old Caroline and ten-year-old Amelia.

The controversial procedure was not without serious criticism from moral environmentalists, who believed that natural catastrophes were evidence of God's displeasure and judgment. In 1756, a French cleric condemned vaccination as a violation of trust in God. He maligned the introduction of infected matter into a healthy body as a barbaric practice evolved by the pagan Turks and imported by English heretics.

1720 An outbreak of smallpox killed 180 or 90 percent of the 200 people living on Foula, a Shetland island northeast of Scotland. To curtail successive bouts of the disease, late in the eighteenth century, two local men established systems of inoculation. The Reverend Mitchell of Tingwall applied standard variolation. His contemporary, John "Johnnie Notions" Williamson, a weaver, blacksmith, clock repairman, and carpenter of Hamnavoe to the northwest, produced a weakened, less dangerous vaccine by collecting matter from pustules, smoking the effluvia, and burying it for extended periods. The hardening of the vaccine protected it during long-term storage. In Williamson's practice, he vaccinated 3,000 people. Neither Mitchell nor Williamson lost a patient to variolation. Williamson's cottage and a plaque preserve his contribution to epidemiology.

1720 Syphilis emerged in Norway, where people called it *radesyge* (bad disease). Doctors treated the sores with mercury injections and Vienna paste, a blend of alcohol, flour, and zinc chloride.

May 25, 1720 The cycle of bubonic plague that started in 1346 ended in Marseilles, France. The last outbreak began when the common *Rattus rattus* infested the city after the docking of the *Grand Saint Antoine* with a shipload of silk and cotton. The contagion might have been avoided if port rules had stopped the ship in Italy. Captain Jean Baptiste Chataud had arrived at Livorno on his way to Marseilles after sailing from Tripoli and Sidon, which was gripped by an epidemic. He duly reported pestilence among passengers and crew. From a quarantine jetty, he observed the heaping of his goods and the removal of crew for sequestering. Additional illness forced the impounding of Chataud's cargo and his isolation at Jarre Island, a harbor contagion barrier where the *Grand Saint Antoine* was burned and sunk to halt infection.

Because Marseilles profited from a trade monopoly with Turkey, entrepreneurs intending to trade imported cottons and silks to the Caribbean and the Americas influenced port authorities to ignore the need for quarantine. A lapse in port authority rules shortened the waiting period from the usual 40 days to 15 or 20 days and allowed infected passengers to proceed ashore to Marseilles with only a perfunctory fumigation. By July, slipshod harbor doctors could no longer deny that plague had reached Marseilles by sea from the eastern Mediterranean. In August, when fatalities were numbered at 1,000 per day, curtailment of grain shipments limited the supply of bread. The disease raged until February 1721 and killed 56 percent of the populace despite the stoking of bonfires and the cleansing of rooms with burning sulfur.

Attempts at halting the plague at the sea-line were crude and ineffective. People armed themselves with *les bâtons de Saint Roch*, long poles with which they warded off the approach of other pedestrians. Officials raised the *mur de la peste* (plague wall), a dense stone barrier over six feet high, across the Plateau de Vaucluse and posted guards to stop violations of quarantine. By the epidemic's end in November, Marseilles was in such a disordered state that enforcers of health regulations enlisted 700 prison inmates to rid the streets of rotting cadavers and purify the city with boiling water, vinegar, lime wash, fumigation, fragrant herbs, and gunpowder. The city's business district lost over half of its hatters, carpenters, and shoemakers and even more masons and cobblers. Only a new round of apprentices or the immigration of skilled labor could replace them.

The suffering from plague over the next two years was so acute in France that the "Good Bishop" H. F. Xavier de Belsunce made a special effort to visit the afflicted. In response to a heavenly message, on October 20, the Venerable Anne-Madeleine Remuzat, a visiting nun, instituted a feast to honor the Sacred Heart of Jesus.

Her pious gesture to relieve suffering from plague blessed Provence, Lyons, Rouen, Constantinople, Cairo, Spain, the Caribbean, Louisiana, Persia, and Syria.

In a population of 80,000, infection killed 39,152 and wiped out 85,000 within Provence, halting travel and commerce. Chronicling events was Jean Baptiste Bertrand, a city physician and inspector of corpses who wrote *A Historical Relation of the Plague at Marseilles in the Year 1720*. Greatly lionized was heroic doctor Pierre Chirac, on staff at the Montpellier School of Medicine, whom Louis XV elevated and named royal physician.

News of the Marseilles outbreak spread fast. When French ships approached the Netherlands, governmental strictures limited crew and cargo from the infected country and later extended the measure to vessels from southwestern Europe and Turkey. At Goeree and Terschelling, Holland, coastal stationmasters refused to let suspect vessels dock. Ships that breached the shore patrol were torched and their passengers and crew isolated for 30 days.

May 1721 When the warship H.M.S. *Seahorse* carried smallpox from Barbados to Boston, as reported by the *Boston Gazette*, the captain reported no sickness. After entering the harbor, one man developed symptoms. The ailing crewman entered quarantine at dockside. Selectmen dispatched 26 free blacks to sweep the thoroughfares.

Although sequestration and street cleaning were immediate, the disease advanced rapidly on Cambridge, Charlestown, and Roxbury, Massachusetts, targeting the very young and elderly. Of 5,579 cases among 11,000 citizens, 844 died at the rate of four per day. As the death toll rose in early winter, the city assembly hired nurses, posted two guards at each infected residence, and hoisted a red flag marked "God have mercy on this house." Except for funeral corteges, the harbor and trade center emptied of traffic. *See also* **June 26, 1721; 1722.**

June 26, 1721 A Congregational minister and health enthusiast, Cotton Mather, recognized the dangers of tuberculosis, which he called the "English disease," and knew about taking citric juice to prevent or cure scurvy, the bane of seagoing ships' crews. He also commented on Turkish methods of inoculation, which he learned from his slave Onesimus, a Coromantee who may have been familiar with the procedure in Sudan. Onesimus explained that he acquired the scar on his arm by dropping small pox matter into a cut in the skin. After a brief and mild case of smallpox, the slave acquired immunity.

The explanation convinced Mather. He published his enthusiasm for inoculation in the *Boston Gazette* and stated in *The Angel of Bethesda* (1722), a treatise on medicine, his personal experience with smallpox vaccination:

> About the Month of May, 1721, the Small-Pox being admitted into the City of Boston, I proposed unto the Physicians of the Town, the unfailing Method of preventing Death, and many other grievous Miseries, from a tremendous Distemper, by receiving and managing the Small-Pox, in the Way of Inoculation. One of the Physicians had the Courage to begin the Practice upon his own Children and Servants; and another expressed his Good Will unto it [Mather, 1972].

Mather professed shock that other Boston doctors "treated the Proposal with an Incivility and an Inhumanity not well to be accounted for" (ibid.).

On Mather's advice, physician Zabdiel Boylston of Brookline, Massachusetts, vaccinated the citizens of Boston. In June 1721, Boylston began with his own son Thomas and two black slaves from his household staff. He reported in the *Boston Gazette* his successful experiment. The public became so emotional that they firebombed Mather's house on November 13, 1721, and forced Boylston into hiding with curses, assault, and threats of hanging. Other respected physicians mocked the notion as inflicting certain death. The Rev. Edmund Massey launched a sermon against Mather for distrusting God and published the text in the *Weekly Mercury* on January 1, 1722.

Of the 286 citizens that Boylston and two other doctors vaccinated, only six died from the procedure. Of the inoculees, most came from the upper social echelons, including well educated, politically conservative, and orthodox religious citizens. For his promotion of the public good, Mather became the first native-born American inducted into London's prestigious Royal Society.

James Franklin, publisher of the *New-England Courant*, initiated America's first media crusade against inoculation, a point of view shared by the town officials. Contributing to the campaign between April and October of 1722 was 15-year-old Benjamin Franklin, James's brother, who imitated the style of the *Spectator* in the 14 "Silence Dogood" essays. The satiric pieces, spoken through the fictional persona of an opinionated widow and submitted anonymously under the pressroom door, expressed Franklin's dislike of the pompous Cotton Mather and his hard-handed indoctrination of the public concerning the benefits of inoculation. *See also* **1753**.

Despite the media blitz, Mather and Boylston's opinion prevailed. In 1726, Boylston published *An Historical Account of the Small-Pox Inoculated in New England* and earned membership in the Royal College of Physicians and the Royal Society of London. His text was so popular that he brought out a new edition in 1730.

September 1721 In Nova Scotia, harbor rules to stem smallpox outbreaks required that no shipper to Annapolis, Maryland, return to port with American cotton wool, which might spread contagion. Only wool that had been aired and inspected was allowed into the country.

1722 Cotton Mather commented in *The Angel of Bethesda* (1722) on pandemic rickets in Massachusetts:

> In our Countrey, it is now a Very Common Malady.... 'Tis a Melancholy Spectacle, which the Rickety Children afford unto us, when we see their Heads growing into an unporportionable Magnitude; with diverse uncomely Protuberances: Their Breasts troublesomely straitened and mishapen; the Sternum Sticking out; Their Bellies enormously Swelled; Their Backs and Bones Crooked; Their Joints tumified; Their Breath short [Mather 1972, p. 276].

August 1722 When Tsar Peter I the Great led Cossack cavalry south to take Constantinople from the Turks, his Russian troopers contracted Saint Anthony's fire or ergotism at Astrakhan on the Volga River delta. The provisioning of horses and men with rye grain and hay from local sources spread the agonizing ailment and killed 20,000 people, weakening the Russian army.

1723 The return of smallpox to southwestern Europe inspired propaganda through articles and verse. In Spain, Benito Jerónimo Feijoo composed a pro-inoculation essay; Manuel José Quintana wrote an epic poem. Because the illiterate were unaware of these writings and because the church did not encourage preventive medicine, the peasants of Spain were slow to seek inoculation.

1725 Under Tsar Peter I the Great, Russians incurred less suffering from endemic and epidemic fever, plague, and typhus. A proponent of progress, the tsar encouraged the medical community to apply scientific analysis and prevention to contagion. In addition to quarantining the sick, medical communities took responsibility for assuring public wellness.

1726 A ship arriving at Philadelphia from Bristol, England, brought passengers who were ill with smallpox. To prevent an epidemic, a humanitarian named Barnes brought them ashore at the Swedish church and dosed them with rum. He led them into the woods to the Blue-house Tavern on the corner of Ninth and South Street. An entry in the *Annals of Philadelphia and Pennsylvania* (1857) states that the victims hid until they recovered.

summer 1726 Residents of Tunis suffered a scourge of endemic malaria after high rainfall increased habitats for disease-bearing mosquitoes. Across Tunisia, as a virulent epidemic caused sickness and death, doctors began prescribing quinine for the first time.

July 1727 Bubonic plague engulfed Astrakhan, Russia, on the Volga River, disrupting the silk trade. First identified among the military returning from Persia, the malady followed residents as they fled Astrakhan. The governor evacuated the area, furthering the spread of disease to the outlands, where migrant workers carried it into rural areas. By June 21, 1728, around 1,300 had died. The disease ended by late summer, when food shortages and malnutrition worsened the state of the people. The final death toll was 15 percent of the populace.

1728 A measles epidemic starting in Veracruz, Mexico, ravaged Mesoamerican natives. Near the year's end, contagion reached the Pima and Tchoowaka living at the mission at San Ignacio de Caburica, where priests baptized victims and buried the dead. Natives burned their houses and moved to new ground. Among the Jemez, 109 died. There were additional fatalities among the Acoma and the Galisteo of New Mexico.

1728 Ruined potato harvests produced so harsh an outbreak of typhus in Ireland that Dubliner Jonathan Swift wrote one of his most biting satires, *A Modest Proposal for Preventing the Children of Poor People in Ireland from Being a Burden to Their Parents or Country, and for Making Them Beneficial to the Public* (1729). By pretending to suggest that poor Irish peasants breed children like livestock for sale as food, Swift typified the predatory attitude of the English, who considered Ireland an ignorant papist outpost and a ready source of land, crops, meat, and wool.

fall 1728 Yellow fever returned to the international port of Charleston, South Carolina, killing many of its victims from secondary pleurisy. Putrefaction caused rapid decay in corpses. Families neglected traditional funeral customs for their kin. Physicians discussed whether the disease emerged from local vectors. Assemblymen failed to discuss regulations because they failed to achieve a quorum until January 1729.

February 1729 Multiple infections spread up the California mission system, a string of Catholic institutions built on a 532-mile stretch called El Camino Real (the King's Highway). The series of outposts was the plan of Franciscan evangelist Father Junípero Serra, Westernizer of the coastal Indians. Called the Apostle of California, he plotted an interconnected series of 21 missions to quarter a campaign for Christianizing area aboriginal tribes conveniently lumped together under the name Mission Indians. In all, measles and venereal disease sliced the native population nearly 80 percent, from 70,000 to 15,000.

May 1730 When news of a smallpox epidemic at Boston reached Marblehead, Massachusetts, townspeople overreacted and erected a fence over the road into town. Into mid-summer, they locked the gate, posted a guard round the clock, and instituted a curfew keeping blacks and Indians off the streets after 9:00 p.m. After Hannah Winters sickened with smallpox in October, infection spread rapidly to most Marblehead families. People vacated their homes and closed shops. To contain the pox, health officials banned nurses from the streets and exterminated stray dogs. Neighboring towns refused to admit citizens of Marblehead. The general uproar lasted until summer 1731. Because only two selectmen were well enough to function, the Justice of the Peace held a special election to fill vacant seats.

1731 *Matlazáhuatl*, an unidentified hemorrhagic fever, assailed Churubusco outside Mexico City, killing many indigenous people. The virus brought on vertigo, headache, fever, dysentery, black tongue, torso pain, and swellings on the neck and head. Following bleeding from eyes, nose, and mouth, patients died on the fourth day.

1731 In the Balkans, a smallpox epidemic at Sarajevo, Yugoslavia, resulted in quarantine of Slavic exports to Venice. The port regulations severely curtailed Slavic trade, increasing the sufferings of residents.

May 1732 The return of yellow fever to Charleston, South Carolina, lasted throughout the hot months, peaking in July with the deaths of up to a dozen white citizens per day. To quell panic, local authorities silenced tolling funeral bells.

August 25, 1733 Smallpox struck virgin-soil territory when natives of Greenland came in contact with European diseases imported after 1721 by Hans Egede and Danish traders to a Scandinavian post at Godthab. Ironically, the carrier among them was a boy named Carl, an islander who had traveled to Copenhagen with five others. All died of smallpox except Carl, who infected kin and well wishers on his return home that spring. The epidemic sickened the boy and hundreds more living within 20 miles of the trading colony. The population of one island fell to only four young siblings whose father had lived long enough to inter his neighbors. So many people died by April 1734 that survivors struggled to buy and store winter provisions for their families.

1734 Epidemic smallpox menaced New Mexicans and overran the Zuñi, killing 200. The scourge appeared in a traditional winter count, a pictographic chronicle recapping events from 1230 C.E. onward. In 1878, tribal keeper Battiste Good documented native history for William Corbusier, a Continental army surgeon at the Rosebud Indian Reservation in South Dakota. Among memorable episodes was the smallpox-used-them-up-again winter among the Sicangu Lakota (or Brulé), a part of the graphic calendar history. The cell depicting a pocked victim is the first evidence of epidemic smallpox on the North American plains.

1734 In southern Poland and Prussia, wars pitting French and German troops along the Rhine resulted in thousands of deaths from typhus, the standard disease among soldiers. Close sleeping quarters and shared blankets that harbored pathogen-bearing body lice worsened the infection.

1735 Don Pedro Casal, royal physician to King Philip V of Spain, visited the Asturias region to observe a local anomaly, an outbreak of *real de la rosa* (king of the rose), which he diagnosed as a form of leprosy. He concluded that the disease sprang from poverty and a preponderance of corn in the diet. French diagnostician de Sauvages of Montpellier declared the leprosy scorbutic in nature, causing tremors of the upper body from a lack of vitamin C.

February 1735 When 47 German Salzburgers settled coastal Georgia north of Savannah, their long sea voyage and limited rations produced beriberi and scurvy. Loose teeth and bleeding gums appeared following their arrival at Ebenezer, where they dined out of a common pot of cooked cornmeal, hominy, or rice. Their expedition apothecary, Andreas Zwiffler, had no clue to the cause of their illness.

Adults tended to die of secondary infection. Infants were likely to succumb shortly after birth, as were mothers. One midwife, Elizabeth Stanley, reported a 31 percent mortality rate among newborns. The total death rate approached 41 percent. Adding to the colonists' misery was endemic dysentery from suspect shallow wells contaminated by privies. The outlook remained grim for their survival until spring 1736, when they moved to Red Bluff on the Savannah River. Gardening supplemented their deficient diet with cabbage, peas, and radishes. *See also* **July 1736.**

May 1735 Over a five-year period, residents of New England identified waves of diphtheria and scarlet fever as "throat distemper." The diseases began in Kingston, New Hampshire, where 40 died, 26 or 65 percent of whom were children. At Hampton Falls, New Hampshire, 20 families lost every child as 210 died of the malady. At Haverhill, Massachusetts, citizens lost half the children from infants to mid-teens. Infection spread to Connecticut, Maine, Maryland, and Massachusetts in late fall. By March 1736, the disease reached Rhode Island. Of the 5,000 who perished, over half were children. By 1741, the death toll had risen to 20,000.

During the Great Awakening, a revival of piety

and religious fervor, the public, in terror of pestilence, reverenced the symbol of the skeleton on tomb markers at Essex County, Massachusetts. Pious physician-preachers used the opportunity to chastise homesteaders for seeking wealth in the New World and for slighting their duty to God. Congregational minister Cotton Mather referred to the advantage of the minister-physician as an "angelical conjunction" that allowed him to minister to the soul while caring for the suffering body.

November 1735 Chapter 21 of William Buchan's *Domestic Medicine; or, the Family Physician* (1749), the prize home medical handbook in Britain and the United States, detailed how an epidemic of *sudor Anglicus*, called *suette miliare* (miliary fever) or sweating disease, extended into January 1736 at Strasbourg, France. The disease baffled doctors and wreaked havoc among both weak and strong. From shivering, yawning, and joint pain, the disease progressed to a lethal high fever, seedlike pustules, delirium, convulsions, and hemorrhaging, which precipitated a quick death.

1736 Pierre Gaultier de Varennes, Sieur de la Vérendrye, a Canadian fur trader and explorer of the plains of southern Manitoba, headed a 50-man party, comprised in part by his three sons, a nephew, a Jesuit cleric, and an Indian guide named Ochagach. While building trading posts in western Canada, the men were the first to report smallpox. The scourge struck the Ojibwa at the Lake of the Woods and along the Winnipeg River. The traders' report speculated that the Hudson Bay Company introduced the disease.

July 1736 On the Atlantic coast in the Georgia colonies, tertian fever beset English and German newcomers, producing a multistage outbreak. The first epidemic of endemic ague—possibly malaria or typhoid fever—struck in July 1736. It killed most of the indentured servants of planter John Mackay. After the fever ebbed in November, settlers experienced widespread dysentery. The ague returned the next summer.

August 1736 In Mexico, a severe *matlazáhuatl* (hemorrhagic fever), or perhaps typhus, struck Tacuba and, within six months, swept into Cuernavaca. Oaxaca and Michoacán reported the highest mortality. Bobbing in and out of populous areas, the disease spared communities from Nochistlan to Teutila and from Guayacocotla to Yagualica and died out by the next summer.

1737 Marguerite d'Youville of Quebec, the first canonized Canadian, applied the nursing model of Saint Vincent de Paul to her own company of visiting nurses, *les Soeurs Grises* (Gray Sisters). During an outbreak of smallpox in Montreal among the Mohawk, the military, and local prostitutes, the sisters set up wards at l'Hôpital Général and visited other victims at home. At *La Crèche d'Youville* (The Cradle of Youville), Mother d'Youville and the Gray Sisters nurtured children orphaned by disease. In 1990, Pope John Paul II proclaimed her Canada's first saint. *See also* **October 1755**.

April 1737 As dysentery, influenza, relapsing fever, and smallpox ravaged Denmark, Finland, Iceland, and Sweden, Swedish shore patrols began surveying incoming vessels, particularly those from Germany, Poland, and Prussia. Epidemic disease elevated death rates and remained active in Iceland into 1739 and in Finland into 1740.

1738 A cargo of African slaves infected Charleston, South Carolina, with smallpox. As a result, James Kilpatrick (or Kirkpatrick) published *An Essay on Inoculation* (1738), which established that only 1 percent of the 800 inoculated died of the disease. He advised physicians to practice the arm-to-arm method to create immunity. In opposition to the practice, the city enacted an ordinance punishing givers and receivers of inoculation within the greater Charleston area with a fine of £500.

As contagion swept west, it killed off half the 15,000 Cherokee living in the Appalachian Mountains and disfigured survivors of the pox. Infection also weakened the Chickamauga, leaving them unfit to defend their homeland from English and French insurgents. Indians fled inland, spreading infection to other tribes.

1738 An upsurge in dysentery cases along the northern coast of Brittany grew into an epidemic over the next four years. The water-borne disease targeted rural French peasants and killed many children at Anjou. Hunger and malnutrition increased the number of fatalities.

spring 1738 At Ochakov in southern Russia, plague struck forces along the Black Sea and moved into the heartland, killing 30,000 soldiers and hastening the resolution of the Russo-Turkish War. Because of quick action to cordon off major cities, the contagion remained localized in rural areas.

summer 1738 Dutch passengers embarking down the Rhine from Rotterdam joined a throng of emigrants to Delaware and Pennsylvania. Of 6,500 passengers, 35 percent died of dysentery and other fevers along the way. At the stopover at Rotterdam, people huddled in tents against cold rain, from which 80 children died. Tightly packed on ships, the sick spread their contagion, which killed large numbers on board. To avoid taxation, captains may have underreported the total buried at sea and kept victims' belongings to compensate for financial loss.

1739 On the San Antonio River, an epidemic struck San José y San Miguel de Aguayo Mission, a Franciscan outpost in San Antonio, Texas, called the Queen of Missions. The scourge reduced to 49 a headcount of Indians from over 20 tribes. To provide a more healthful environment, the Catholic hierarchy moved the frontier mission to higher ground.

1740 When typhus overran the French at Anjou and Brittany, the disease killed 55,000 over a two-year period. Worsening the toll were the combined forces of unfavorable weather, poverty, and a forced migration of homeless and jobless peasants. The poor endured a horrific time by crowding extended families into small huts, thus spreading infection and body lice, the vector for typhus. Within a year, the disease erupted in Lorraine, killing inmates of prisons in an era rife with dysentery, influenza, malaria, and typhoid.

In Sweden, a hard winter preceded an unprofitable agricultural year and the migration of peasants in search of jobs and food. Living in close quarters huddled together to keep warm, they spread the body lice that carried typhus. Mass migration across the Finnish border carried the infection to a wider area, creating an epidemic that burgeoned into 1742 and paralleled outbreaks of dysentery and relapsing fever. *See also* **spring 1742.**

January 1740 After a poor potato harvest and dismal winter of 1739, Ireland suffered famine accompanied by the two-pronged pestilence of dysentery and typhus. From his home, the Irish Anglican philosopher and anti-materialist George Berkeley of Kilkenny, the bishop of Cloyne, offered charity, supplies, food, and doses of tar water that he decanted from his own stock. By 1741, the death toll stood at 4 percent of the nation's population.

January 1740 In western France, lethal influenza during the late winter raised the death toll two or three times among older citizens. At the Hôtel-Dieu in Paris, fatalities rose by 50 percent as the disease plus famine and a simultaneous outbreak of dysentery and typhus felled the poor. Many collapsed because they were too riddled by tuberculosis and chronic respiratory ailments to survive additional trauma.

summer 1740 When yellow fever reached Guayaquil, Ecuador, via ships from Santo Domingo, it spread from one site to another. At first, the disease centered on the port, infecting both Huancavilca Indians and Spanish insurgents. The fever returned two years later over the same Caribbean route, but killed fewer people because of lifetime immunity that survivors acquired from the initial outbreak.

1741 At Cartagena, Colombia, Admiral Edward "Old Grog" Vernon failed in his assault, which followed the British declaration of war on Spain in 1739. He began with an ill-advised harbor landing, led by Major-General Thomas Wentworth against an enemy raining shot from superior locations. Contaminated well water spread dysentery, which debilitated amphibious forces. Yellow fever struck 8,431 of those marines and sailors who lacked immunity. Heavy loss of life reduced the two leaders to bitter quarrels and led to their subsequent withdrawal from Cartagena.

1741 Austrian forces abandoned Prague, Czechoslovakia, to the French because typhus killed 30,000 Austrian soldiers, leaving too few standing to defend the city.

winter 1741 In London, resurgent typhus killed 8,700 between its beginning and the end of contagion in February 1742. Worsening the death toll were failed harvests, unemployment, unremitting cold, and worsening conditions for the urban poor.

May 1741 In southern Argentina, diarist Richard Walker commented on outbreaks of scurvy during a voyage through the Straits of Le Maire aboard the *Centurion*. By June, mortality increased to a total of 200 deaths, leaving a handful of crew healthy enough to perform watch and sea duty. He wrote:

> This disease so frequently attending all long voyages, and, so particularly destructive to us, is surely the most singular and unaccountable of any that affects the human body. For its symptoms are inconstant and innumerable, and its progress and effects extremely irregular; for scarcely any two persons have the same complaints [Carey 1987, p. 222].

He listed swollen limbs, discolorations, putrid gums, labored breathing, trembling and shivering, skin ulceration, and lassitude and swooning from the least exertion. Some dropped dead upon stirring from their hammocks.

June 1741 Philadelphians fell sick from yellow fever, which forced urbanites to flee to the country and rural people to avoid the town. Some blamed the epidemic on German and Irish immigrants. Because of the outbreak among German newcomers, the fever acquired the name "Palatine distemper." Lexicographer Noah Webster, author of *A Brief History of Epidemic and Pestilential Diseases* (1799), preferred the term "American plague." By October, the death count rose to 250. Two years later, the malady again ravaged Philadelphia and New York.

1742 Traveling west from Mexico by overland trade routes, a severe *matlazáhuatl* (hemorrhagic fever) or perhaps typhus, struck Baja California. At Sierra Gorda, the scourge killed 200,000 as it raged for another year.

1742 Among the Piro of the upper Ucayali River of Peru, Jesuit missionaries attempted to gather potential converts into a mission. After epidemic disease terrorized the Indians, they recoiled from the touch of whites and scattered to the forests. The ingathering of tribes was not complete until 1788, when Father Pedro García welcomed Peruvian natives to the mission of Nuestra Señora del Pilar de Bepuano.

spring 1742 The emergence of "hot fever" in southwestern Sweden appears to refer to epidemic relapsing fever, which ravaged the citizenry simultaneously with typhus and dysentery (Cohen 2001, p. 329). Around 12 percent of the populace died at Varmland, which suffered the brunt of contagion. Perpetuating sickness throughout the land was extensive troop movement by land and sea at the conclusion of Sweden's unsuccessful war with Russia, which ended with the Peace of Abo and the ceding of parts of Finland to the victors.

March 1743 A trader from Corfu carried goods and plague to the port of Messina, Sicily, which prided itself on successful harbor quarantine measures. Although agents burned both the ship and its cargo, they failed to halt contagion. Over three months of widespread illness, authorities launched preventive measures against a pestilence that historians named the Great Plague of 1743. Priests led prayers and street processions to no avail. Around 50,000 of the city's 62,755 citizens perished. The loss virtually emptied the area. Grateful survivors displayed candles at each side of the Sanctuary of Gesù Ecce Homo.

July 1743 When yellow fever returned to New York City, the death roll rose to 217. Residents, including revered physician and historian Cadwallader D. Colden of Dunse, Scotland, later lieutenant governor of New York and author of *Observations on the Climate and Diseases of New York* (1738), began to make a connection between the "bilious plague" and the harbor. On January 11, 1744, as a member of the governor's council, he published an essay in the *New York Weekly Post Boy* recommending purifying water and cleaning sewers, drains, and piers to prevent fever.

1745 The establishment of the London Smallpox and Inoculation Hospital at Coldbath Fields fought epidemics through vaccination of citizens. However, the profiteering of freelance agents at the rate of two to five pounds per patient placed the prophylactic measure out of range for the majority of people needing protection. Another factor inhibiting universal inoculation was the time required for recovering from the mild case of pox that the procedure caused.

1746 Epidemic disease in Spain placed a high demand for medicine at all social levels. At Buendía, pharmacist Gaspar Narro refused to lower prices, even though medication could relieve the high death rate of the poor. After citizens forced the closure of his dispensary, the town hired José Antonio de Canora as his replacement and offered him a monopoly on pharmaceuticals.

June 1746 After threats to French holdings in North America, Jean-Baptiste-Louis-Frederic de la Rochefoucauld de Roye, duc d'Anville, an incompetent commander, led a fleet of ten warships, 45 troop transports, and 11,000 soldiers to Chubucto to lay siege to Louisburg, Nova Scotia. A rocky voyage via the West Indies extended to three months because of delays. The men suffered an outbreak of typhus, complicated by scurvy and exhaustion from battling Atlantic storms. Before completion of the landing at Acadia, 1,000 died at sea and produced a death rate nearing 20 percent on some vessels. Among the dead was d'Anville, who suffered fatal seizures. On shore, starvation lessened the chance of surviving disease, which killed an additional 135. In October, after burning the worst of their ships, the French limped home on the remaining 42 ships.

The aftermath of the failed invasion devastated Acadian Indians, who had supported the French against the British. Local Micmac traders collected the garments and blankets left by the departing fleet, spreading so deadly a scourge that the natives were virtually annihilated. Huge burial mounds at Chubucto received the dead. The Indians believed that the French intended the infection as a form of biological warfare.

1747 Emergence of angina maligna, a common name for diphtheria, struck Cremona, Italy, inhibiting swallowing and breathing. The lethal disease asphyxiated many children, a piteous sight described in detail in the treatise *Angina Maligna* (1749) by respected physician Martino Ghisi. He filled his text with clinical observations as well as his examination of corpses.

June 6, 1747 At Albany, New York Governor George Clinton announced an ordinance forbidding inoculation, which had become an explosive topic during pandemic outbreaks of smallpox.

1748 After the arrival of Franciscan missionary Francisco Ano de los Dolores in 1748 to Texas, the Tonkawa and other central Texas tribes died in great numbers from smallpox at Mission of San Francisco Xavier de Horcasitas on the San Xavier River. The Lipan Apache took advantage of the weakened mission Indians. Their attack plus a second epidemic forced the evacuation of the surviving Tonkawa in 1755 to a frontier mission on Guadalupe River.

February 1748 Pennsylvania's provinces suffered a pandemic of pleurisy that felled many servants. To fill vacancies in homes and institutions, employers imported great numbers of indentured replacements. Because of the weakened state of the colony, treaty negotiations with the Six Nations—Mohawk, Oneida, Onondaga, Cayuga, Seneca, and Tuscarora—ceased until summer, when the contagion ended.

October 1750 At Philadelphia, epidemic smallpox targeted urban dwellers. Quaker quilter, horticulturist, and diarist Hannah Callender Sansom, at age thirteen, began a daily journal to reflect on events. Among other challenges, she accepted her own facial disfigurement from smallpox. During the American Revolution, she faced additional losses of kin and friends to diphtheria, dysentery, and measles.

ca. 1750 According to the diaries and journals of travelers in Ethiopia, medical communities built around sources of hydrotherapy turned into healing villages. Their reputation for cures enticed syphilitics, lepers, and their families, who sought treatment with natural thermal baths. It was not until 1901 and 1934 that the Ethiopian government built separate leprosaria as a means of restricting the travel of lepers among the general population.

ca. 1750 In Europe, doctors believed that mothers brought about endemic puerperal fever by hiring wet-nurses to feed their infants. Male doctors shifted blame from contagion to the suppression of breast milk in post-parturient women. By charging mothers with weak maternal instincts and unnatural attitudes to their bodies, doctors exonerated themselves as bearers of disease.

1750 In London during the Black Assize, the crowding of unwashed prisoners awaiting trial in the passageway leading from Newgate Prison into Old Bailey spread typhus, called "putrid fever" or "jail fever." The disease killed one-third of the prison's inmates as well as two judges, the lord mayor, aldermen, and other court officials. The sheriff of London, Sir Stephen Theodore Jansen, issued a monograph, *The Jail Fever* (1767), to the Lord Mayor that summarized the cause and nature of jail fevers. He warned:

> Several judges, sheriffs, magistrates, juries, and whole courts of judicature, have been infected by those contagious diseases, which caused the loss of many valuable lives, particularly at the Old

Bailey, and formerly at the assizes at Oxford, all owing to the horrid neglect of gaolers, and even of the sheriffs and magistrates, whose office it is to compel the gaolers, to the most rigorous repeated orders and attention to their duty, without the least indulgence or remission [Jansen].

Among Jansen's complaints was the construction of the prison on only three-quarters of an acre.

December 1751 A lethal onslaught of smallpox struck London and moved across England, killing around one-third of its victims in Chelmsford. The severity of the outbreak spurred citizens to seek variolation. The demand encouraged Robert Sutton of Debenham, Suffolk, to limit the invasive nature of inoculation by introducing matter from pustules through tiny incisions into the outer layers of skin.

1752 During an eight-year typhus epidemic in the Netherlands, Sir John Pringle, British army surgeon and founder of modern military medicine, reformed clinical care and camp sanitation as a means of halting the spread of camp fever, a general term encompassing typhus and typhoid fever. A champion of preventive medicine, he was an experienced army doctor and the author of *Observations on Diseases of the Army in Camp and Garrison* (1752), which described outbreaks of typhus, then called "gaol fever." He advocated barracks ventilation and improved sanitation around army encampments as a deterrent to perennial outbreaks of disease among soldiers. He established the need for personal hygiene, ventilation of hospital wards, proper clothing, boiling of drinking water, and relief from overcrowding, extremes of temperature, dampness, and fatigue. The work, which stayed in demand for over a half century, contained a segment previously issued as "Observations upon the Nature and Cure of Hospital and Jayl-Fevers," an impetus to the development of a public health service to stem epidemics.

1752 In the Georgia colonies along the Atlantic coast, German and English settlers depended on flour and salt meat for food. Because of the lack of fruits and vegetables, they tended to suffer beriberi and scurvy from dietary deficiencies until the arrival of the first African slaves to work the land and grow crops. To the detriment of Europeans, the newcomers bore new infections of dysentery, malaria, and typhoid fever.

1753 During an onslaught of smallpox in Boston, Benjamin Franklin, who had opposed inoculation in 1721 and lost his four-year-old son Francis to the pox, reexamined the worth of immunizing the public. After he conducted a statistical analysis of the immunity of people receiving inoculation, he determined that the method worked.

1753 In a period when 5,000 British sailors died each year of scurvy, ship's doctor James Lind, an expert on diet and hygiene among English seamen, studied the incidence of scurvy during a round-the-world voyage aboard the H.M.S. *Salisbury*, which embarked in 1748 under the command of Commodore George Anson. Lind conducted a study known as the Salisbury Experiment by testing anti-scorbutics on twelve sailors through dietary supplements of citrus juice, herbal barley water, vinegar, cider, and seawater. From his findings, he categorized scurvy as a nutritional disorder, which had plagued humankind in the time of Hippocrates and during the Crusades.

Lind took immediate action to prevent scurvy. He augmented shipboard diet of biscuit, meat, and salt with fresh fruits and vegetables. His report, *A Treatise on Scurvy: In Three Parts. Containing an Inquiry into the Nature, Causes, and Cure, of the Disease* (1753) noted his surmise about shortages of vitamin C in the diet:

> Oranges and lemons were the most effectual remedies for this distemper at sea. I am apt to think oranges preferable to lemons, though it was principally oranges which so speedily and surprisingly recovered Lord Anson's people at the Island of Tinian, of which that noble, brave and experienced commander was so sensible that before he left the island one man was ordered on shore from each mess to lay in a stock of them for their future security [Lind, 1753].

Lind's study impacted England's naval provisioning, as did his second treatise, *On the Most Effectual Means of Preserving the Health of Seamen* (1757). For his concern for sailors, he was named the father of nautical medicine. *See also* **1758.**

Concurring with Lind was John Huxham, an English expert on infectious disease and the author of "A Method for Preserving the Health of Seamen in Long Cruises and Voyages" in *An Essay on Fevers* (1757). Huxham observed how quickly arrival in port cured ailing sailors with fresh air, clean water, and fresh food, particularly greens and fruit. In his opinion, these benefits cleansed the blood and restored wellness. He advocated wine, cider, apples, lemons, and oranges as a cure for scurvy. He added that, if these fruit sources cured, they would also prevent outbreaks of the disease.

April 1753 Within eight months, a severe epidemic of measles blanketed Japan. Traveling from Kyushu southwest to northeast, the pestilence killed many people, even on the far reaches of Hokkaido.

1754 In Philadelphia, an outbreak of "Dutch distemper" caused citizens to blame servants emigrating from Germany and Holland. The following year, the disease afflicted whole neighborhoods, acquiring the name "jail fever," a slang term for typhus. To ward off contagion, enforcers of health regulations began inspecting ships and settlements of German immigrants.

May 1755 The emergence of "gall fever" (smallpox) among Hottentots or Khoikhoi of Cape Colony and Cape Town, South Africa, contributed to the destruction of the indigenous culture as far away as Namibia. Borne to the continent from India, the disease killed over 1,000 whites and their slaves, who were treated at separate hospitals. As graveyards received more burials than space accommodated, failure to bury corpses a sufficient depth raised a sickening odor. When the onslaught began to ease ten months later, survivors gave thanks, fasted, and prayed.

October 1755 The heaviest loss of Canadian Indians to smallpox occurred over a two-year span, beginning with the onset of the Great Smallpox Epidemic. Carried by British and French soldiers, the disease struck white westerners from Niagara traveling north to Quebec and Montreal. At age 64, Mother Marguerite d'Youville, hero of the smallpox epidemic of 1737, began rebuilding the hospital founded by *les Soeurs Grises* (Gray Sisters). Indians who admired and respected her for previous nurse care pooled their keepsakes, bolts of cloth, knives, and blankets to donate to the cause.

Mutual contacts resulted in additional infections of the pox. The Seneca spread contagion to the Montaignais. Combat during the French and Indian War passed infection to Fort William Henry, New York, and Fort Edward, Annapolis Royal, Nova Scotia, and limited military engagements. In August 1757, when Indians waylaid the British at Fort William Henry, the pillage of personal effects spread the pox to attackers, killing them before they could transport the loot to their villages.

1757 During the Seven Years' War, syphilis ravaged Lithuania after the arrival of Russian forces. The next year at Breslau, Poland, some dysentery and more typhus killed 19,000 people, evenly striking locals and the military. *See also* **1817.**

1758 While employed at Portsmouth's Haslar Hospital on England's southern coast, James Lind, a Scottish doctor and naval surgeon, studied outbreaks of dysentery, scurvy, and typhus. To prevent loss of manpower and crippling or death to sailors, he proposed frequent delousing, distillation of seawater into potable water, and a way to preserve citrus fruit juice to prevent scurvy from a vitamin C deficiency.

March 22, 1758 Contributing to New Englanders' mistrust of smallpox inoculation was the death of Jonathan Edwards, revered founder of New England theology and author of the stirring sermon *Sinners in the Hands of an Angry God* (1741). He succumbed to a secondary fever after variolation during his first weeks in office as president of Princeton University.

May 1758 When smallpox struck the susceptible population of Burford, England, 110 people— 14 percent of the populace—succumbed to the disease before it died out in August.

1759 In the Caribbean, William Hillary, an English physician and author of *Observations on the Changes of the Air and the Concomitant Epidemical Diseases in the Island of Barbados* (1759) commented on the symptoms of endemic sprue or diarrhea alba, a tropical intestinal complaint that robbed islanders of fatty nutrients. He explained the misery of burning mouth and stomach, small pustules on the tongue, and intense diarrhea that saps victims of their strength.

summer 1759 During a large building project at the new village of Bethania, severe fever invaded the Moravian settlers of Bethabara, North Carolina. The disease killed eight Moravians and other local settlers. After the death of her husband, Hans Martin Kalberlahn, Sister Anna Catharina began a mission to newly orphaned children. Epidemiologists later identified the disease as the first outbreak of Rocky Mountain spotted fever, which was not formally recognized on the Atlantic seaboard for another 170 years.

winter 1759-1760 During the French and Indian War, the British garrison at Quebec lost 700 soldiers to scurvy and contagious diseases. By April, nearly 32 percent of the Redcoats were too ill for military duty. In an attempt to recapture Quebec, Brigadier General Gaston Lévis took the advantage by mobilizing a strike force of 3,900 French troops and 3,000 Canadian militia. The battle of Sainte Foy gave the advantage to the French until the arrival of 497 British warships with reinforcements. The total dead from disease and combat reached 1,000.

September 4, 1760 Smallpox ravaged colonials, slaves, and Indians throughout the Cherokee Wars of 1759–1760. As reported in the *Pennsylvania Gazette*, during an epidemic at Keowee, South Carolina, Cherokee of the upper township isolated themselves from the sick in the lower township. At Charleston, South Carolina, a dense population pattern made the citizens more vulnerable to contagion brought in by returning soldiers the previous January. The infection continued into November.

A mass inoculation drive was the British colonies' last public health effort preceding a general vaccination. South Carolina Huguenot Ann Ashby Manigault kept a diary that included details of inoculation of her children. She surveyed domestic health measures to ward off summer fever (malaria) through purges, emetics, bleeding, and bark curative, a source of quinine.

The issue of smallpox among Africans figured in posters announcing slave auctions. Austin, Laurens, & Appleby advertised the arrival of the slaver *Bance Island* carrying 250 healthy prospects from the "Windward & Rice Coast" (Steinberg 2002, p. 56). To boost sales from leery buyers, the text promised:

> The utmost care has already been taken, and shall be continued, to keep them free from the least danger of being infected with the SMALL-POX, no boat having been on board, and all other communication with people from *Charles-Town* prevented. N. B. Full one Half of the above Negroes have had the SMALL-POX in their own Country [ibid.].

On each side of the poster, stereotypical sketches of blacks in grass skirts and feather crowns illustrated their vigor.

1761 Over a four-year period in the Spanish missions of Mexico and Baja California, the Chumash contracted from their conquerors gonorrhea and *mal galico* (syphilis) and died from measles epidemics. Missionaries feared that European diseases would annihilate the Indians, but continued building missions north along the Pacific coast. To the south, typhus and smallpox felled many in Mexico City and San Luis Potosí, a Spanish mining center in the Bolivian mountains.

1761 When a ship from Veracruz, Mexico, infected Havana harbor, Spain's main Caribbean port, with yellow fever, British forces suffered thousands of casualties. The next year, when the infection returned to Cuba, the British lost over 53 percent of a total of 15,000 men. From Havana, a ship bore the fever north to Philadelphia, where, in 1762, Benjamin Rush, the city's humanitarian epidemiologist, observed an outbreak that enabled him to isolate symptoms and the course of the fatal disease.

1762 After observing large numbers of the poor suffering *mal de la rosa* (pellagra) at Oviedo, Spanish court physician Gaspar Casal y Julian, author of *Historia Natural y Medica de el Principado de Asturias* (*Biology and Medical History of Asturias Province*) (1762), remarked on the piteous cranial tremor, burning lips and mouth, coated tongue, and diffuse weakness as well as crusting of the skin on the neck. From his professional writing came a nickname for skin lesions on the neck, called the Casal necklace. Because his patients were poor, he recognized that their diet relied too heavily on corn and not enough on fresh meat, a source of niacin or B-complex vitamins.

1762 In Sweden, soldiers mustering out after the Seven Years' War spread syphilis, a disease common to men who had been away from wives and home for long periods.

1762 At Cayenne, Guiana, yellow fever struck virgin-soil territory, felling thousands and hampering European settlement of the lush coastline.

1763 After the spread of smallpox over Texas and Central Mexico for two years, Father Ignaz

Pfefferkorn, a German Jesuit missionary at Guebavi Mission, witnessed epidemic disease in Sonora. He reported that more Yaqui and Seri than Hispanics died of the pox. Most adults perished from infection complicated by starvation and dehydration. The priest himself suffered a fever and had to be transported to the mission at Cucurpe.

1763 An immunization drive in Boston lowered the death toll when smallpox struck 170 out of 20,000 citizens. Chronicling events, James Gordon, a wealthy Boston merchant and landholder active in the Anglican Church, noted the high death rate in the first 13 cases, of which 11 died. Although the assembly quarantined infected residences, the virulence of the disease stampeded the populace to Cambridge and sent 5,000, including some Acadian immigrants, in search of variolators for inoculation. To assure the colony's welfare, the Overseers of the Poor, a civic benevolence modeled on English systems, underwrote the cost of immunization for the underclass.

The next year, the epidemic of smallpox struck Philadelphia, killing 300, primarily indigent children. The city's doctors formed the Society for Inoculating the Poor Gratis, the first benevolent association launched to fight smallpox among those lacking the funds to pay for vaccination. Among volunteers soliciting funds for the task was Benjamin Franklin, whose son Francis had died of the disease at age four in 1736.

July 7, 1763 At Fort Pitt (modern Pittsburgh) during Pontiac's rebellion, commander-in-chief General Jeffrey Amherst urged Colonel Henry Bouquet, the commander in the west, to distribute smallpox-contaminated blankets to area Lenni Lenape or Delaware. Amherst plotted genocide through biological warfare to rob the French of their Indian allies. Militia commander William Trent had anticipated the suggestion the previous May by dispatching a smallpox-permeated handkerchief and blankets, which effectively introduced lethal disease to natives of the area. The spread of the pox contributed to the collapse of the tribal alliance brokered by Pontiac, an Ottawa chief. Contagion also devastated the Iroquois, Potawatomi, Wea, Kickapoo, Miami, and Shawnee.

1764 An Italian, Angelo Gatti of Pisa, author of *Reflexions on Variolation* (1764), credited vaccination as a means of preventing smallpox.

1764 Ethan Allen stood on the meetinghouse steps at Salisbury, Connecticut, to allow Thomas Young to ingraft tissue from a victim of smallpox into Allen's arm. The act challenged laws and penalties as well as the New England clergy's condemnation of inoculation.

1764 During an upsurge of smallpox in Boston, the court moved its sessions to Cambridge. They built huge fires in the hearths to ward off infection and unintentionally burned down Harvard Hall. With the building went the most complete library in the Americas and a well-equipped scientific laboratory.

John Adams and his brothers, Elihu and Peter Adams, allowed Continental Army surgeon John Warren to treat them with milk, mercury, and ipecac for a week preparatory to variolation against smallpox, which Adams called the "king of terrors" ("More to Dread"). The procedure required making a quarter-inch slit in the left arm with a lancet and inserting a virus-bearing thread. Adams came down with the disease and stopped writing to Abigail, his wife, lest he spread contagion. The disease left him with pockmarks.

1764 When typhus sickened the people of El Paso, Texas, their symptoms ranged from headache, weakness, and rash to a high fever. The disease killed around 230, primarily adults of the Del Norte district. A Benedictine monk, Diego Zapata, maintained a burial registry in which he recorded the deaths from typhus with a T.

1766 Jonathan Carver reported in *Travels Through the Interior Parts of North America, in the Years 1766, 1767, and 1768* (1781) that smallpox seemed a fitting punishment to Canadian Indians for their cruelties to whites:

> The small pox, by means of their communication with the Europeans, found its way among them, and made an equal havock to what they themselves had done. The methods they pursued on the first attack of that malignant disorder, to abate the fever attending it, rendered it fatal. Whilst their blood was in a state of fermentation, and nature was striving to throw out the peccant matter, they checked her operations by plunging into the water: the consequence was, that they died by hundreds [p. 326].

He added that survivors carried serious pockmarks for the rest of their lives.

February 1766 The Pima of Fort Tubac, Arizona, in the upper Santa Cruz River Valley suffered an infectious epidemic, possibly typhus, that spread north from Sonora, Mexico, the previous June. The scourge wiped out families and depopulated villages at Calabazas, Guebavi, Sonoita, and Tumacácori. Throughout March, the infection caused complications that killed many survivors of the initial viral attack. The great loss of indigenous people compromised Spanish military operations into Apache territory because many Indian scouts died or were deafened, blinded, or weakened by disease.

1767 John Smith began inoculating residents of Yorktown, Virginia, against smallpox, generating objections that he would cause an epidemic. A year later, some college boys left quarantine too early and spread virulent disease in Williamsburg, where two victims died. The mishap caused other colonists to doubt the wisdom of inoculation.

1767 Captain Samuel Wallis and his English sailors aboard the H.M.S. *Dolphin*, the first English ship to dock in Tahiti, apparently introduced syphilis to islanders, who named it *apa no Pretane* (the British disease). (H. Smith 1975, p. 43). Corroborating the diagnosis was a journal entry by ship's master George Robertson, author of *The Discovery of Tahiti*, in which he noted the condition of the crew:

> We carried Ten Men ashoar to the Sick Tent, three of them was Very Bad, and has Been so ever since we Left England, with Damn'd Inveterate Poxes and Claps; the Rest has but Trifling Complaints, mostly of the Same Kind [ibid, p. 40].

Robertson added that, except for these men on sick call, all other crew members were fit for duty as well as for "trade with the Young [Tahitian] Girls" (ibid., p. 42).

Within eight months, Louis Antoine de Bougainville, author of *Voyage autour du Monde* (*Voyage Round the World*) (1771), arrived on April 12, 1767, aboard the *La Boudeuse* along with *L'Etoile*. For the enjoyment of his crew, he confessed that he "offered them young girls; the hut was immediately filled with a curious crowd of men and women" (ibid., p. 42). The captain also mused, "I am yet ignorant whether the people of Tahiti, as they owe the first knowledge of iron to the English, may not likewise be indebted to them for the venereal disease which we found had been naturalized among them" (ibid.).

The time span between dockings of English and French ships leaves in question who was responsible for infecting Tahitians with venereal disease, which joined other maladies in dramatically reducing populations of native Polynesians. By the time English sea captain James Cook sailed the *Endeavour* to the Society Islands in 1769, he found syphilis rampant among islanders. Local contacts passed gonorrhea and syphilis to 40 of his sailors, who transported the pathogens to New Zealand.

1768 As measles engulfed Mexico City for a year, a dearth of grain weakened citizens. The scourge targeted infants and children, who were the most likely to die of contagion.

1768 Tsarina Catherine II of Russia in 1768 sought variolation for herself and Grand Duke Paul from English physician Thomas Dimsdale of Hertsford. Because Dimsdale helped suppress endemic smallpox in St. Petersburg, he received awards and accolades. The tsarina set up a public inoculation clinic for the needy and extended the campaign into rural Russia. Her thinking helped enlighten the Russian medical community on epidemiology. *See also* **September 24, 1771**.

June 24, 1768 While smallpox ravaged Williamsburg, Virginia, anti-inoculation riots in Norfolk protested the practice of Archibald Campbell and John Dalgliesh, who variolated friends in Campbell's home. The protesters fired the house, forcing Campbell to remove patients on a stormy night to the safety of a lazaretto. Three days later, the mob destroyed Campbell's home.

Norfolk newspapers bristled with the pros and cons of inoculation and the danger of a smallpox epidemic. The pro-inoculation forces hired a young attorney, Thomas Jefferson, who accepted inoculation as scientifically sound, had himself variolated at Philadelphia around 1760, and successfully pled the case of Campbell, Dalgliesh, and their supporters. In 1770, a law regulating inoculation in Virginia required a license and allowed the procedure only for people infected with the disease. Seven years later, an amendment broadened the application of the law if a majority of the patient's neighbors in a two-mile radius gave written consent.

1769 James Latham, a British military surgeon, learned in England from inoculators Robert and Daniel Sutton how to protect people from smallpox without actually infecting them with live pathogens. Latham variolated 1,250 Canadians, including prominent Quebec City citizens.

1769 During a diphtheria epidemic in New York, Samuel Bard wrote a classic monograph, *An Enquiry into the Nature, Cause, and Cure, of the Angina Suffocativa, or, Sore Throat Distemper* (1769), concerning his treatment of patients and autopsy of the dead.

early 1769 Following drought, flood, and famine, an upsurge of smallpox in Bengal reached most of India. Contagion ravaged Murshidabad in March, killing 63,000; Bengal posted death tolls of three million. Street cleaners were unable to cope with the heaps of corpses outside dwellings where whole families died. The next year, the disease killed Syefuddowla, the Muslim viceroy.

April 27, 1769 The threat of smallpox to Southern plantations raised caution in buyers considering a new crop of slaves from Africa. When Captain Thomas Davies disembarked at the port of Charleston, South Carolina, from the *Countess* on a voyage from Gambia, representatives of the John A. Chapman Company tried to allay fears in prospective slave buyers. An illustrated broadside announced,

> This is the Vessel that had the Small-Pox on Board at the Time of her Arrival the 31st of March last: Every necessary Precaution hath since been taken to cleanse both Ship and Cargo thoroughly, so that those who may be inclined to purchase need not be under the least Apprehension of Danger from Infliction ["Broadside"].

The soothing text promised that this lot was the "likeliest Parcel that have been imported this Season" [ibid.].

fall 1769 During the three-ship expedition of Gaspar de Portolá from La Paz, Bolivia, to San Diego, California, the men suffered shipboard exhaustion and scurvy. Most of the sailors and half the soldiers were bedridden. The four crewmen healthy enough to perform their duties required the aid of the remaining soldiers. When the expedition set out on October 1, 1769, scurvy sickened 17 men and killed army surgeon Don Pedro Prat. Additional infection from enteritis responded to meals of leafy greens, which benefited victims of scurvy. Indians contributed to the starving party with seeds and *pinole*, a hot cereal common to Southwestern natives.

1770s In a teaching hospital in Lombardy, every woman delivering a baby during the 1770s died of puerperal fever.

1770s In Sweden, syphilis and other venereal diseases created a demand for medical understanding of the cause and spread of infection. In eastern Götland, patients sought treatment at the county hospitals of Linköping, Norrköping, and Vadstena. The curative methods ranged from severe diet controls to the *Röknings kur*, consisting of fumigating patient lesions with arsenic and mercury vapors.

To exterminate syphilis, Peter Hernquist of Skara studied syphilis and surgery in Paris. In his medical practice, he treated syphilitics at his 16-bed hospital with mercury ointments and drugs. Backed by Abraham Bäck, president of the Collegium Medicum in Stockholm, Hernquist made regular reports on symptoms of syphilis and his curative methods, including control of nocturnal pain with opium.

1770 An observer of relapsing fever in Ireland, John Rutty, author of *A Chronological History of the Weather and Seasons on the Prevailing Diseases in Dublin* (1770) commented on an infestation of fever in late July through October. The persistent illness caused fever, headache, and sweating.

1770 During a smallpox epidemic on the Pacific coast of North America spread perhaps by Spanish explorers, nearly 30 percent of native peoples died after their first contact with Europeans. The outbreak killed over 11,000 in western Washington State, reducing the indigenous population from 37,000 to 26,000. Additional mortality from influenza, measles, and other infectious diseases over the next 80 years further reduced the headcount to 9,000. The expedition of British navigator George Vancouver, which arrived in Juan de Fuca Straits on May 12, 1792, witnessed the mortality from deserted villages and bones scattered about. Among pockmarked survivors, some were blinded by smallpox, which he described in *A Voyage of Discovery to the North Pacific Ocean and Round the World, 1791–1795*.

1770 During the Russo-Turkish War, the tenth in a series of 15 plague epidemics in the Ukraine caused the most damage. Typically spread from the Ottoman Empire, this five-year outbreak extended to Moscow. Health workers, directed by epidemiologist Ivan A. Poletika, vigorously quarantined the sick, built treatment hospitals, and initiated more modern methods than doctors had previously applied. Through the new curative measures introduced by Danilo S. Samoilovich, a noted author on plague control, Ukrainians made significant progress in annihilating the disease.

1770 To stem epidemics at the port slave market in Rio de Janeiro, Brazil, a royal decree established harbor statutes. Incoming vessels had to supply documentation including a passenger list and ship's death book, which the port authority compared with embarkation data from African ports. The highest rate of death occurred in winter during the long voyages of slaves from Mozambique as compared to Guinea slaves from the Bight of Benin in West Africa.

January 21, 1770 A month-long outbreak of measles struck the Indians of Santa María Magdalena south of Fort Tubac and blanketed Sonora, Mexico, and the Gulf coast of Texas. Until February 18, the disease continued to kill. At San Ignacio Mission, the percentage of fatalities was higher as the disease spread throughout Sonora and into the San Miguel River Valley, weakening native warriors battling the insurgent army of Gaspar de Portolá and forcing the Indians to surrender. A lesser outbreak hit the Saint Ignatius and Guebavi missions.

1771 During an epidemic of pellagra among Italian peasants, Francesco Frapoli (or Frapolli) wrote a treatise that named the disease *vulgo pelagrain* from the Italian words for "common sour skin."

March 1771 In *Diary of David Zeisberger, a Moravian Missionary among the Indians of Ohio* (1870), the immigrant evangelist explained how he accepted assignments among native people in New York and Pennsylvania and learned their languages. He regretted the effects of smallpox on natives of the Tuscarawas Valley. Because medicine men could do nothing to heal the sick, he accepted an invitation from the Delaware. After he delivered one sermon, the disease abated. The Delaware invited him to live among them. On May 3, he and 28 Delaware converts built a village that he named Schoenbrunn. Natives trusted Zeisberger so thoroughly that they made him a chief and gave him a seat on their council.

September 24, 1771 Russian Tsarina Catherine II, a champion of public health, decided to move large factories out of Moscow as a means of ridding the city of industry's noise, smoke, and contagion from diseased workers. Spurring her decision was the bubonic plague that robbed Russian factories of laborers. Because of deaths and work slow-downs, her edict was unnecessary.

In the face of harsh tsarist reaction to the epidemic, Moscow erupted in a two-day riot called the Plague Rebellion, which was marked by destruction of property and looting. Leading the revolt were the poorest citizens who suffered the greatest loss from economic stagnation. Because instigators summoned forces by ringing the Kremlin bell, the tsarina had the clapper removed. A prelude to the Peasant War of 1773, the uprising demonstrated mounting peasant anger at the brutality and excesses of feudalism.

1772 James Lind, a Scottish doctor and naval surgeon, issued a third edition of *An Essay on Diseases Incidental to Europeans in Hot Climates* (1768), the first handbook on tropical medicine. His addendum explained the German method of preserving fresh cabbage as *zoorkool*, a compressed, brineless sauerkraut. The purpose of his recipe was the addition of an anti-scorbutic to the diet of sailors to prevent the scurvy that was rampant throughout the navy.

1772 For a year, a recurrence of an unidentified epidemic called *matlazáhuatl* (hemorrhagic fever) raised the death toll in the valley of Mexico.

1772 Tahitians suffered their first outbreak of gastric influenza, which reached a virgin-soil population in French Polynesia in November from infected crew aboard the *Aguila,* the vessel of Spanish explorer Domingo de Boenechea. During the month-long visit, island traders experienced the unprecedented symptoms of fever, head and stomach pain, and sore throat. The return of the vessel in 1774 generated a second infection.

October 1772 When English forces attacked Saint Vincent during the Carib Wars, they faced 1,500 black Carib, former slaves who had escaped or survived the shipwreck of slavers from Africa to the Caribbean. In home territory, the natives battled six battalions of redcoats. The British completed pacification of the island by February 27,

1773, but at the cost of a treaty with islanders. Some 150 fell in combat and 500 incurred yellow fever. Of the hospitalized men, 20 percent died.

October 18, 1772 When the Rev. William Martin led an emigrant band of Covenanters from Ulster, Ireland, to South Carolina, the party filled five ships that set out from the northeastern port of Larne. On August 25, the first to embark, the H.M.S. *James and Mary*, carried smallpox. Upon approaching Charleston harbor, the new-comers were obliged to anchor off Sullivan's Island until December to prevent an epidemic.

winter 1772 Plague proved so lethal in Persia that it slew two million citizens as it spread from Baghdad to Basra, killing 1,000 per day. Hospital filled over capacity. At Basra in April 1773, a quarter million succumbed to the scourge. Authorities at Shiraz limited all travel from the south, from which the scourge spread to Bahrain, raising the death rate to two million. To spare occupational forces, the British cut off all contact with peasants. The head of the East India Company fled by ship to Bombay, India.

1773 In lower Quebec, Canada, the outbreak of *mal de la Baie Saint-Paul* or Molbay disease, a non-venereal form of syphilis, spread a miserable ulcerative scourge allegedly introduced by a detachment of Scottish troops. In its last stage, it caused pain in the chest, labored breathing, and diminished hearing, sight, and smell. Government investigation of the outbreak disclosed that Canadians had unsanitary habits, including drinking from common cups, sharing smoking pipes, and feeding infants by chewing food and spitting it into their mouths.

In 1785, James Bowman published a 16-page monograph, *Direction pour la Guérison du Mal de la Baie Saint-Paul* (*Methods for Treating Baie Saint-Paul Disease*), Canada's first printed medical monograph. Additional writings dealt with the severity of the epidemic, which eventually infected 1 percent of the populace. In 1783, Bishop Jean-Olivier Briand of Quebec urged his curés to dispatch the afflicted to medical treatment centers. Another early treatise author, Robert Jones, issued *Remarks on the Distemper Generally Known by the Name of the Molbay Disease* (1786).

October 16, 1773 At the busy ports of Marblehead and Salem, Massachusetts, where West Indies traders often bore contagion, smallpox killed 14 of 59 victims. Marblehead's three selectmen ignored the suggestion of Captain John Glover, his brother Jonathan, Azor Orne, and Elbridge Gerry to build an inoculation clinic, so the trio undertook the project. They completed a two-story edifice superintended by Hall Jackson, a prominent clinician from Portsmouth, New Hampshire.

Immunization put hardships on Massachusetts citizens. Because cowpox inoculation caused a mild form of the disease, patients entered isolation at Cat Island. After a few of them died from variolation, angry mobs destroyed boats ferrying patients to treatment. Vigilantes captured four men stealing clothing from the hospital, tarred and feathered the culprits, and paraded them on a cart through Salem to the music of fife and drums. One brigand received a lashing at the public whipping post.

Opponents of inoculation stormed the hospital after rumors of 22 new cases of smallpox. Threatening multiple lynchings, the rebels demanded that staff padlock the facility. On January 26, 1774, arsonists burned the building. When the sheriff arrested and jailed them, men from Salem commandeered the jail and set them free. Opposing bands then armed for public confrontation, causing the hospital's owners to back down before the set-to turned into an anti-inoculation war.

spring 1774 Because wounded men were likely to contract infectious diseases at military hospitals, the British quartered in Massachusetts established smallpox vessels as harbor contagion wards for soldiers with the pox.

March 12, 1774 At Geneva, New York, Ote-tiani or Red Jacket, a Seneca notable respected by white military leaders as a peacemaker and negotiator, characterized an epidemic of smallpox as the Great Spirit's vengeance against elders who withheld from him the rank of sachem. After his elevation, he took the name Sagoyewatha (He Keeps Them Awake).

summer 1774 Benjamin Jesty, an English farmer at Yetminster, Dorset, determined to protect his pregnant wife Elizabeth and their three children from a smallpox epidemic. He knew that his two dairymaids, Ann Notley and Mary Reade, were immune to smallpox because they had survived cowpox. To obtain injectable material, he took his family to Elford's farm at Chetnole and removed pus from the udder of a diseased cow with a darning needle. He chafed Elizabeth's arm to admit the pus, then repeated the test on his two sons.

The procedure, the first documented smallpox inoculation, was successful. After the boys, Robert and Benjamin, suffered a few days of cowpox, they were immune to smallpox. Elizabeth became sicker than the children, but survived. Jesty performed the procedures on residents of Worth Matravers on the Isle of Purbeck. After villagers learned of Jesty's method, on market day, they scorned and ridiculed him and pelted him with rubbish. His tombstone bore testimony to his wise assumptions about cowpox.

late 1774 Aboard the *Ouwerkerk*, crew of the Dutch East India Company en route to the Netherlands colonies suffered epidemic phagedaena (or phagedena), a gangrenous ulceration or infection of wounds. Upon approach to Macassar in southwestern Celebes, Indonesia, the men conspired to charge ship's surgeon Adriaan van Brakel with neglecting to treat the sick. In an early case of conflict between law and medicine, the judge acquitted van Brakel. Proceedings produced the first description of tropical phagedaena.

1775 An explorer of Utah and Colorado, Father Silvestre Vélez de Escalante, a Franciscan priest and missionary to Santa Fe, New Mexico, noted that disease, hunger, and violence reduced some 7,500 Hopi in seven pueblos to 798. To protect themselves from the smallpox emanating from Mexico City, the Hopi at Oraibi Pueblo ejected Spaniards from their environs.

1775 A Spanish sailor, Bruno de Hezetaon, an expeditioner with Juan Francisco de la Bodega y Quadra, introduced smallpox to northern Pacific coast tribes. A population of 200,000 indigenous people fell to 125,000. Exacerbating the decline were outbreaks of measles and pertussis. As disease invaded the Columbia Plateau into Oregon and Idaho, fervid prophetic activity among tribes ignored the colonial incursion and stressed spiritual unease among individual Indians. Prophets attempted to interpret myths and apply ritual to a purification and renewal of the unbalanced native world. Terrified by multiple health crises, native people responded to the call.

1775 During the American Revolution, according to Elizabeth A. Fenn, an assistant professor of history at George Washington University and author of *Pox Americana: The Great Smallpox Epidemic of 1775–82* (2001), the scourge spread across the entire United States. It affected soldiers in April at the battles of Lexington and Concord, Massachusetts, and kept General George Washington's Continental Army from confronting the Redcoats, who were routinely variolated upon enlistment. Fortunately for the patriots, Washington was immune to the disease, which he had contracted at age 19 in Barbados in early November 1751.

On February 5, 1775, Jemima Condict Harrison, a daily journal keeper in Pleasantdale, New Jersey, stated her hesitance to be variolated during the smallpox scourge:

> Was my Cousins Knockulated I am apt to think they will repent there Undertaking before they Done with it for I am Shure tis a great venter. But Sence they are gone I wish them Success And I think they have Had good luck So far for they have all Got home Alive But I fear Cousin N Dod Wont get over it well [Harrison, 1930, 43].

Jemima's qualms captured the vacillation of colonists who anticipated a mild case of the pox, but feared pockmarks, scarring, blindness, or dying of secondary infection.

June 1775 During the general inoculation of the people of Philadelphia, Martha Washington submitted to the procedure. She needed protection from contagion to free her for travel with her husband and the Continental Army.

July 10, 1775 As smallpox beset Boston during constant military maneuvers, funerals numbered up to 30 daily. To prevent panic, city ordinances forbade tolling bells. In October, Massachusetts legislators suspended ferry service from Boston to Chelsea to contain the outbreak. On August 24, 1775, in addition to other trials and treachery facing Massachusetts colonists, satiric letter writer and playwright Mercy Otis Warren of Plymouth commented on epidemic contagion in Boston. In Newton, New Hampshire, letter writer Mary Bartlett believed that the scourge during the faceoff against England would decide the fate of patriots.

fall 1775 The smallpox epidemic halted General Benedict Arnold's invaders of Quebec,

Canada, which colonial patriots hoped to claim as the 14th colony. Close contact in crowded billets and unclean living conditions spread the pathogen throughout the winter. With only 4,500 of Arnold's original 10,000 troops still functioning after the march from Kennebec, Maine, he couldn't trust their health and stamina for a siege. Connecticut Governor Jonathan Trumble witnessed the tents full of moaning, dying soldiers, who needed professional care. The last week of December, officers attempted to contain the epidemic at a smallpox hospital.

On December 31, General Richard Montgomery attacked Quebec during a blizzard and fell dead as his men routed. The 400 American prisoners taken by the Canadians died in large numbers and spread smallpox in close and uncomfortable quarters, where there was no drinking water. Those soldiers who arrived back at the American camp departed when their tour of duty ended. General Arnold tried to tighten the isolation of the sick and reported to Washington a count of 100 down with the pox. Reinforcements quickly contracted the disease. See also **February 15, 1776**.

Thomas Dickson Reide, surgeon to the British army's first battalion and author of A View of the Diseases of the Army in Great Britain, America, the West Indies, and on Board of King's Ships and Transports (1793), recognized the significance of the American rout in Quebec. In June 1776, he inoculated around fourteen unprotected Redcoats.

November 18, 1775 As disease overwhelmed the British military quartered in Massachusetts, General Lord William Howe obtained a headcount of his soldiers who had no immunity to smallpox and ordered them inoculated. He later reduced the order from obligatory to voluntary.

1776 At Socorro in north central Colombia, the onslaught of smallpox in a prosperous district killed 6,000 or 18 percent of the populace, many of them babies and children. Most of the victims were servants. The poor who could not afford burial abandoned corpses of babies on church steps. Worsened by inadequate harvests, the disease was endemic into 1780, when the government began inoculating citizens.

1776 Among loyalist Scots and English fleeing the American colonies and settling at Belleville, Ontario, infant mortality reached 160 per 1000

live births primarily because of widespread acute infections, iron-deficiency anemia, enteritis, scarlet fever, and tuberculosis. Fatalities resulted from poor urban sanitation and the low nutrition of weaning foods, which indigent parents diluted with contaminated water. In contrast to Anglican and Methodist mothers, Catholic women produced a higher survival rate by breastfeeding babies.

January 4, 1776 Commodore Esek Hopkins of Rhode Island, a veteran of the French and Indian Wars, moved the Continental fleet out of Philadelphia along the Delaware River to rid coastal waters of British attack forces. During the maneuver, his ships edged slowly through heavy ice. As the convoy slipped into the Atlantic Ocean and south to take Nassau and seize the gunpowder and cannon from Fort Montagu on March 3, the crew of four ships suffered heavy loss of manpower from smallpox. Because Captain Nicholas Biddle of the brig Andrew Doria had ordered his men variolated, his was the only vessel of the fleet spared the disease. Thus, it became the Continental hospital ship.

February 1776 So many black and white soldiers caught smallpox at Tucker's Point outside Portsmouth, Virginia, that a camp cemetery acquired 300 new burials, a disproportionate number of whom were black soldiers. The navy surgeon advised a mass inoculation. Captain Andrew Snape Hamond of the Roebuck chose to decamp, burn the victims' brush huts, and, on May 27, move the fleet north to the Piankatank River at Gwynn's Island on the lower Chesapeake Bay.

In the new setting, more soldiers fell ill with smallpox and epidemic typhus. The crew of one boat exposed residents of Saint Inigoes, Maryland, to the pox. When John Murray, Earl of Dunmore and governor of Virginia, fled Gwynn's Island for Saint Augustine, New York, he burned some sick men in their huts and left unburied rotting corpses at the Cherry Point battery. A few survivors dragged themselves to the shore to plead for help. The total dead neared 500. On July 4, 1776, Thomas Jefferson used Dunmore's cruelty as one of the issues named in the Declaration of Independence, which charged that King George III had "ravaged our coasts, burned our towns, & destroyed the lives of our people."

February 15, 1776 To forces in the village of Sorel outside Quebec, General Benedict Arnold issued stern orders against amateur inoculations with needles and pins. He vowed court martials for enlistees, ousting for officers who disobeyed, and death to doctors who inoculated patients. In spite of Arnold's attempts to control his men, smallpox terrified them more than a firing squad. In March, Colonel Seth Warner brought the Green Mountain Boys of Vermont and quietly countenanced inoculations concealed on the thigh. One of the illicit inoculators, Josiah Sabin, received blindfolded patients who could not identify him as the violator of Arnold's order.

March 1776 In a year that brought widespread influenza, Japan faced another in its long history of epidemic measles. A lethal pathogen, the disease beset citizens over the entire island chain.

May 1776 At Quebec, the Continental Army's camp count had fallen from 2,505 to 1,900, of whom 786 were too weak to fight when fresh British troops overran the Canadian plain along the Saint Lawrence. American soldiers fled, abandoning the sick to the ministrations of the Redcoats. When the withdrawing colonials reached Sorel on May 11, three civilian observers—Charles Carroll, Samuel Chase, and Benjamin Franklin—studied the frazzled men and realized that susceptible reinforcements would soon incur the same fate.

On June 2, Major General John Thomas, the new army commander at Quebec, succumbed to smallpox. When Brigadier General Horatio Brig traveled north to take charge of the patriots, he brought Jonathan Potts, author of *Utility of Vaccination* (1771), whom the army had hired to tend the sick. At Fort George, Niagara, Ontario, Potts found four surgeons and four aides attending 1,000 sick men. In mid–July, the number tripled as the army regrouped at Ticonderoga, New York.

July 1776 In Boston, Abigail Adams and her four children submitted to smallpox variolation by the Sutton method, which the originator, Robert Sutton, a Suffolk surgeon, had used in England on 300,000 with few fatalities. Less expensive and less time consuming than earlier procedures, Suttonian inoculation required no dietary or medical preparation and no cutting of the skin. For serum, they drew from an inoculated subject rather than from a smallpox survivor. In 1796, Sutton's son, Daniel Sutton, published *The Inoculator*, which explained the Suttonian method of vaccination.

July 3, 1776 Members of the Second Continental Congress were wary of the smallpox epidemic in Philadelphia, where Virginia orator Patrick Henry received inoculation from Benjamin Rush, the city's eminent epidemiologist. Philadelphia health officials rescinded the ban on variolation and allowed inoculation through July 15. The shift in attitude toward the procedure spurred Bostonians to demand protection from smallpox. By December, the pox so engulfed Virginia that the colonial authorities also lifted statutes governing inoculation to allow families already hit with the disease to seek protection.

late in 1776 British Vice Admiral James Young attempted to abandon 100 ailing American prisoners of war on the Caribbean island of Antigua. The local governor, William Matthew Burt, declined to accept the men because of an epidemic of smallpox, which could kill the men in their weakened condition. He also feared importation of dysentery, typhus, and other diseases rampant in military prisons. By 1778, the British government forced the governors of Barbados, the Bahamas, and other Caribbean islands to house prisoners of war. In March 1782, Admiral George Rodney began anchoring British prison ships off Antigua, Jamaica, and Saint Lucia and stocking them with 5,000 French captives.

late winter 1776 At Boston, the state council ordered the corralling of smallpox victims in quarantined quarters to limit spread of the scourge. Town authorities set guards at ferry docks and main roads, limiting free trade with outsiders. In March and April, the disease continued to infect more people and spread to Lord Dunmore's fleet on the Chesapeake. At the insistence of John Morgan, Director General of the Hospitals and Physician in Chief to the American Army, staff of the Continental Army began to favor inoculation.

1777 One of the first contagious and lethal diseases to afflict Oceania was scrofula. Called *lila* (wasting disease), it infected the South Pacific islanders of Taiarapu (Cliff et al. 2000, p. 144). The disease appears to have arrived aboard a Spanish ship that anchored near Otaheite (modern Tahiti) or from the crew of the British frigate H.M.S. *Pandora*, which carried the disease to Fiji in 1779.

1777 An outbreak of Saint Anthony's fire or gangrenous ergotism in Sologne southwest of

Paris, France, killed 8,000 from gangrene. The Abbaye of Saint Antoine received pilgrims at the cloister's infirmary, where veneration of the saint's relics and bones relieved symptoms, called *gangrène des Solognots* (gangrene of those from Sologne), *mal ardent* (the burning evil), and *ignis sacer* (sacred fire). Those cured of the malady left their crutches in the chapel as a thank offering.

1777 At Mission Santa Clara de Asís, California's eighth mission founded on January 12 by Father Junipero Serra, Indians succumbed to a measles epidemic. Disease gave priests an opportunity to demonstrate the powers of baptism as a means of restoring dying children to health. Because of the severity of the outbreak, Mission Santa Clara led California religious sites in baptisms and Christian burials in hallowed grounds. A Dominican priest, Father Luis Sales, author of *Noticias de la Provincia de Californias en Tres Cartas* (*Observations on the Province of California in Three Letters*) (1794), surmised that natives of Baja California died in large numbers from European diseases because they lacked experienced nurse care and were too weakened to tend and feed the sick, who died of dehydration and malnutrition.

1777 An outbreak of gangrenous petechial quinsy at Moivron, Lorraine, targeted French children and teens. Modern epidemiologists conclude that the disease was scarlet fever.

January 1, 1777 General George Washington, commander in chief of the Continental Army, realized that, for every man killed in combat, he was losing nine men to camp fever (typhus), dysentery, scurvy, typhoid, and typhus. Those men furloughed home returned with fresh bacterial and viral contaminants, including gonorrhea and syphilis. Worsening the situation for the patriots was a smallpox epidemic at the military hospital in Bethlehem, Pennsylvania, where sanitation was primitive. To halt infection, Washington debated the evidence of physician Zabdiel Boylston's success in the 1720s at variolation, the early term for transfer of smallpox serum to the arm.

To protect colonial soldiers and assure the outcome of the American Revolution, Washington ordered physician and Continental Congressman William Shippen to begin the variolation procedure in Philadelphia. Washington vacillated on the order from January 28 to February 5, but returned to his original decision, which Dr. Benjamin Rush corroborated. John Cochran, the army's physician and surgeon general, superintended the mass inoculation at Newtown, Pennsylvania. Other stations at Alexandria, Dumfries, and Fairfax, Virginia; Georgetown, Maryland; Morristown, New Jersey; Hudson Highlands and Ticonderoga, New York; and Bethlehem and Philadelphia, Pennsylvania, completed the task for veterans and new recruits with the improved Suttonian method. Of those vaccinated, the army lost one of every 1,000 to the disease.

January 13, 1777 After the battle of Long Island on August 27, 1776, Vice Admiral James Young anchored the 64-gun H.M.S. *Jersey*, an infamous British prison hulk wrested from privateers, some 100 yards off the Brooklyn Navy Yard at Wallabout Bay, New York. The ship joined a fleet of twelve prison vessels, including the *Falmouth, Good Hope, Hunter, Prince of Wales, Scorpion, Stromboli,* and *Whitby*. The largest of the flotilla, the *Jersey* filled rapidly with 1,100 prisoners of war in barred lockups belowdecks, the dark, dank holds reserved for French and Spanish prisoners.

According to Henry Reed Stiles's three-volume *A History of the City of Brooklyn* (1867), starvation, vermin, typhus, dysentery, typhoid, pneumonia, and smallpox rapidly depleted inmates, as did the random bayoneting by sentries. Stiles particularized

> loathsome dungeons, denied the light and air of heaven; scantily fed on poor, putrid, and sometimes even uncooked food; obliged to endure the companionship of the most abandoned criminals, and those sick with small-pox and other infectious diseases; worn out by the groans and complaints of their suffering fellows, and subjected to every conceivable insult and indignity by their inhuman keepers, thousands of Americans sickened and died [p. 332].

Some who survived physical contagion went mad with delirium and raged like beasts.

Each day, the British collected a dozen corpses, which guards lowered on ropes from the forecastle. Burial parties ferried the dead by handbarrows to the shoreline for crude interment in shallow pits. Authorities left the living to starve in a penetrating stench that kept intruders at bay. Historians estimate the prison ship mortality at 12,500. On January 13, 1777, General George Washington was so incensed by reports of the high death rate of prisoners that he dispatched a stinging letter to General Lord William Howe. Washington protested "shocking accounts of their barbarous usage" and promised retaliation (Petriello, 101). In February 1778, the last of the original prison ships was burned, but more took their places. *See also* **July 4, 1782.**

April 1777 A rumor terrorized patriots at Kingston, New Hampshire, that the British were plotting biological warfare through the spread of smallpox among civilians.

May 3, 1777 Smallpox had spread so rapidly through Virginia that Martha Washington feared for the staff at Mount Vernon. George Washington petitioned William Shippen, medical supervisor of the Continental Army, for aid, including stocks of calomel and jalap to treat the sick. In all, 130,000 colonials died; of that total, 25,000 were soldiers.

mid–September 1777 Under orders from the Continental command, soldiers conscripted from North Carolina as provisioners and laborers underwent inoculation by William Rickman, Director of the Continental hospitals in Virginia. When smallpox symptoms felled the men, Rickman and his aides deserted the troops, leaving them cold and hungry. Disillusioned by lack of supplies and support, he retired in 1780.

winter 1777–1778 During a miserable winter encampment at Valley Forge northwest of Philadelphia, General George Washington discovered that some 4,000 soldiers had avoided inoculation. When smallpox once more raged along with typhus and typhoid fever, in January 1778, with troop strength down from 14,000 to 8,100, he forced the men to accept variolation. In March, he stopped receiving enlistees at inoculation clinics in Alexandria and Georgetown and directed them immediately to Valley Forge, where troop numbers hovered slightly above 7,300.

By May, judicious drafting of men boosted the army to over 15,000. When the Continental forces once more engaged the British, the military hospital still floundered under a patient load of 3,800, comprised of the sick and recent inoculees. Treating them as the army pressed forward over the Delaware River to the battle of Monmouth, New Jersey, in June 1778 were aides, military surgeons, and the Women of the Army, a volunteer corps of nurses, seamstresses, cooks,

and launderers who numbered 20,000 during the American Revolution.

1778 Three "Dunkard" surgeons and two mates staffed the Brothers House Hospital of Ephrata, Pennsylvania, where they treated about 500 patients. After a heavy toll of wounded from the battle of Brandywine on September 11, 1777, the rise in typhus cases included patients, nurses, and local volunteers. Gravediggers made places for their corpses at a hill behind the facility. The high rate of contagion provoked military enforcers of health regulations to burn the building.

1778 In French West Africa, yellow fever felled British soldiers at Fort Saint Louis and spread to Gorée, Senegal, where around 68 percent of white victims died of the infection.

1778 Elizabeth "Molly" Page Stark of Derryfield, New Hampshire, served as medical officer for the troops of her husband, General John Stark, during a smallpox epidemic. To accommodate patients, she turned her residence into a hospital. When she petitioned the General Court to allow her to inoculate her family and house staff, a judge denied her plea. Only when her husband, General Stark, the state's most distinguished hero of the Revolutionary War, submitted the request to vaccinate soldiers and prisoners of war did he receive approval.

June 10, 1778 A scourge, probably smallpox, gripped the prison at Fort Moultrie, a palmetto log installation on Sullivan's Island, South Carolina. To guarantee security, Colonel Francis "Swamp Fox" Marion chose as guards only men who were immune.

July 1778 During the War of the Bavarian Succession, also called the Potato War, 152,000 Prussian troops and 20,000 Saxons under Frederick II the Great invaded Bohemia. The three armies saw no action after Frederick negotiated hostilities with Joseph II of Austria. Beset by dysentery and typhus, the idle soldiers sickened at an alarming rate. Some 14,550 Prussians and 12,625 Austrians died from contagion contracted in filthy billets. Another 16,052 deserted, in part because of appalling encampments. Chastened by the debacle, Frederick determined to reform his military medical service.

fall 1778 As Maryland patriots sailed to the Caribbean along with mercenaries from Waldeck, Germany, smallpox afflicted the colonists in port at Jamaica. When the ship reached Pensacola, Florida, on January 30, 1779, the Germans, who were immune, survived. The colonial sick required some three weeks to recover. Among the pro–British Creek, the disease infected Indian traders already weakened by famine and dimmed their enthusiasm for the American Revolution. Spread from the Chickamauga Cherokee and the Creek of Georgia, the disease advanced through the Tennessee hills and moved as far south as Mexico City.

December 1778 After the British took Savannah, Georgia, 400 colonial prisoners of war incarcerated on prison hulks in the harbor suffered dysentery, smallpox, yellow fever, starvation, vermin, and exposure. In *Memoirs of a Revolutionary Soldier* (1950), Pierre Colomb, a captain in the Georgia regiment of the Continental Army, described conditions in the hold of one vessel:

> Scurvy ran rampant, at least three or four wretches a day died, and this number was soon increased to ten or twelve, the brutal nature of their diseases leaving to others only the task of throwing the bodies overboard. The number of sick increased every day. The poisoned air we breathed affected even the healthiest among us, and yet no help was forthcoming, no medicines, no fresh supplies of any sort [Ranley, 2000].

After retrieving the day's dead, sentries tossed their remains in the sea. Colomb evaded their fate for three months of incarceration. At his release, he was covered in sores and vermin. That same year, the British disposed of their prison ships.

1778–1779 English explorer Captain James Cook and his crew aboard the H.M.S. *Discovery* and the H.M.S. *Resolution* spread venereal disease among Hawaii's one million native inhabitants. Lieutenant James King, an officer aboard Cook's expedition and author of *Voyage to the South Pacific* (1784), commented on endemic boils, scaly skin, skeletal deformities, and syphilis in the Sandwich Islands with attendant blindness. He noted, "A squinting sight was pretty common" (Bushnell 1993, p. 27). In King's estimation, "We were the authors of the disease in this place" (ibid., p. 37).

Surgeon John Law, on the return of the H.M.S. *Resolution* in March 1779, confirmed native reports that they acquired venereal disease during Cook's previous landfall in the islands. When

Captain Jean-François Galaup de La Pérouse visited Hawaii in May 1786, he and his ship's surgeon, M. Rollin, observed scantily clad women. La Pérouse noted "traces of the ravages occasioned by the venereal disease," which Hawaiians called *ka'oka'o* (red rot) (ibid., pp. 41, 76). A century after the first infection, fewer than 50,000 of the race survived.

1779 An epidemic of dengue fever at Java was the first reported in medical history. British medical officer David Bylon, who contracted the disease at Batavia (modern Jakarta), Indonesia, identified it as *knokkol koorts* (knuckle fever). During a simultaneous outbreak of dengue fever in Egypt, a medical historian, Abdel Rahman al Gaberti, composed a similar description in *Al-Jabarti's History of Egypt*.

1779 Largely because of an outbreak of smallpox, the Western Comanche, who maintained heavy trade into the upper Arkansas River Valley in guns and bullets, tools, fabric, dray animals, skins, and slaves, faced a decline in their fortunes. Because of their quick-witted shift of focus from the whites of the Missouri valley to the Spanish of the Southwest, they managed to overcome their losses and recover by cooperating with European settlers. By 1787, the Comanche had formed a symbiotic relationship with the Spanish by joining them in wars against the Apache in exchange for weapons and ammunition.

1779 In the Austrian Netherlands, epidemic dysentery, called the *rode dood* (red death), entered villages as Dutch soldiers marched home from combat in Germany. At Brabant, citizens suffered needlessly out of ignorance of the disease. In this same period, the French postponed an invasion of England after *la grande dysenterie* sickened soldiers stationed in western France.

April 1779 During colonial troop movements led by Colonel Casimir Pulaski over the Carolinas and Georgia, an epidemic of smallpox paralyzed the state capital of New Bern, North Carolina, and extended west to the mid-state Moravian enclave of Salem. Because the Moravians chose not to inoculate, their community suffered through October. Prophylactic measures consisted of tar daubed on the forehead and tobacco stuffed into nostrils. For the next two years, nothing halted the waves of smallpox that killed Moravians at Bethabara, Bethania, Friedberg, and Salem.

October 1779 Black slaves seeking to serve the British in exchange for freedom succumbed at a rapid rate to smallpox and typhus. In Charleston, South Carolina, according to preacher and memoirist Boston King, a former plantation slave:

> I was unable to march with the army, I expected to be taken by the enemy. However when they came, and understood that we were ill of the smallpox, they precipitately left us for fear of the infection ["Memoirs of Boston King"].

British officials corralled black patients a mile from the military encampment, immuring them in unsanitary surroundings without water or rations. King and 25 others sheltered in a hut a quarter mile from the post hospital.

October 4, 1779 A double epidemic of smallpox and measles spread over land routes to Sonora, Alta, and Baja California, and to Texas, halting European empire builders in the Southwest. At Mexico City, Esteban Morel was *la primera inoculada* (the first professional inoculator) in the Spanish colonies and his patient, Barbara Rodriguez de Velasco, the first inoculee. Because of Morel's success, Viceroy Martín de Mayoraga supported a general immunization of the populace.

The governing council set up a clinic in the Convent of San Hipólito and distributed handbills urging variolation, but few took advantage of the prophylactic measure. Morel inoculated only 14 people. For his work, he was paid 200 pesos. By December 27, over 8,821 of the 44,000 cases of smallpox in Mexico City died. The majority were young people. *See also* **November 1779**.

November 1779 After smallpox engulfed loyalists in Charleston, South Carolina, according to General William Moultrie, the militia failed to defend the city from the British. Because of widespread fear of infection, only 5,000 patriots turned out to fight the advance of General Henry Clinton's 10,000, who triumphed on May 12, 1780. That summer, large numbers of children suffered infection. The British, smug in their immunity to smallpox, met their own dilemma with mounting malaria. Those prisoners of war whom the British interned at Saint Augustine, Florida, encountered widespread pestilence and observed slaves too sick to harvest the fall rice crop.

1780 According to the winter count of American Horse, chronicler of the Oglala Lakota, smallpox began killing his people in an epidemic that stretched into 1781 across the northern plains. The disease was a greater catastrophe for Hopi, Paiute, and Pueblo on the Colorado Plateau than for the Navaho. Combined with the number killed by measles, smallpox deaths, which numbered 5,025 among Pueblo Indians alone, contributed to the permanent weakening of tribal strength.

The epidemic lowered the defenses of the Mandan and Hidatsa. To stem Sioux attacks, the Mandan dug ditches inside their fortification in the Nuptadi village on the Missouri River in modern North Dakota. Because the maneuver failed to protect the Mandan, they moved closer to the Hidatsa, their allies, and built three villages. The Arikara joined them, but squabbles forced the Mandan out of the alliance to new quarters in the Knife River district to the west of Dakota Territory.

To the north into Canada, Ojibwa, Chipewyan, and Sioux suffered a three-year pestilence that killed 60 percent of indigenous people. Those clans fleeing to the prairie carried the smallpox virus that killed and maimed. Trade with the Arikara, Cree, Hidatsa, and Mandan extended contagion along the Rocky Mountains. The corpses left unburied drew predators from the woods. The wolves that contracted the virus looked too shabby and denuded to be valuable to trappers. Disease also sapped the Maliseet of the Maritime Provinces, where diminution of the population allowed Loyalist settlers to overrun ancestral hunting and fishing grounds.

1780 The French developed a unique two-for-one system of treating congenital syphilis and researched a pilot program in an institutional setting. In the neonatal ward of Vaugirard Hospital in Paris, physicians treated syphilitic wetnurses with mercury as a means of passing the drug through breast milk to infants. The system, which was in operation for a decade, had a greater curative result on women than on their charges. A parallel campaign to rid poor and abandoned women of sexual promiscuity was unsuccessful.

April 1780 According to eyewitness accounts by military officer Stephen Kemble, the first command of 20-year-old Horatio Nelson, captain of the frigate *Hinchinbroke*, placed Nelson in command of 1,400 men, whom he led up the San Juan River along Nicaragua's mosquito coast and across Lake Nicaragua to Granada within view of the Pacific Ocean. The purpose was to split Spanish possessions midway in Central America. During the rainy season, the British traveled into the interior through heavy infestations of disease-bearing mosquitoes to assault Fort San Juan, a western redoubt on the boundary of Costa Rica.

Nelson lost few to battle, but by October, 1,080 English soldiers were dead of malaria and yellow fever. Nelson's own health was so impaired by dysentery and fever that his men evacuated him by canoe to the coast, then had him transported to Jamaica. He had to return to England to recuperate. Meanwhile, Spain recovered Fort San Juan.

summer 1780 Benjamin Rush, a physician and professor of chemistry at the College of Philadelphia, summarized data from an outbreak of "bilious remitting fever" or dengue fever (also breakbone fever). Half of the residents acquired the pestilence from passengers and mosquitoes arriving on ships from the Caribbean. Patients suffered fever, nausea, vomiting, rash, intense pain, and hemorrhagic symptoms.

1781 Settlers led by Lieutenant José de Zúñiga from Alamos, Sonora, to Alta California spread smallpox along their route. On their overland progress from the Gulf of California to Loreto, they infected the missions of Baja California, where Indian residents died. A Dominican priest, Father Luis Sales, author of *Noticias de la Provincia de Californias en Tres Cartas* (*Observations on the Province of California in Three Letters*) (1794), reported on Indian retreats to caves and the rapid demise of adults, who left their children without food. At three missions—San Fernando de Velicatá, San Francisco de Borja, and San Ignacio—Dominican priests successfully variolated Indians. Additional pox infections of the Alameda, Bernalillo, Corrales, Hopi, Sandia, and Santo Domingo caused great loss of life.

January 1781 Smallpox struck Salisbury in south-central North Carolina among the colonial militia and prisoners of war. General Nathaniel Greene rejected a proposal to inoculate the populace. By April, the disease had reached Fayetteville to the east. Pestilence raging among Georgia Indians moved over trade routes to Texas and Santa Fe, New Mexico. Disease decimated Indians and their Apache and Comanche slaves at Pecos, Picuris, Santa Clara, San Ildefonso, and San Juan pueblos and killed soldiers at the Santa Fe presidio.

By February 21, 1782, Father Juan Bermejo was so overwhelmed with burials that he stopped entering names in the parish register. That summer among the Pima of Tubac in northern Sonora, a severe outbreak precipitated their abandonment of the area. At the Tumacácori Mission, the disease killed around 16 percent of residents. At Mission San Ignacio, the disease targeted small children.

March 1781 When cholera struck Ganjam, India, it felled 1,143 of a regiment of 5,000 British soldiers under command of Colonel Pearse and spread down the coast to Calcutta before dying out. *See also* **October 1782.**

spring 1781 At Camden, South Carolina, two of the victims of rampant smallpox aboard prison vessels were teenaged brothers, Andrew and Robert Jackson. With no bedding or clean clothing and meager food and water, they seemed doomed to perish until their mother, Elizabeth Hutchinson Jackson, arranged a prisoner exchange. The trio departed in the rain. Robert died of the pox, but Andrew survived severe delirium. Late into his presidency, Andrew Jackson extolled women like his selfless mother, who died nursing the fever-ridden prisoners of war at Charleston.

summer 1781 Along the Strait of Georgia in British Columbia and Puget Sound, Washington, smallpox sapped the Salish as contagion moved along the plains into the lower Columbia basin. Details from fur trader Samuel Hearne, author of *A Journey from Prince of Wales's Fort in Hudson's Bay to the Northern Ocean* (1772), and a Scots eyewitness, Mitchell Oman, along with the map of the Red River and Missouri River Basin drawn by fur trader Donald Mackay in 1791 attested that the Assiniboin, Atsina or Gros Ventre, Cree, Chipewyan, Kainai, Peigan, and Siksika of west central Canada died out in vast numbers from smallpox. At native encampments and trading posts along the Saskatchewan River, enclaves were littered with reeking corpses. Native survivors hunched together in disbelief.

Nebraska historian Mari Sandoz recorded in *The Beaver Men: Spearheads of the Empire* (1964) that the devastation was shattering to the

Assiniboin, who transported the pox in a clutch of scalps they had taken in the upper Missouri country:

> Twice now this stinking spotted disease had come up the ladder of rivers, following at the heels of the white man like a hungry wolf but going around him like the arrow that goes around the white buffalo, and then striking the Indian down.… The disease ran through [them] as it had the eastern peoples almost a hundred fifty years ago, and many, many tribes since. Scarcely an Indian struck by smallpox lived [p. 158].

They were aware of the "small cuts," their term for variolation, but considered the technique "just another foolish story," yet they could not deny that the whites survived while Indians died (ibid.). The Assiniboin shamans sang, chanted, danced, drummed, painted sacred symbols, and administered herbs, "often until they too, fell before this powerful enemy whose throat none could grasp" (ibid.).

The Assiniboin tried to outrun the disease, but found the stalker to be relentless. Only after the dread winter passed could they return to claim skeletal remains, tie them into tree forks or on scaffolds, or reclaim bones left by feeding wolves. Reverently, the Assiniboin collected skulls and settled them in circles at holy places on the prairie. In coming years, the sad survivors came to visit, smoke, and "throw their minds back to the time before the spotted disease" (ibid., p. 159).

Battiste Good, winter count-keeper of the Brulé Sioux, named both 1779–1780 and 1780–1781 as the smallpox-used-them-up winter. The contagion reached the Nez Percé across the Rockies along the lower Snake, Salmon, and Clearwater rivers in central Idaho, Oregon, and Washington. Those who survived were dubbed the "rough-face," a term that clung to the pock-marked evidence of smallpox.

The Assiniboin survived at one-fourth their former numbers; the Chipewyan appear to have lost 90 percent of their population to the epidemic. Diarist David Thompson, a cartographer working for the North West Company, cited Sokumapi, an Indian who approached a camp of the Snake Indians and found them all dead or dying. An eyewitness, William Walker, journal-keeper for Hudson House trading post, surmised that many weakened and died of starvation because there was no one to feed the sick. The loss permanently altered the fur trade, halting the flow of goods to Montreal until 1782. *See also* **June 10, 1782, May 1792.**

July 13, 1781 As a diversion, General Alexander Leslie, a British military leader following General Lord Charles Cornwallis, plotted an act of bioterrorism. He proposed to distribute 700 black slaves exposed to smallpox among loyalist estates at Portsmouth, Virginia. At Philadelphia, Quaker author Lowry Jones Wister recorded the agony of watching her toddler, William Wister, expire from the pox.

October 1781 Shut in at Yorktown, Virginia, General Lord Charles Cornwallis fought off the Continental Army by land and battled the French at the shore. He depended on black laborers to perform menial service, but found them highly susceptible to smallpox and typhus, which killed them in great numbers. To rid himself of the burden of nursing and feeding ex-slaves, he expelled them from the town, leaving them to die in unspeakable misery and filth and rot unburied. Loyalists charged the British with ejecting the sick as a way of infecting the enemy. Thomas Jefferson surmised that only one tenth of the 30,000 black conscripts survived. In 1786, Benjamin Franklin charged the British with heinous and inhumane acts after they

> inoculated some of the negroes they took as prisoners … and then let them escape, or sent them, covered with the pock, to mix with and spread the distemper among the others of their colour, as well as among the white country people; which occasioned a great mortality of both [Ranlet 217].

late 1781 An assessment of losses during the American Revolution corroborated earlier claims that epidemic smallpox, typhus, measles, and other communicable diseases were far deadlier than battle (Bayne-Jones 1968, p. 56):

Group	Combat	Disease	Total rates per 1,000 deaths per year
Colonials	20	180	200
British	18	100	118
Hessians	18.75	62.5	81.25

November 29, 1781 Two months into a voyage from São Tomé in West Africa's Gulf of Guinea, Captain Luke Collingwood sailed the slaver *Zong* on its way toward Jamaica with a cargo of 415 black slaves. En route, he tried to cover up a deadly pestilence aboard ship before he docked at slave depots in the Caribbean. To guarantee that the ship's owners in Liverpool, England, would collect insurance, he drowned 133 Africans, whom he chained together before hurling them overboard in mid-ocean. His reasoning was calculated and cold-blooded: if the slaves died naturally, the loss was his; if he threw them overboard while still alive for the safety of the ship, the underwriters would have to reimburse the company for the loss.

The owner of the *Zong*, James Gregson, filed a claim, which abolitionist Olaudah Equiano, an Ibo freedman from Benin, refuted with eyewitness testimony in his autobiography *The Interesting Narration of the Life of Olaudah Equiano, or Gustavus Vassa* (1789). The case came to trial in March 1783 and resulted in a landmark decision elevating the status of blacks from cargo to human beings. Reported by abolition leader Granville Sharpe, the incident provoked strong abolitionism throughout the Caribbean and England, particularly among Quakers.

June 10, 1782 As described in the Hudson Bay Company's *Cumberland House Journals and Inland Journals, 1779–1782*, smallpox could kill Indians before it reached the pustule stage. The disease reached the Great Lakes and infected the Cree, Dakota, Menominee, and other fur-trading tribes. Forest expeditioner and company bookkeeper Matthew Cocking, author of *Journal of Matthew Cocking, from York Factory to the Blackfeet Country, 1772–73*, reported that when Ojibwa traders approached the Hudson Bay, they observed an epidemic on Lake Winnipeg among the Cree. On June 23, the six pockmarked survivors traveling on the Saskatchewan River to the York Factory in Manitoba were all who remained of five families infected by Canadians from Cumberland House.

The epidemic was unique in that it arrived from the interior rather than entering the region upstream by white boat parties. When William Tomison reached York Factory on July 2, 1782, and corroborated news of the pox in the interior, he preceded six canoes of Cree fur traders. Nine days later, three of the Indians died of the scourge. To save lives, Cocking warded off other Indians who frequented the outpost. When three more sick Cree arrived on August 6, he sequestered them upstream with supplies and medicine.

Although Cocking could not save their lives, the quarantine kept the disease from spreading. The smallpox epidemic revealed the compassion of Tomison and Cocking, who received dying natives, fed and sheltered them, and tended them in their last hours. The men displayed respect for natives by hacking graves from frozen turf. Cocking wrote to headquarters:

> I believe never Letter in Hudson's Bay conveyed more doleful Tidings than this. Much the greatest part of the Indians whose Furrs have been formerly & hitherto brought to this Place are now no more, having been carried off by that cruel disorder the Small Pox ... the whole tribe of U'Basquiou Indians ... are extinct except one Child [Mcintyre & Houston, 1999].

Reduced by half, the Ojibwa resettled permanently to the west, causing conflict with the Dakota, but living peaceably with the surviving Cree and Assiniboin.

Despite the successful isolation of smallpox, the impact on the fur industry was severe because European buyers enlisted far fewer native trappers to supply pelts. Many of those who worked trap lines were still recovering from grave illness. The disease raged out of control until William Todd, an officer at Fort Pelly on the upper Assiniboine River, began vaccinating natives. **See September 21, 1837.**

July 4, 1782 Loyalist Captain Thomas Dring of Rhode Island, author of *Recollections of the Jersey Prison-Ship* (1829), entered the confines of a British prison hulk under the wardenship of David Sproat, a loathed tormenter of captives. Dring recognized inmates in various stages of smallpox in an airless room fetid with corruption. He immediately self-inoculated with a pin, choosing the web between his thumb and forefinger. After binding the wound, he found the disease in progress by the next day.

On July 4, 1782, as the war reached its close, guards savaged the frail prisoners, who cheered the arrival of independence for the American colonies. The angry British sealed the hatches, abandoning survivors without fresh air or water. When victorious Americans liberated the flotilla in 1783, they found only 1,400 prisoners alive. Liberators abandoned the H.M.S. *Jersey* to rot and sink into the bay.

Poets Joel Barlow and Captain Philip Freneau commemorated the courage of survivors. Freneau's "The British Prison-Ships" (1780) charged colonial Tories with inhumanity to their neighbors. Of the patriot sufferings he wrote:

> No masts or sails these crowded ships adorn,
> Dismal to view, neglected and forlorn!
> Here nightly ills oppress the imprison'd throng-
> Dull were our slumbers, and our nights too long-
> From morn to eve along the decks we lay,
> Scorch'd into fevers by the solar ray;
> No friendly awning cast a welcome shade;
> Once was it promis'd, and was never made.
> No favors could these sons of death bestow,
> 'Twas endless cursing, and continual woe;
> Immortal hatred doth their breasts engage,
> And this lost empire swells their soul with rage
> ["Jersey"].

A monument honors the martyrs of British prison ships at Fort Greene Park, New York, named for General Nathanael Greene.

October 1782 Cholera inundated India's east coast from Madras to Gan-jam. At Madras, the sickness sapped British soldiers. In 72 hours, contagion killed some 50 men; in the first 30 days, 1,000 Redcoats came down with the disease.

1783 Weakened by a famine, the Dalmatians on the Adriatic Sea suffered an outbreak of bubonic plague that raged into the Balkans. Guards posted on the Bosnian border stopped the influx of sick migrants and quarantined them in pest huts. By sea that September, the Venetian navy blockaded the Croatian seaport of Split. Inland, the sick crowded into the Luzaz lazaretto, dying at the rate of one of every ten citizens. With the closing of houses of worship, establishment of a second hospital, supervision of graveyards, and disinfection of incoming goods, officials began to rein in the disease.

April 1783 Cholera recurred in Burma and continued to dog India, striking the central highlands and Uttar Pradesh to the north. When pilgrims gathered along the Ganges River, a Hindu holy site, the pestilence raced through their number, killing 20,000. General Hari Pant, leader of the Maratha forces, noted a serious loss of manpower during the British battles against Tipu (or Tippoo) Sahib, sultan of Mysore.

July 25, 1783 At Lancaster, Pennsylvania, letter writer Sarah Burd Yeates summarized a surge of measles, which killed three of her children during a three-month epidemic. As the pathogen advanced on Philadelphia and New York, it developed secondary pneumonia, a lethal threat of handicap or death to infants and weanlings.

1784 Obstetrician and pediatrician Michael Underwood of London observed cases of infantile paralysis. In the article "Debility of the Lower Extremities" in *Treatise on Diseases of Children* (1789), he remarked on lameness that developed in children following an unidentified fever and debility. He evolved the term "poliomyelitis" to describe the effects on myelin, the tissue that enshrouds nerves. After his career progressed from surgery to child welfare, he earned the position as court physician to the Prince of Wales, the future Hanoverian King George IV. In 1793, Underwood collected data on diagnosing and treating palsy and paralysis for a subsequent edition of his medical text, the most widely consulted Anglo-American pediatrics compendium of its time.

1784 In Bas-Poitou, France, quick action by the medical community reduced panic, suffering, and loss of life from epidemic infectious pneumonia. The disease brought on chills, fever, cough, shortness of break, chest pain, and panting. Children tended to incur blue skin, convulsions, and unconsciousness; the elder displayed confusion.

1784 To determine the cause of rampant puerperal sepsis and maternal death, the staff of the Vienna General Hospital established the Vienna Maternity Hospital, a teaching institution treating poor women. Patients agreed to accept free labor, delivery, and post-partum care from midwives and obstetricians and to surrender their infants to the state orphanage.

May 1784 At Kherson, Ukraine, Danilo S. Samoilovich (or Samoylovich), author of *Description of the Plague Which Attacked Kherson* (1784), was a physician of the Koporsky regiment and founder of the region's first scientific medical society. After battling outbreaks of plague, he became the Ukraine's first epidemiologist by establishing plague control. His microscopic study of the pathogen determined whether the disease was epizootic or limited to human victims.

1785 Some 32 years after James Lind connected scurvy to a lack of vitamin C in the diet, Sir Gilbert Blane, a Scottish physician and compiler of *Observations on the Diseases Incident to Seamen* (1785) found sailors still weakened from poor diet. He observed that scurvy worsened from the compounded assault of cold, damp, unsanitary conditions, boredom, and inactivity. He was outraged by the number of cases of scurvy and wrote to the admiralty, "I beg leave to call to mind that 1,518 deaths from disease, besides 350 invalids, in 12,109 men, in the course of one year is an alarming waste of British seamen, being a number that would man three of his Majesty's ships of the line" ("Historical Maritime").

Composing from the point of view of a doctor as well as financial adviser, Blane counseled the admiralty:

> Scurvy is one of the principal diseases with which seamen are afflicted; and this may be infallibly prevented, or cured by vegetables and fruit, particularly oranges, lemons and limes. I am well convinced that more men would be saved by a purveyance of fruit and vegetables than could be raised by double the expense and trouble employed on the imprest service; so that policy, as well as humanity, concur in recommending it [ibid.].

For the entire British navy, Blane recommended classifying citrus fruit as disease prevention rather than rations.

After his appointment as fleet physician, Blane proposed shipping lemon juice laced with wine or rum as a preservative on long voyages. The blend was to be dispensed in one-ounce servings after the first two weeks at sea. Sailors could sweeten it with sugar or mix it with the regular ration of rum or grog. The regimen diminished sick days from scurvy by half. For his wisdom, Blane became royal physician to George IV and William IV. See also **1795.**

1786 As a result of a frigid winter, warm spring, and serious famine, epidemic pneumonia decimated the poor of Mexico City. Folk healers advocated eliminating icy cold sour drinks made from citrus fruit. For the 12 percent of victims who died, Viceroy Bernardo de Gálvez instituted immediate deep burial of cadavers. City officials sought outside assistance in preventing and containing illness. They enlisted priests and monks to set up a free public education program, which stressed vocational training. The gesture generated a protest from the private teachers' guild, which lost pupils to the free classes.

1787 In New Orleans, an international port of the southern United States, Don Esteban Rodri-guez Miró of Catalonia took seriously the frequent outbreaks of disease. In his fifth year as Spanish governor of Louisiana, he assessed the community's hospital services and supported the rebuilding of the Hospice of Saint Charles. As a means of preventing epidemics, he placed municipal authorities in charge of draining standing water, ridding streets of untended dogs and farm animals, and cleaning and repairing gutters. He also initiated above-ground burial to shield the city from contagion when flood waters caused coffins to float out of cemeteries. To protect the citizenry, Miró erected a small lodge to isolate slaves suffering from smallpox.

1788 Shortly after the arrival of the first English fleet of 11 ships carrying 1,300 whites to Sydney, Australia, an outbreak of smallpox or possibly chicken pox overwhelmed aborigines. Raging for months, the disease killed one white sailor, Joseph Jeffries, from the crew of the *Supply*, but wiped out more than half of the indigenous population. One clan was reduced to three survivors. Mass sickness created panic in natives, who fled over the land, infecting others. Those left untended died of disease complicated by dehydration and hunger.

1788 At Mixteca Alta, a humid mountainous region in Oaxaca, Mexico, plague was rampant and deadly among indigenous people. Rapid loss of victims reaffirmed folk beliefs of an imminent apocalypse.

fall 1789 In early winter, sickness in Maine, Massachusetts, and Rhode Island preceded an outbreak of respiratory infection in Nova Scotia. When influenza overwhelmed New England, New York, and Nova Scotia, thousands of victims survived the initial infection, but succumbed to secondary pneumonia. Arriving through posts in Connecticut, the pathogen spread to Hartford, felling Noah Webster, compiler of *A Grammatical Institute of the English Language* (1784). A recrudescence occurred in spring 1790, when George Washington, then age 58, nearly expired of the disease.

winter 1789 Anatomist and surgeon Jacques-René Tenon, a member of the French Royal Academy and author of *Mémoires sur les Hôpitaux de Paris* (*Memoirs of Paris Hospitals*) (1788), considered wholesome medical institutions a measure of a people's level of civilization. To prove a point about the need for isolation wards, he reported on the unsanitary conditions in the maternity ward of the Hôtel de Dieu in Paris, where women were housed three to a bed in an overheated, poorly ventilated room. The staff made no separation of women in labor from those sick and dying of puerperal sepsis.

December 1789 Scottish obstetrician Alexander Gordon's *Treatise on the Epidemic Puerperal Fever of Aberdeen* (1795) summarized an outbreak of childbed fever or puerperal septicemia that raged in the city's lying-in wards for three months. It struck women of all classes, afflicting four of his patients. He concluded that the disease resembled erysipelas and was spread by human contact with midwives, nurses, and physicians. Sadly, he concluded that he himself had unknowingly infected many new mothers.

1790 At the end of the Russo-Swedish War of 1788, a second epidemic of syphilis in Sweden derived from soldiers coming home from combat in Finland.

1790–1791 The South Seas suffered an era of infection by European pathogens. As a result of contact with new maladies, the island of Nuku-hiva declined in native population from 17,700 in 1804 to 5,000 in 1848. In Fiji, the emergence of an unidentified disease similar to a severe cold killed whole villages. According to Father Ildefonse Alazard, author of *Réponse de Calomnies de M. Dejeante* (*Reply to the Lies of Mr. Dejeante*) (1905), *pakoko* (tuberculosis) was unknown in Polynesia until a Hawaiian named Tama, one of the first islanders to travel to Boston, carried the pathogen from Massachusetts to Vaitahu in the Marquesas Islands.

With the arrival of the first European sailors and seal hunters, New Zealanders suffered their first epidemic in the modern age. Sickness began among indigenous Maori with what locals called *tikotiko toto* (bloody feces), which may have been dysentery. Spread from an English vessel docked at Mercury Bay, the lethal illness preceded an outbreak of flu. Within three years, an unknown illness decimated the Bay of Islands. *See also* **1792.**

1791 On Spain's military frontier in Chile, endemic smallpox among the Mapuche and Araucana Indians along the Bío-bío River produced cultural conflict between conquerors and conquered, animists and Jesuits. The Spanish failed to enforce a quarantine.

December 5, 1791 During an epidemic of rheumatic fever in Vienna, composer Wolfgang Amadeus Mozart was one of the victims. Medical historians link his fever and delirium to traits of the infection, which include a delayed reaction to strep throat.

1792 In Tahiti, dysentery devastated natives from contact with the crew of the *Anguila*, the ship of English navigator George Vancouver. When Captain William Bligh reached the island in April to load breadfruit cuttings, he listened to the complaints of the widow of Chief Hitia'a, who died from the scourge. She blamed Vancouver's men for her loss.

1792 A slaver from southern India docking in the Indian Ocean at Île de France carried smallpox, which killed many black Mauritians. Heavily involved in revolt, the people debated widespread inoculation and their rights as citizens to refuse preventive treatment for themselves and for their slaves. Others promoted a general inoculation among their work force, much as they would protect farm livestock. The slaves themselves practiced a form of inoculation along with accumulated folk treatments derived from Africa, India, and Madagascar.

1792 Shortly after departing medical training at the University of Glasgow, Thomas Masterman Winterbottom, an explorer of Rio Nuñez and Senegambia, saw symptoms of endemic sleeping sickness or trypanosomiasis among the natives of Sierra Leone. In 1803, he issued a monograph, *An Account of the Native Africans in the Neighborhood of Sierra Leone*, containing clinical observations of the disease.

1792 To halt epidemic typhoid fever and typhus among the poor, John Ferriar, a Scots physician at the Manchester Infirmary and author of *Medical Histories and Reflections* (1792), urged the establishment of fever wards to quarantine infected patients and proposed sanitation reforms. He appointed a board of health in 1796 that recommended ventilation of homes, whitewashing residences every six months, widening streets to admit daylight to fetid drainage ditches, street cleaning and removal of dung, judicious location of privies, and regulation of boarding houses. The suggestions never went beyond the preliminaries.

1792 A recurrence of yellow fever, spread by the *Aedes aegypti* mosquito, infected residents of New Haven, Connecticut, and New York City. At the port city of Charleston, South Carolina, the disease remained active over a seven-year cycle. Rampant contagion ruled out crowded entertainment venues and festivals. At Baltimore, recurrent fever stymied the cultural efforts of promoter William Godwin, a member of the American Company and producer at the New Theatre, who wanted to found a resident city acting corps.

1792 At Pancevo, Serbia, epidemic smallpox curtailed trade with the nearby regions under Turkish control. For seven years, health authorities maintained border controls and quarantines as a means of containing infection.

May 1792 When Captain George Vancouver, a British navigator and author of *Voyage of Discovery to the North Pacific Ocean and round the World* (1795), explored the north Pacific coast of the Americas aboard the *Anguila*, he found an extensive cache of skeletons on the Strait of Juan de Fuca between Canada and Washington State. The crumbling residences and absence of living inhabitants at Puget Sound suggested a catastrophe brought on by smallpox in a virgin soil population. Two clues were the pockmarked faces and blind eyes of local natives. On May 21, Archibald Menzies, a botanist and cartographer of Puget Sound, corroborated Vancouver's findings with his own observations of native losses to pestilence.

August 1792 King Frederick William II of Prussia lost 29 percent of his fighting force after dysentery struck his 42,000 soldiers during an invasion of France. Within weeks, the French overcame the insurgents at the battle of Valmy, France, and forced the Prussians back to Germany over muddy roads littered with the dead and dying. On October 1, as the Duke of Brunswick led the survivors from Grand-Pré and Verdun east across the Rhine, the disease besieged civilians across northeastern France.

1793 As tuberculosis infections increased, medical writer Thomas Beddoes, a radical Oxford-trained physician and chemist, set up a treatment center at Bristol, England, where he acquired funds from Erasmus Darwin, Josiah Wedgwood,

and William Reynolds to apply nitrous oxide, hydrogen, oxygen, carbon dioxide, and other gases as a cure. Beddoes first tried to interest landlords in setting up rooms where tubercular patients could sleep while cows breathed on them. Aided by Humphry Davy as superintendent of the Bristol Pneumatic Institute for Inhalation Gas Therapy, Beddoes tended victims of pulmonary tuberculosis free of charge using equipment designed by engineer James Watt, whose daughter died from TB.

May 1793 Spread from Pennsylvania simultaneously with epidemic cholera, an outbreak of scarlet fever assailed Connecticut, centering in Litchfield and New Fairfield. At Hartford and New Haven, a recrudescence in February 1794 killed many children. New Haven lost 52 white citizens to the disease. The fever ravaged an unknown number of Algonquian, Miami, Potawatomi, and other Great Lakes tribes, killing babies, children, and village elders.

July 1793 Yellow fever reached Pennsylvania via infected refugees aboard transport vessels from Haiti and Santo Domingo, where 44 percent of English occupation forces died of the disease. On the voyage north, disease-bearing mosquitoes appear to have flourished in water casks on deck. When the ships docked with hordes of refugees fleeing revolution in the Caribbean, many were already sick.

As summer downpours increased mosquito incubation sites, the disease broke out in the low part of town among French seamen quartered at Denny's Lodging House. In the words of *Watson's Annals of Philadelphia and Pennsylvania* (1857):

> Then the haunts of vice were shut up; drunkenness and revelling found no companions; tavern doors grew rusty on their hinges; the lewd or merry song was hushed; lewdness perished or was banished, and men generally called upon God. Men saluted each other as if doubting to be met again, and their conversation for the moment was about their several losses and sufferings ["Watson's Annals"].

Fever killed one-tenth of the people of Philadelphia, including three-month-old William Temple Todd and Quaker attorney John Todd, the first husband of Dolly Madison, and halted U.S. government staffs at the customs office, treasury, and post office, which were housed temporarily in the city during the building of Washington, D.C.

A prominent physician, Benjamin Rush, a signer of the Declaration of Independence, voiced the medical community's concerns. Backed by Andrew Brown, publisher of the *Federal Gazette*, Rush advised citizens to rest, stay out of the sun, avoid night air and liquor, and use vinegar and camphor to rid sickrooms of infection. He encouraged the healthy to retreat to the country. Among the notables who took his advice were President George Washington and Martha Washington, who retreated to Mount Vernon; Secretary of State Thomas Jefferson; Secretary of War John Knox; and Secretary of the Treasury Alexander Hamilton and his wife, Elizabeth Schuyler Hamilton, both of whom came down with fever and fled to New York outside Albany. En route, Hamilton encountered a roadblock, where a patrol torched his belongings and disinfected his family's conveyance. An initial toll of 140 mortalities rose in one month to 1,400. By October, 4,044 had died.

The onset quickly depleted the body via malodorous diarrhea, hemorrhaging, and vomiting, and left a putrefying corpse. Workers sprinkled vinegar and scattered shredded pungent artemisia and burning tar in the streets. As witnessed in the journal of Quaker carpenter and volunteer Edward Garrigues, member of a hospital committee and one of the Guardians of the Poor, the privileged fled the city, leaving the city overseers and guardians like Garrigues to convey the suffering to the Spruce Street poorhouse, to Ricketts' Circus, and to the Bush Hill pesthouse, a mansion turned into an emergency hospital. Families left new patients in holes dug on the grounds until the staff could admit them. In all, Jean Devèze, Stephen Girard, Peter Helm, head matron Mary Saville, and 19 nurses attended 807 fever victims without adequate supplies or running water.

Officials had difficulty hiring cooks and medics to treat the sick and carters and gravediggers to bury the dead. Distinguishing themselves were five medical students of Benjamin Rush, along with female prisoners and two former slaves. Influenced by John Lining's comments on the immunity of blacks to the 1742 yellow fever epidemic in Charleston, Rush requested the help of two blacks, cobbler Richard Allen and minister Absalom Jones, who knocked on doors to determine who was still alive. The aides collected orphaned children and summoned burial crews to haul out fetid corpses. Other black heroes of the epidemic included nurses Sarah Bass and Mary Scott. A third black caregiver, Caesar Cranchal, perished in service to the poor.

Contagion peaked on October 11 and fell rapidly with the arrival of cool autumn weather and the first frosts. On November 14, 1793, historian Matthew Carey published *A Short Account of the Malignant Fever Lately Prevalent in Philadelphia* (1793), which he updated with addenda and a list of 4,000 victims interred in local cemeteries. The text described the collapse of the business and financial district and blamed coffee and putrid animal matter for spreading fever. In response to Carey's claim that white health workers saved the day, Allen and Jones published a monograph entitled *Narrative of the Proceedings of the Black People, During the Late Awful Calamity in Philadelphia* (1793). Other authors characterizing the epidemic and speculating on its causes included gothic novelist Charles Brockden Brown, architect Benjamin Latrobe, and lexicographer Noah Webster, author of *A Brief History of Epidemic and Pestilential Diseases* (1799). Another writer, poet Philip Freneau, composed the poem "Pestilence" (1793), which muttered ominously,

> Nature's poisons here collected,
> Water, earth, and air infected—
> O, what a pity,
> Such a City,
> Was in such a place erected! [Freneau].

Because Haitian refugees bore the brunt of the contagion, white citizens blamed the disease on Caribbean immorality. When German newcomers died in record numbers, established Philadelphians claimed that the disease should be called Palatine fever. *See also 1797.*

1794 A bilingual smallpox edict carved on stone and issued between the Tibetan King Tri Ralpchen and the Chinese Emperor Wen Wu Hsiaote Wang-ti intended to reduce contagion at Lhasa. Among precautions against the pathogen, workers built smallpox hospitals. The influence of the edict illustrated the strength of the Manchu dynasty in governing the Asian interior and in forcing Tibetans to adopt East Asian preventive measures, notably, mixing lymph from a smallpox victim with camphor for inhaling through the nose.

June 1794 In Santo Domingo, yellow fever arrived from Grenada, Jamaica, and Martinique during military disquiet produced by French wars. At the coast city of Port-au-Prince, fresh troops strengthening the British occupational force carried the contagion that spread the fever. When the enemy took the highlands and curtailed the supply of mountain streams to the English in the harbor, the garrison had to drink from polluted vats. The total loss to the British was 2,200 or 40 percent of its manpower. Contagion continued felling new waves of soldiers, forcing the British to decamp three years later to healthier climes.

In Haiti, pandemic yellow fever cost the British military many deaths and additional loss of island labor. Contributing to the incidence of disease was the faulty placement of barracks, inadequate medical treatment, and general troop debility, exacerbated by alcohol abuse. Doctors, who were uncertain about distinguishing between malaria and yellow fever, made efforts to reform weak elements of the Army medical corps.

1794 Epidemic plague swept Oran, Algeria. The scourge arrived with Muslims pilgrims returning from a hadj to Mecca, Arabia.

June 10, 1794 Yellow fever emerged at New Haven, Connecticut's largest harbor receiving Caribbean trade. The outbreak prompted one New York newspaper to muse that contagion arrived from the West Indies in a trunk of clothes contaminated by a sick sailor. Reportedly introduced by a sloop from Martinique, the disease spread to the Isaac Gorham family in the dockyard, killing Elizabeth Gorham, her son, and an infant. After the first deaths, port authorities at New York and Philadelphia quarantined all ships and goods arriving from New Haven.

Drawing on the experience of Philadelphians the previous summer, New Haven's health authorities suppressed news of the rising deaths. In secret, they set up a hospital for fever patients and hired nurses with money collected from donors. On August 20, the city staff released a report claiming that they had contained the fever. The diary of Ezra Stiles, Yale College president, told the sad truth—that up to 90 percent or 18 out of every 20 victims died of yellow fever.

August 7, 1794 When yellow fever struck Baltimore, people died at the rate of 23 per day. The epidemic emerged in April as fever among children and paralleled an outbreak of smallpox and dysentery. The Baltimore media suppressed news

of the local epidemic, but some issues published letters from Philadelphians about their sufferings from yellow fever. Thomas Drysdale, a young doctor, blamed the scourge on putrid air blowing from rotting vegetables on wharves and in houses. The populace recalled the sufferings in Philadelphia the previous year and panicked; all who could leave headed for the country. Intercity rivalry provoked open enmity between Baltimore and Philadelphia concerning where the disease was more virulent. Before October weather ended the contagion, several overworked doctors had died of the fever.

1795 A Scottish doctor, Sir Gilbert Blane, physician to the English navy and personal physician of George IV and William IV of England, caused naval administrators to upgrade shipboard cookery and to supplement the crew diet with citrus fruit. Naval surgeon James Lind originally proposed the idea in 1753 as an antidote to rampant scurvy. The addition of lemons from Mediterranean groves and Caribbean limes resulted in the nickname "limeys" for British seamen and "lime-juicers" for naval vessels (Kemp, 1988, p. 485). The concept of more vitamin C for sailors passed to the British merchant marine under the Merchant Shipping Act of 1854.

March 1795 The Tatca and Sacla Indians who entered Mission Dolores in San Francisco, the sixth in California's 21-mission chain, suffered a devastating epidemic that may have been typhus. The outbreak made an immediate change in native attitudes toward the Spanish. In less than two months, the depleted natives fled the mission to merge with other East Bay tribes to attempt a stand against Hispanic settlers.

summer 1795 Dual epidemics terrified residents of Virginia. When Norfolk residents suffered the first of five major yellow fever epidemics, coastal authorities quarantined shipping all the way to Alexandria. *See also* June.

June 1795 An employee of the Missouri Company, Jean-Baptiste Truteau was a Canadian-born explorer and fur trader whom the Spanish licensed to establish contact with the Arikara of South Dakota. After traveling north to the Dakotas, wintering at Fort Randall, and learning the local language, he compiled facts about Nebraska and Dakota Indians in *Journal of Truteau on the Missouri River, 1794–1795*. He reported that smallpox had so reduced the Arikara that their 32 villages had shrunk to two. The Arikara merged tribes or departed to join the Omaha or Mandan, who had settled on the Knife River.

late July 1795 A new yellow fever outbreak in New York City arrived from the Caribbean on the brig *Zephyr*, where the ship's boy succumbed to "bilious remitting fever," which had begun in two other crewmen. The harbor health inspector, Malachi Treat, recognized the symptoms of yellow fever. By the time it had run its course, the fever had killed 732. Of that number, 500 were immigrants.

To maintain trade, the captain of the *Zephyr* and city officials refuted Treat's diagnosis and allowed passengers to land. Undeterred, Treat rowed to Nutten Island to supervise the burial of the ship's boy. A second ship, the *William*, docked and reported fever among the crew, but response from the health committee remained lax. On July 29, Treat died of fever, yet newspapers suppressed the cause.

The health committee supervised rapid burial of victims throughout August and watched for more infected sailors in port. They ordered cleaning of markets, yards, streets, and basements along the East River. By August 15, the death toll rose to 14. At Bellevue Fever Hospital, an attending physician was stricken with fever. When seven more victims sickened, citizens surmised that they were not learning the truth about the fever. On August 28, the *Philadelphia Gazette* warned New Yorkers of the danger of letting the malady spread. Three days later, to the consternation of John Jay, New York's recently elected governor, Philadelphia's harbor quarantined all ships from New York.

Physician Alexander Anderson, educated in Edinburgh, took charge of Bellevue Fever Hospital and witnessed the death of a nurse from the fever. Short-staffed, he pondered resigning from a hopeless situation. When the death count rose to 25 per day, panicky citizens discussed the mounting epidemic and began evacuating their homes. Quietly redirecting patients from Bellevue to the offices of city doctors, municipal agents played on community prejudice by blaming immigrants and the homeless for increased incidence of fever and dispatched observers to determine if Anderson were competent to treat the sick.

Meanwhile, Philadelphians posted watchers to observe stage lines and to bar New York refugees. Benjamin Rush, the East Coast expert on yellow fever, advised on methods of ridding the city of contagion. New York doctors began discussing methods of treatment as prominent citizens fell ill. In the absence of benevolences from New York City coffers, Philadelphians sent donations to help the poor. By early November, the epidemic subsided.

October 2, 1795 During the Brigands' War, the second phase of the Carib Wars, General Sir Ralph Abercromby, commander-in-chief of the British forces in the West Indies, and his 4,000 British regulars faced down Black Caribs led by Martin Padre, Du Valle (also Duval or Duvalier), and Chief Chatoyer, who died in battle. Upon the resettlement of 5,080 black natives to Roatan in the Bay Islands off Belize and Honduras, over 2,500 died of plague.

1796 A smallpox epidemic in London killed 3,500, a death toll that was one-tenth of the fatalities from pox throughout Britain. To end millennia of epidemics, Edward Jenner, a medical student of John Hunter at Saint George's Hospital, London, and later general practitioner in Berkeley, Gloucestershire, demonstrated the efficacy of vaccination against smallpox. To establish immunity, he injected pus from a cowpox lesion on the hand of dairymaid Sarah Nilmes into the arm of eight-year-old James Phipps. After six weeks, Jenner tested Phipps's resilience by injecting him with fluid from a smallpox lesion. Because of the success of the experiment, which Jenner reported in a monograph, *An Inquiry into the Causes and Effects of the Variolae Vaccinae* (1798), Europeans began demanding the procedure to protect them from contagion.

1796 During author Eugenio Texeira's observation of medical treatment in the seven parishes of Cuzco, Peru, he discovered that syphilis targeted the Quechua and Aymara Indian populations.

May 1796 When Napoleon's troops overcame Austrian forces in Italy, the French settled in Lombardy, spreading typhus among Mantuans. Within a year, the disease migrated across Italy and into Sicily.

1797 In Mexico, 10,000 died of smallpox that engulfed Mexico City, Oaxaca, Orizaba, and Puebla from 1793 to 1795. By the time the contagion reached Guatemala, the death toll rose to 25,000 or 16.7 percent out of 150,000 infections. Families who concealed their sick kin fought off soldiers who enforced mandatory hospitalization. To spare Spanish colonies the pandemic, surgeon Francisco Xavier de Balmis set up a smallpox vaccination program that drew on 25 Spanish orphans as antibody reserves. He continued until 1813 in Cuba, Venezuela, Mexico City, Texas, the Philippines, Ecuador, Peru, Chile, Colombia, Paraguay, Panama, Bolivia, and Puerto Rico. His initiative won the support of the benevolent Charles IV of Spain, whose three-year-old daughter Maria Teresa died of the pox in 1794, and of Viceroy José de Iturrigaray, but provoked the jealousy of philanthropic doctor Francisco Oller. As a result of medical infighting, governor Ramón de Castro rejected de Balmis's crusade from Puerto Rico.

1797 A yellow fever epidemic at Providence, Rhode Island, appeared to derive from the trading vessel *Betsey* arriving from the Caribbean in August, killing 36 patients and a physician, Ephraim Comestock. Another theory proposed that rotting garbage was a source of virulent miasma causing "ship fever." Victims developed fever, delirium, head and torso pain, and jaundice and vomited a black substance. To ward off panic, the Town council advocated immediate interment of corpses in tarred sheets and burial of bedding and personal effects. Limited funeral attendance and conveyance of caskets through back alleys suppressed general knowledge of the scourge.

Because the city derived its wealth from commerce, enforcers of health regulations placed no quarantine on ships. They also took no action requiring refuse cleanup in the dock area lest they incur discrimination against ships from their harbor or give the city a reputation for filth. On his own, activist and medical historian Moses Brown began investigating domestic sources of contagion from swampy ground. Survivors of the 1793 outbreak in Philadelphia collected $1,500 for relief of the poor at Providence.

When contagion struck Philadelphia, it terrorized survivors of the massive epidemic of 1793. One observer, French-born physician Felix Pascalis-Ouvière, a graduate of the University of Montpellier and experienced doctor of tropical medicine at Santo Domingo, settled in Philadelphia to study the methods of Benjamin Rush, who created a regimen requiring purging and blood-letting, limited diet, cold baths, and cold drinks. Pascalis-Ouvière compiled an eyewitness memoir, *An Account of the Contagious Epidemic Yellow Fever, Which Prevailed in Philadelphia in the Summer and Autumn of 1797* (1798), in which he lauded his mentor for offering citizens good advice during troubled times.

1798 After European colonists enslaved the Puri Indians of São Paulo, Brazil, who occupied ancestral lands between the Mantiqueira Mountains and the Paraíba River, the natives resided at São João de Queluz under unfamiliar and artificial living conditions. Exposed to a smallpox epidemic, they died in record numbers, losing their cultural and racial heritage as fewer survived to perpetuate it. Meanwhile, colonists settled the Puri at the Queluz mission and usurped their land. By 1831, only six Puri survived.

late July 1798 Yellow fever raged in New York City, killing upwards of 3,000. One alleged source of contagion was the schooner *Fox* out of Jeremie, Haiti. After the fever emerged at Coenties wharf on the southern tip of Manhattan, the first victims were prominent merchants. One hero of the outbreak was a medical student, Walter Jonas Judah, one of the first Jews to study medicine in the United States. He offered free medical care and distributed money from his pocket to patients who couldn't afford medicine. He died of the fever at age 21.

One victim, Philadelphia-born Quaker novelist Charles Brockden Brown, felt confident that he lived far enough away from contagion to be safe. He commented in his journal:

> Heavy rains, uncleansed sinks, and a continuance of unexampled heat, have within these ten days given birth to the yellow fever among us in its epidemical form. Death and alarms have rapidly multiplied, but it is hoped that now, as formerly, its influence will be limited to one place…. My mode of living, from which animal food and spirituous liquors are wholly excluded, gives the utmost security ["Charles Brockden Brown"].

After Brown and his friend, physician Elihu Hubbard Smith, evacuated a sick friend, Joseph Scandella of Venice, to their home for treatment, both Brown and Smith caught the fever. Brown survived, but Smith and Scandella did not. Brown used his recuperation as an opportunity to write four novels. In *Arthur Mervyn, or Memoirs of the Year 1793* (1799), which contains the events surrounding Scandella's death, and again in *Ormond* (1799), Brown referred to epidemics with graphic detail and commented on society's loss from yellow fever outbreaks.

summer 1798 As yellow fever advanced unabated summer after summer, the loss of 22,000 of General Victor Emmanuel LeClerc's 25,000 French forces to fever in Haiti forced Napoleon to concentrate his military might in Egypt and Malta. Napoleon chose to give up a huge parcel of territory to President Thomas Jefferson, who was glad to acquire the Louisiana Purchase, negotiated on April 10, 1803, by Secretary of State James Monroe for $11,250,000. Within four years, 24,000 French soldiers in Haiti died of yellow fever, with another 8,000 hospitalized. LeClerc himself succumbed to the fever in Santo Domingo on November 2, 1802. *See also* **mid-May 1802**.

September 10, 1798 At war with Spaniards in British Honduras, British colonists won the battle of Saint George's Cay in part because they were healthier, knew the territory, were well supplied with materials and leaders, and had no choice but to fight to survive. Spanish insurgents, who were overwhelmed by yellow fever and lacked motivation, were easily defeated.

December 1798 After 38,000 French troops invaded Egypt and Syria in Napoleon's grab of eastern Mediterranean lands, they risked losing all to bubonic plague, from which 2,000 French expeditioners perished. The military medical corps followed hygienic standards in laundries and showers, launched a general decontamination, and distributed preventive medicine while boosting morale. The regimen, carried out into 1801, saved the French army from severe loss to disease. *See also* **March 11, 1799**.

1799 To halt endemic smallpox throughout the Spanish realm, the periodical press in Madrid and Latin America disseminated monographs and 19 books on the value of vaccination. Printed in Bilboa, Cádiz, Catalonia, Madrid, Montpellier, Pamplona, and Zaragoza, the texts provided a valuable tutorial to colonists, who began to request vaccination.

March 11, 1799 When plague assailed Napoleon's garrisons in Egypt at Abukir, Alexandria, Damietta, and Rosetta, it posed a small danger. At Jaffa, Israel, however, contagion compromised the entire force. A diarist, Major T. K. Detroye, remarked on violent fever, buboes, and sudden death. He described the military doctors' attempt to ward off panic by declaring that the 14 dead, four suicides, and 31 ailing men did not have plague. Nonetheless, the Armenian convent filled with patients, from whom the monks and French medical staff fled.

By keeping himself and his soldiers focused on the campaign, Napoleon refused to give in to fear. According to René-Nicolas Desgenettes, Napoleon inspected the foul-smelling hospital, addressed the sick, and gave the appearance of an unhurried visit. He even helped remove the body of a soldier soaked in the effluvia of an abscessed bubo, which Napoleon insisted should be lanced. Because of the general's courage, he raised morale for the move up the Mediterranean coast to Acre and Haifa three days later. Among the 300 plague victims he left behind was Adjutant-General Pierre Joseph Grézieu, who died treating the sick. With some 30 dying each day, only one-twelfth of the patients survived.

Napoleon's forces carried the plague first to Haifa, then Acre, where the Turks suffered their own outbreak. The French medical staff received plague victims at a special pesthouse on Mount Carmel outside Haifa. By the time the army set siege to Acre, the increase in plague victims reduced the likelihood of victory. The general wisely chose to return to Alexandria, but concealed his failure under the excuse that Acre was overrun with plague and thus might endanger French troop strength. His forces abandoned the moribund by the wayside. Napoleon advised Desgenettes to euthanize the suffering with large doses of opiates, but his medical officer considered the act unprofessional. *See also* **May 1801**.

May 2, 1799 As sickness prevailed at Acre, Israel, after Napoleon's siege, Turkish forces suffered an outbreak of plague, which was endemic in part because of climate and poor standards of hygiene. The scourge enveloped the Armenian convent with patients. Plague infected Colonel Antoine le Picard de Phélippeaux, a military engineer at Jaffa who boarded the *Theseus* for departure. When he died of fever, crew buried him at sea and jettisoned his belongings. Despite fumigation of his cabin, the disease spread to the crew.

1800 In response to a nine-year epidemic of scarlet fever in Germany, Samuel Christian Hahnemann of Saxony, founder of homeopathy, published a monograph, *The Cure and Prevention of Scarlet Fever* (1801), which booksellers distributed with a vial of belladonna, a drug derived from the berries of the poisonous nightshade. His cure, which also prevented infection, was the forerunner of homeopathic treatment, which applies minute doses of natural substances to stimulate the body to heal itself. Near the end of the scourge, he recommended treating fever with aconite or monk's hood in his treatise *Observations on the Scarlet Fever* (1808).

1800 During the Austrian siege of the French in Genoa, Italy, typhus devastated inhabitants. A dedicated English physician, William Batt, happened to be lodging in the city and volunteered to diagnose and treat the sick. The retreating French marched west along the Mediterranean toward Nice, spreading fever in San Remo and causing a recrudescence in Genoa.

1800 In Persia, health authorities initiated quarantine as plague struck Mosul (Iraq) and swept the Persian Gulf as far north as Constantinople. The fumigation of travelers and the closing of roads and ports halted tourists, merchant trains, and British vessels, but did not stop the disease from progressing to Baghdad, where the scourge killed many natives, but no Europeans.

early 1800s At Quangdong, a trading nexus, Chinese epidemiologist Qui Xi worked to suppress outbreaks of endemic smallpox by practicing the European method of variolation. He devoted his career to immunization by taking advantage of China's reception of intellectual ferment from Europe.

July 6, 1800 At Cádiz, Spain, yellow fever felled thousands after a ship from Cuba brought three moribund victims to the harbor. The rise of the death toll to 200 per day or 15 percent of the citizenry halted trade, killed many physicians, and overwhelmed priests and burial crews. In one of five outbreaks to rock the area in the 19th century, the contagion ranged out to Jerez, Medina Sidonia, and Seville. In Andalusia, senior naval surgeon Domingo Vidal proposed control of harbors and frontiers, quarantine, and scrutiny of provisions and public sanitation to spare Catalonia from the epidemic. Another hero, navy doctor Miguel Cabanellas Cladera of Majorca, treated patients at Cádiz, Cartagena, and Seville.

The disease was a mystery to medical experts. As explained by Mexican botanist José María Mociño, a volunteer for the High Council for Health Preservation, in his *Disertación de la Fiebre Epidémica, Que Padecio Cádiz* (*Treatise on the Epidemic Fever That Assailed Cadiz*) (1804), hindering local doctors was an academic disagreement on transmission of infection. A Cádiz city decontamination program consisting of cleaning drains, hosing streets, burning fragrant resins, and firing cannon paralleled the feeding of the poor in city barrios. Individuals venturing onto the streets covered their mouths and noses and carried nosegays and garlic bunches in the same fashion as victims of the Black Death in the late Middle Ages. *See also* **1804**.

The application of fumigation to infected dwellings derived from a system pioneered by Louis Bernard Guyton-Morveau, a colleague of Antoine-Laurent Lavoisier. After he perfected acid fumigation in Dijon, France, in 1773, seven years later, the Académie des Sciences proposed cleansing prisons with the fumes, a preventive that the Conseil de Santé had adopted for sanitizing military hospitals in 1794. Adapted by British physician James Carmichael Smyth with the addition of saltpeter smoke, the method was applied to Spain's yellow fever epidemic. By 1815, health experts—notably Juan Manuel de Aréjula, a Spanish scientist—realized that fumigation had some value in killing the lice that cause typhus, but no worth in quelling yellow fever. Because Ferdinand VII insisted that the government not appear ignorant, his regime suppressed a chapter of Aréjula's work.

August 1800 A yellow fever epidemic in Baltimore targeted the poor at Fells Point, Maryland, killing 1,197. Health workers disinfected streets with ash, lime, and lye. Dr. John Crawford imported vaccine from London; navy physician John J. Giraud advocated ipecac to prevent and treat the scourge. Addressing the needs of the underclass and immigrants, philanthropist and physician James Smith charged Joseph Townsend, member of the board of health, with profiting from his appointment while neglecting his duty to the poor. As a result of Smith's opening his house to the afflicted, city expenditures for the needy increased in 1801, when a pesthouse opened near Fort McHenry. Eleanor Rogers promoted formation of the Benevolent Society of Baltimore.

fall 1800 When Marblehead, Massachusetts, developed an outbreak of smallpox, Elisha Story brought live virus from England and inoculated his children and their friends. Because he had received a virulent stock, he spread the disease, arousing panic that led to an anti-inoculation riot. During a town meeting, ignorant folk threatened Story with lynching. The disease proliferated among the poor and killed 68. The outbreak waned in mid–January 1801. Benjamin Waterhouse, author of *A Prospect of Exterminating the Smallpox* (1800), alerted citizens to the danger of allowing uncertified personnel to perform the procedure. He advised Boston's board of health to set up free public health clinics and disseminated pamphlets on safe inoculation.

November 22, 1800 After 181 days at sea, the East Indiaman *Royal Admiral* transported English prisoners to Port Jackson to serve convict contractors at New South Wales, Australia. Among the 300 inmates and eleven missionaries, 43 prisoners, two members of a convict's family, and four crew died on the voyage of "gaol fever (typhus)," which also killed the ship's surgeon, Samuel Turner. In port, one Anglican missionary died and another quit the mission. On October 30, 1802, the provincial governor declared that many of the victims would never fully recover.

1801 Acadian midwife Marie-Henriette Ross of Bras d'Or, Cape Breton Island, Nova Scotia, distinguished herself as "Granny Ross," a birthing expert and homeopathic healer during a smallpox epidemic. By isolating the sick in a small cabin and inoculating others via the Turkish method of scratching the skin with a needle bearing matter from a smallpox blister or scab, she established her own immunization system. Some 70 years later, her grandson, Thomas Ross, applied the same method during an upsurge in the disease.

1801 In California, the Chumash and Tongva remained centered in their worldview during an diphtheria epidemic introduced by the Spanish. Tribal prophets guided the Chumash in a backlash against the newcomers. The Tongva resorted to witchcraft and executed two shamans for failure to contain the disease.

The following year, California Indians living in missions from San Luis Obispo to Mission San Carlos suffered an outbreak of diphtheria and pneumonia, which targeted the young. Another wave of 1,600 victims died from measles, which skewed the native population as it lost children aged ten and younger. Altogether, epidemics reduced California's Mission Indians in number by around 45 percent. *See also* **1804**.

early 1801 Smallpox blanketed the Great Plains from Canada to the Tex-Mex border. As described by the log of William Clark and Meriwether Lewis and by Nebraska historian Mari Sandoz in *The Beaver Men: Spearheads of Empire* (1964), the pox generated hostility between Native Americans and European insurgents. The fur trade between the French of Louisiana and the Iowa, Quapaw, Salish, Wichita, Pend-d'Oreille, Caddo, and Spokane tribes virtually ceased after an epidemic rendered natives too weak to trap and process pelts. The Osage lost 2,000 of their nation. The battered Missouria joined the Otoe. The 250 remaining Ponca fell victim of the Dakota. Pawnee numbers declined by 75 percent.

Tribes lost as much as 90 percent of their former population. The Omaha, at half their former strength, burned their villages and became nomads and raiders. Parents feared that disfigured survivors would produce the same pock markings on the unborn.

To woo traders back to the pelt business, native trappers heaped skins on an island for smoking to kill pathogens. The French attempted to halt the spread of contagion less out of humanistic concern than on behalf of their dwindling income. The Omaha were so wary of Europeans that they halted a convoy of Spanish who tried to establish business in the upper Missouri region. Along the Pacific coast a hemorrhagic form of the pestilence struck the Flathead, Lummi, Nootka, Tsimshian, Tlingit, Haida, Tillamook, and Nez Perce, leaving severe disfiguration and disabilities and turning corpses black. Lewis and Clark noted the extinction of the Clatsop and Chinook. *See also* **January 1, 1802**.

May 1801 At the battle of Abukir, Egypt, British commander Sir Ralph Abercromby's invasion force faced less danger from combat than from bubonic plague, which decimated the French and Turks. Thomas Robertson, surgeon on the H.M.S. *Leopard*, reported that an ailing ship's baker was the first of many stricken with plague. Similarly, the men of the H.M.S. *Adam Smith* reported fever and buboes. On May 9, officials established a tent contagion ward on shore, with a total loss of 14. In contrast, the French lost up to 40 men per day to the scourge, which cost Napoleon a proposed conquest of Egypt.

summer 1801 The Atsina or Gros Ventre, prairie dwellers of southern Canada, caught smallpox from the Arapaho of the Missouri River region. An epidemic extended across the Columbia Plateau and reached the Salish of Vancouver Island on Canada's Pacific coast. In the early 1900s, Mourning Dove, also known as Christine Quintasket, co-founder of the Colville Indian Association and the first woman elected to the Colville Tribal Council, explained how native medical care worsened smallpox in the previous century. She described her grandmother's memories of its first appearance in the Northwest. Because people tried to sweat out the disease in steamy lodges and then plunged into cold streams, they quickly worsened and perished, sometimes by drowning or pneumonia.

1802 A Quaker physician, Benjamin Waterhouse, a professor of theory and physics at Harvard Medical School, supported vaccination of the public against smallpox in a monograph, *A Prospect of Eliminating the Smallpox* (1800). When Harvard and Boston doctors demanded a public exhibition of vaccination, the Boston Board of Health supported Waterhouse and his campaign to stamp out the pox. After Waterhouse acquired vaccine from John Haygarth of Bath, England, Thomas Jefferson, a proponent of the venture, inoculated 80 members of his household and the families of his daughters. He distributed supplies in Virginia, Pennsylvania, and west with the Lewis and Clark expedition to inoculate Indians along the Pacific Coast. In 1806, Jefferson proclaimed in a letter to Edward Jenner that the extermination of smallpox was a great service to humankind.

1802 During widespread sufferings from tuberculosis, William Heberden, a London-born lecturer at Cambridge University, issued *Commentarii de Morborum Historia et Curatione* (*Commentaries on the History and Cure of Diseases*) (1802), the last major medical treatise composed in Latin. The text recognized the severe contagion of the disease in companions, particularly people who shared beds with consumptive patients. The most threatened were parturient women, who enjoyed a remission of symptoms before giving birth. Afterward, the disease quickly overcame them.

January 1, 1802 Only a third of the Omaha of northeastern Nebraska survived an epidemic of smallpox, which they encountered from trade with whites. Tribe members shuddered at the advance of the unknown scourge from lodge to lodge. Some were crazed by fever and fled to the prairie to perish alone. The faithful honored their 50-year-old Chief Blackbird and mounted his body on a horse under a burial mound atop a ridge overlooking the Missouri River. Braves suspended scalps that Blackbird had taken from a pole extending from the cairn, which became a Nebraska landmark known as Blackbird Hill. In their weakened state, the surviving Omaha were unable to fight off invading Sioux from the Dakotas.

On August 11, 1804, William Clark, one of the two leaders President Thomas Jefferson dispatched on the Lewis and Clark Expedition, wrote in his diary:

> We landed at the foot of the hill on which Black Bird I, the late King of the Mahar who Died 4 years ago, & 400 of his nation with the Smallpox was buried, and went up and fixed a white flag bound with Blue, white & red on the Grave, which was about 12 foot Base & circular, on the top of a Pinnacle about 300 foot above the water of the river. From the top of this hill may be Seen the bends or meandering of the river for 60 or 70 miles round.... Above the Bluff on this Creek the Mahars had the Smallpox & 400 of them Died 4 years ago [*Journals*, 1987].

Clark, who dated the epidemic to 1800, commented on Blackbird's equestrian burial, which had already become a frontier legend. Clark later remarked that the Omaha suffered so severely from disease that they abandoned ancestral lands.

spring 1802 When smallpox struck the Mississippi Territory, Governor William Charles Cole Claiborne took bold action to avert an epidemic. To save the town of Natchez, he initiated the territory's first mass vaccination against the pox.

mid–May 1802 In Santo Domingo during the struggle between Napoleon and black freedom fighter François Dominique Toussaint L'Ouverture, French sailors and occupational forces at Le Cap and Port-au-Prince suffered 40,000 deaths over an eight-month period. Stymied by loss of manpower, General Charles Victor Leclerc's army and the French navy under Rear Admiral Louis Villaret-Joyeuse incurred a decline in command as officers and enlisted men perished in a steady advance of contagion.

The aggressive infection struck new replacements shortly after they landed on the island. General Leclerc himself was felled by yellow fever, a disease linked to the Caribbean sugar trade. Although Napoleon dispatched additional troops, the collaboration of black rebels with British Redcoats drove the surviving 10 percent of French soldiers off the island. Loss of manpower also strapped the fleet of enough materiel to renew the siege.

1803 An outbreak of smallpox in Manger, Norway, resulted from peasant rejection of variolation, which was first performed in 1765. Of 3,500 residents, 208 died of smallpox. The locals held out against vaccination until 1829. Thirty years later, another epidemic killed 27.

1803 In an era of war and starvation, a year-long yellow fever epidemic in La Mancha, Spain, disclosed faulty leadership, which failed to reduce stagnant water or to improve public sanitation. Spanish peasants suffered a shortage of provisions and a lack of concern for basic citizen rights. Doctors tended to quibble over details of treatment and delay applying controversial new medicines.

The disease moved through much of Andalus, striking Cádiz, Écija, Gibraltar, Granada, and Seville. At Málaga, over one-third of the population died from the scourge. In contrast to other cities, Córdoba suffered fewer deaths because of municipal containment measures. In addition to a sanitary cordon and military supervision at the entrance to the city, health authorities cleaned and fumigated infected residences and sequestered visitors.

March 7, 1803 To ward off outbreaks of smallpox in Canada, a missionary, Père Jean Le Noir of the Abenaki village at Saint Francis, requested arm-to-arm vaccinations for himself and the natives. By the next year, many Canadian Indians sought preventive treatment and survived.

spring 1803 Measles returned to Japan, where it devastated the populace. The pathogen, introduced by the docking of a Korean vessel, possibly at Nagasaki, swept over the island chain. The scourge tended to kill those who contracted pneumonia or other complications.

summer 1803 At Ruggell, Liechtenstein, citizens and their livestock suffered epidemic anthrax. Debate as to the diagnosis and type of infection occupied the medical community for two decades. In 1822, Gebhard Schädler issued a report to the Graubünden medical society at Chur, Switzerland, proposing an accurate analysis of a pathogen that could infect both humans and animals.

September 1, 1803 During rampant smallpox in the Western Hemisphere, Charles IV of Spain organized an altruistic expedition to Asian and American settlers to vaccinate them free of charge. Sailing aboard the *María Pita*, his messengers, supervised by doctors Francisco Xavier de Balmis and Jóse Salvany Lleopart, bore the vaccine from Puerto Rico, the first port of call, through unmapped territory in Argentina, Bolivia, Chile, Colombia, Ecuador, Panama, Paraguay, Peru, and Venezuela. The medics set up tutorials on how to prepare and administer variolation, which local health authorities recorded.

1804 Mission La Purisima Concepcion, the 11th coastal facility in Father Junipero Serra's 21-part California mission chain, reached its peak occupation when disease began reducing the hapless population. Epidemic measles and smallpox inundated the 1,522 Chumash living on the grounds under the guidance of Father Mariano Payeras. According to cemetery records, over a period of three years, some 500 Indians received Christian interment. The worst of the epidemic occurred in 1806, when 220 died.

The phenomenon of epidemic pox impressed the Chumash with disturbing insights. As the disease culled the young and old from their number, the nation lost its repositories of cultural knowledge and its hope for the future. Survivors surmised that their leaders were losing power and withdrew into a lethargic state. They suffered an erosion of their belief system, experienced social disruption, and forgot how to conduct traditional rituals. Because the Franciscan fathers were immune to smallpox, the Chumash accepted Catholicism as a superior supernatural source of wellness.

1804 When West Indian yellow fever struck Cartagena, Spain, it produced prodigious numbers of sick. Over 9,000 patients crowded the Spanish naval hospital, which was designed to accommodate 4,000. At Ecija, Andalusia, Mexican botanist José María Mociño, a qualified physician, aided the High Council for Health in quelling the disease, which he described in *Memoir on the Yellow Fever* (ca. 1805). In Tuscany, Italy, arrival of a trader from Havana by way of Barcelona to Livorno spread infection from tainted water butts on board. Only 30 percent of 1,600 patients at Livorno and Lucca survived the disease.

1804 The first understanding of tuberculosis as a complex of diseases came from René Théophile Hyacinthe Laënnec of Quimper, Brittany. While treating an epidemic of tuberculosis cases in the Necker Hospital of Paris, he drew on the work of Baron Jean Nicolas Corvisart des Marets, Napoleon's personal physician, and of French army surgeon Gaspard Laurent Bayle. Laënnec, inventor of the stethoscope, published *De l'Auscultation Médiate, ou Traité du Diagnostic des Maladies du Coeur* (*Mediate Auscultation or a Treatise on the Diagnosis of Heart Disease*) (1818), which summarized his analysis of chest sounds. He contracted tuberculosis and, in 1826, fell dead at home at age 45 after devoting his last efforts to conquering the disease.

1804 Over a four-year period, an outsider to Cuba, Tomás Romay-Chacón, a Basque physician, campaigned among peasant communities and island health workers to extend inoculation against smallpox.

1804 The first great scourge to strike Hawaii in modern times occurred as Kamehameha I mobilized troops at his court at Waikiki to assail Kauai, the next stage in his plan of total island conquest. According to Samuel Manaiakalani Kamakau's *Ruling Chiefs of Hawaii* (1961), a prophet warned the king not to embark, but he continued on his way. At the staging area, soldiers came down with severe *ma'i oku'u* (squatting sickness), perhaps typhoidal enteritis or cholera, that fouled water and provisions. The ensuing epidemic was so virulent that victims perished within hours of the onset of sickness. Many of the king's advisers died; he also contracted the disease, but recovered. Because survivors lost their hair, natives called the disease *po'o-kole* (head stripped bare). In all, the Polynesians declined by over half or possibly as many as two-thirds.

May 20, 1804 After the introduction of inoculation in Mexico, Texas governor José Antonio Cavellero followed the orders of the Spanish King Charles IV in battling contagion. The governor ordered health care workers to

> set aside a hospital room in that capital, and another in each Province in your district, where the fluid will be kept fresh and communicated from arm to arm, to as many as may need it, free of charge if they are poor, the hospital staff periodically and constantly performing operations in rotation, a few at a time [Nixon 1946, p. 57].

By 1806, the reduction in outbreaks of smallpox proved the Spanish king correct in his methods of protecting colonial health.

June 4, 1804 When yellow fever broke out adjacent to the U.S. Navy Yard in Wallabout, New York, officials traced contagion to the *La Ruse*, a brig from the infected port of Guadaloupe. Several of the crew contracted the fever; the cook died of it. The ship jettisoned its bilge in the harbor, which spread disease to neighboring houses. Because of the severity of the epidemic, the harbor authority ordered the vessel and the schooner *Greyhound* into quarantine. The next year, enforcers of health regulations banned traffic of ships sailing from ports inundated with yellow fever.

August 28, 1804 At the British island of Gibraltar, disease was a constant problem because of a compromised water system, crowded residential areas, inadequate sanitation and garbage removal, and a constant stream of travelers and immigrants from the active ports. A four-month outbreak of yellow fever from a merchant named Santo attacked the colony and spread north over much of Andalusia, Spain. The civilian population lost 4,854 to disease. Among British military stationed on the west end of the Mediterranean, the epidemic killed 1,082. The disease flourished on the island over the next quarter century.

spring 1805 An epidemic of cerebral meningitis struck Geneva, Switzerland, killing 33. The authorities chose not to class the outbreak as a true epidemic. The following year, Gaspard Vieusseux, a native physician, reported on the occurrence in the *Journal de Medicien Chirurgie Pharmacie* and characterized symptoms as violent headache and abdominal pain, chills, vomiting, white tongue, vacant gaze, high fever, and purplish patches on the skin.

April 21, 1805 Because of endemic smallpox and yellow fever in Spain, a royal decree set up test hospitals and equipped them with inoculation rooms. Beginning at Seville and Barcelona, the nationwide program offered a sliding scale of fees for peasants and nobles. Nonetheless, the test hospitals found few people seeking vaccination because of the high death rate and bad reputation of institutional care.

November 1805 During the French occupation of Vienna, the overcrowding of hospitals with foreign troops spread dysentery and typhus. As noted in the *Memoirs of Baron de Marbot, Late Lieutenant-General in the French Army* (1892), written by an aide-de-camp to six of the general's marshals, after the battle of Austerlitz on December 2, typhus took more lives than combat. Some 12,000 wounded men died of typhus at Brünn, Czechoslovakia, where the epidemic infected civilians and advanced to Austria, East Prussia, Galicia, Hungary, Moravia, and Silesia. Following the battle of Jena in October 1806, soldiers bound their wounds in the field to avoid transport to an aid station in unsanitary ambulances.

1806 Along the Missouri River, George Clark and Meriwether Lewis witnessed the decimation of tribes that had no knowledge of treatment for European diseases. According to the explorers' journal, of a tribe of 700 Chinook, 400 died of smallpox, which reduced them to a frenzy. Survivors burned the village and killed their families in hopes of a better life after death.

1806 During a measles epidemic among California's Mission Indians, infirmaries tended up to 300 patients daily. At Mission Dolores, children living on the grounds died at the rate of 335 per thousand. In San Francisco, the rate was 880 per thousand, nearly obliterating the generation under ten years of age. Surprisingly, between 1776 and 1825, Catholic priests engaged only one medical doctor within the 21-mission chain.

1807 In gratitude for successful vaccination against smallpox, chiefs of the Iroquois League sent Edward Jenner a wampum belt and thank-you note. In recognition of his contribution to health, they promised to teach their children about his gift to humanity. He also received monetary rewards from the British government totaling £30,000 and appointment as the personal doctor for George IV.

1807 In Tahiti, dysentery spread among natives from contact with the crew of the whaler *Britannia*.

March 1807 The spread of typhus through both sides—38,000 French and 15,000 Prussians—impeded the Napoleonic campaign at Gdansk, Poland. The besieged city, tottering from sickness and starvation, gave in to French Marshal Pierre François-Joseph Lefebvre on April 27, leaving the way clear for Napoleon. As a reward, Lefebvre advanced to Duke of Danzig and commander of the imperial guard. *See also* **January 11, 1813.**

1808 Scots historical and romantic novelist Sir Walter Scott's *Memoirs of the Life of Sir Walter Scott* (1837–1838) speak of his personal experience with polio in early childhood:

> One night, I have been told, I showed great reluctance to be caught and put to bed, and after being chased about the room, was apprehended and consigned to my dormitory with some difficulty. It was the last time I was to show such personal agility. In the morning I was discovered to be affected with the fever which often accompanies the cutting of larger teeth. It held me three days. On the fourth, when they went to bathe me as usual, they discovered that I had lost the power of my right leg [Lockhart 1838, p. 14].

1808 In Gothenburg, Sweden, a small outbreak of poliomyelitis was the subject of writings by the local health officer, Christopher Carlander.

July 30, 1809 During the Napoleonic Wars, the British Navy entrusted 352 transports and 264 warships to the command of Viscount Robert Stewart Castlereagh to take Antwerp, Belgium, and boost the Austrians in their bid to overthrow Napoleon. Based at Walcheren Island, a French naval depot on the Scheldt River estuary on the North Sea, the expedition was an abysmal failure. Defeating the effort of 40,000 British soldiers was steamy weather, mediocre strategy, mosquito-infested swampland, and epidemic malaria. According to one rifleman, pestilence generated some ten or twelve burials at sea per day from his ship alone. In addition to the 217 slain in combat, 21,000 came down with "Walcheren fever," which killed 7,000. The First Division retained only 93 soldiers fit for duty.

Over the eight-week assault, British military physicians lacked the quinine needed to battle an epidemic. Dutch locals and religious volunteers augmented military fever hospitals, which were understaffed, unprepared, and jammed with the sick. Those men who returned to English bases quickly challenged facilities with more cases of malaria than the military could accommodate. Meanwhile, the French waited out the siege, allowing malaria to do their work for them.

The fiasco cost the British government £2,000,000 and much more in manpower. As of February 1, 1810, a total of 11,513 officers and men were still absent from active duty because of malaria. In 1812, Lord Wellington was so leery of depending on fever-weakened men that he rejected them from service. Because of the post-war undercurrent generated by George Canning, minister of foreign affairs, Castlereagh engaged in a duel at which Canning triumphed by wounding his opponent. Generally disliked and castigated as a failure, Castlereagh committed suicide in 1822.

1810 In Saint Louis, Missouri, Elizabeth Ann Bayley Seton, the first canonized citizen of the United States, opened the first U.S. Catholic hospital, where she treated lepers of all races as well as the elderly and deaf-mutes.

1812 So great an outbreak of fever ravaged the Monk's Mound brotherhood in Illinois for two years that it impeded the Cistercian brothers from their labors. Dom Urbain chose to abandon the mission and transfer his colony to Maryland.

October 19, 1812 General Napoleon Bonaparte ended his Russian campaign after 80,000 of his 450,000 men sickened from dysentery and typhus following the battle of Ostrowo, Poland. After finding the city of Moscow stripped of provisions and supplies, France's Grand Army had to retreat, leaving several thousand victims behind. In the melee, 62,000 Russians died of typhus. On the long, cold march west to France, the supply train suffered the loss of many pack horses and their load of food. Pursued by Cossacks, the men slogged their way home in worn boots and shredded uniforms and pillaged farmhouses for what they could steal.

Heading for Paris, Napoleon had to abandon 30,000 men at Vilna, Lithuania. Most of them died of typhus, frostbite, hunger, and exhaustion. In 2001, a chance find of a burial pit turned up 2,000 skeletons. Many were curled into fetal positions as they surrendered to disease and the elements. The survivors plodded westward, spreading contagion in Prussia and Germany. In all, only 6,000 French soldiers survived the combined onslaught of exposure, malnutrition, and disease. On the way, they spread typhus in Austria, Switzerland, and France.

1813 Italian surgeon Giovanni Battista Monteggia of Lavenea, a professor of anatomy at Milan's Great Hospital and author of the multi-volume *Istituzioni di Chirurgiche* (*The Profession of Surgery*) (1802), clarified the symptoms and permanent afflictions resulting from polio. The paralytic disease emerged in baffling epidemics throughout Europe, usually in summer.

1813 The Austrian military brought smallpox to Milan. Naturalist and surgeon Luigi Sacco, author of *Trattato di Vaccinazione* (*Treatise on Vaccination*) (1809) and a proponent of epidemiologist Edward Jenner, was skilled in cowpox variolation. He immunized himself and thousands of others with matter he took from patients in Vares, Lombardy. His method unfortunately spread erysipelas and syphilis.

Northwest of Venice at Rivalta, Italy, the tainted cowpox vaccination of 63 children spread syphilis to 70 percent. The disease killed some patients shortly after the procedure and infected parents and wet-nurses. When smallpox invaded Rome, Pope Pius VII authorized a general immunization program. A teaching hospital at Milan, Ospedale Luigi Sacco, and a clinic at the University of Milan preserve Sacco's name and honor his push for national inoculation.

January 1813 Physician Luis José Montaña observed epidemic spotted fever (typhus) in Mexico. The scourge drained the nation of money and food to aid the poor as 65,500 fell ill with the disease. A friend and colleague, Samuel Latham Mitchill, editor of *The Medical Repository*, America's first medical journal, published Montaña's findings. One outcome of widespread sickness and death was the swamping of a form of social security initiated in 1770, when tobacco factory management created a fund for workers. As a result of poor management and huge demands for loans and financial aid, the fund collapsed in November 1814.

January 11, 1813 A revival of the horrors of typhus that had ravaged Gdansk, Poland, in 1807 followed the retreat of Napoleon's men from Russia. Around 70 percent of French occupation forces came down with the disease during a siege by Prussians and their Russian allies. Sapped by cold and hunger, the French died in huge numbers, taking with them the local peasants they infected. By May, 11,000 infantrymen had fallen onto the roadsides and perished. In passing the corpses, the healthy hurried on their way to avoid contagion. As the French marched home, they quartered with civilians, spreading typhus among their caregivers. At Berlin, Brandenburg, Dresden, Frankfurt, Hamburg, Leipzig, Weimar, and Wittenberg, hospitals and lazarettes overflowed with wounded French, Prussian, and Russian soldiers, who were doubly threatened by scurvy and typhus.

At Torgau, Saxony, 30,000 soldiers died from epidemic dysentery and typhus, which overwhelmed the medical services of a town of 5,000 people. City officials adapted homes, churches, schools, and public buildings into hospitals to accommodate the sick and dying. Among the dead was Louis Marie Jacques Amalric, the Comte de Narbonne, commander in chief of the French national guard. The director of one of the hospitals was a pioneering homeopathic physician, Moritz Wilhelm Müller, author of *De Febre in Inflammatoria* (*On Fever in Inflammations*) (1810).

From Germany the still-healthy soldiers passed into the northeast corner of France and carried typhus and dysentery into Alsace, Burgundy, Champagne, and Lorraine. As marching men succumbed to infection, they quickly filled available hospital space. In November, a resurgence of the epidemic throughout the garrison and civilian population raised the total of patients to alarming numbers. In Metz, 60 troopers perished each day. At Freiburg, sick men lay shoulder-to-shoulder on foul straw without bedding or nightshirts. As military enforcers evacuated the sick from Trier by boat for transport down the Moselle and Saar rivers, over Germany, the death count rose to 300,000. *See also* **November 1, 1813**.

July 1813 Around 50 expired per day in Malta during an epidemic of plague imported by sea vessels. The outbreak forced the evacuation of Manderaggio outside Valletta. In all, the island lost approximately 4,500 or five percent of its population. To treat the massive number of sick, officials adapted the Villa Bighi into a plague hospital directed by physician Gio Batta Saydon and Luigi Pisani, the island's medical practitioner for the poor. Another health professional, Agostino Naude, recorded particulars of the infection in Latin.

The disease, called *il Pesta i Kbira* (the Great Plague) moved to smaller parts of the island cluster before reaching the Ionian islands. Municipal outlay strapped finances as islanders required more prevention and care. For example, the cost of fumigating mail arriving to the island raised the price of stamps. At the end of the scourge, in an act of thanksgiving, the devout erected statues of Saint Sebastian, the Virgin Mary, and Saint George, the intercessor for Gozo Island. In 2001, film director Salvu Mallia re-enacted the era in a made-for-TV film, *Pesta 1813*, starring actors and extras from the island.

November 1, 1813 Near the end of Napoleon's wars, among 30,000 French soldiers quartered at Mainz, Germany, typhus infiltrated 18,000 victims comprised of both the military and civilian populations. Inadequate hospital staff left the sick unwashed and untended on straw pallets. As medical workers and gravediggers died, survivors piled corpses outside the city. The epidemic worsened over six months. On the homeward march, soldiers spread the lice-borne pathogen into the countryside. *See also* **January 1814**.

late winter 1813 During the War of 1812, the impact of shifts in commerce and loss to disease affected life in New England and into Canada. Across the Northeast, John Jacob Astor's fur trading cartel lost profits after native trappers contracted smallpox. At Burlington, Vermont, war profiteering on provisions for some 4,000 soldiers boosted trade with Canada, but citizens failed to support the war effort despite the encouragement of the weekly newspaper, the Burlington *Sentinel*. Residents despised troop rowdyism and stealing and regretted the late-winter epidemic of pneumonia. At war's end, long-suffering New Englanders cheered at Thomas Macdonough's triumph at the battle of Plattsburgh Bay, New York, the official end of the Redcoat menace.

1814 Because Swedish health authorities feared an outbreak of smallpox, they crusaded for general inoculation in Uppsala. Assuring the success of preventive measures were the clergy and parish clerks of the Church of Sweden, some of whom performed the inoculations personally on congregants.

1814 When the convict transport *Surrey* left England on February 22 with 201 prisoners to colonize New South Wales, epidemic typhoid struck crew and passengers, killing the captain, surgeon general, both mates, and two guards. Without a navigator, the notorious "plague ship" drifted off the Australian shore until one sailor aboard the *Roxberry* volunteered to board the ship. On July 28, the *Surrey* docked at a Sydney quarantine station, the first authorized in Australia. The cemetery on Jeffrey Street received the area's first European burials. An investigation convinced naval authorities that the epidemic resulted from the captain's negligence in requiring cleanliness in the convict quarters, where only 164 survived the pestilence. Subsequent convict transports carried naval surgeons to suppress pestilence.

January 1814 As German forces pursued Napoleon's retreating men, they spread typhus, encouraging an epidemic that sickened 30,000 and killed thousands. The April death toll stood at 9,100. Paris hospitals reported 7,600 dead by June. Military evacuation plans called for riverboat transport down the Loire and Seine to contagion wards at Rouen and Tours.

1815 A spring outbreak of cerebrospinal meningitis reached Liguria, perhaps from trading vessels docking in the Gulf of Genoa, and killed 70 percent of victims in the isolated coastal town of Albenga, Italy, and a cluster of surrounding villages.

1815 Thomas Young, an English doctor, surmised that 25 percent of Europeans were tubercular. Some fifteen years later, 30 percent of England's laboring class succumbed to consumption. In this period, a much-studied case of the tubercular family resulted from the deaths of the wife and children of the Rev. Patrick Brontë of Haworth, a Yorkshire clergyman who apparently infected the family from his chronic cough. Although his wife Maria apparently died from puerperal fever, her depleted state suggested to medical historians a chronic illness from tuberculosis. One by one, Brontë's children sickened with the disease: 12-year-old Maria in May 1825, 11-year-old Elizabeth the following month, Branwell and Emily in 1848, and Anne in 1849. Charlotte Brontë's death in 1855 at age 38 raises the question of her own depletion from the family curse of tuberculosis.

1815 At the end of Anglo-Nepalese War, smallpox and chicken pox epidemics invaded a virgin population in western Nepal, killing 17-year-old Raja Girvana Yuddha Vikrama on November 26, 1816. After the prime minister committed suicide, the regency passed to the raja's three-year-old son.

1815 On the shores of Sierra Leone, where yellow fever was endemic, the disease targeted European teachers, soldiers, sailors, and missionaries. Over the next nine years, the disease decimated Freetown with suffering from jaundice, body aches, and black vomit. Nearly 60 percent of white residents died, but most of the native Temne survived. Medical historians considered Sierra Leone the home of yellow fever in West Africa. When a 90-member back-to-Africa pilgrimage reached the docks aboard the ship *Elizabeth* in spring 1820, fever attacked the black Americans, destroying their settlement.

1816 The first pandemic of cholera began in Bengal (modern Bangladesh), India, and, in one decade, advanced on China, Japan, and the Caspian Sea. In 1821, English soldiers transported the disease for the first time beyond the borders of Hindustan when they departed from Bombay to the Arabian Peninsula. In the fifth year, cholera overran the Chinese at Canton, Ningpo, Wenchow, and the Yangtze valley. In the seventh year, it moved north to Peking and west along trade routes to Russia. By the winter of 1823–1824, the scourge waned in southern Russia before it reached Europe.

When disease struck eastern Galicia, Rabbi Tsvi Hirsh of Zydaczów, Ukraine, an influential *tsadik* (model of righteousness) since the previous century, applied Hasidic piety to the question of treatment. In opposition to the Austrians and the Jewish *maskilim* (intellectuals), he defied rationalism by rejecting modern science. To the Ukrainian Jews of Munkács stricken with cholera, he urged rejection of Austrian physicians in favor of trust in God, the fount of divine grace.

1816 Widespread hunger, unemployment, and homelessness contributed to a three-year outbreak of smallpox in England, Scotland, and Ireland. Hardest hit were peasant children who had not been vaccinated by the Jenner method. The disease recurred in 1837, 1871, and 1901 before mandatory inoculation and stronger public health measures controlled the disease. Simultaneously, the British Isles suffered an outbreak of bubonic plague and a typhus epidemic that some medical historians identify as relapsing fever or typhoid.

In 1840, a lengthy discussion of predisposition involved epidemiologists and sanitarians of the medical community. The focus of their debate was the effects of natural susceptibility in generating disease. William Pulteney Alison, a proponent of social medicine, named poverty as a contributing factor in the 1817–19 Irish fever epidemic. Opposing Alison's argument was public health pioneer Edwin Chadwick, who contended that one cause triggered epidemics. He cited as corroboration Neil Arnott and Thomas Southwood Smith's 1837–38 Reports to the Poor Law Commission.

1816 Among Tuscans, a two-year epidemic of typhus coincided with a post–Napoleonic famine resulting from war. Holy Roman Emperor Ferdinand III promoted the planting of potatoes in Italy to relieve hunger and set up temporary pesthouses. He initiated public works to provide economic relief to the underclass, whom unemployment forced into a nomadic existence that spread disease. The scourge advanced on Croatia and south and east to Venice, Trieste, and Ancona. Within a year, the disease surfaced as far south and west as Florence, Rome, and Naples and into Ragusa, Sicily, infesting naval crews that carried infectious body lice about the western Mediterranean.

1816 Following Napoleon's triumphs in Spain in 1808, the nation lost most of its New World holdings, except for Cuba, the Philippines, and Puerto Rico. After General Pablo Morillo's army restored New Granada (modern Colombia, Ecuador, Panama, and Venezuela) to Spanish control, he lost 3,000, or one-third of his men to pandemic dysentery, malaria, smallpox, and yellow fever. Rebel leader Simón Bolívar concluded that Morillo's troops lacked the natural immunity that protected local Latinos.

January 1816 Following a failed potato harvest, the return of dysentery and typhus to Ireland struck one-quarter of the populace, killing a total of 110,000. As the poor migrated across the land in search of work, medical treatment, and charity, they spread disease at hospices and inns. Citizens nailed up fever sheds to receive those vagrants with nowhere to turn for help.

1817 The Jutland syphiloid received its first public notice in Denmark after spreading for over 65 years following infection by the Russian military throughout Norway and Sweden. Officials connected the outbreak to laborers building the Crown Prince Dyke and to a pandemic in the marshlands that afflicted whole villages. Author Isak Dinesen used the location and disease in a popular religious fable, *Babette's Feast* (1959), which was filmed in 1987, starring Stéphane Audran as Babette Hersant.

1817 In an effort to regain control of Spanish colonies, King Ferdinand VII amassed a battleship, brig, and two frigates into an expeditionary convoy assembled at Cádiz, Spain, which he postponed because of an outbreak of yellow fever introduced by returning colonial soldiers. Although the navy and army mustered troops and bought the ships from Russia to carry the assembly to the Río de Plata in Argentina and Uruguay, a rebellion planned by disgruntled crew interrupted the planned voyage, which the king cancelled on January 1, 1820.

August 1817 A six-year pandemic of Asiatic cholera involved most of the continent of Asia during heavy colonizing activity by the British. The first of three major cholera pandemics, the disease started outside Purneah, India, the previous year and spread to Jessore and Calcutta, killing thousands of Bengalis and hundreds of British occupation troops. The disease advanced into the Ganges basin, killing some victims in less than 24 hours. Contagion fanned out in all directions, traveling over Ceylon, Nepal, Southeast Asia, Japan, Java, Russia, and Afghanistan. At Oman on the southern tip of the Arabian Peninsula, British forces infected slavers, who bore the disease to Zanzibar. *See* **March 1818.**

late 1817 Exploitation of German and Swiss immigrants fleeing post–Napoleonic Europe gave Dutch ship owners an opportunity for price gauging and shortchanging passengers aboard filthy, overcrowded transatlantic vessels. According to the diary of Dirk Cornelis de Groot, captain of the three-master *April*, 400 or 57 percent of 700 passengers under his command died of a shipboard typhus epidemic on their way from Amsterdam, Holland, to New Castle, Pennsylvania. After the ship's docking on January 3, 1818, the U.S. Congress legislated the first health and sanitation regulations for passenger ships.

1818 A diphtheria outbreak at Tours, France, intrigued French physician Pierre-Fidèle Bretonneau, who disproved notions that the disease was either gangrenous scurvy or the second stage of bronchitis. Imported by French soldiers, symptoms included malodorous breath, ulcers and grayish-green patches on the face, inflamed gums, and swollen glands. From studies of civilian and military corpses, he discovered the membranous obstruction to the throat and deduced that the disease was contagious. He named it *diphtheritis* from the Greek for "leather disease" and countered the suffocating effects of throat obstruction by becoming the first physician to perform a tracheotomy.

1818 Following the death of Dey Ali Khoji VI at Algeria from plague, merchants or voyagers carried the pestilence to Tunisia, which suffered the loss of one-quarter of the citizenry and its place among Mediterranean power wielders. Continuing into 1820, it was the last great outbreak of the disease on Africa's northern coast. Muslims used the scourge and accompanying famine as an opportunity to proselytize Jews. After the execution of a Jew, the worsening of infection caused Muslims to interpret the disease as Allah's disapproval of burning Jews at the stake.

ca. 1818 Norwegian health officials combatted rampant puerperal fever by building maternity hospitals at Christiana and Bergen. The purpose of these facilities was the training of midwives in aseptic examination and delivery techniques.

March 1818 Against a backdrop of flooding and hunger, epidemic cholera, held in check during the previous winter, struck Allahabad, India. The pestilence ranged into the heartland and northward as well as south to the island of Ceylon. British occupation troops infected the enemy in Afghanistan and Nepal.

summer 1818 Endemic syphilis in Inner Carniola and Kras, Slovenia, puzzled doctors, who thought they were observing a new disease. One theory claimed that the pathogen derived from syphilis complicated by leprosy, scabies, or tuberculosis. Spread by asexual, extragenital contamination among the unhygienic, illiterate poor by Turkish soldiers and merchants, the disease suggested a hybrid of syphilis and another form of contagion. Under Baron Freiherr von Stifft, a Viennese physician and politician, the government required medical exams of residents of infected communities and made compulsory a regimen of domestic decontamination and treatment with mercury at a special hospital in Postojna. Within a decade, the outbreak ebbed and the Postojna hospital closed, in part because civil authorities and the clergy upgraded living standards and understanding among peasants.

1819 To curtail the smallpox outbreaks that killed off indigenous peoples, Archbishop Pedro José Fonte Hernández, the last European archbishop in Mexico, campaigned to the northeast for vaccination among the Huastec Indians living east of the Sierra Madre. In a well orchestrated crusade that he financed personally, he offered rewards to willing patients.

1819 Upon arrival in Hawaii, explorer Captain Louis Claude Desaulses de Freycinet, a physician and author of *Voyage autour du Monde* (*Voyage around the World*) (1839), recorded ophthalmia and widespread scabies:

> Men, women, children, from the poorest to the most sovereign chiefs, none seemed to be exempt; and pestilence of the same nature even sullied the skin of several Europeans who had been living for a long time among them…. I do not believe it to be contagious, and it does not seem to be caught except through prolonged cohabitation with an infected person [Bushnell, 1993, p. 231].

In explanation of the spread of infectious disease, Jacques Arago, an artist who accompanied the expedition, remarked on the promiscuity of native women in Honolulu. He declared that they sold their bodies for "a handkerchief, a necklace of glass beads, one or two shining buttons, or similar trifles" (ibid., p. 189).

1819 After yellow fever emerged in southern Spain at Cadiz, researchers linked it to a merchantman carrying sugar cane from Brazil. Within two years, soldiers, Jewish peddlers, fishermen, and travelers spread the scourge to Jerez, Minorca, Seville, and Barcelona as well as Martinique, Guadaloupe, Jamaica, Cuba, and Cayenne, French Guiana. Medical observers generated insights describing the hardships of peasants who were still recovering from an era of famine.

April 1819 Aboard the French slaver *La Rodeur*, an outbreak of ophthalmia (trachoma) produced temporary blindness in bondsmen and crew. The captain thought he could save money on his financial loss by shackling the ailing slaves and drowning them in the sea. After murdering 39 Africans, he encountered the Spanish slaver *Leon*, on which the whole crew was blinded by shipboard ophthalmia. On arrival at Guadaloupe on June 21, the captain of the *La Rodeur* and all his crew except one had developed ophthalmia. To express the disgust of abolitionists, Quaker poet John Greenleaf Whittier published the dramatic poem "The Slave Ship" in the October 11, 1834, issue of *The Liberator*, the powerful emancipationist journal of William Lloyd Garrison.

winter 1819 During the "little smallpox winter," measles crossed the northern plains states by Northwest Company canoes as far north as the Mackenzie River Valley in Canada's Northwest Territories. During the wild rice harvest on the Winnipeg River, the virus and starvation killed many Lac Seul Ojibwa, who were too weak to hunt or gather. The death of two-thirds of the Ojibwa left only 76 at Lac Seul. To the south in the United States, the "great red skin" disease reached the Menominee the following year along the Saint Croix River in Wisconsin. Crazed by the failure of healing songs and chanting shamans, the Indians experienced social and cultural disintegration. By 1821, French missionaries successfully enforced quarantine among diseased natives.

winter 1819 Epidemic scurvy threatened the frontier soldiers at Cantonment, Missouri. The first recorded epidemic in the state of Nebraska, it struck one-tenth of the roster. Officers tended to ignore the degeneration of strength and health among the ranks, who fed daily on individual ratios of three-quarter pound of salted, dried, or smoked beef or pork per man. Amid inadequate disposal of sewage and garbage, by January 1820, the men incurred dysentery and respiratory ills, which worsened ulcerated gums, edema, fetid breath and urine, swollen joints, and bloat. Few survived isolation in the three-room barrack hospital; 20 died during their conveyance by keelboat through pack ice down the Missouri River to Fort Osage. By March 10, 1820, the death rate reached 100 of 345 patients. Survivors fed on wild onions and raised corn, potatoes, and turnips in their garden.

1820 Daniel Williams Harmon, Vermont-born explorer and fur trader for the Northwest Company and author of *A Journal of Voyages and Travels in the Interior of North America* (1820), shed light on the high death rate of Native Americans from epidemics. While on a 19-year expedition between Montreal and the Pacific Ocean over Saskatchewan, Manitoba, Alberta, and British Columbia, he deduced that the Cree worsened the danger of disease by resorting to sweat baths, a treatment involving a sweat lodge or sauna tent heated by a small fire and moistened with steam emitting an herbal essence.

1820 When yellow fever invaded Savannah, Georgia, it sickened over 25 percent of the populace. A bold physician, William Coffee Daniell, recognized the lack of worthy cures and urged local doctors to abandon mercury and purgatives, which did more harm than good. In response to the upsurge of yellow fever in the eastern and southeastern United States, Nicolas Chervin, a French physician and anti-contagionist, set out on a four-year trek to observe local response to the mysterious disease. From New Orleans he traveled to Charleston, Norfolk, Washington, D.C., Baltimore, Philadelphia, New York, Boston, and Portland, Maine, collecting data via a questionnaire. He used the data to impress on his European contemporaries that the fever was not contagious.

1820 When a virulent strain of smallpox struck Piedmont, Sardinia, the government distributed free inoculations. As a result, the infection rate for the area was only 1 percent, with a small death toll. The test of vaccination during an epidemic proved to doubters that inoculation was worth the risk.

1820 Settlers in Michigan created their own miasmas that spread ague, their term for malaria. By digging millponds and plowing fields flat with no drain lines, farmers encouraged the propagation of disease-bearing mosquitoes. Because roughly half the pioneers were felled by fever, many mothers died in childbirth. Over the next decades, they developed methods of improving their investment by screening windows and draining croplands.

1820 In northern Auckland, New Zealand, an influenza epidemic struck the Thames or Waihou River region and spread with Maori troops as they marched to the south. The scourge wiped out whole villages.

March 1820 As British imperialism gripped Asia following the Napoleonic wars, pandemic cholera at Songkla, Siam, derived from Penang in the Malay Peninsula. After a warm April, the scourge gradually advanced on Vietnam and Manila in the Philippines. By May, cholera killed one-fifth or 30,000 of the residents of Bangkok. River-borne corpses spread contagion in drinking water.

King Rama II ordered the firing of weapons to deter evil and called for a return to piety, including the recitation of holy verses, donations to mendicant monks, and city processions by land and water from the Emerald Buddha and sacred relics in Wat Phra Keo. Refuting peasant notions of demonic malice, the king concluded that the pathogen originated in poisonous volcanic gas. He ordered the collection of herbal medicine formulas and traditional remedies for disease at Wat Chetuphon and also released the royal guard from service, opened prisons, and initiated a national fast. As a result of the most terrifying epidemic in the nation's history, citizens welcomed Western medical practice.

September 1820 In the Philippines, a six-month outbreak of cholera erupted in Manila. Among the sick was Carlos Luis Benoit, a physician with the national medical corps. The high rate of infection required a cart service to remove corpses. In the Paco district, citizens constructed an octagonal cemetery to accommodate the dead and placed statues of deceased children in niches.

Natives suspected that foreigners who enriched themselves on colonial trade poisoned the water to rid the islands of its indigenous people. As a result, on October 9, angry Indio hordes assailed Europeans at Binondo, Cavite, Manila, and Tondo, slaying 27. To restore control, the governor placed artillery at the plaza opposite Binondo Bridge in Manila. When rioting broke out the next day against the Chinese, Archbishop Zulaybar of Manila displayed sacramental elements, but made no inroads against frenzied agitators. When the violence ended, city authorities took seriously the need for a sanitary waterworks.

1821 The onslaught of putrid fever or diphtheria in Mariefred, Sweden, resulted from a mystifying contagion. According to the treatises of city surgeon Johan Gabriel Collin, the source was a miasma arising in the bogs. He advocated training the populace in hygiene and moderation.

1821 The introduction of cholera in Persia proved immediately lethal to victims. Carried by British forces traveling to Oman by way of Bombay, India, the disease killed 10,000 at Muscat and around 18,000 at Basra. Contagion spread across the nation, reaching the Caspian Sea by the next year. At Baghdad, Iraq, General Muhammad 'Ali Mirza died of cholera along with a portion of his army. Another outbreak among the military struck Persian troops as they assaulted Turks at Khoi, Iran. At Tauris (modern Tabriz, Iran), the disease killed 5,000. Contagion spread to Astrakhan, Russia, and to Syria via cameleers journeying from the Persian Gulf to Aleppo in November 1822. By mid–1823, the disease advanced from Antioch and Laodicea, Turkey, north to the Caspian Sea.

April 1821 In Indonesia, widespread Asiatic cholera in Java emerged at the harbor of Semarang. Over eight months, the outbreak resulted in 125,000 fatalities. The contagion reached Borneo and Sumatra to a lesser degree.

summer 1821 Arriving aboard the *Gran Turco* from Cuba, yellow fever hit spread from the port and docks to peasants of Barcelona, Spain. Health authorities failed to take positive action as disease advanced into the fall months, killing around 20,000 or one-sixth of the citizenry and migrating to the Balearics and the coastal Mediterranean towns of Tarragona and Tortosa. A border quarantine and the scuttling of infected ships preceded a state of martial law, but control of riots and looting was ineffective. Famine worsened the situation for those who were trapped in the city. Aiding them were nuns, volunteers operating a soup kitchen, and a five-member medical team including an expert, Jean Andre Rochoux, author of *Recherches sur le Fievre Jaune* (*Research on Yellow Fever*) (1820). Survivors doubted that the city would recover.

late July 1821 The spread of Asiatic cholera from China reached Korea, killing 1,000 at Pyongyang within ten days. The superstitious placated angry gods with ritual and feasting. Their piety appeared to control the disease during the winter months, but a spring recrudescence revived panic as the disease moved toward Japan.

1822 After the start of the Industrial Revolution, increased smog, malnutrition, and child labor in European factories encouraged rampant rickets, a skeletal disease resulting from inadequate vitamin D. A Polish doctor observed that children in Warsaw incurred rickets, but rural children remained healthy and ricket-free. Diseased toddlers retained the soft, pliant frame of infancy and were slow in acquiring motor skills. They developed bowed legs, knock-knees, and scoliosis. Some suffered seizures severe enough to kill them. The doctor discovered that sunbathing provided relief from the anomaly.

1823 A cholera outbreak in Astrakhan, Russia, along the Volga River infected 392 people in two months, killing over 52 percent.

August 22, 1823 A month-long yellow fever epidemic afflicted Brooklyn, New York, allegedly brought by the ship *Diana* from New Orleans, or the brig *Trio* and spread by consumption of foul-smelling fish sold in local stores. In one city block of Furman Street facing the East River, health officials reported 19 cases. Among the ten people killed by the contagion was John Wells, a survivor of an Indian massacre during the American Revolution.

November 1823 Measles re-emerged in Japan, moving from the western shores to Tokyo by the next calendar year.

1824 At Jessore, Bengal (modern Bangladesh), 75,000 Indians died of leishmaniasis, which locals called *kala-azar*. The epidemic yielded the first data on the infection, a protozoan wasting disease caused by the bite of pathogen-bearing sandflies and inflicting a 95 percent mortality on victims. Because leishmaniasis resembled malaria, it went undiagnosed for decades.

1825 A renewed spread of smallpox over England engulfed the country with illness. Increased admissions to the London Smallpox Hospital, Europe's first smallpox institution, founded in 1745, nearly equaled the total number of sick from the 1750s. The poor, who were least likely to be inoculated, suffered the highest fatality rate.

1825 According to the collected writings of an Italian physician, Cesare Bressa, completed in 1825, beriberi was a serious problem among slaves of the Louisiana planters living on the Mississippi Delta outside New Orleans. The result of thiamine deficiency, the disease resulted from a diet of cornmeal and salt meat. Symptoms ranged from torso swelling, bowed limbs, and skin lesions to blindness and convulsions.

mid–February 1825 While transporting the bodies of Hawaii's King Kamehameha II and Queen Kamamalu, who had died the previous winter of measles, Captain George Anson Byron, commander of the H.M.S. *Blonde*, encountered an outbreak of smallpox aboard ship. He was forced to abandon his original route and retreat to Valparaiso, Chile, to seek help for his crew. His intent was to spare Hawaiians an introduction to smallpox.

1826 The second great Asiatic cholera pandemic, which originated along the Ganges River Delta in Bengal (modern Bangladesh), India, raged over Asia and Europe for 11 years. It spread to the Punjab and Gujarat and over trade routes to Kabul, Afghanistan, and, within a year, into Bukhara, Russia. At Banaras, up to 300 Indians died each day. In summer, 172 soldiers stationed at Kanpur perished. After a late-summer lull, the disease renewed its virulence in early winter at Agra, Delhi, and Mathura. *See also* **early 1827**.

1826 The First Burma War resulted from Burma's attempt to annex Bengal (modern Bangladesh). The Rangoon army of 11,500 sepoys and British regulars, led by Sir Archibald Campbell, lost 75 percent of its manpower to dysentery, malaria, and scurvy. General Joseph Morrison's army of 11,000 suffered 1,495 deaths from dysentery and malaria brought on by heavy rain. At war's end, the combined Anglo-Indian force lost 15,000, of which 600 were combat deaths and 14,400 were fatalities from epidemic disease.

1826 In New Zealand, the arrival of the H.M.S. *Coromandel* from Sydney over the Tasman Sea to the Bay of Islands introduced pandemic influenza. On a war mission to Akaroa, two chiefs, Towiwi and Totoes, died of the virus. The Maori, a virgin-soil population, reverted to wrapping the head in leaves, heating the body in sweat huts, and plunging into cold streams to reduce fever. Sealer John Boultbee estimated that pestilence killed as many as 80 percent of natives and turned Okahu into a ghost town.

January 1, 1826 A three-year pandemic of dengue fever spread from Savannah, Georgia, down the southeastern Atlantic seaboard through

Charleston, South Carolina, and Pensacola, Florida, to New Orleans on the Gulf Coast and into the American Virgin Islands, Bermuda, Cuba, and Jamaica. The disease progressed south through the Caribbean, striking the French islands, Barbados, and Tobago and advancing to Curaçao north of Venezuela and to Cartagena, Colombia, and Veracruz, Mexico. During the spread of disease, Spanish health workers named the fever from the Swahili term *dinga* or *dyenga*, referring to a cramp from the surprise attack of an evil spirit. The next year, Samuel Henry Dickson of Charleston, a tropical disease specialist and founder of the Medical College of South Carolina, issued *On Dengue: Its History, Pathology, and Treatment* (1839), a report on the intense pain, skin eruptions, and fever caused by the disease. He noted that few victims needed medication or died from the contagion.

1827 A smallpox epidemic swept northern Mexico and greatly reduced the population of Alta California's Mission Indians. Numbers are conjectural because no one kept a record of their deaths. At Santa Barbara, measles contributed to the loss of life. Confusing the history of the era's sufferings is the *Personal Narrative of James O. Pattie of Kentucky: The True Wild West of New Mexico and California* (1831), composed by fur trapper James Ohio Pattie. He claimed to have petitioned the territorial governor for permission to vaccinate 22,000 Indians against disease on a route that took him from San Diego to Fort Ross north of San Francisco.

early 1827 A recurrent juggernaut of cholera emerged in the Ganges basin and reached into Chittagong and Calcutta, India, and east to Moulmein, Burma. By late spring, medical authorities reported worsening scenarios at Agra and Delhi and sickness as far north as the lower Himalayan Mountains and west into Sind (modern Pakistan), Afghanistan, and Persia.

1828 During the Greek Revolution, Louis-André Gosse, a visiting Swiss doctor from Geneva, became the Greek naval commander, state medical officer, and supervisor of sanitary measures during a bout of plague. Introduced by the Egyptian army during a prisoner exchange, the pestilence arrived on the island of Aegina on September 16 from Navarin, France, with the first 5,500 soldiers. The pathogen flourished among starving Greek slave women and children at Methoni on the Peloponnesus. The ruins of Ottoman fortresses contained putrid waste in their environs that had caused fever and plague among the Turks for the past seven years. By mucking out ditches, the French set standards of hygiene that impeded the scourge. Meanwhile, disease-ridden passengers aboard a ship from Constantinople spread plague to Damietta on the Nile Delta and to Alexandria, Egypt, and Beirut, Lebanon.

September 1828 The initial assault of pertussis among New Zealanders devastated infants, children, and the elderly. Almost all children at Anglican missions incurred the dire sickness and secondary pneumonia, influenza, and respiratory illness, which closed schools. After thriving for a year at Port Jackson and Sydney, Australia, the disease ranged north to the port of the Bay of Islands, probably spread from a foreign ship. Infection overtook both indigenous Maori and Europeans and raged into October. Consumption ravaged Maori females, leaving them skeletal and weak.

1829 At Astoria, Oregon, Northwest Company trader Duncan McDougal, "the great smallpox chief," brandished a vial of liquid that he claimed contained the disease. His bluff terrified chiefs of the Chinook, a powerful nation that lived along the Columbia River near Seattle, Washington, as far south as the Willamette Valley. After the pox and "ague fever" (malaria) engulfed up to 90 percent of the tribe by summer, the Chehali, Salish, and Tillamook absorbed survivors into their villages. By the time that waves of pioneers arrived over the Oregon Trail in 1841, few Pacific coast Indians could rally a protest.

1829 Malaria invaded the warriors involved in the Spanish expedition from Cuba to Mexico. Dispatched by Spain's King Ferdinand VII, General Isidro Barradas led 3,000 insurgents on a fleet of 15 ships. Traveling to Tampico, Mexico, the Spanish army of 7,000 faced the troops of General Antonio López de Santa Anna, who demanded surrender from the suffering Spaniards. On its last attempt at New World conquest, Spain lost 215 to battle and 1,500 to malaria.

February 1829 Over ten months of combat in the Balkans and the Caucasus during the Russo-Turkish War, Russia lost 12,857 men in the field and 30,000 more from dysentery, malaria, fever, and plague, especially in winter quarters at Jassy and Moldau, Romania. More than 210,000 of the infected languished in lazarettes and regimental sick bays. By March, a new upsurge of disease sapped the military south of the Danube and in urban sections of Bulgaria and Moldavia. In May, a tent city at Varna, Bulgaria, sheltered the ailing infantry, who died at the rate of 100 per day. Of 28 physicians, only 13 remained healthy. During the aftermath, another 96,722 Russian soldiers succumbed to contagion, contrasted with 80,000 fatalities among the Ottoman Turks.

February 1829 Multiple diseases, notably malaria, slew around 150,000 Indians of the Pacific Northwest after New England Captain John Dominis sailed the brig *Owhyhee* from the Juan Fernández Islands off Chile to Oregon. The ship, which was the first oceangoing vessel to enter the Willamette River, arrived with mosquito larvae in the water butts carrying malarial protozoa. Stagnant pools created by the overflowing Columbia River gave pathogen-bearing insects a new breeding ground for an epidemic that lasted over four years.

First to suffer the "cold sick" disease, the Multnomah of the Willamette River Valley virtually disappeared by summer, as did the Clackamas, who lost 90 percent of their number. After the Indians had helped free the *Owhyhee* when it went aground at Deer Island, they contracted disease from close contact with sailors sick with fever and ague. The indigenous people suspected Dominis of deliberately infecting them. More Indians died at Fort Vancouver, Washington, and on the Lower Columbia River and Klamath Lakes. Along with measles and smallpox, malaria prevailed over the next three years, striking the Kootenay and Thompson people near the Multnomah and spreading up the seacoast to the Nootka and Salish, killing up to 95 percent in each outbreak. *See also* **August 1830.**

August 1829 According to the Sydney, Australia, *Gazette,* when the *America* docked at New Holland, Papua New Guinea, harbor authorities quarantined the ship because of an outbreak of measles among 169 male prisoners. The disease killed seven inmates and one guard, but did not spread to the island's virgin soil population.

August 26, 1829 Cholera that overran India spread into Orenburg, Russia, in the Ural Mountains. Some 3,100 died in Moscow. By fall, nearly 56 percent of 68,091 cases died in Russia as the unusually lethal contagion moved west to Poland and south to Tehran, Iran, killing princes in Herat, Afghanistan.

During the war in the Balkans, Caucasus, and Transcaucasia, health authorities cordoned off and quarantined Sevastopol, Russia, before the disease arrived. The stringency of house arrest and resultant hoarding of food by health authorities produced hunger and deaths from exposure and privation of food, fuel, and clothing. In June 1830, female citizens and sailors rebelled against sequestration. The revolt, which resulted in the murder of Governor Nikolai A. Stolypin and other enforcers, netted seven executions, sentences of hard labor against 1,000 insurrectionists, and 4,200 deportations.

October 1829 An influenza pandemic begun in Guangzhou, China, spread to the Philippines, Borneo, Sumatra, Java, and Japan. When it emerged in Europe, it appears to have traveled over Siberia and struck Moscow in November 1830, involving the entire European continent before journeying to the Americas. Flu reached Australia on Macassan fishing boats from Indonesia along the north shore and beset virgin-soil territory. Borne by the Murray River expedition of Captain Charles Sturt, father of Australian exploration, the disease moved from Raffles Bay and into Queensland and the Murray-Darling basin. Combined with measles and smallpox, the disease killed the island's aborigines, leaving acreage open for squatters to settle.

1830 When physician George Bennett of Plymouth, Massachusetts, visited Rótuma in the Society Islands of the South Pacific, he discovered endemic dysentery, which killed many islanders each year. Sickness so troubled Chief Ufangnot of Saflé that he offered his visitors inducements to stay and set up a medical practice:

> You stay at Rótuma, make people well, as too many people die, and you have made some well, and know how to cure all people, you will have plenty wife, plenty yam and pig, plenty land, and be all the same as one king [Bennett 1831, p. 478].

Bennett also diagnosed rampant purulent ophthalmia or trachoma in infants, which mothers refused to treat with the lotions he provided. Folk cures involved vigorous massage with coconut oil, cutting, or roasting over a slow fire.

1830 The *Messenger of Peace*, a schooner dispatched by the London Missionary Society, introduced influenza to a virgin-soil population in Samoa. The least likely to survive were the elderly and patients weakened by lung disease. Some expired long after the initial flu epidemic. Islanders blamed the Congregationalist missionaries and the new religion for a disease the recurred annually.

1830 The gradual depopulation of the remote island of Saint Kilda in the Scottish Hebrides derived in part from lack of mates for marriages, emigration, natural disasters, and the 80 percent death rate among male children from *tetanus infantum*. The training of midwives in germ theory introduced methods of controlling neonatal infections. In 2004, new research exonerated birthing techniques and blamed the local choice of living on bacteria-rich soil.

1830 Smallpox spreading over the plains states wiped out 75 percent of the Blackfoot and threatened the Multnomah and other indigenous people of the lower Missouri, Platte, and Kansas river valleys. Although President Thomas Jefferson had publicly advocated inoculation to Indian leaders in Washington in 1801, a public health initiative had limited success among the Delaware and Shawnee migrating to Kansas City. In 1832, the disease struck the Potawatomi of Illinois and Wisconsin and infected up to one-third of the Menominee and Winnebago. Of the depredations of the disease, George Catlin's *Letters and Notes on the Manners, Customs, and Condition of the North American Indians* (1841) reported:

> The smallpox, whose ravages have now pretty nearly subsided, has taken off a great many of the Winnebago and Sioux. The famous Wa-be-sha, of the Sioux, and more than half of his band, have fallen victim to it within a few weeks, and the remainder of them, blackened with its frightful distortions, look as if they had just emerged from the sulphurous regions below. At Prairie du Chien, a considerable number of the half-breeds, and French also, suffered death by this baneful disease [1989, p. 420].

He added that last-minute inoculation did not save those already infected with smallpox. "In almost every instance of such," he remarked, "death ensued." (ibid., p. 421).

When the disease reached the mouth of the Columbia River in Oregon, it sickened the 60-year-old Chinook leader Comcomly, who had welcomed the Lewis and Clark expedition and aided John Jacob Astor's fur traders. The natives provided a ritual cedar bark canoe burial at Point Ellice. Afterward, Meredith Gairdner, a doctor employed by the Hudson Bay Company, acquired Comcomly's sloped skull and, in 1835, carried it to Edinburgh, Scotland, for scientific study by phrenologists. John Richardson exhibited the curiosity at the Royal Naval Hospital in Portsmouth, England. The Chinook retrieved the skull in 1972 for burial.

August 1830 Along the Pacific Northwest, the introduction of malaria, which locals called "fever and ague" or "intermittent fever," was borne by the *Anopheles malculipennis* mosquito, a native of western Oregon's swamps and the Cascade Mountains. The disease spread from Fort Vancouver, Washington, down the lower Columbia River and the Willamette Valley. At Fort Vancouver on October 11, John McLoughlin reported the deaths of 75 percent of local Indians and the burning of a village to cleanse the area of disease. The epidemic affected the native Chinook and Kalapuya cultures and severely disrupted the mission of the Hudson Bay Company by depriving it of native laborers.

fall 1830 A recurrence of plague in Persia blanketed the nation with suffering and death. Traveling from Tabriz, the disease claimed 30,000 lives. Terrified residents abandoned the city, carrying contagion to outlying areas. The infection advanced on Baghdad and Basra, infecting the pasha and killing two of his wives. The governor retreated, leaving residents to fend for themselves as corpses littered the streets and alleys.

November 1830 Hampering the uprising of Jews and the military against Russian occupation troops at Bialystok and Podlasie, Poland, was an epidemic of cholera. A priest, Jan Dolinowski, visited peasant cottages to bring aid and comfort. The Jewish Choleric Cemetery on Bema Street in Bialystok received 1,000 corpses in mass graves.

early 1830s Among Australia's 100,000 aborigines, smallpox arrived from Moreton Bay in the north and advanced to Adelaide in the south. In virgin-soil territory along the Murray River, the virus generated a 45 percent fatality rate that targeted women and children. The loss left twice as many males alive as females and increased the value of women and the availability of food and water to the 10,000 survivors.

1830s Although vaccination reached Latin America in 1805, Hondurans suffered terrifying upsurges in cholera and smallpox. Historians blame the failure of official immunization campaigns on lack of funding, political instability, and the low level of literacy among peasants.

1831 At Mangareva in the Marquesas Islands, a missionary brought disease from Rapa in the Austral Archipelago, where 90 percent of islanders died of an unidentified epidemic. To Polynesians, the man appeared to spread an epidemic of European origin among natives. Outraged, they blamed his god for attacking them and forced the outsider to flee back to Rapa.

1831 When the Russian military invaded Eastern Europe, soldiers spread pandemic cholera to Poland, Hungary, Germany, and Baltic ports. At Lublin, Polish Jews suffered the brunt of the epidemic, during which Jewish medical workers treated civilian and military victims. In Bohemia, local mayors were forced to reorganize the public health services, yet Viennese officials refused to establish a state-subsidized board of regional physicians. When Asiatic cholera invaded Germany, the two-year onslaught gave politicians a scapegoat on which to project rampant xenophobia. They alternately described the disease as foreign or alien, insidious and treacherous, the result of sinful or degenerate behaviors, and the result of liberal thinking.

Health authorities in Austria, England, Finland, France, Russia, Scotland, Sweden, and Wales reported the pandemic. In England, eyewitnesses described a crisis unlike any caused by foreign invaders. A royal proclamation set up the Central Board of Health in June 1831 to guard public health. Some 50,000 Muscovites fled their city; Parisians poured into the countryside at the rate of 700 a day. East Prussian overreaction produced a rigid sanitary corridor to the east as well as quarantine quarters and isolation wards. Sanitary commissions administered village cleansing and superintended special graveyards for victims. The harsh measures annoyed residents, but could not halt the disease, which killed over 14,000 and generated panic and rioting in Königsberg.

During the emergence of cholera in Alexandria, Egypt, Mehemet 'Ali Pasha established an epidemiological study center. The institute developed into a laboratory favored by European bacteriologists studying Middle Eastern scourges. At Basra, Iraq, Mandaean priest Yahia Bihram survived the pestilence, of which priests perished in large numbers. To perpetuate traditional religious practice, he and Ram Zihrun, the son of a priest, conferred an emergency ordination on each other. Their dedication preserved the last extant Middle Eastern Gnostic cell.

Islamic doctors and health officials contributed to pestilence control. In Persia, Mirzâ Mohammad-Taqi Shirâzi and Sâveji wrote about outbreaks of *heyzah*, a seasonal intestinal complaint that may or may not have been cholera. Shirâzi issued three monographs arguing that *heyzeh* was a sporadic diarrhea caused by pervasive hunger. He also clarified the concept of *vaba* as contagion.

1831 Health sleuths linked an unidentified malady in eastern Denmark in 1831 to three seasons of heavy rains that mildewed hay and rye in Falster, Lolland, and Sjaelland. The resulting mold and contaminated grain in the marshlands precipitated achy joints, delirium, diarrhea, head pain, sweating, thirst, vertigo, and vomiting, symptoms of malaria. Old men in rural areas suffered the highest death rate. The official response to the epidemic prepared the nation for a bout of cholera in 1853.

June 1831 Transported by Russian soldiers to Poland, an outbreak of cholera struck Bohemia and Hungary, killing a quarter of the population. As a result, peasants abandoned farms and moved into towns, bringing their culture and the Slav and Magyar languages. A hero of the epidemic, David Didier Roth, a Vienna-trained homeopath who had practiced in Paris, elevated his branch of medical science, earning appointment as official staff doctor of Hungary. By fall, the disease traveled on to Berlin and Hamburg, Germany.

To the south, cholera migrated to Africa from holy sites in Arabia via Moroccan pilgrims returning home from a hadj to the sacred compound in Mecca. As the faithful made their way South, they spread bacteria through the Islamic world. At Alexandria, Egypt, and Istanbul, Turkey, the disease took hold and moved into the northern shores of Africa, where European sanitary measures contained the disease and set a standard for Moroccan health care.

The scourge progressed up the Balkans and

the Danube, and into Eastern Europe. One eyewitness, Belgian businessman Aristide Dethier, issued letters to his father that autumn about the effects of cholera in Izmir, Turkey. For the next 70 years, European nations pressed the Ottoman government to enact health precautions to halt the epidemic.

October 26, 1831 The first great wave of cholera, which traveled from Bengal, India, across Europe, found England largely unprepared despite the creation of a health bureaucracy the previous June. The disease arrived at Sunderland, England, on a ship from Hamburg, Germany. After passenger William Sproat died of the scourge, health authorities denied the existence of the disease and thus allowed it to proliferate north in Scotland, Ireland, and south to London.

Victims suffered hideous retching and diarrhea, dehydration, joint and muscle pain, and a graying of the skin. Sir William Tennant Gairdner, Glasgow's medical officer and professor at the University of Glasgow, called cholera monstrous and insidious because of its intensity and contagion, which mystified the medical community. In all, the epidemic claimed 52,000 lives, bringing to mind the Black Death of the Middle Ages. Exacerbating the disease was a parallel epidemic of influenza, which weakened the immune system, leaving it incapable of fighting off cholera.

The threat of disease in England produced observable shifts in thinking. James Phillips Kay-Shuttleworth, a Manchester physician at the Ardwick and Ancoats Dispensary, issued a monograph, *The Moral and Physical Condition of the Working Classes Employed in the Cotton Manufacture in Manchester* (1832), which singled out Irish laborers as sources of rampant Catholicism, spreaders of immorality among English workers, and causes of epidemics. Conservative Anglican ministers used the catastrophe to their advantage and declared cholera a national retribution caused by sin and debauchery. Evangelicals scheduled a nationwide day of fasting that carried the support of William IV.

The vituperative rhetoric that Kay and his followers generated worsened the local disrespect for doctors. To bolster their reputation and authority and halt the oppression of the poor, health experts rebutted pulpit oratory by focusing on the physical needs of the British. Their insistence on sane discussions of disease initi-

ated a modern public health movement. *See also* **May 17, 1832**.

October 30, 1831 After cholera advanced on Romania, rebellion in Transylvania along the Tisza River derived from Slovak rumors that nobles had deliberately spread the disease among peasants. At Slovakia, a municipal deputation in Liptov County earned credit for curtailing the epidemic. The cooperation of citizens with the government health agency prefaced the acceptance of Europe's public health initiative.

While Russians battled the disease, paranoia made health workers suspicious and uncooperative. In the estimation of Roderick E. McGrew, author of *Russia and the Cholera 1823–1832* (1965):

> The English delegation which had visited the military hospitals under the guidance of Sir James Wylie offered to take charge of a number of cholera cases, but their offer was refused because of the violent excitement of the people against all foreigners, more particularly against medical men, whom they lately looked on as emissaries employed by their enemies to poison them [pp. 4–5].

When the disease reemerged in 1892, the former paranoia resurged and kept pace with contagion. *See also* **1892**.

One upshot of the *Cholera asiatica* epidemic was the test of homeopathy among the nobility, who witnessed the deaths of thousands. Because the mysterious pestilence proved both Russian medicine and the Russian government powerless to save a quarter million victims, Tsar Nicholas I lost half his subjects. He was unable to recruit men for the military, his trading centers halted operations, serfs died on the untilled land, and quarantine virtually imprisoned the peasantry. The mounting unrest and rioting caused the Ministry of War to warn General Ivan Dibich that no past threat had so seized the motherland.

To ease the sick in a time when people had lost faith in standard medicine, the Russian statistician and landlord Semen Korsakov invented a set of homeopathic dilutions. He welcomed peasants to his practice and served as district inspector of cholera hospitals. As actuary, he collected data proving that homeopathy saved 69.8 percent of the 480 patients he treated. Because of his skill as a healer, landlords and physicians flocked to Korsakov for supplies and instructions.

To meet the needs of the epidemic, Korsakov's

wife worked days at a time manufacturing curative powders. During the rise of homeopathy into respectable practice, the Polish physician Valenty Cherminsky, who became the minister of war in October 1831, converted from standard medicine to Korsakov's methodology. He pursued the acceptance of homeopathy at cholera hospitals because it was inexpensive, beneficial, fast, and reliable.

winter 1831 War and epidemic smallpox struck the Pawnee with a mortality rate of 50 percent. The population suffered irreparable loss. Caught between encroaching whites and enemy Dakota to the north, they barely hung onto ancestral homelands in the Platte Valley of Nebraska. Worsening their situation was famine from the depletion of bison, grass, and woodlands along the Oregon Trail.

early 1832 The second pandemic of cholera moved through Europe and into the far reaches of the British Isles. Carriers were often itinerant laborers, masons, and wet nurses journeying from Paris to rural areas of France. At York, Ontario, where officials of the York Medical Society had prepared for the epidemic for months, residents felt confident that the contagion would spare their city of 25,000. By June, the number of reported illnesses in tenements reached epidemic level. Advancing rapidly, the infection rate peaked in mid-summer and dropped steadily through August. In all, 1.7 percent of the populace sickened with cholera; 185 died.

The fear of death from the scourge pressed citizens into a new awareness of religion and municipal responsibility toward the public. In Wales, ministers formulated appropriate prayers in Welsh and urged piety and fasting among churchgoers. In March 1832, the Irish suffered 25,000 deaths, forcing the conversion of the defunct Richmond General Penitentiary into a lazaretto. Irish emigrants carried infection west across the Atlantic to Canada and New York.

The outbreak of cholera in eastern Canada provoked curtailment of incoming passengers from England, Ireland, and other cholera-infected areas. Immigrants entered quarantine at the Grosse Isle reception center outside Quebec, but in early June the system failed to stop contagion from the brig *Carricks*, on which over 31 percent died of the disease while crossing the Atlantic from Ireland. Additional sick from the

ship *Voyageur* boosted the outbreak, which filled beds at Quebec's Hôtel de Dieu, filled tent wards on the Plains of Abraham, and spread misery to Beauport, Montreal, and Pointe Levi.

After moving into Upper Canada, cholera felled soldiers as well as the 134 Mohawk at the Jesuit mission at Caughnawaga south of Montreal. Diarist James Lesslie recorded the suffering at York. By spring 1832, the disease sped along with emigrés to Lower Canada, killing 5,820, and north to Upper Canada, killing over 1,000. At Montreal, Indians living in crowded conditions incurred 42 deaths. Also stricken were members of the Potawatomi, Winnebago, Menominee, and Ojibwa living on ancestral lands. *See also* **June 1832**.

March 29, 1832 Details of pestilence that swept Europe emerged in the religious pamphlet of Oxford scholar Vaughan Thomas; in letters and diaries, including the journal of Johann Mathias Wellenstein, a high court official at Luxembourg; and the personal papers of Guillame Dupuytren, chief surgeon of the Hôtel de Dieu in Paris. The French confronted the collapse of traditional values and the failure of the church to provide emotional and spiritual support. At Seine-et-Oise, hampering the survival rate were despair, terror, and a general distrust of doctors and medicine.

To halt panic, enforcers of health regulations buried some victims immediately in hastily dug graves. Lines of corpses sewn in sacks awaited burial. At Père Lachaise, a cemetery in the city's suburbs, hearses lined up awaiting entrance to the crowded burial ground. In a letter to a friend, Heinrich Heine, a German lyric poet, described a costume ball in Paris at carnival time where maskers collapsed, turned blue, and died, causing a riot and an immediate exodus of 120,000 from the city.

The epidemic influenced the artistic presentation of deadly disease as a demon. After reading Heine's letter, Alfred Rethel pictured the personified Death as a cutthroat. Composer Franz Liszt became obsessed with death. While visiting novelist Victor Hugo, Liszt repeatedly played the "Funeral March" from Beethoven's *Sonata in A-flat* as funeral corteges passed by on their way to Notre Dame Cathedral. A countess who resided near Liszt's apartment recalled that, throughout the night, he played variations of "Dies Irae" (Day of Wrath), a musical image of Judgment Day. By the end of the outbreak, cholera had claimed 100,000 in France.

May 17, 1832 At Manchester, England, Sir James Phillips Kay-Shuttleworth practiced medicine at the Ardwick and Ancoats Dispensary during the cholera scourge. Twenty years before science discovered the cause of cholera, he examined the pathology of the Asiatic strain to determine how it spread. He studied the sick and dying of Irish Town, a colony of Irish workers living at Oxford Road near a polluted stream. To receive the huge influx of Irish patients, local people had stripped a cotton factory of machines to create Knott Hill Hospital.

While managing 14 district boards, Kay-Shuttleworth compiled significant data. He warned the comfortable middle class that the state of health in Little Ireland impacted all factories and the livelihood of the English. To district magistrates he proposed:

> to remove the evils enumerated; and offer the following suggestions with a view to their partial amelioration:
>
> *First,* to open the main sewer from the bottom, and to relay it.
>
> *Secondly,* to open and unchoke the lateral drains, and secure a regular discharge of the water, &c., into the main sewer.
>
> *Thirdly,* to enforce the weekly cleansing and purification of the privies.
>
> *Fourthly,* if practicable, to fill up the cellars.
>
> *Fifthly,* to provide the inhabitants with quick–lime, and induce them to whitewash their rooms, where it can be done with safety.
>
> *Sixthly,* if possible, to induce the inhabitants to observe greater cleanliness in their houses and persons [Kay-Shuttleworth].

June 1832 As famine and hardship worsened living conditions in Ireland, outbreaks of cholera, dysentery, scurvy, and typhus killed the youngest, oldest, and weakest. Hysteria caused self-appointed messengers to distribute blessed turf. Ostensibly sanctified by priests, bags of dirt assured the receivers of protection from the disease. Some 45,000 immigrants chose to flee Ireland by crowding aboard seagoing vessels at £3 each. After incubating lice and disease in the unhealthful atmosphere below decks, the newcomers carried cholera to Montreal and Quebec, killing 3,347. *See also* **November 10, 1836**.

July 1832 In New York, a cholera epidemic killed over 3,000 between July and August. Outside Princeton, New Jersey, contagion targeted laborers on the Delaware and Raritan Canal. In Providence, Maryland, a Cuban nun known only as Mother Mary extended community care to victims of the disease. In Saint Louis, Missouri, the disease proved so damaging to families that Mrs. Ann Perry formed the Saint Louis Association of Ladies for the Relief of Orphan Children. By October, the scourge moved far to the south and took 4,340 people in New Orleans.

During the Black Hawk War, when Chief Black Hawk headed a war party of Fox and Sauk, cholera engulfed tribes of the area. As General Winfield Scott's soldiers moved cautiously from Buffalo, New York, to Fort Dearborn, Indiana, they introduced cholera in Illinois. A new outbreak hit Indian Territory in 1833 and struck the Great Lakes Indians and Halifax, Nova Scotia, the following year.

In the West, where contagion crept along trails and waterways, epidemic cholera proved useful to Mormons seeking to convert unbelievers and to proselytize members of other denominations. During an era of anti–Mormon persecution, members stood firm in the belief that the scourge derived from an angry god who punished them for lack of faith. Because the Mormons as a group incurred a lower rate of deaths than the rest of the population, doubters were impressed and embraced righteousness.

1833 A malaria outbreak that began at Fort Vancouver, Washington, engulfed the Maidu, Miwok, and other California and Oregon Indians and white settlers. The scourge claimed up to 75 percent of Indians, who had developed no immunity to the disease. An eyewitness, trapper George Yount, reported on thousands of withered corpses lying on the plains and in desolate villages. Tenaya (or Tenieya), the son of a Miwok leader, led 200 surviving Mono, Paiute, and Yosemite from the Fresno reservation back to their homeland in Yosemite Valley, where he served as chief until his death in 1853. Within a decade of the malaria epidemic, the native population of Central California had declined 92 percent.

1833 A Japanese physician and *rangakusha* (Western scholar), Takano Choei, who was educated in Dutch medicine by Phillip Franz von Siebold, wrote articles on disease in Japan. He informed the outside world of the malnutrition and pestilence that devastated the island nation during the Tenpo Famine of 1833, a five-year period of rice crop failure, revolt, and death among 300,000 refugees in northeastern Japan. Takano summarized precautions against contagion and listed foods that would rescue the starving from death.

1833 During a cholera epidemic in Kentucky, medical author John Esten Cooke of Virginia, a professor at Transylvania University Medical School and the Louisville Medical Institute and author of the three-volume *Treatise of Pathology and Therapeutics* (1828), expressed his opinion on the use of calomel as a curative in the *Transylvania Journal of Medicine*, which he edited. He selflessly treated the sick during the outbreak by purging the blood of impurities with large doses of calomel, which reduced fever.

January 1833 In Spain, cholera, called the "blue plague," killed 300,000 (Kohn 1995, p. 317). Response to the scourge uncloaked the elitism of the medical community. In reports from medical academies of Cádiz, Granada, and Seville, the health establishment clearly favored protecting the aristocracy while ignoring the abysmal filth of the lowest social class. At the nation's capital, the Sanitation Junta of Madrid, which King Ferdinand VII established on April 19, 1832, listed hospitals and aid stations, doctors, surgeons, and pharmacists. The registry assured medical supplies, noted expenses, and disseminated suggested home remedies for symptoms. Describing the despair and unease of the period are eyewitness accounts in letters to the administration at Le Rioja from Don Cipriano Palafox Portocarrero, the Count of Montijo.

January 7, 1833 During rampant infections from cholera, London doubled its usual burial schedule; two weeks later, burials quadrupled. The poor died in record number along with public officials, actors in theater troupes, and financiers at the Bank of England. A wave of "cholera-phobia" permeated the writings of Alfred, Lord Tennyson, Arthur Hallam, Thomas Babington Macauley, Thomas Carlyle, Charles Dickens, Elizabeth Gaskell, Rudyard Kipling, and Florence Nightingale. The disease returned three more times—1848 to 1849, 1853 to 1854, and 1865 to 1866—before containment measures and strict harbor regulations prevented the importation of contagion.

spring 1833 An unidentified epidemic—possibly cholera, measles, or malaria—invaded California, annihilating whole communities along the San Joaquin river. In autumn, frontiersman and trapper Colonel James J. Warner reported that hundreds of Indians camped along the Tuolumne and Stanislaus rivers the previous year in tents and lodges had been reduced to only six or eight native inhabitants.

summer 1833 Widespread flu in Persia traveled south from Constantinople and Syria, infecting nearly all the citizenry of Tehran, Iran, and killing many. Panicking the populace were food shortages and the infection of Shah Fath (or Feth) Ali, who left the nation leaderless during his illness. He recovered, but mourned the death of his heir, Abbas Mirza.

July 1833 When Cuba lost 8,000 to cholera, largely slaves, Mexico suffered over 15,000 deaths. The epidemic arrived from Tampico, then ravaged San Luis Potosí, Bolivia. One resident, Enrique Androis, a local French tailor, was taken for deceased and carried out on the daily run of the dead cart. Heaped in a burial pit, he regained consciousness and struggled back home on foot. Because of his weak knock at the door, he climbed into his first-floor room and knocked to summon his staff. The frightened servants alerted police and a padre, who examined the man and determined that he had been erroneously removed with the day's corpses. He recovered and went back to work.

The disease pressed on to Mexico City and north to New Orleans and San Antonio. In Mexico City, sickness rapidly destabilized the presidency of Valentín Gómez Farías. At Guadalajara, Mexico, authorities reported 3,000 deaths. In Texas, health officials at Brazorio met to discuss how to clean public thoroughfares and built six carts to carry away human excrement that people regularly piled in gutters. By August, 200 victims per day died of cholera.

On August 3 in Celaya, Guanajuato, Francisco Eduardo Tresguerras, Mexico's leading 19th-century architect, died when cholera swept the country. By mid–August, the epidemic threatened the country's survival. Prevention and disease containment measures enabled Guanajuato to expand despite intermittent wars.

1834 According to a report by a Scottish neurologist, Sir Charles Bell, author of *An Exposition* *of the Natural System of the Nerves of the Human Body* (1824), polio reached epidemic proportions on the island of Saint Helena off the southwestern coast of Africa. Bell chronicled the spread of fever among toddlers and infants and noted that the disease inhibited development of their limbs and motor functions.

1834 Plague returned to Egypt, slaying 30,000 or one-third of the sick as it moved from the port at Alexandria up the Nile to Cairo. After Muslim officials took control, Mehemet 'Ali Pasha instituted European-style health controls. By listening to infidel Italian physicians, autopsying remains, and limiting the number of mourners at burials, he aroused a backlash among Muslims. Especially offensive was the work of Russian doctors at Mansura who tested plague transmission by paying volunteers five piastres per day to wear the garments of deceased plague victims. The volunteers remained healthy, but one of the Russian physicians caught plague and died.

A Frenchman, Antoine Barthelemy Clot of Grenoble, chief surgeon to the Egyptian army and personal surgeon to the pasha, was skilled in forensic medicine. He tested human-to-human transmission by injecting himself with the blood of a plague victim and surviving. Honored as a founder of modern medicine in Egypt and its first minister of health, Clot—called "Clot-Bey," "Clot Bek," or "Klute Bey"—was beloved for founding the nation's first medical school, for organizing sanitation measures, and for carrying back to France some of Egypt's earliest medical instruments. A thoroughfare in old Cairo bearing his nickname preserves his memory.

September 1834 Cholera inundated Sweden late after emergence of the disease in India in the Ganges and a rapid spread over much of the globe. In Finland and Sweden, King Karl XIV Johan and Governor Jacob Wilhelm Sprengtporten of Stockholm prepared for the onslaught of disease that had been moving westward for three years. The small outbreak disarmed officials before swelling to a mass of cases. Advancing to the northeast along watercourses, cholera produced a death rate of 14.5 percent among peasants who condoned the worst sanitation.

Gothenburg data recorded 4,600 fatalities. Twenty temporary contagion wards added 625 beds to the two hospitals, where staff treated the sick with bismuth, calomel, camphor, and menthol. City staff employed burial crews from poorhouses to inter the dead in special graveyards. As a result of initial confusion and uncertainty in the health bureaucracy, the governor centralized administration for the eight parishes and established a strict health policy improving sanitation and controlling contagion. In this same period, virulent cholera inundated Algeria from Gibraltar, killing 967 in Oran and 1,457 in Mostaganem and Mascara.

October 25, 1834 Migration from the British Isles to New South Wales carried hazards for shipboard epidemics. Conversion of the *John Barry* from convict carrier to passenger ship packed the cabins with 400 passengers. On passage out of Dundee, Scotland, 30 died of fever before reaching Australia. When charterer John Marshall, the chief operator of the bounty system, took 350 emigrants from Gravesend, England, to Australia aboard the *David Scott*, measles ravaged the ship. Many children died of complications from pneumonia by the time the vessel reached Sydney on October 25. *See also* **March 1835**.

1835 In Canada's Northwest Territories, a ship harboring at Fort Simpson spread smallpox to Pacific tribes, killing up to a third of their number. Over four years, the pestilence ravaged the Tlingit, formerly the strongest of British Columbia's coastal Indian groups. Robbed of strength, wealth, and number, the Tlingit ceased resisting white settlers. The Haida escaped contagion; others vaccinated by the Hudson Bay Company survived.

1835 Upon his visit to Egypt, journeyman Alexander William Kinglake of Brittany, author of *Eothen, or Traces of Travel Brought Home from the East* (1844), learned that infantry trainees died at a high rate. The predominant diseases were dysentery, plague, and trachoma, a major cause of blindness.

1835 As white investors seized land and gradually subsumed power in the Hawaiian Islands and overthrew Queen Liliuokalani, they also characterized epidemic leprosy as a sign of ethnic weakness and moral degeneracy. The bacilli first emerged in 1835 after Hawaiians came in contact with Chinese laborers on sugar plantations. Patient Zero, Kamuli, lived on Kauai Island. Hawaiians, who were susceptible to the illness, named it *Ma'i Pake* (Chinese sickness).

1835 A recrudescence of cholera struck southern France. In Nice, the populace benefited from a plan of action formulated by the military, municipal authorities, doctors, and hospital administrators during the initial outbreak of disease three years before.

January 1835 An epidemic of measles, introduced the previous October 25 with the docking at Sydney of the migrant charter *David Scott*, spread from Australia to Hobart, Tasmania, and, by whalers, to New Zealand's South Island, where 4,000 Maori died. One grievous loss, the death of Te Whakataupuka, the paramount Maori chief and peacemaker over Canterbury and Otago, occurred in December 1835.

March 1835 On one of bounty manager John Marshall's charter ventures out of Cork, Ireland, aboard the H.M.S. *Lady McNaghten*, no one was permitted to wash during the four-month passage. Pertussis and measles killed 56 children; 50 adult emigrants and crew died of typhus. Captain George Hustwick signaled distress. Nearing Australia off Cape Howe, Governor Richard Bourke dispatched a surgeon and medical chest to aid ship's surgeon George Forman. When the *Lady McNaghten* docked at Sydney, another 80 people lay dying. Both Hustwick and Forman were seriously ill.

In this same period, measles flourished in a virgin-soil population when it invaded New Zealand and Tasmania. The disease arrived with the docking of the *Children*, a passenger brig from Sydney. The carrier, a Maori on his way home from Australia, introduced the disease in South Island, New Zealand.

November 1835 John Badham, a young doctor from Nottinghamshire, England, commented in the *London Medical Gazette* on multiple occurrences of paralysis in children that stemmed from a small polio epidemic.

November 1835 When cholera arrived from France to the Italian ports of Genoa and Turin, Italy, it initiated a three-year outbreak. In November, the scourge infected Trieste and Venice to the northeast and moved on to other urban areas. Worst hit were the working poor, who lived in squalor. Among the 69,211 dead were citizens of Genoa, Milan, Naples, Palermo, and Rome. In reference to the era's sufferings, writings by Giacomo Tommasini, Gianandrea Giacomini, and Pirondi proposed a variety of points of view on contagion, prevention, and treatment of cholera.

1836 Jamaican-Scottish herbalist Mary Jane Seacole, a widow living in Kingston, Jamaica, and author of *The Wonderful Adventures of Mrs. Seacole in Many Lands* (1857), studied treatment of infectious diseases throughout the Caribbean. She opened a boarding house in Gorgona, Panama, and received native launderers, boatmen, and muleteers during a cholera epidemic. When the devout begged for priests to process through town with statues of saints to remove contagion, she ridiculed their superstition. Instead of prayers, she proposed mustard plasters, herbal beverages, and measured doses of calomel. With all her nurse-care, many died of the fever. She survived a mild case.

1836 After an outbreak of smallpox, the Winnebago living on the Fox and Wisconsin rivers ceded their homeland to the United States government and resettled farther west on Iowa Neutral Ground.

1836 In Udine, Italy, epidemic cholera generated rumors that linked cholera to the immorality and unwise diet of the poor. The populace compelled civic officials to adopt hygienic and sanitary measures to protect the water supply. To treat the underclass, health workers adapted a 15th-century convent into a temporary cholera hospital offering free treatment for the poor. To assure complaining citizens that patients would not live there permanently, administrators insisted that they would close the facility after the epidemic ended and designate the structure as a military depot.

spring 1836 In Pennsylvania, William Wood Gerhard, a Moravian author of *On the Typhus Fever, Which Occurred at Philadelphia in the Spring and Summer of 1836* (1837), commented on the rampant spread of typhoid fever. He was the first to report the characteristics that differentiated typhoid from typhus.

October 1836 A lethal outbreak of flu across Asia and Europe began in Sydney, Australia, and Cape Town, South Africa. After ravaging urban centers across Europe, it reached North Africa and the Middle East within three months. The high rate of infection was pronounced in Florence, Italy, where only one-quarter of the city escaped infection, and in Copenhagen, Denmark,

Geneva, Switzerland, and Lyons and Paris, France, where half the populace caught the disease. In Milan, contagion peaked in mid-summer, targeting females and older citizens, largely on the city's outskirts. Epidemiologists surmised that contact with sewage and contaminated water during market gardening as well as peddling door to door may have heightened infection.

October 7, 1836 Because Europeans connected cholera with ignorance and filth, outbreaks in Germany sparked heated denials that the disease existed in its environs. The *Augsburger Allgemeine Zeitung* (*Augsburg General Newspaper*) linked its denial to the economy:

> According to letters from many corners of Germany, people are, much to our surprise, under the delusion that cholera is in Munich. As a result, many visitors are being prevented from coming to Munich, accommodation for the October Festival has been cancelled, and the number of visitors at this year's festival is strikingly low. One cannot repeat often enough, therefore, that there is no trace of cholera in Munich or in the surrounding areas and there is not the slightest reason to believe that this dreaded sickness will occur here [Bastian 1994].

By winter, hundreds of Munich citizens came down with cholera and died.

November 10, 1836 The worsening of endemic typhus in Ireland derived its virulence from the mass migration of people looking for work and shelter. Crowded into workhouses, they shared body lice, the carriers of contagion. At Swineford workhouse on November 10, an extra 120 packed the facility, where 367 died of the fever. Another overstuffed shelter at Ballina, County Mayo, added 200 newcomers to the base population of 1,200 inmates. In all, 1,138 at Ballina died, including the health supervisor. The same unhealthful condition at Ballinasloe, County Galway, killed 254, including all medical officers.

At Ennis, County Clare, where there was no bread to be had, so many pounded on the workhouse doors that the staff summoned the police. On December 24, 1846, Captain Edmund Wynne, the District Inspector for Clare, described the depths of despair among the poor. During heavy December sleet, starving, ill-clad women and their children scoured turnip patches, yanking up raw tubers and gnawing them in the row, dirt and all. It was another year

before the government could organize soup kitchens, an outreach of the Temporary Relief for Destitute Persons of Ireland Act.

1837 The spread of smallpox among the Aleut, in addition to ill treatment of natives by Russian fur traders, reduced the number of indigenous Alaskans by 90 percent in a few weeks. Over half of the residents of Bristol Bay region died, leaving whole villages abandoned. The Dena'ina Athabascan of Tyonek lost half their members, as did the Tlingit of Wrangell, Alaska.

1837 Influenza struck New Zealanders over a six-week period, infecting 800 Maori and killing the children of a British physician. Within months, a streptococcal malady infected area people, reaching residents of the Matamata mission. A recrudescence of flu assaulted the Maori at the same time that the disease emerged at Samoa, reputedly introduced by English preacher John Williams on a soul-saving expedition from the London Missionary Society. After he arrived aboard the H.M.S. *Messenger of Peace*, the spread of disease and subsequent deaths aroused suspicion in tribal chiefs. As a result, Maori warriors killed some of the whites who brought suffering and death to the island. In December 1838, according to a pioneering missionary, Bishop William Williams of Waiapu on the north Island of Aotearoa, New Zealand, a fresh outbreak of influenza engulfed the Maori. The virus infected every native, killing large numbers.

1837 While in dock in Freetown, Sierra Leone, the crew of the H.M.S. *Curlew* incurred yellow fever, which they carried to Gambia in the larvae of infected mosquitoes living in their water supply.

June 5, 1837 With the docking of the steamboat *Saint Peter* at Fort Clark, North Dakota, Francis A. Chardon, a French chronicler and trader with the American Fur Company, warned the Mandan of a mulatto deckhand ill with smallpox. Despite Chardon's good intentions, three Arikara women arrived at Fort Clark with smallpox and survived to travel along the Missouri River to a Mandan village. Because a few infections burgeoned into an epidemic, the Mandan population shrank from 1,600 to only 31 survivors (some historians report 125 survivors). Arikara, Hidatsa, and Sioux raiders carried the disease back to their villages. Chardon's journal recorded the natives' sufferings, including the death of his own two-year-old son, Andrew Jackson Chardon, on September 22. By the end of September, Chardon estimated the Mandan at one-eighth their former number and the Arikara and Hidatsa reduced by half.

Among the Mandan dead was their second chief, Mah-to-toh-pa or Four Bears, whom artists Karl Bodmer and George Catlin painted. Chardon immortalized the 42-year-old chief as a beloved leader and peacemaker. In his dying words, Mah-to-toh-pa faced death with grace, but regretted leaving a corpse with a visage grotesquely mottled by a white man's disease. In the description of George Catlin's *Letters and Notes on the Manners, Customs, and Condition of the North American Indians* (1841):

> This fine fellow sat in his lodge and watched every one of his family die about him, his wives and his children ... when he walked out, around the village, and wept over the final destruction of his tribe; his braves and warriors all laid low; when he came back to his lodge, where he covered his whole family with a number of robes, and wrapping another around himself, went out upon a hill at a little distance, where he laid for several days ... resolved to starve himself to death. He remained there until the sixth day, when he had just strength enough to creep back to the village, when he entered the horrid gloom of his own wigwam, and laying his body alongside of the group of his family, drew his robe over him, and died on the ninth day.... So have perished the friendly and hospitable Mandan [1965, pp. 257–259].

The villagers were so quickly overwhelmed that few remained to perform death rituals. Wolves, dogs, and rats devoured the remains of the unburied.

Permanently weakened, the remaining Mandan were forced to leave the Knife River area and to assimilate themselves among the Hidatsa and Arikara at a reservation up the Missouri River near Fort Berthold, North Dakota. According to Chardon's journal, one Mandan chose murder-suicide of his wife and himself over the desolation that remained. After acculturation, few pure Mandan survived. *See also* **June 24, 1837.**

June 24, 1837 When the steamer *Saint Peter* continued along its route up the Missouri River, disease spread to Fort Union, North Dakota, on the Montana border, John Jacob Astor's American Fur Company trading post in Assiniboin territory at a juncture with the Yellowstone River.

As described in Charles Lepenteur's *Forty Years a Fur Trader on the Upper Missouri* (1872), station manager Jacob Halsey vaccinated those exposed to disease:

> Our only apprehensions were that the disease might spread among the Indians.... Prompt measures were adopted to prevent an epidemic. As we had no vaccine matter we decided to inoculate with the smallpox itself; and after the systems of those who were to be inoculated had been prepared according to Dr. Thomas' medical book, the operation was performed upon about 30 Indians and a few white men. This was all done with the view to have it all over and everything cleaned up before any Indians should come in, on their fall trade, which commenced early in September [1933, pp. 110–111].

Unfortunately for natives, the procedure was too late to save them from contagion.

After smallpox emerged in the fort's native population, over half of the 40 Assiniboin died. When staff turned Fort William, Wyoming, into a hospital, native women served as nurses; John Brazo was the undertaker. Lepenteur reported,

> Some went crazy, and others were half eaten up by maggots before they died; yet, singular to say, not a single bad expression was ever uttered by a sick Indian. Many died, that those who recovered were so much disfigured that one could scarcely recognize them [ibid., p. 111].

Artist George Catlin reported suicides by knife, gun, and plunging onto rocks.

As the *Saint Peter* moved on to Fort McKenzie, Montana, near the confluence of the Marias and the Missouri rivers, according to Francis Chardon, the disease spread from 5,000 Blackfoot and Piegan at the fort to the Arikara, Atsina, Hidatsa, Kainai, Sarcee, Siksika, and Sioux and into tribes of the Canadian prairie. Vaccination clinics conducted by the Hudson Bay Company saved the Cree from infection. Around 4,000 Blackfoot perished. The Assiniboin, also devastated, merged with the Crow, Atsina, Santee, and Yankton Sioux, whom the remaining Assiniboin joined in battle against the U.S. cavalry at the Little Big Horn in 1876.

July 1837 An 18-month epidemic of typhus invaded Britain's industrial centers. Novelist Elizabeth Gaskell, who succored the poor of Manchester, England, typified the disease among the underclass. In Chapter VI of *Mary Barton* (1848), Gaskell claims the disease is "brought on by miserable living, filthy neighborhoods, and great depression of mind and body" (Gaskell). Her championing of slum dwellers won the praise of a contemporary, social novelist Charles Dickens.

September 21, 1837 Upon hearing of terrible contagion among tribes along the Missouri River, William Todd, a surgeon in the Swan River District, Montana, anticipated an outbreak of smallpox and immunized with cowpox vaccine the 60 Indians at Fort Pelly, Saskatchewan. He dispatched fresh serum to Carlton House, Île-à-la-Crosse, Edmonton House, Fort Chipewyan, and Fort Ellice. For all plains Indians over a series of epidemics around Fort Clark, North Dakota, the estimated survival rate was sobering.

1838 In the Canadian Yukon, the Inuit contracted smallpox, which killed 67 percent. Because Russian traders were vaccinated and immune to the disease, natives concluded that the outsiders were poisoning them. Inuit attackers murdered the employees of a trading post and stole bales of pelts.

1838 French professor of hygiene and internal pathology Gabriel Andral, founder of the science of hematology and author of the first hematology text, *Précis d'Anatomie Pathologique* (*Summary of Pathological Anatomy*) (1829), and of *Clinique Medicale* (Medical Clinic) (1833), was the first medical technologist to make a chemical analysis of blood albumin and fibrin. While examining the sick, he preserved clinical details as a record of epidemics. In 1838, he observed that an outbreak of grippe or influenza in Paris was curiously varied in symptomology, an early acknowledgment of mutations in infectious disease. For his work, he was named a commander of the Legion of Honor.

1838 In Switzerland, Johann Jakob Guggenbühl claimed to have observed an alpine epidemic of pneumonia that he called Alpenstich. In the next decade, epidemiological analysis disclosed errors in his logic. He appears to have compiled data on anthrax, erysipelas, typhus, influenza, and plague, treating the diseases as one demographic phenomenon.

1838 On a French expedition off the coast of Chile, Captain Dumont d'Urville faced a serious loss of manpower from scurvy in sailors aboard the *Astrolabe* and its sister ship, the *Zélée*. When the men grew discouraged, nine crewmen deserted at Talcahuano. The captain left the sickest behind and sailed on to Valparaiso. When the ships entered the Pacific Ocean, sailors encountered dysentery and fever, which killed 20 by the time d'Urville anchored in Hobart, Tasmania.

May 1838 After the U.S. government decided to uproot the Cherokee from their Appalachian settlements at New Echota and Ridge, Georgia, over 7,000 troops invaded Cherokee homelands, joining 2,000 state militiamen. They erected stockades and, late in May, began corralling some 8,000 Indians whom they seized from ancestral lands so white homesteaders could usurp both acreage and goods. During a summer drought, cholera and dysentery weakened the internees, forcing leaders to settle with whites over the long walk west.

After the Cherokee buried their dead, the military forced 16,000 off native land and loaded them on 645 wagons. Of the total departing in October, one-quarter died either in holding pens or along the trek west to Oklahoma. According to cavalryman John G. Burnett:

> I saw the helpless Cherokees arrested and dragged from their homes, and driven at the bayonet point into the stockades…. The sufferings of the Cherokees were awful. The trail of the exiles was a trail of death. They had to sleep in the wagons and on the ground without fire. And I have known as many as twenty-two of them to die in one night of pneumonia due to ill treatment, cold and exposure ["The Trail Where They Cried"].

The wagon train slowed along the way just long enough for Indians to inter children and the elderly who died of hunger and weakness from the effects of pellagra, pneumonia, and tuberculosis. The Cherokee diaspora earned the name *oosti ganuhnuh dunaclohiluh* or Trail of Tears ("Native American Tribes"). *See also* **August 14, 1842.**

summer 1838 Rockwell, Illinois—a planned temperance community founded in 1836 by investor and state legislator John Rockwell—provided a terminus for the Illinois and Michigan Canal. A series of financial upsets and an unprecedented malaria epidemic squelched interest in the project. Into the fall months, inexplicable contagion spread over the West and proved especially virulent in the Illinois bottomland. Many of the colony's newcomers died of the fever; survivors moved on to more salubrious climates.

November 5, 1838 On a miserable 18-week voyage from Gravesend, England, to Sydney, Australia, 315 passengers aboard the ship *Maitland* suffered a serious assault of scarlet fever. Ship's surgeon John Smith treated 286 patients, of whom five adults and 29 children died. According to the Sydney *Gazette*, the *Maitland* suffered more misfortune than previous emigrant ships and reached harbor at Watson's Bay with many people still bedfast. The local medical board quarantined the ship and crew.

1839 During a decade of typhus, cholera, scarlet fever, pertussis, and influenza waylaying England, Edwin Chadwick's *Report on the Sanitary Condition of the Labouring Population of Great Britain* (1839) noted that for every death from age and mishap, eight occurred from contagious disease. Because of epidemics, polluted air, overcrowding of the poor in unplumbed housing, contaminated meat and fish, and famine, it was not surprising to Chadwick that one-third of England's children never reached age five.

Chadwick's alarming writings influenced Friedrich Engels's *Condition of the Working Class in England* (1844). In an analysis of Manchester, the text reported that Irish Town was a clutch of damp, unhygienic cellar rooms along unpaved lanes. Privies and pigsties complemented ankle-deep muck around windowless mud huts to create a miasma of squalor and despair. The 18,000 people willing to live in these hovels were the lowest-paid workers, often employed in dangerous, degrading jobs in construction and canal engineering. From sleeping three to a bed or on the floor, residents spread disease among family members. In Chapter VII, Engels remarked, "The children are puny, weak, and in many cases, severely crippled" (Engels).

1839 The Kiowa winter count reported an epidemic, which natives called the "smallpox winter." Disease endangered the Hidatsa and Mandan and killed 8,000 Blackfoot, 2,000 Pawnee, and 1,000 Crow. Too weak to bury all their families, survivors placed corpses in burial pits or threw them into rivers. Some people killed themselves after most of their community died.

1839 An onslaught of *tito apoplettico* (meningitis) overwhelmed Italy, traveling from French military billets at Ancona into Calabria and along the Amalfi coast. The disease migrated west to Sicily and as far away as Corfu. Only 20 percent of victims survived the bacteria.

1839 From 1839, the people of Oceania confronted the contagious diseases imported from the Western world. A disastrous series of epidemics savaged Tahiti, beginning with pertussis in 1840, smallpox the following year, dysentery in 1843, scarlet fever in 1847, and a virulent three-year run of measles from 1852 to 1854. These and earlier outbreaks, along with alcoholism, opium addiction, and violence, helped to depopulate the Society Islands. *See also* **June 1841**.

Cliff et al. (2000, pp. 146–147) give the following breakdown of diseases:

Date	Epidemic	Island	Death Rate
1839	influenza	Cook Islands	2.5%
1841	smallpox	Tahiti	unknown
1843	dysentery	Bora Bora	3%
		Rarotonga	unknown
	measles	Leeward	3%
1848	pertussis	Samoa	5%
	smallpox	Tahiti	10%
1854	measles	Bora Bora	3%
		Moorea	unknown
		Tahiti	10%
1865	dysentery	Leeward	33%
1875	measles	Fiji	25%
1893	measles	Samoa	10%
		Tonga	5%
1911	measles	Samoa	5%
1918	influenza	Leeward	25%
		Samoa	25%
		Tonga	10%

1840 After opening the first orthopedic hospital, orthopedist Jakob von Heine of Stuttgart, Germany, the discoverer of "Heine-Medin disease," studied an outbreak of poliomyelitis. Thirty-five years later, he reported on "Spinale Kinderlähmung" (infantile spinal paralysis) at a convention of the German Society of Natural Scientists and Physicians in Freiburg, Germany. At the Roosevelt Warm Springs Institute for Rehabilitation at Warm Springs, Georgia, a bronze bust in the Polio Hall of Fame honors Heine's work.

1840 When Asiatic cholera overran Havana, Cuba, health authorities feared that the disease derived from miasma or stale air. To increase air circulation and fumigate streets, agents spread fragrant resin and fired cannon throughout the city.

1840 On Kodiak Island, Alaska, a smallpox epidemic produced an irreparable tragedy. Disease reduced the population of Eskimo at Koniag village in Karluk and hastened the social disintegration of natives and acculturation by whites. Assisting the disoriented natives were Russian traders, who helped survivors at 30 Kodiak villages to consolidate into seven communities, the largest of which they built on Woody Island. Russian volunteers constructed community centers, storehouses, and residences and supported the Eskimo during their migration from defunct villages. Missionaries completed the task of replacing native traditions with Anglo-Christian values and beliefs.

ca. 1840 Syphilis presented a number of ethical, religious, and social dilemmas to Europe. From Iberia to Russia, fear of venereal disease affected the set-up of foundling homes. In Bologna, Italy, staff of the local *ospizio* (receiving home) feared that the admittance of syphilitic babies would infect the children's wet nurses. From the women who fed infants, the disease could spread to their husbands and their own nurslings. The marginalization of suspect abandoned infants compromised their health and increased death rates, which were over twice that of babies reared at home by their parents.

To prevent and treat syphilis, experimenters in England, France, Italy, Norway, and the United States sought a viable vaccine. With live pathogens, they inoculated a test population drawn from hospital patients as well as from orphans, infants, the elderly, soldiers, female prostitutes, prisoners, the insane, and lepers. Hospital officials violated patient rights by conducting the study without patient consent. By 1852, the Paris Academy of Medicine found human experimentation unjustified and inhumane, but the testing remained in use for another 16 years in England and Norway.

summer 1840 During the French conquest of Algeria, Louis-Philippe, King of France, established the French Foreign Legion. While defending Miliana on the Mediterranean coast in a drought, half the 1,232-man garrison died of North African fevers, contributing to the identification of Africa as the "white man's grave." Army doctor François Maillot reduced suffering by prescribing quinine to control malaria. Around 70 of the 600 survivors died by fall. Within the whole province, the French lost 8,000 soldiers, most to disease. Added to the 3,200 who died or were invalided out of service between 1831 and 1835, the total illustrates the high price that France paid to subdue Algeria.

1840s Among slave infants born in plantation quarters in the American South, *trismus nascentium* (neonatal tetanus) was endemic. Added to deaths from convulsions, rickets, malnutrition, pica, and parasites, tetanus fatalities threatened owner investment in prize workers. Spread by bacteria infecting the umbilicus, the disease was generally fatal. Medical experts William O. Baldwin and John M. Watson proposed sterilization of the navel to prevent the disease.

In 1845, J. Marion Sims, the father of gynecology and designer of the speculum, set up an experimental clinic on his property in Lancaster, South Carolina, to perfect methods of treating women's ills. Before his colleagues in septic conditions without anesthesia, he operated on black slave women and treated neonatal tetanus in their infants. Rather than blame the infant disease on unhygienic lodgings, he faulted blacks for inadequate morals.

1840s The absence of licensing requirements in California allowed two Irish brothers, Richard S. Den and Nicholas A. Den, to treat victims of a smallpox epidemic. While serving with the Californios during the Mexican-American War, the two also treated outbreaks among soldiers. Richard Den established a medical practice in Los Angeles. *See also* **June 4, 1847; 1848.**

1841 When Edward Lucatt, a traveler and author of *Rovings in the Pacific* (1851), observed the people of Rotuma Island north of Fiji, he found endemic disease:

> They are subject to huge swellings of the members, called by us elephantiasis, but by them *fe-fe;* to scorbutic eruptions, and to the breaking out of virulent tumours, which eat into and decay the bone. I beheld some shocking spectacles. There is also a blight, which at seasons affects the atmosphere, and many are apt to lose sight of one or both their eyes [p. 161].

Islanders, who had no understanding of yaws and other infections, attributed sickness to heavy indulgence in kava, the favorite fermented drink, which tended to dull thinking and roughen the skin with squamous patches.

1841 In Ceylon, a dearth of laborers at central plantations forced British colonial overseers to import workers from India. The rise in hookworm infestation created a life cycle generating larvae that migrated to the lungs. Pneumonia often killed the immigrants, who lived in unhygienic quarters, used unsanitary latrines, and walked barefoot on soil infested with nematodes, a type of intestinal parasite. The Rockefeller Foundation investigated the situation and suggested cures, but offered no intervention in unhealthful lifestyles.

1841 An inexplicable rise in Mangareva's death rate from 1841 to 1845 may indicate rampant tuberculosis in the South Pacific. Accompanying the scourge was famine from food shortages.

1841 During an epidemic of plague in Egypt, the viceroy, Mehemet 'Ali Pasha, posed stringent health measures enforced by the military. He established examination procedures and imposed strip searches and the washing of patients to rid them of contagion. After each examination, the subject received new sanitary garments to replace infected clothes and shoes.

June 1841 With the docking of a U.S. vessel from Valparaiso, Chile, via Hawaii, smallpox reached Matavai Bay at Papeete, the capital of Tahiti. Before docking, the captain concealed from harbor authorities the deaths of his brother and five Hawaiians. After the ship's two-week stay, contagion gripped Papeete and claimed 200 islanders. A chief and a native priest arriving from Moorea carried infection back to their people, of whom 67 sickened and 42 died. After the fortuitous arrival of the American warship *Yorktown* to Papeete with vaccine aboard, a general inoculation saved many islanders at Tahiti and other islands in the group.

September 1841 An emergence of yellow fever in New Orleans martyred Methodist evangelist and nursing volunteer Elijah Steele, who had preached through Louisiana and Mississippi. A self-educated minister and follower of pulpit preacher William Winans, Steele had staffed the Seniasha Mission and congregations in Port Gibson and Woodville, Mississippi. Upon appointment to New Orleans to serve the Poydras Street Church, he battled the yellow fever epidemic and died caring for others.

August 14, 1842 By the end of the Second Seminole War, a seven-year conflict, deaths from swamp fever (malaria) outweighed the loss of life from combat. The last major Indian war on the Atlantic coast, the conflict drew U.S. cavalry and some 5,000 Seminole who refused to move into trans–Mississippi Indian Territory on the infamous Trail of Tears. Under chiefs Alligator, Jumper, Osceola, and Wildcat, the Indians fought Colonel Zachary Taylor's regulars and militia through fever-ridden wetlands. The total carnage was the highest for any U.S. battle with indigenous people. Florida militiamen suffered 55 combat deaths and hundreds from fever.

1843 In the Cook Islands and French Polynesia, dysentery ran rampant for six months, particularly on Aitutaki, Bora Bora, Huahine, Mangaia, Raiatea, Tahaa, and Tahiti. The Rev. Henry Albert Royle, an agent of the London Missionary Society, accounted for infection at Aitutaki from some 35 whalers and traders in port during the year. Medical historians made a direct connection between infected men on a whaler and 130 deaths among natives on Rarotonga, where the vessel docked.

1844 Adventurer Alexander William Kinglake of Brittany, barrister and chronicler of the Crimean War, compiled *Eothen, or Traces of Travel Brought Home from the East* (1844), a classic travelogue of Cairo, Constantinople, Cyprus, Damascus, Gaza, Lebanon, Nablus, Smyrna, the Suez, and the Troad in northwestern Turkey. Upon experiencing the strictures of routine harbor quarantine at the border separating Serbs from Turks, he noted the severity of violations:

> It is plague, and the dread of the plague, that divide the one people from the other. All coming and going stands forbidden by the terrors of the yellow flag. If you dare to break the laws of the quarantine, you will be tried with military haste; the court will scream out your sentence to you from a tribunal some fifty yards off; the priest instead of whispering to you the sweet hopes of religion will console you at duelling distance, and after that you will find yourself carefully shot, and carelessly buried in the ground of the Lazaretto [Kinglake 1844].

Kinglake noted suspicions of "compromised" persons, those who had been in contact with infected people or objects. At Cairo, he especially avoided the relaxing drying beds of Turkish baths, where diseased people rested after bathing when they contracted plague.

Kinglake advanced theories of disease, particularly the cause of plague in Turkey, which he blamed on touching infected objects, including furs and garments:

> It is held safer to breathe the same air with a man sick of the plague, and even to come into contact with his skin, than to be touched by the smallest particle of woollen or of thread which may have been within the reach of possible infection. If this be a right notion, the spread of the malady must be materially aided by the observance of a custom prevailing amongst the people of Stamboul. It is this: when an Osmanli dies, one of his dresses is cut up, and a small piece of it is sent to each of his friends as a memorial of the departed—a fatal present [Kinglake 1844].

His squeamishness suggested the difficulties of a fastidious European in an area rife with plague.

Of the Turkish dread of plague, Kinglake observed their meticulous rules of personal and residential quarantine:

> It is a part of their faith that metals, and hempen rope, and also, I fancy, one or two other substances, will not carry the infection; and they likewise believe that the germ of pestilence lying in an infected substance may be destroyed by submersion in water, or by the action of smoke. They, therefore, guard the doors of their houses with the utmost care against intrusion, and condemn themselves, with all members of their family, including European servants, to a strict imprisonment within the walls of their dwelling. Their native attendants are not allowed to enter at all, but they make the necessary purchases of provisions; these are hauled up through one of the windows by means of a rope, and are afterwards soaked in water [Kinglake 1844].

1845 In the heart of the Amazon rainforest at Loreto in northeastern Peru, epidemic smallpox reduced the Ucayali tribes living at the Franciscan Sarayacú Mission to only 1,200 members.

1845 Immigrant Germans traveling to Galveston or Indianola, Texas, faced a 12–14 week sea voyage from Bremen. Because promoters crammed up to 300 people in steerage and fed them mainly dried beef and hardtack, many developed scurvy, pellagra, enteritis, and dysentery.

April 3, 1845 On a voyage from Plymouth, England, to Freetown, Sierra Leone, the crew of the H.M.S. *Clair* contracted yellow fever, which infected 13 and killed seven. On a subsequent leg

to Gambia, half the ship's crew died of the disease. As the ship traveled to Boa Vista, Cape Verde, in August, 39 on board died of the fever, which afflicted islanders. The Royal Navy dispatched a military surgeon, James Ormiston McWilliam of Edinburgh, author of *Medical History of the Niger Expedition* (1843), to investigate the cause of the outbreak.

February 1846 Following two years of disappointing island harvests and a season of heavy floods, typhoid fever inundated Java, striking most of the populace. Spread from the central mountains, infection raged into 1850, with a death rate as high as 47 percent.

March 28, 1846 Danish officials sent a medical health officer, 26-year-old Peter Ludwig Panum of Rønne, Denmark, the first professor of physiology at the University of Copenhagen, to investigate an epidemic of 6,000 cases of measles. Disease spread among 7,782 Faroe Islanders by a carpenter returning from a job in Copenhagen after he came in contact with a sailor suffering from measles. Panum wrote a report, *Observations Made during the Epidemic of Measles on the Faeroe Islands in the Year 1846*, one of the first epidemiological analyses ever published.

In the text, Panum deduced that the disease struck people younger than 65 because the 98 members of the older population had developed immunity from an outbreak in 1781. He postulated transmission through human exhalations and established the incubation period and lifelong immunity to the measles virus. Since 1986, the Panum Institute, which houses the university's faculty of medicine, has preserved his name.

May 1846 The third in a series of great Asiatic cholera pandemics broke out in the Middle East after simmering for months in Bengal (modern Bangladesh). At Mecca, Arabia, a hub of epidemics, Islamic pilgrims were among the 15,000 fatalities. From Tehran, where 12,000 died, the disease struck Persia and inundated the Arabian peninsula. When Shah Mohammad panicked in mid-summer and packed up his household for a hasty move to outlying quarters, he spread fear across the land. His flight did not save his minister, two of the royal wives, and a daughter and son.

The contagion spread overland to southern Russia and Siberia before penetrating Moscow and Riga in fall 1847. Over the seven-year span of the pandemic, the scourge reached the far islands of Scandinavia and Britain and returned to the Mideast. Cyclic cholera outbreaks rocked the Mississippi Valley, New Orleans, Canada, Central America, and the Pacific coast. The interwoven pattern and reinfections made the disease difficult for epidemiologists to trace.

The fight against cholera produced unforeseen heroes. In Slovenia, the writings of Janez Bleiweis, a veterinarian of Carniola, improved public health service. By issuing articles in the vernacular in *Novice* (*The News*), which he edited, he aided in disease prevention among the uneducated. He also served on medical councils and upgraded medical institutions in Ljubljana.

early June 1846 In the United States, a virulent outbreak of measles spread from a canoe convoy from the Red River Valley in the Dakotas to the northwestern trading posts. Respiratory complications killed natives, costing the Hudson Bay Company a full complement of workers. At Norway House, a fur depot, the Ojibwa suffered a heavy onslaught. Returns from the summer's business moved contagion into the interior, striking the Chipewyan in Saskatchewan with heavy losses.

winter 1846 As Brigham Young led a phalanx of Mormons west to form a colony in Salt Lake City, Utah, his followers wintered along the Platte River at Omaha, Nebraska, where 600 died of malnutrition and cholera. The arrival of survivors in Utah among the Numic-speaking Ute, Paiute, and Shoshone of the Great Basin generated demographic catastrophe over the next seven years. Rapidly, European pathogens spread epidemics along Indian trade networks. *See also* **1853.**

1847 After a decade-long hiatus, typhus returned to northern Europe and inundated Great Britain during "Black '47," killing 30,000 in England and Wales. The English blamed Irish immigrants who crowded into Liverpool, spreading "Irish fever," which had devastated Ireland the previous year and depopulated Belfast in March 1847. In summer, a doctor from Lowtherstown, Ireland, observed the sick in 40 beds at the Union Fever Hospital. She stated that

> persons coming in while in the state of starvation were often taken with fever, and in many instances had tedious recoveries. I think that in the better classes the head was much more affected than in the poor. The medical profes-

sion has suffered more from this fever in this county than any of the better classes.... The worst class of fever we had was that which was imported from Scotland, and the common people were well aware of it, and called it the Scotch, or black fever, from the great darkness of the countenance during the attack ["Irish Fever Hospitals," 1849, p. 125].

Dillon of County Mayo blamed crowded jails and prison hospitals for spreading contagion; at Ballinrobe, Pemberton observed the huddled inmates of the workhouse, who passed fatal illness to the staff.

At Innis Boffin, Alexander Fry tried to increase strength with food. Of his efforts, he reported shifts in attitude toward the medical profession:

> I bought up all the milk I could get, and distributed it among the sick, most of whom had nothing to drink but water, I also obtained some rice which after a few days I was able to allow my patients.... I had numerous enemies to encounter, and many difficulties to be overcome, owing to the—fairy doctors—and the ignorance and superstitions of the people. However, after some time, when they saw the patients whom I had in hospital recovering, and some who could not be persuaded to come into hospital, dying, their prejudices gave way, and they began no longer to look upon me as a dangerous emissary from the Government [ibid., p. 374].

When English hospitals could hold no more patients, authorities nailed together isolation sheds and riverboat lazarettos.

When epidemic typhus reached Ghent, Belgium, and Oudemaarde in East Flanders, poor health and malnutrition made the underclass more susceptible to disease. Like Alexander Fry, health workers, doctors, and local priests succored the hungry with handouts of food. Because of lethal contagion, high death rates for doctors and ministers precipitated shortages of professional care.

1847 At the upscale Vienna General Hospital, Hungarian obstetrician Ignaz Philipp Semmelweis concluded that medical students who dissected cadavers spread contagious puerperal fever to maternity cases at the rate of one out of every five mothers admitted to give birth. The ward relegated to the poor gained a despicable reputation for infection, causing women to plead for transfer to a better location, even an alleyway. Semmelweis made the connection between the dissecting tables and fever after a teacher performing an autopsy cut himself and died of the same disease that killed mothers in the wards.

On the basis of his theory about microscopic contaminants, Semmelweis ordered students to wash away the remnants and smell of dissections in chlorine solution before examining parturient women. He recognized additional sources of sepsis in contaminated cuffs, instruments, sheets, and towels. Obedience to his rule dropped the mortality rate to 2 percent. Because he was an outsider speaking with an accent, he was fired and his antiseptic measures dismissed as unsubstantiated.

1847 On the way to New Zealand to settle the village of Panmure on the Tamaki River, 312 passengers boarded the emigrant ship *Clifton* out of Gravesend, England, during its stop at Galway, Ireland. In an era that saw four million Irish flee the potato famine, most of the passengers were from Ireland. On the four-month sea journey, 432 cases of disease—measles, scurvy, smallpox, and typhoid fever—weakened the settlers. Officials at Auckland feared the newcomers would set off an epidemic, but survivors disembarked without carrying disease to the islands.

1847 The arrival of agents of the Hudson Bay Company at British Columbia introduced technological advancement, cultural change, and infectious diseases, all of which permanently altered native life. The influx of outsiders in the Pacific Northwest introduced the nation's first epidemic of measles, which rapidly increased native suspicions of European outsiders. Among those devastated by illness were the Cayuse. *See also* **fall 1847.**

February 1847 London master chef Alexis Benoît Soyer, who cooked for 1,500 aristocrats at the Reform Club in Pall Mall, abandoned his career to help the hungry during the Irish Potato Famine. In addition to designing a soup boiler, he erected feeding stations at the Royal Barracks in Dublin, where 1,680,500 poor crowded city shelters. Soyer equipped his food factories with coal ovens, *bain-maries*, cutting tables, and steam boilers on wheels. To speed the process of feeding the famished Irish, he paralleled a row of tables holding food basins with another row offering washbasins and sponges to cleanse dishes and reduce contagion from typhus. To keep diners from infecting each other, he zigzagged the line to eliminate crowding and fed each seating in six minutes. Each hour, his system accommodated 1,000 diners.

spring 1847 Eastern Canada had its share of misery from rampant typhus. Contagion arrived in Nova Scotia from Liverpool aboard the English ship *Lady Constable*, which docked at Charlottetown, Prince Edward Island. Unfortunately for local people, port authorities moved too slowly to contain a typhus epidemic. After isolating passengers, health agents discovered that the infection had already gripped the harbor community. Of 140 hospitalized victims, 23 died.

In May, the docking of 30 "coffin ships" at the Grosse Isle quarantine station at Quebec overwhelmed authorities with epidemic typhus that had beset the doomed Irish passengers in steerage during the 45-day crossing. Without adequate food, water, or sanitation facilities, they died in large numbers without seeing North America. When the *Syria* from Liverpool docked, George M. Douglas, Medical Superintendent of the quarantine station, counted nine corpses and the remaining 202 passengers "all were wretched and poor—disease—fever and dysentery—broke out a few days after leaving port" (O'Driscoll and Reynolds, 1988).

As the survivors crowded into an isolation area measuring 3.0 by 3.5 miles, sickness spread to subsequent waves of immigrants, killing 5,400. At Montreal, additional contagion required desperate measures to house the sick. Officials linked three tiers of wooden boats from Victoria Pier to Wind Mill Point, turning the space into much-needed hospital wards. At Saint John, New Brunswick, a makeshift hospital on Partridge Island warehoused sick immigrants.

On June 25, the heavy toll on newcomers to Canada caused the Assembly to petition Queen Victoria for a more humane strategy for receiving and housing arrivals:

> We humbly venture to state, that the arrangements for the reception of the sick at Grosse Isle, the quarantine station, although made on an extensive scale, have proved wholly inadequate to the unexpected emergency, that the entire range of buildings intended for the use of emigrants generally, at the station, have been converted into hospitals, and are still insufficient for the numerous and increasing sick; but the island itself ... has been reported as not sufficiently extensive to receive all those who by the regulation of the health officers are required to perform quarantine [Marr and Kirocofe, 2000].

The emotional speech informed the queen that new arrivals to Canada were frequently "the help-less, the starving, the sick and diseased, unequal and unfit as they are to face the hardships of a settler's life" (ibid.). In 1996, Canada raised an official memorial to some 3,000 Irish famine refugees whom cholera claimed at Grosse Isle. *See also* **late summer 1849**.

June 4, 1847 During the Mexican-American War, 1,000 of General Winfield Scott's 12,000 men suffered epidemic enteritis at Puebla, Mexico, 100 miles southwest of Jalapa. The loss followed a previous downing of 1,000 by disease at Veracruz soon after landing. Field surgeons suspected water and food contamination, weather, and exhaustion as the culprits. When the army moved on in mid–August, 3,000 sick and convalescing soldiers remained behind. Because more collapsed on the march to Mexico City, officers established makeshift hospitals until supply trains could evacuate the sick.

At the end of the two-year war in February 1848, the losses of Taylor's American forces were carefully tabulated and scrutinized under a health management system advanced during the tenure of Joseph Lovell, Surgeon General of the U.S. Army from 1818 to 1836. The totals reflected the same rampant epidemics that colonials had met in the American Revolution. Of the 100,454 who fought in Mexico, there were 12,519 casualties. A total of 1,549 or 12.3 percent died in combat; 10,970 or 87.7 percent of the fallen died of disease, chiefly dysentery, when spread to civilians after the army demobilized. One plus for the army's inoculation program was the absence of smallpox among epidemic diseases.

fall 1847 After a large wagon train passed through Washington's Walla Walla Valley, an outbreak of dysentery and measles killed half the Cayuse and almost all their children. While white children recovered from disease, the Indians worsened their weakened condition with sweat baths and plunges into cold rivers, which killed half their number. Already outraged by high traffic of whites along the Oregon Trail, the survivors blamed the local Presbyterian missionaries for the scourge and punished them in the Indian tradition, which required the assassination of any medicine man who failed to cure illness.

On November 29, the Cayuse, led by Chief Tiloukaikt, converged on the mission at Wai-latpu, which was sponsored by the American Board of Commissioners for Foreign Missions. In a face-to-face meeting, irate natives blamed the white healers of poisoning Indians. The chief and his braves then retaliated against Marcus Whitman, his wife Narcissa, and eleven others by bludgeoning them with tomahawks, taking 53 captives, and burning the mission.

The attack incited Colonel Cornelius Gilliam, a fundamentalist preacher, to lead 550 militiamen against the Cayuse in February. Their vengeful retaliation provoked the Cayuse War in the lower Columbia and Willamette valleys. Tiloukaikt and Tomahas surrendered, stood trial, and faced hanging at Oregon City. Tiloukaikt compared himself to Jesus, who accepted martyrdom to save people. Indian agents dispatched the remaining Cayuse to live with the Umatilla on a reservation at Thorn Hollow, Oregon. Joseph Meek, speaking on behalf of settlers, convinced Congress to establish the Oregon Territory in August 1848, making it the first official territory west of the Rocky Mountains.

October 1847 Scarlet fever invaded Tahiti, causing illness in islanders and the Rev. John Rodgerson, a volunteer from the London Missionary Society. The disease beset Raiateans for a decade. At Bora Bora, natives located a medicine to halt infection. *See also* **1848**.

December 17, 1847 Because of general ineptitude, Governor Mariano Ustáriz had to answer charges made by auditors from the District of Caracas, Venezuela. In addition to nepotism and mishandling of public monies, the governor had to explain his failure to contain a measles epidemic. Upon his ouster, Marcelino de la Plaza took charge and launched an investigation of malfeasance in public office.

1848 As more people from the North American coast crossed the Pacific, islanders experienced calamitous epidemics among virgin-soil populations. Scarlet fever emerged in Australia, New Zealand, Tahiti, and Tasmania. Simultaneously, influenza, measles, and pertussis threatened to annihilate native Hawaiians, killing 40,000 or 26.7 percent of a population of 150,000, notably, a whole generation of infants and children. Within October and November 1848, Honolulu lost 769 to pestilence imported by the American naval frigate *Independence* from Mazatlan, Mexico, to Hilo. Secondary bronchial pneumonia, cardiac failure, encephalitis, and diarrhea increased the mortality rate in July 1849. Burials without coffins became the norm. In 1853, Oahu shrank to 20 percent its former population from epidemic measles and smallpox. James Michener immortalized the struggle to save indigenous people in *Hawaii* (1966), source of a 1966 film starring Julie Andrews and Gene Hackman in the roles of caregivers.

1848 In Paris, a cholera epidemic produced a new sensitivity in Catholic officials. Less eager to blame sin and waywardness of the lower class for the outbreak, church officials supported the government's programs of hygiene rather than prayer as a means of quelling the outbreak. Emerging from the era was a respect for the urban poor and a serious analysis of social problems contributing to social unrest. Among those who died in the twelve-month siege was a heroic Prussian general, the Count de la Roche-Aymon, who served Louis XVIII of France as field marshal.

February 1848 Berlin's minister of culture posted anatomic pathologist Rudolf Ludwig Karl Virchow of the University of Würzburg to Upper Silesia to observe an epidemic that may have been typhus or relapsing fever. He analyzed the deplorable squalor of several million ignorant Polish peasants who refused to bathe and acquired crusts of dirt that hosted body lice, which produced "crowd diseases." Virchow added that a meager diet of potatoes and vodka, overcrowded dwellings, and a lifestyle that encouraged kwashiorkor, dysentery, and malaria contributed to sufferings during widespread famine from a failed potato harvest.

Virchow's conclusion, *Reports on the Typhus Epidemics of Upper Silesia*, completed on March 10, 1848, noted that, if diseases result from hardships of individuals, then epidemics indicate mass disturbance of populations. He stated:

> There can no longer be any doubt that the epidemic spread of typhus is only possible under living conditions such as those caused by poverty and lack of culture as witnessed in Upper Silesia. Remove the conditions and I am convinced that the typhus epidemic will not return [Bastian 1994].

Virchow blamed reactionary Prussian authorities for doing nothing to prevent disease among the illiterate poor. The government suspended him from his post as medical examiner for two weeks and denied him room and board at the Charité Hospital for publishing provocative liberal essays in the journal *The Medical Reform* advocating social consciousness. His work promoted the public health movement in Europe.

December 1848 An epidemic of cholera hit Bergen, Norway, and spread northwest to Manger. After the local doctor collapsed, he was too weak to serve the public cause. A priest, Father Michael Sars, took over local care. To end the epidemic and minister to the peasantry, he collected medicine, set up a contagion hospital, hired a doctor from Bergen, assembled wood and nails for coffins, and established burial grounds. After leaving the church in 1854, he studied marine biology, taught zoology, and became one of the most respected scientists of his time.

1849 An estimated 350,000 Europeans traveling over the 2,000-mile California Trail lost somewhere between 20,000 and 30,000 or 8.6 percent along the way to cholera as well as to measles, mumps, pneumonia, tuberculosis, pertussis, and accidents. George Bent, the biracial son of a Southern Cheyenne woman named Owl Woman and of frontier guide and trader Colonel William Bent, was a trusted friend of North American Plains Indians. Bent described Asiatic cholera epidemics as cramps, the term the Cheyenne used to name a scourge that had killed hundreds. As summarized in his autobiography, the sickness struck during an encounter with the great migration of prospectors west to the California gold fields in 1849. In Indian villages, braves made medicine and lifted their weapons in vain against an invader they could neither see nor kill.

During the cholera epidemic, the Cheyenne and Sioux suffered the most fatalities, which coincided with malnutrition and depression caused by the disappearance of the buffalo. In deserted villages along the Platte River, tepees held the corpses of whole families; the class system weakened. In Bent's words,

> Our tribe suffered very heavy loss, half of the tribe died, some old people say. A war party of about one hundred Cheyennes had been down the Platte, hunting for Pawnees, and on their way home they stopped in an emigrant camp and saw white men dying of cholera in the wagons. When the Cheyennes saw these sick white men, they rushed out of the camp and started for home on the run, scattering as they went, but the terrible

disease had them already in its grip [Sajna 2000, p. 71].

Some of the braves died en route, including the speaker's uncle and aunt.

The outbreak of cholera in multiple camps panicked the Indians, who broke up into small families and clans and fled to safer territory. Discouraged by events that made such a vast change in life on the Santa Fe Trail, on August 21, Colonel William Bent chose to blow up and torch Bent's Fort, the famed terminus that he, his brother Charles Bent, and partner Ceran Saint Vrain had built in 1833 in southeastern Colorado on the Arkansas River. Their despair attested to the end of an era.

1849 When cholera reached Paris, France, Joseph Désiré Tholozan, the clinical assistant to physician Michel Lévy at the Val-de-Grâce military hospital, was editing the *Gazette Médicale de Paris*. To isolate the cause and nature of the disease, he determined its pathology and incidence. Nine years later, as physician to the shah of Persia, he advised that the government take charge of in-country quarantines and suppress the annual hadj to Mecca, Arabia, as a control of cholera outbreaks.

1849 A year after the California Gold Rush brought smallpox to the southern plains Indians, a cholera epidemic savaged the Brule-Sioux, Maricopa, Comanche, Kiowa, Menominee, Ojibwa, Papago, Pawnee, and Pima. The Comanche, whose population dropped 40 percent to 12,000, lost to disease the peacemaker Mopechucope (Old Owl) and Chief Santa Anna (or Santana), a great leader and peacemaker who was the first of his people to visit the white president in Washington, D.C. Contributing to the devastation of First Peoples was the shrinkage of native buffalo herds upon which the people depended for food, hides for tepees and robes, and cultural stability. Subsequent outbreaks of smallpox in 1862 and more cholera five years later permanently debilitated the Comanche, forcing bands to merge.

1849 One of the nutritional diseases to emerge from the Irish potato famine was xerophthalmia, a drying and thickening of the eye from lack of vitamin A. Large numbers of homeless, aged, and destitute Irish people sheltered in workhouses, where faulty diet caused the epidemic eye disease among children whose bodies had inadequate stores of vitamin A.

1849 During a civil war in Samoa, the arrival of pertussis via a sailing crew from Tahiti spread to outlying islands. Only months after a storm leveled crops, the disease beset the children of Savaii and Upolu islands. Within 18 months, five percent of the citizens died, primarily the weak and very young.

January 20, 1849 After 150 pauper children died from the cholera outbreak at Peter Drouet's Infant Poor Establishment, a children's farm or public foster home for 1,400 boys in Tooting, England, journalist Charles Dickens became outraged at the loss. He wrote four angry articles for *The Examiner*, a liberal weekly journal, protesting the deaths. In his closing line of the first article, he stormed: "[Drouet's farm] was brutally conducted, vilely kept, preposterously inspected, dishonestly defended, a disgrace to a Christian community, and a stain upon a civilized land" (Diedrick, 1987).

A week later, the *Illustrated London News* reported on a public inquest, at which Drouet verbally abused a child, Henry Hartshorne, and threatened reprisals. Testimony further excoriated Drouet for starving the children on too little solid food and too much liquid. Examination of remains found distended bellies, common symptoms of kwashiorkor. The boys were poorly clothed for the winter and slept three to a bed in a cold, unventilated dormitory. Contributing to contagion were sluggish drains and an open sewer. The coroner compared the atmosphere at the farm to the Black Hole at Calcutta. The jury found Drouet guilty of manslaughter.

early spring 1849 On his journey from New York to the California gold fields, Augustus Campbell wrote six letters to his family describing the overland Panama route, which was beset by pandemic malaria and yellow fever. Hiking and riding muleback from Chagres to Panama City, he survived disease, brigands, and a travel hiatus and set out for the California coast, which he reached in May. Although he sought quick wealth from mining, he earned better pay as a doctor to diseased miners.

March 1849 When the cholera epidemic spread from India to Europe and the New World, it plunged heavily into Mexico. American pioneers died by the dozens; Mexicans suffered a much higher fatality rate. By May, 60 people died per day in Saltillo; in Durango, the rate was 500 per day. Church bells rang as residents of Monterrey held a ritual procession to plead for divine intervention against the pestilence.

To the north, bands of travelers crossing overland trails to Oregon and California experienced cholera outbreaks over a three-year period. Worsening their chances of survival were unsanitary conditions, few opportunities to bathe or wash clothes and bedding, sleeping in close quarters in an enclosed wagon bed, and impure food and water. As the wagon trains passed west from Fort Laramie, Wyoming, they met with better water sources, which increased their chances of reaching the Pacific Coast alive.

April 1849 Asiatic cholera ransacked residents of the Missouri River Valley into 1850, spreading panic. At Cincinnati, Ohio, because the worst epidemic in the city's history targeted German, Irish, and Jewish immigrants, city officials moved slowly to counter contagion. Quarantine and terror reduced commerce from the thousands of immigrant trains that typically converged in Saint Joseph and Independence to stock up on necessities before crossing the Rockies. To extend boom times, Missourians concealed from the outside world the danger of contracting disease.

Cholera struck river steamers indiscriminately. Those boats carrying contagion burned smudge pots and firebands and posted criers to yell warnings. Passenger Isaac Wistar reported 18 dead on one boat, where crew mummied corpses in blankets, stretched them out on deck, and carried them ashore for burial. At Saint Louis, Missouri, cholera hit in May, killing 34; by mid–June, the death rate was 84 per day. The dead filled cemeteries, requiring burial in the city's natural limestone caves and tunnels. The total for a population of 63,000 was 4,557 deaths, or 7.2 percent. Enforcers of health regulations posted quarantines, fumigated alleyways, drained sinkholes, removed industrial waste from Chouteau's Pond in the Mill Creek Valley, and began construction of an upgraded sewer system.

Infection spread across the plains as wagoneers moved their families steadily westward. Pioneers treasured their bibles, stores of matches, quinine, an opiate called bluemass, whiskey, hartshorn for snakebites, and barrels of pickles and citric acid, two cures for scurvy. Nonetheless, when cholera struck, some died of diarrhea and fever within ten hours of taking sick. Up the Platte Valley, an unbroken line of graves, marked and unmarked, dotted the land from Independence, Missouri, to Fort Laramie, Wyoming.

Swedish immigrant Gustaf Unonius, a lay Episcopal minister and graduate of the Nashota House seminary, fought cholera among Scandinavians of Saint Ansgarius parish in Chicago. His experience among squabbling Swedes and Norwegians and Episcopalians and Lutherans was so unpleasant that he gave up the ministry. In 1858, he returned to Sweden and dissuaded others from emigrating. He published his experiences in an autobiography, *A Pioneer in Northwest America: 1841–1858* (1861).

April 1849 Letters and journals from the California mining camps reported the malnutrition among people reduced to stirring water and a little grease into flour to make dough cakes that they baked in hearth ash. Because of vitamin deficiencies from an absence of vegetables and fruit in their diets, many bled from the gums, lost teeth, developed purplish skin and swelling and cramps in the legs, and weakened as scurvy usurped their strength. On the American River at Middle Fork, Edward Buffam dispatched a friend to buy salts. When Buffum's legs blackened, he feared the end.

A friend's discovery of wild beans produced a meal of boiled sprouts. That plus spruce bark kept Buffum alive until he could buy fresh potatoes in Coloma at the inflated price of three dollars per pound. A better source of beneficial vitamin C was lemons, which rose to a dollar each. The price encouraged California farmers to plant citrus groves, which became more lucrative than prospecting.

The experience of near starvation on the California gold fields was so harrowing to Buffum that he moved to San Francisco, where the steamer *Brooklyn* arrived from New York City bearing passengers suffering from the same disease. Scurvy became an endemic ill, especially for those traveling the eight-month passage around Cape Horn. For those passengers who died in transit, their ill fate concluded with a perfunctory wrapping in canvas and burial at sea. One victim left enough money to pay for embalming in a barrel of rum to preserve his remains for the passage back east and a cemetery burial.

summer 1849 Avian artist and journalist John Woodhouse Audubon led an expedition of 100 men from New York City to New Orleans to the mouth of the Rio Grande by water and overland across Mexico to San Diego, California. As the men reached the Sonora uplands, cholera struck amid the dry heat. At Altar, he wrote,

> Half of us are on foot, our clothes are ragged and torn, and we have lived on half rations, often less, of beans and what we call bread. Several days we were 20 to 24 hours without water, no grass for our horses, and inexpressibly weary always [Jackson 1980, p. 179].

Their health had improved late in September, when the Pima welcomed them to a village on the Gila River near Phoenix, Arizona, where the men fed on corn and melons. By their arrival in California on October 30, the expedition consisted of only 50 men.

late summer 1849 The arrival of 38,000 European immigrants to Quebec initiated a new epidemic of cholera, emanating from the quarantine center on Grosse Isle. The deaths of 1,185 newcomers paralleled a similarly desperate situation in Montreal. At Toronto, the fatalities reached 18,000.

mid–October 1849 Cholera arriving by boat from Algeria evaded the harbor blockade imposed by Ahmed Bey at Tabarca, Tunisia. Within ten months, nearly 48 percent of 118,500 victims died. The disease spread to Tunis and the Sahel, killing desert dwellers. At Tunis, the combined losses to Jewish and Muslim sectors raised the death toll to 7,300.

October 21, 1849 Ironically, an outbreak of cholera in Sacramento, California, began with the docking of the steamer *Carolina* bearing the good news of statehood. Arriving from Panama, the ship's passengers suffered 14 deaths, yet port authorities in San Francisco did not quarantine the survivors. The disease killed 20 Californians the first day. An emergency hospital set up at the Odd Fellows lodge took in patients. Within ten days, 28 more fatalities subdued jubilation over California's admittance to the Union.

Thousands fled into the hills, carrying cholera with them. When contagion killed 150 per day in early November, city appointees began burning garbage in the streets. The disease killed 16 doctors as it spread to Placerville and into mining camps. By December, the toll rose to 900.

November 1849 In the crowded city of Rio de Janeiro, Brazil, a long bout with yellow fever appeared to derive from ships traveling from Louisiana and Cuba and later from a Danish vessel from Bahía. The scourge slew over 100,000. The outbreak forced Brazilians to re-examine folk customs. Because of the seriousness of contagion, officials and medical experts persuaded citizens to abandon mythic concepts of death and to accept mortuary and burial services as a sanitary means of disposing of corpses.

1850 When prospectors overran the Northern Paiute homeland in Nevada during the 1849 Gold Rush, they carried sickness into a virgin-soil population. Six-year-old Sarah Winnemucca witnessed an epidemic of typhus. In adulthood, she drew on memories of sickness and death during her lectures promoting the preservation of native Paiute. *See also* **July 1868**.

1850 The docking of a California vessel in the Cook Islands introduced mumps, a disease that rarely troubled Polynesians. The infection apparently migrated to Samoa and other islands in the cluster.

1850 The soldiers at Camp Yuma, the forerunner of Fort Yuma, California, depended on game and a barrel of flour for meals. When the garrison moved out and left a nine-man squad to maintain the camp, they fell under siege by hundreds of Indians. The soldiers, compromised by scurvy from a lack of fresh fruits and vegetables, were so demoralized by news that their relief had turned back that they buried ammunition and equipment and abandoned the spot. The military revived the outpost at Camp Independence, a nearby ferry crossing on the Colorado River.

1850 A second cholera epidemic assaulted Cuba, killing up to 34,000 slaves. The disease spread through the Caribbean later in the decade, generating similar disproportionate deaths among Africans in the American Virgin Islands, Barbados, Jamaica, and Puerto Rico. Worsening the chances of slave survival were malnutrition, impure water, crowded hovels, and concentration of laborers near ports, the source of contamination.

1850 Theodor Bilharz, a 25-year-old German physician and researcher educated at Freiburg and Tübingen, settled in Egypt to work at Kasr el Aini hospital and teach at Cairo University. To overcome endemic schistosomiasis, he discovered and described the parasite causing the disease, which scientists named bilharzia in recognition of his studies. During an expedition on the upper Nile in 1862, his attempt to rid Egypt of the disease ended when he died of typhoid fever at age 37.

1850 During a period of low incidence of rabies in France, citizens became obsessed by fear of the disease. Hysterical pedestrians recoiled from contact with dogs and panicked over animal bites and scratches. Some developed delusions called hysterical rabies.

1850 In southeastern China, the 15-year Taiping Rebellion, triggered by religious fanaticism and starvation, was alleged by the *Guinness Book of World Records* to be the bloodiest civil war in history. A threat to the ruling Qing dynasty of China, the revolt caused the deaths of around 30 million people before eventually succumbing to Manchu forces. A barbaric war lacking the usual glorious goals and rhetoric, combat massacred people by the tens of thousands. Others succumbed to malnutrition, epidemic cholera, and suicide.

March 31, 1850 Cherokee healer and herbalist Sarah "Sallie" Ridge Paschal of Galveston, Texas, turned a two-story home into a yellow fever clinic, where she alleviated fever with freshly brewed native teas. City officials enacted a quarantine, which they strengthened in 1853 by erecting the state's first quarantine station. It was still possible for ships to circumvent restrictions by landing on the Gulf shores to offload passengers and crew.

July 1850 Along the Atlantic seaboard, a two-year outbreak of dengue fever hugged the coast as it moved from Charleston, South Carolina south to Savannah and Augusta, Georgia. It spread to the Gulf Coast at Mobile, Alabama; Galveston and Brownsville, Texas; New Orleans; and Matamoros, Mexico. Into the Caribbean, the disease felled Cubans at Havana and crept into Lima, Peru, as contagion moved inland from the Pacific Coast.

July 24, 1850 The cargo vessel *Abby Baker* embarked for the California gold fields on November 7, 1849, carrying the wife and four sons of Captain Timothy Pratt. After a three-month voyage via the Straits of Magellan, the captain died. His crew journeyed on to San Francisco and Sacramento, where a cholera epidemic infected the captain's family, killing his widow and three of their sons. Aiding the sick were Jewish volunteer efforts, led by Samuel I. Neustadt. That fall, cholera victims became the first corpses interred at the Hart Cemetery, the nation's first Jewish graveyard in the Far West.

August 7, 1850 At Guadalajara, Mexico, officials fought Asiatic cholera, which overran the Hospital Belén and the substandard, unhygienic San Juan de Díos hospital. Increasing infection were piles of human feces on the streets awaiting removal by night cart and sewage running through town into the San Juan de Díos River. Wealthier residents deposited their waste in latrines, which consisted of open holes in the ground.

Council members advocated more latrines, fewer night cart collections, restriction of fertilizing gardens with sewage, and ignition of fireworks and artillery to stir the air and rid it of miasma or stale air. Ordinances closed tanneries and moved soap works beyond the city limits. Police restricted the size of gatherings and closed pool parlors, lotteries, and social halls. To halt the spread of disease from poor to rich, authorities assigned doctors by districts.

1850s To rid Southwestern natives of mumps imported by Navaho traders, the Zuñi began performing a ritual dance in yei (spirit) masks. The dance was similar to the Navaho yebichai (or yebeichai) and incorporated a water sprinkler and clown. Typically, on the final night of the nine-day Nightway healing ceremony, Navaho dancers followed a revered masked leader, a medicine man called the Talking God, who chanted and sang.

The Zuñi version of the ceremony, which impersonated a line of benevolent gods, concluded with an all-night procession through their pueblo. The sick spat on performers and begged for the end of the epidemic. The spiritual yebichai dance became a standard part of Zuñi healing rites.

1850s When cholera devastated France, the town of Saint-Etienne instituted a model of modern epidemic containment. Officials based municipal action on the theory expounded in *Des Airs, des Eaux, et des Lieux* (*On Air, Water, and Occasions*) (ca. 310 B.C.E.), a French translation of Hippocrates's treatise, which advised community sanitation to prevent pathogens from permeating the atmosphere and water supply. In 1862, Philippe Béroud reported on the era's experience in *Étude sur l'Hygiène et la Topographie Médicale de la Ville de Saint-Etienne* (*Study on Hygiene and Medical Topography of the City of Saint-Etienne*) (1862).

1850s At Copenhagen's *Kommunehospitalet* (General Hospital), built in 1769 as a poorhouse, staff retained Denmark's poor in an isolated area. The growing number of patients, mostly the working poor, who were largely young and male, died at the rate of one in ten, usually on the day of admittance. While sheltering and nourishing the sick, the Danish medical staff examined them thoroughly, recording rampant venereal diseases, scabies, and fatal pulmonary tuberculosis.

1851 Count Edgar Bourée de Corberon, an intellectual living in Croatia, urged the Austrian monarchy to combat epidemic cholera and typhus by establishing a medical facility in Zagreb. The planned structure of Zagreb Central Hospital was not complete until 1917.

1852 The third pandemic of cholera spread from India to Tehran, Iran, where the death rate reached 700 per week. When the disease swept through the heart of Persia and north over the Caspian Sea, it killed one million Russians.

May 27, 1852 During the settlement of New Zealand by the "Fencibles," the British crown recruited retired soldiers for the Royal New Zealand Fencibles Corps to defend settlement camps, which quickly filled with immigrants. On January 14, the ship *Inchinnen* left London by way of Portland, England, with 78 male pensioners and 181 women and children. On their arrival at Auckland on May 27, 1852, for transportation to Howick, the ship's surgeon Henry Richards and his hospital assistant reported 22 deaths, mostly from an upsurge of measles and chicken pox during the voyage.

October 29, 1852 Count Gaston Raousset-Boulbon of Avignon, France, a freelancing exploiter, agreed to represent mining investors in a company called La Restauradora. He led 600 mercenaries from California to Mexico, where he faced 1,200 Mexican regulars at Hermosillo, Sonora. In combat, dysentery proved deadlier to insurgents than bullets. After five miserable months, Raousset abandoned the idea of hunting

gold and setting up a French settlement after he and his officers became too ill to carry out their initial expedition.

1853 After surveying medical techniques throughout the Caribbean and Panama, herbalist and innkeeper Mary Jane Seacole returned to her home in Kingston, Jamaica. She fought a yellow fever outbreak at Up-Park Camp and at her residence, where medics conveyed victims from ships in port.

1853 Fort Thorn, near Hatch, New Mexico, earned the nickname "the sickliest post in the territory" after travelers introduced malaria to the region (Kraemer 1996, p. 221). Recurrent fevers robbed the men of strength to perform their duties. After six years of fighting disease, military authorities abandoned the fort.

1853 Yellow fever carried on British vessels assailed a virgin-soil population in the Andes. In the high country, typhus complicated public health. At its worst, the emergent fever enveloped poor Peruvians with nasal and skin hemorrhaging and total prostration. In airless, dark huts, peasants had no chance of surviving.

1853 During the Mormon migration from Kansas to Salt Lake City, Utah, parties of converts from England and Scandinavia began their crossing of the Rocky Mountains. At Atchison, Kansas, the lead party of 500 Latter Day Saints immigrating from Liverpool, England, suffered severe cholera as it headed over the plains. Around one-third died along the way.

spring 1853 A sick sailor aboard the American trader *Charles Mallory* introduced smallpox in Hawaii, initiating a terrifying epidemic that greatly reduced the population of native Hawaiians. Harbor authorities sequestered the ship at Kalihi and transferred the sick to Waikiki, where 3,000 died. Medical authorities quarantined victims, burned their clothing and grass huts, roped off streets, and posted guards to halt the influx of disease. After Mormon proselytizer William Uaua mobilized a religious campaign on Maui and Molokai in March 1853, he pressed on to Honolulu. Elders from Utah began ministering to hundreds of people sick with festering scabies and ulceration and blinded by syphilis and smallpox, which also infected lungs, heart, and brain.

After a public health official, Richard Armstrong, ordered vaccination, fearful natives fled to the mountains or other islands. Only Lanai, Molokai, and Niihau remained free of contagion. Those islanders terrified of the inoculation procedure picked at the injection site and sucked out the vaccine. On June 15, islanders converged for a public ceremony honoring the dead. One survivor, Kamahiai, bore hideous marks on his face. His appearance was wholly changed and his throat so scarred that he could barely speak.

Into the summer, the epidemic worsened, killing some 1,500 in Honolulu. By July 1, a clinical doctor and professional inoculator, George Albert Lathrop, was treating 568 patients at one time. After 1,500 received a defective vaccine, corpses accrued in shacks, alleys, and hillsides. Those interred in shallow graves tempted prowling dogs and hungry pigs, which natives had left behind in their flight from contagion. To dispel the stench at the Kakaako pest hospital administered by Gerrit Judd, an altruistic medical missionary, attendants purified rooms with burning tar.

Despite precautions and fines for those avoiding vaccination, smallpox killed 80 percent of the natives of Oahu as well as 275 nuns and preachers. In October, the Mormons abandoned their mission. By 1857, Brigham Young ordered them all home. In early 1854, the death toll reached 5,750, completely disrupting the island development as fearful Hawaiians hid in remote locales and avoided white settlements, missions, and harbors. Blame focused on white authorities and forced the resignation of Gerrit Judd, but did not deter the island government from requiring inoculation of newcomers and Hawaiians. Assisting in the immunization program at Lahaina harbor was missionary Dwight Baldwin.

August 11, 1853 At New Orleans, the nation's "death capital," 8,000 people or 7 percent of the populace died of yellow fever, which claimed 2.5 times more whites than blacks (Pritchett and Tunali 1995). The sickness earned the name "strangers' disease" because it struck German and Irish immigrants more frequently than it felled long-term residents or native-born Louisianians, who had built an immunity (ibid.). The *Daily Crescent* reported how victims were buried 50 to a pit:

> What a feast of horrors! Inside, corpses piled in pyramids, and without the gates, old and withered crones and fat huxter [huckster] women ... dispensing ice creams and confections, and brushing away ... the green bottle-flies that hovered on their merchandise, and that anon buzzed away to drink dainty inhalations from the green and festering corpses [Hanger, 1997].

Widespread misery sent the elite and moneyed classes scurrying to northern climes until the contagion passed. In 1855, the Louisiana legislature set up a quarantine system and the nation's first board of health, ably led by Samuel Choppin and Forster Axson.

early November 1853 A cholera epidemic in Spain arrived with the docking of the battleship *Isabel la Católica* in Vigo harbor. The pathogen targeted the poor of Galicia, who were battling crop failure and malnutrition. By late summer of the following year, of 672 stricken with the disease at Vic northeast of Barcelona, 370 perished. In 1855, Madrid lost 4,200 to the scourge. By the end of the infection, over 28 percent of 830,000 victims had died. From the demographic data recorded by city and province, D. Nicasio Landa compiled for the General Directorate of Health and Charity a valuable epidemiological study of contagion and morbidity.

1854 At Savannah, Georgia, attorney John Elliott Ward used his one year in office as mayor to complete a waterworks, initiate a sewer line, and upgrade the police force and city paving. His term also called for action against a hurricane and yellow fever, a constant threat in Southern ports.

1854 After centuries of defeat by malaria, African explorers used prophylactic quinine to assure the success of overland expeditions along inland waterways. Promoted by Alexander Bryson's 1847 report, the drug accompanied Scots explorer William Balfour Baikie in 1854, when his company mapped 700 miles of the Niger River without loss of life. The successful venture along the Benue Tributary from Nigeria into Cameroon marked the beginning of European penetration of Africa's dark heart and introduction of British trade.

1854 Joseph Désiré Tholozan, in service to the French military medical corps during the Crimean War, observed battlefield acrodynia, cholera, dysentery, scurvy, and typhoid fever. He chronicled a pestilence that may have been murine typhus, a form borne by mouse fleas that causes rash, fever, and headache. The disease spread from an American transport vessel carrying horses from the Crimea. In Persia, he continued his career in epidemiology as personal physician to Nasreddin Shah, for whom he superintended treatment of cholera and plague. *See also* **1856.**

spring 1854 In the South Seas, epidemics of measles proved deadly among virgin-soil populations. With the arrival of immigrant ships, simultaneous outbreaks of scarlet fever and measles swept New Zealand, endangering indigenous tribes. Introduced by crew aboard a Tasmanian vessel, the paired scourges killed 4,000 Maori.

The H.M.S. *Beejapore*, which docked at Sydney in March, also bore measles and scarlet fever, which were common among large numbers of young and infant passengers. On the way from Liverpool, England, over 12 percent of the people on board died of contagion. Australian harbormasters immediately quarantined the vessel, but could not stop the infection from spreading to mainland children.

To the north in French Polynesia, a ship traveling from New Castle, Australia, docked at Tahiti in April on its way to San Francisco and spread measles. Infection and resultant complications endangered Polynesians, who had no immunity to the disease. From the main island, measles advanced to Moorea, claiming one-tenth of the islanders. When contagion died out in fall, some 800 Tahitians had perished, primarily the weak and elderly. The disease migrated to the Cook Islands aboard a Tahitian schooner in early 1855, felling most of the populace. At Tonga, mortality figures claim that the scourge killed five percent of islanders.

mid–June 1854 After serious epidemics of cholera in Quebec in 1849, 1851, 1852, and 1853, the next year, the disease returned to the Saint Lawrence Seaway as outsiders entered the city by ship, bringing the pathogen. Again, Montreal suffered over two months of contagion and death. The disease crept over Hamilton, Toronto, and Kingston, and into Saint John, New Brunswick, where water and food incubated bacterial contaminants. Worsening the situation in Saint John were wells that absorbed sewage. The outbreak moved south, striking Portland, Oregon, and as far inland as Manitoba's Red River Valley.

Canada's first indigenous English-speaking women's order, the Sisters of Charity of Saint John (later called the Sisters of Charity of the Immaculate Conception) got their start after the adult deaths from cholera left many Irish-Catholic orphans in the area. The sisters built a care center and erected more receiving homes to the west. An Acadian house, the Religieuses de Notre-Dame du Sacré-Coeur, extended the order's outreach, which thrived into the mid–1960s.

August 1854 In Stradella, Sardinia, city officials had to act fast to halt the advance of cholera. Mayor Agostino Depretis solicited financial aid from wealthy citizens and used the money to improve sanitation and containment methods.

August 31, 1854 After an outbreak of *Vibrio cholerae* in London in 1831, the epidemic remained virulent in England for 23 years, particularly at Newcastle and Gateshead, where it slew tens of thousands. In 1853 alone, 10,675 died in London and 1,527 at Newcastle. Hector Gavin, author and editor of the *Journal of Public Health*, served as secretary of the General Board of Health, which authorized a local study of contagion. A British surgeon-anesthesiologist, John Snow of York, studied the 800 houses at Golden Square of Saint James parish in the Soho region of London, where 500 fell sick by September 10 and, by the end of the month, 616 died. Snow isolated a major source of disease by linking sewage-contaminated water from a single pump to victims of the outbreak.

Although the Vauxhall Water Company declined to take blame, Snow's house-by-house demographics proved that victims drank from the same source, into which the soak of diapers from an infected infant had been poured. The local governing board hesitated to accept Snow's findings, but agreed to remove the pump handle. To their surprise, no new cases emerged. Illogically, officials continued to dismiss Snow's charge that the water caused the epidemic and to champion moral environmentalism, an assumption that sin and degeneracy caused disease.

Snow refused to let know-nothings ignore his discovery. In reference to *On the Mode of Communication of Cholera* (1855), a report he issued to Parliament, he stated,

> I have satisfied myself completely, that the chief mode of propagation of cholera in the South district of London, throughout the late outbreak, was by the water of the Southwark and Vauxhall Water Company containing the sewage of London; and containing consequently whatever

might come from the cholera patients in the crowded habitations of the poor; and I am satisfied that it spread directly from individual to individual, sometimes in the same family, but by similar means; that is, by their swallowing accidentally what came from a previous sick patient ["Snow's Testimony"].

His demographic and cartographic method of studying sources of drinking water, which he published in the *Medical Times and Gazette,* became a model of epidemiological investigation.

In contrast to Snow's pure scientific method was *Memoir on the Cholera at Oxford* (1856) by Sir Henry Wentworth Acland, the Regius Professor of medicine at Oxford University. He produced a demographic map of southern London and located cholera cases over past years and their contiguity to low-lying bogs, polluted water, and contaminated rivers and streams. In addition to scholarly reflection, he blended data on sanitation with moral and religious analyses of the epidemic.

October 21, 1854 When Florence Nightingale left England for Scutari, Turkey, on the Bosporus, British soldiers fighting the Crimean War contended with wounds, cholera, dysentery, scurvy, and typhus and died at the rate of one out of two. She found them lying on straw mats in stone cells at the Barrack Hospital, an abandoned military quarters near Istanbul. The Sisters of Mercy had already dispatched 50 nurses to treat the 4,000 ailing British.

On November 5, the Nightingale cadre of 38 began filling four miles of beds with casualties from the battle of Inkerman. The women set up a laundry and dietary center functioning round the clock and organized medical supplies, bedclothes, utensils, soap, and sickroom essentials. Supporting their efforts were Hector Gavin, the Crimea's sanitary commissioner, and volunteer dietitian and chef Alexis Soyer, who applied restaurant management to military kitchens and improved survival rates by upgrading nutrition and cleanliness. By contrast, the French and Russian armies continued to suffer huge losses from typhus.

When Nightingale returned home in July 1856, a personal bout with typhoid and battle fatigue threatened her health. A year later during the Sepoy Rebellion in India, she outlined by mail from her home sufficient sanitary guidelines to stem epidemics of cholera and plague. Outraged at wasted lives, she scolded the military for

tolerating a mortality rate six times higher than that of British civilians. Under her orders, the military medical corps lowered the death rate from 6.9 to 1.8 percent.

1855 Epidemic cholera returned to southwestern Europe, killing 180,000 in Spain. When contagion reached Istria from northeastern Italy, a priest, Antonio Fachinetti of Savicenta, compiled *Memoria in Occasione del Cholèra-Morbus nell'Anno 1855* (*Recalling the Cholera Epidemic of the Year 1855*) concerning the spread of the third cholera epidemic, which began in 1849. In addition to epidemiological details, he gave an ethnic breakdown of the citizenry drawn from parish archives.

1855 The introduction of epidemic disease in Brazil divided health and civic leaders over preventive measures. Contributing to the public forum were articles in city newspapers on sanitation and the causes and transmission of yellow fever, which was pandemic in the coffee estates. Of the 200,000 victims, a large share were African slaves. The loss of blacks as laborers precipitated economic disaster among planters and investors. At Bahía, navy surgeon Freitas Albuquerque observed the epidemic and later used his experience as a healer during an outbreak at Belém.

1855 Diphtheria blanketed Europe with disease, which then infected Africa, Asia, Australia, and North America. The British recorded a death rate of 517 per one million citizens. Hard hit in Sweden, children collapsed by the thousands. At Sundsvall and Göteborg, disinfection of residences, isolation at epidemic hospitals, and serum treatment improved mortality rates.

1855 To stop endemic yellow fever in Havana, Guillermo Lambert de Humboldt evolved a homeopathic form of vaccination. It gained the support of the Cuban governor, José Guiterrez de la Concha, who wanted some method of protecting soldiers from disease. The method figured in *Pneumopathies Infectieuses Communautaires chez la Personne Agée* (*Infectious Lung Disease in the Aged*) (1858) by Nicolás Manzini, who blamed poor sanitation as a source of contagion.

June 7, 1855 The 1855 yellow fever epidemic in Norfolk, Virginia, earned the nickname "The Death Storm." The loss of 2,000 residents produced one of the city's worst disasters. The last of five outbreaks, the scourge arrived at Hampton

Roads during an emergency stop of the steamer *Ben Franklin*, which was traveling from Saint Thomas in the Virgin Islands up the Atlantic coast to New York City. Over a 12-day quarantine of the ship, the captain violated harbor regulations by emptying his bilges and opening the hatches. The first victim, a machinist laboring in the hold, died on July 8.

The disease rapidly engulfed a slum known as Irish Row and spread to Portsmouth. Health agents evacuated the tenement and received residents at Barry's Row. Citizens in flight from the "yellow jack" attempted a retreat from the pestilence, but found themselves unwanted at Old Point, where guards shouldered weapons to keep them out. By summer's end, the Norfolk business community was moribund as volunteers arrived to treat the fallen.

Within 90 days, the fever infected nearly every resident. One steamer, the *J. E. Coffee*, delivered mail and coffins. Eventually, the number of corpses overran accommodations and required burials in other containers or simple wrappings. George D. Armstrong, a pastor at Norfolk Presbyterian Church, compiled epidemic data in *A History of the Ravages of the Yellow Fever in Norfolk, Virginia, A.D. 1855*.

summer 1855 The emergence of cholera at Fort Leavenworth, Kansas, demanded additional medical care and ward space to relieve the crowded hospital. Some of the 115 sick soldiers were billeted on the floor of a hot, fly- and rat-infested stable. The men drank from a common container—water barrels filled at the Missouri River. Sick men could observe the loading of coffins with 24 corpses, one of which was too large to fit and had to be wedged in by force.

1856 When tuberculosis invaded Toronto's immigrant population, patients took solace from the Sisters of Saint Joseph, a company founded in Montreal in 1843 by Bishop Ignace Bourget. The nuns established a benevolent shelter and food bank, the House of Providence, one of four Catholic hospitals built by the sisters.

1856 Amid rampant dysentery in India, Edward John Waring, an English physician with the Madras Medical Service and personal attendant on the Rajah of Travancore, studied racial differences that caused a higher rate of infection in European soldiers than in local sepoys.

1856 Norwegian officials directed victims of epidemic leprosy to the district health officer for identification and recording on the national leprosy register. Public health initiatives also provided examination and treatment of patients' families. With safeguards of privacy, some people avoided being labeled as lepers.

1856 At the end of the Crimean War, the combined losses from wounds, frostbite, and disease devastated the allied forces of England, France, Russia, Sardinia, and Turkey. Deaths were overwhelmingly the result of contagion from scurvy, typhus, cholera, typhoid, malaria, and dysentery, although data vary by source:

country	mortality	combat deaths	contagion deaths
England	22,000	4,000	18,000
France	95,000	30,240	6,000
Russia	475,000	128,000	347,000
Sardinia	21,000	2,194	1,972
Turkey	165,000	45,400	20,900
Total	750,000	approx. 20%	approx. 80%

1856 In Costa Rica, Nicaraguan insurgents carried cholera that killed a total of 10,000 in Alajuela, San José, and Cartago. Disease inflicted more misery and loss than natural disasters or combat. As soldiers returned home, they carried cholera to their communities, where one-tenth died. Treating the sick was immigrant physician Carlos Hoffmann, a German biologist educated by Alexander von Humboldt, who also served as chief surgeon to the Costa Rican military.

April 1, 1856 Louis Vallery-Radot, biographer of epidemiologist Louis Pasteur, reported on a six-week period in which deaths from puerperal fever overran the Paris Maternity Hospital. Of 347 patients, 64 or 18.4 percent died of childbed fever. The alarmed staff closed the facility and transferred remaining patients to another locale. Unfortunately for the new mothers, they carried contagion with them and all died of it.

November 15, 1856 Upon observing an epidemic of typhoid fever at a Clergy Orphan School at Saint John's Wood, London, William Budd of Devon, a lecturer at the Bristol Medical School, published in the November 15, 1856, issue of *Lancet* a summary treatise, *Typhoid Fever: On Its Nature, Mode of Spreading, and Infection*. He established the severe contagious nature of the disease and recognized that infection spread from fecal matter to water, milk, and mouths. For control of infectious body fluids, he urged strong solutions of zinc chloride for washing and disinfecting the patient and sickroom.

1857 In the marshes of England, unidentified ague was an endemic misery that had beset the populace from 1826 to 1829. Fenlanders assumed that the disease was malaria derived from exudations of local bogs. Although few perished of the tertian marsh fever, most suffered pervasive chronic sickness, which they treated with white poppyhead tea. The disease weakened workers and limited economic advancement because traders avoided the area. The most ambitious fenlanders migrated to safer climes. Local healers evolved their own deterrents and curatives for bog ague. Mosquitoes carrying the pathogen became extinct, ending the ailment.

spring 1857 As a 2,500-man cavalry expedition advanced on Mormons at Utah from Fort Leavenworth, Kansas, soldiers suffered a mix of ailments, including enteritis, poison oak, and syphilis. Contaminated water and inedible food exacerbated symptoms. The men encountered Pawnee suffering a smallpox outbreak. Some soldiers acquired syphilis from contact with female Pawnee. As the column approached the Platte River in Nebraska in late summer, outbreaks of malaria and scurvy added to the misery.

June 1858 Disease traveled with Commodore Matthew C. Perry when he sailed his flagship, the sidewheel steamer U.S.S. *Mississippi*, to Japan. When his fleet of four ships arrived from Shanghai to a treaty-signing in Nagasaki, one shipboard case of cholera spread throughout the port and as far north as Tokyo. As contagion gripped the island chain, it killed three million people.

Ogata Koan of Bitchu, a physician and medical instructor trained in Nagasaki in the Dutch tradition, collected data on the epidemic, which he published in *Korori Chijun* (*Treatise on Cholera*) (1858). As a result of his accumulated knowledge of contagion, in 1862, he advanced to the post of physician to the shogun and director of Igakusho, the Tokugawa shogunate's institute of Western medicine. Medical historians honor Ogata as the modernizer of Japanese medicine.

1860 Physician Thomas Collett began treating endemic scabies among the peasants of western Norway by community regiments rather than

individual protocols. The board of health provided barrels of sulfur ointment for classroom teachers to distribute to pupils infected with scabies and also to their families and playmates. Collett's experiment encouraged self-care and the formulation of sulfur medicines.

1860 In the agricultural South—Alabama, Florida, Louisiana, Mississippi—productivity declined over two decades in part because of endemic hookworm among U.S. citizens living in unhygienic conditions. Medical investigators deduced from skeletal remains and mortality figures that workers were incapable of maintaining a vigorous day's work. Because of the ravages of *Necator americanus*, "the vampire of the South," the laboring class suffered chronic iron-deficiency anemia, impaired brain function, and one million deaths annually. Hardest hit were some 40 percent of school-age children.

1860 Measles again robbed Australia of its young children after the disease blanketed Sydney. Over 270 died. Others were too drained of strength and immunity to fight off dysentery, enteritis, and pneumonia.

July 1860 Cholera, the frequent ravager of India, recurred at Bengal (modern Bangladesh) before sweeping Agra and Mathura in midsummer. *See also* **May 1861.**

December 1860 The arrival of measles at Perth in western Australia caught the attention of reporters for the *Independent Journal*. They noted with relief that only residents of York were infected. However, by December, the disease had afflicted nearly all of Perth's households and spread into Fremantle. In all, 57 people died of the scourge.

1860s Emergent kala azar, a malaria-like fever that enlarges the spleen, spread through Bengal (modern Bangladesh). Although it had been active in that part of India for four decades, it remained unidentified until 1903 when it was named leishmaniasis and the vector parasite labeled *Leishmania donovani*.

1860s According to a doctoral dissertation written by Sophia Jex-Blake in 1877, the New England Hospital for Women and Children in Boston, Massachusetts, contributed to the rise in deaths from puerperal fever by failing to uphold standards of disinfection. She also charged that the hospital placed undue stress on unwed mothers by isolating them from the institution's general population.

1860s At Merthyr Tydfil in south Wales, epidemic cholera threatened investment in coal and ironworking. To save workers for service to these operations, industrial management provided the area with clean water and a sewage system and a cadre of medical officers.

1861 The medical history of Oceania derived in large part from the writings of missionaries. When the ship *Hirondelle* carried measles to Anatom, Vanuatu, within three months, Pacific islanders died at the rate of one in three. The contagion advanced on Erromango, where a missionary, George Gordon, interpreted the terrible loss as God's retribution for the people's crimes and sins. Natives were so angry at his self-righteousness that they dismembered him with axes.

According to *Histoire de Rotuma* (ca. 1873), a log kept at the Sumi Catholic Mission, Father B. Joseph Trouillet compiled the oral history of Rotuma in the Society Islands. He surmised that in 1861 during the reign of Kaunufuek, the 87th high chief, dysentery reached epidemic proportions. The disease returned a decade later, claiming a total of 68 whites and islanders.

1861 A frontier legend and ballad from Alma, Colorado, credited a dance hall employee nicknamed Silver Heels with nursing sheepherders from Bayou Salado who were stricken with smallpox. Because she contracted the disease and incurred scarring and disfigurement, she crept out of town. Miners named a local peak at Fairplay, Colorado, Mount Silverheels in her honor. A ghost story reported that a veiled woman garbed in black wandered the Buckskin Joe cemetery and wept at the graves of the dead.

1861 A squadron of four ships—the *Jason, Firebrand, Spiteful,* and *Racer*—from Nassau in the Bahamas, Gulf of Mexico, Jamaica—imported to Halifax, Nova Scotia, passengers and crew ill with the fever, catarrh, bloody vomit, and diarrhea of yellow fever. Out of a total of 855 on the ships, 499 caught the contagion, which killed 162. Preserving the hardships of the era are the records of dockyard hospitals and a naval cemetery.

1861 During a scourge of yellow fever at Saint-Nazaire, a port city in Brittany, François Melier and his colleagues deduced that the harbor was the source of contagion. Melier's understanding of the movement of pathogens led to port authority quarantines, ship inspections, and other sanitary procedures.

April 12, 1861 Shortly after the outbreak of the U.S. Civil War, military camps required improvised hospitals in churches, hotels, schools, government buildings, and homes to house and treat the wounded. The makeshift clinics also received soldiers suffering from pneumonia and outbreaks of dengue fever, dysentery, enteritis, measles, mumps, and typhoid. An immediate call for volunteers preceded organized patient transfer and the set-up of kitchens preparing invalid foods and tonics. Noncombatants collected an array of beds and mattresses, pillows and bedding, bandages, dishes, and cutlery.

In Washington, D.C., volunteers laundered linens and patient clothing, fought lice, and hand-fed the sick. Poet and medic Walt Whitman, author of *Memoranda During the War* (1876), made himself welcome with gifts of citrus fruit, jams and fruit syrup, pickles, dried fruit, and rice pudding, a food easily digested by feverish patients. Another literary volunteer, novelist Louisa May Alcott, caught typhoid fever and had to return home for treatment and recuperation.

May 1861 When a renewed outbreak of cholera hit India, it struck English barracks as well as Bengali prisons. Within two months, contagion advanced from Allahabad and Kanpur into Gwalior and Jabalpur before reaching Delhi at the height of summer. Outside Lahore, heavy losses among British troops required repositioning of forces to safer ground.

As local people collapsed with sickness and hospital staff became too ill to attend patients, natives turned to folk fetishes as a deterrent to infection. The dubious debated the possibility that cholera originated in European settlers. The British, fearing for their investment in the Indian subcontinent, named Sir John Strachey to head a board of inquiry and initiated standards of sanitation for army barracks and government compounds.

The next summer, the scourge swept Persia, killing up to 840 people weekly at Meshed, Iran, including a royal prince. A contingent of pilgrims from Meshed carried the infection throughout Tehran's slums, where officials made no effort at containment. The disease spread across the Arabian Peninsula and north across the Caspian Sea into Russia.

In 1865, long-lived pockets of cholera refreshed the scourge in India. Rapidly, disease spread along the Malabar coast, felling 40,000 on its way east to Karachi, Pakistan, and into Persia. Overall, India suffered 84,000 reported deaths from the pestilence.

July 21, 1861 After the battle of Manassas, Virginia, Captain Sally Louisa Tompkins, the Confederacy's only official female officer, turned her Richmond home into the 22-bed Robertson Hospital. Through her makeshift clinic passed combat wounds, pneumonia, dysentery, and typhus. Her nephew reported that "As medicines were contraband of war, her treatment for all diseases was air, light, turpentine, and whiskey, all home products. If these failed, her panacea was prayer and the Bible" (Boardman 1916, 75). Of the 1,333 patients she treated, only 73 perished. When she died in 1916, her amazingly high standards of care and cleanliness earned a funeral with military honors and a tombstone erected by the Daughters of the Confederacy.

1862 The fourth pandemic of cholera moved from Bengal and Mauritius to Japan, Indonesia, Europe, North America, and Africa in 12 years. *See also* **November 1863; May 1865; October 1865.**

1862 During the Civil War, to escape the tedium of a Virginia camp, soldiers in Vermont's Second Volunteer Infantry gladly took part in the Peninsular Campaign. From the combined effect of contaminated water, exposure, and fatigue in filthy camp hospitals, they died in record numbers from chicken pox, dysentery, enteritis, malaria, measles, mumps, pertussis, pneumonia, and typhoid fever. The U.S. Surgeon-General, William Alexander Hammond, had few cures or palliatives to offer beyond coffee, quinine, and whisky.

1862 A three-year epidemic of scarlet fever struck the Tlingit of British Columbia and the Mackenzie-Yukon tribes of northwestern Canada. Simultaneously, a pandemic of scarlet fever and acute rheumatic fever in western Norway remained virulent until 1884. Of 1,155 victims, mostly children, 13.3 percent died, primarily from kidney failure. Simultaneously, 76 adult patients contracted acute rheumatic fever. Health officials surmised that separate strains of streptococci caused the outbreaks.

1862 The start of a four-year epidemic of typhus in London resulted in more than 3,100 hospital admissions and 10,000 deaths, striking a disproportionate number of slum dwellers.

February 1862 Japan suffered its worst onslaught from measles, leaving only one-third of the citizenry untouched. Imported from a vessel arriving at Nagasaki, the disease pummeled Tokyo, infecting children and teens, including the popular young Shogun Tokugawa Iemochi, husband of the Princess Kazunomiya.

March 12, 1862 An outbreak of smallpox among Pacific Coast and Puget Sound Indians began with the docking of the *Brother Jonathan*, a steamer carrying 350 gold-seekers from San Francisco to Victoria, British Columbia. Newcomers spread disease as they filled grog shops and brothels staffed by native women. The *Daily British Colonist* reported the first case and blamed Indians for "filthy habits." White medical officials vaccinated whites, but few Indians. In the opinion of Robert Boyd, author of *The Coming of the Spirit of Pestilence: Introduced Infectious Diseases and Population Decline Among Northwest Coast Indians, 1774–1874* (1999), the epidemic was solely the fault of exploiters and racists.

The first deaths were reported outside Victoria, where Vancouver Islanders began ousting the native population and burning their dwellings. The exodus spread disease into Sitka, Alaska, and Washington state, killing tens of thousands who had responded to promoters seeking hearty women, sturdy woodsmen, and devoted families as settlers. In April 1863 at Fort Victoria, John Sebastian Helmcken used an ivory blade to vaccinate 500 natives. In May, a missionary, Father Leon Fouquet, vaccinated 3,400 Indians, but the clinic was too late and too limited.

As the disease worsened, the Haida and Tsimshian fled contagion by canoe. In the lead was the Rev. William Duncan, a Church of England missionary, who guided the Tsimshian toward his compound at Metlakhatla. Within five days, 300 Tsimshian developed symptoms. Two days later, with the aid of local officials, the Songhee rushed to the Discovery and San Juan Islands, carrying contagion with them.

The death toll was devastating. Helmcken

reported that nearly every Indian infected with the pox died,

> whether he was taken care of in the Indian small pox hospital or not—and it was also said whether he had been vaccinated or not. I do not believe the last assertion because the Songish [Songhees] Indians kept comparatively free from the disease and many of them at various times had been successfully vaccinated by me—arm to arm [Blakey-Smith 1975, p. 187].

The serious loss of life and the maiming of survivors by scarring and disfiguration were typical of a virgin-soil population encountering European disease.

Native medicine was partially to blame for the high mortality. Joseph Crow was an eyewitness to a sweating-and-cooling regimen at the Tomanus House, which concluded with family tossing their sick into the bay, where they died from shock. Workers segregated the sick in the woods. They tied delirious children to trees and left them to die alone with only a blanket for warmth, water to drink, and salmon to eat. As the dead rotted unburied, a putrid odor hovered.

By December, 14,000 deaths were reported among the Bella Bella, Haida, Kwakiutl, Makah, Skallam, Tlingit, and Tsimshian. Some feared for the survival of the Bella Bella and Tsimshian nations. The Songhees, who received the greatest number of vaccinations and profited from isolation, lost the fewest members.

April 8, 1862 Following defeat at Fort Donelson and Fort Henry, Tennessee, Confederate soldiers lost Island No. 10, which General P.G.T. Beauregard, commander of the Confederate Army of the Mississippi, had selected as the bulwark for defending the Mississippi River. Sixty miles below Columbus, Kentucky, rebel soldiers feared the heavy artillery of a U.S. Navy flotilla. Added to their problems with poor leadership and rainy weather was a three-pronged outbreak of dysentery, measles, and mumps. As described by letters written by Confederate regimental surgeon Samuel H. Caldwell to his wife, the men despaired and caved in rapidly under Union assault.

summer 1862 In an eyewitness account, privateer Thomas E. Taylor, author of *Running the Blockade: A Personal Narrative of Adventures, Risks, and Escapes during the Civil War* (1896), claimed that yellow fever engulfed Nassau, filling the Caribbean port with sickness, death, and mourning. The disease beset ports along the Atlantic, especially Charleston, South Carolina, and Wilmington, North Carolina, where 42 percent of the populace died.

1863 After an outbreak of leprosy swept Hawaii, King Kamehameha V criminalized the sick and exiled sufferers into permanent quarantine at the settlement at Kalawao on the Kalaupapa Peninsula of Molokai, a virtual stockade made inaccessible by steep escarpments. At the medical colony, they were further segregated by race. Joseph de Veuster of Tremeloo, Belgium, studied the Hawaiian language and received ordination at Sacred Hearts Congregation in Louvain as Father Damien, a devoted nurse-missionary to the island's hopeless lepers and their families.

In 1873, Father Damien began ministering to the outcasts with medicines and bandages. For the dead, he constructed coffins and dug graves. In 1884, when he, too, was infected, Franciscan sisters and Vermont native Ira Barnes Dutton of the Sacred Hearts Brotherhood traveled to the commune and treated Father Damien until he died at age 49 in mid–April 1889. Hawaiians raised a statue to his memory in Honolulu and in National Statuary Hall in Washington, D.C.

July 1–3, 1863 At the battle of Gettysburg, Pennsylvania, General George E. Meade's Union forces suffered 23,000 casualties to the Confederacy's 31,000. Some 5,000 victims of dysentery or enteritis accompanied the ambulance train. As explained in Colonel William W. Blackford's memoir *The War Years with Jeb Stuart* (1945), General Robert E. Lee was among the number so weakened by enteritis that he couldn't sit a saddle.

August 1863 At the beginning of the third year of the Civil War, the Union Army established a pesthouse at Ellis Island on the Mississippi River outside Alton, Illinois. During an outbreak of smallpox, prison staff quarantined Confederate prisoners of war and some civilians. Eyewitness accounts contained in *A Camp and Prison Journal* (1867), compiled by Confederate Captain Griffin Frost, a newspaper publisher from Palmyra, Missouri, provide glimpses of the suffering of inmates and the dying, most of whom came from Arkansas, Mississippi, Missouri, and Tennessee. Staff buried the dead in unmarked graves and interred others at the Confederate Soldiers'

Cemetery in North Alton. The island acquired the folk name Smallpox Island.

November 1863 For three years, the fourth cholera pandemic spread over India and to the Middle East and Mediterranean. At Pandharpur, Hindus visiting the holy site suffered heavy loss to cholera. Those departing for home infected other Indian cities as they sickened and died along the way. By December, Bombay was seriously engulfed with pestilence, which killed 4,588 within the next year. *See also* **May 1865**.

November 1863 At Dunedin on New Zealand's South Island, a heavy toll of dysentery, enteritis, and typhoid worsened with an epidemic of scarlet fever. Officials reported 119 deaths. To ward off future outbreaks from incoming passenger vessels, the Port Chalmers authority established two islands as quarantine stations.

1864 Sisters of Charity from Leavenworth, Kansas, set up the first civilian hospital. They sheltered cholera victims at Fort Harker, where 200 died of disease. The nuns found homes for children whose parents died during the epidemic. Superintending the effort was George M. Sternberg, who upgraded sanitation and personally treated the sick while compiling data from the era for the Army Medical Museum.

1864 When hundreds of Union soldiers slowly starved on inadequate rations at the Confederate prison in Andersonville, Georgia, they died of an undiagnosed affliction. Staff assumed that the cause was typhoid fever. In 1916, W. J. Kerr, the prison's former chief surgeon, surmised from the symptoms of epidemic pellagra in the South that this disease probably had killed the ill-fed prisoners.

January 1864 General James Henry Carleton and Colonel Christopher "Kit" Carson rounded up 9,000 Navaho from their pastures and orchards at Canyon de Chelly, Arizona. Their soldiers destroyed hogans, flocks, and peach trees to prevent the Indians from returning. In March, the resulting Long Walk took the unwilling Navaho plus 500 Mescalero Apache some 300 miles southeast to Fort Sumner at Bosque Redondo, New Mexico. Those who collapsed along the way the soldiers shot and hastily buried.

In virtual imprisonment, the Indians starved from crop failure, sickened on brackish water, and died of outbreaks of chicken pox, dysentery, malaria, pneumonia, and smallpox. In a cultural stalemate, military doctors refused to acknowledge Navaho healing methods; the Navaho rejected Army medical procedures. Hundreds of native women also contracted venereal disease from federal soldiers. It took eyewitness accounts from Navaho Chief Barboncito and military leader Manuelito to awaken government officials in Washington to the plight of Fort Sumner's internees. When the U.S. army reversed the migration with the Treaty of 1868, the remaining 7,000 Navaho occupied a 3.5 million-acre reservation, much of which consisted of their ancestral homeland.

February 1864 Allegedly spread from a Portuguese vessel, epidemic smallpox hit northern Angola at Ambriz, causing an exodus to the north. Contagion traveled on a caravan route along with loads of copper, cotton, ivory, and wax, causing outlanders to halt trade. Two famous victims of the epidemic were Portuguese explorer Joaquim Rodríguez Graca and Hungarian adventurer-cartographer László Magyar, who had searched for the headwaters of the Congo River and had become a local chief after marrying Princess Ozora, the daughter of the chief of Bihe.

April 1864 During three years of South Pacific raids by Peruvian blackbirders, the kidnap of 360 islanders from Rapa, a remote southern island of Austral Archipelago in French Polynesia, resulted in enslavement and transportation to Peru aboard the infamous "snatch-snatch" ships. Slavers advertised most of the lot as manual laborers and sold them at Callao for £2 each to plantation owners. Of the snatched islanders, 344 died of dysentery and smallpox.

Within months of the raid on Rapa, the Peruvian government ended the brutal slave trade and repatriated captives to Polynesia. On return of the 16 survivors to Rapa on August 18, 1864, the captain of the ship *Barbara Gomez* refused to disembark at the island, where fewer than 66 percent of islanders survived epidemic smallpox and tuberculosis. That same year, Father Eugène Eyraud, Friar Roussel, and a band of priest-missionaries began Christianizing and baptizing islanders. By mid–1877, Rapa's population sank to 111 and its culture and written language virtually disappeared.

spring 1864 While yellow fever raged for the third year along the Atlantic seaboard during the

Civil War, Southern sympathizers headquartered in Canada hatched a plot to spread the disease to Northern cities. A prominent Kentucky cavalryman from Woodford County, Luke Pryor Blackburn, a field surgeon under General Sterling Price, claimed to be a New Orleans physician. In mid–April 1864, he traveled from Halifax, Nova Scotia, to Bermuda to treat fever victims and won a citation from the Queen's Admiralty for altruism.

A month later, Blackburn shipped to Halifax eight steamer trunks packed with vomit-soaked bedding, gowns, and soiled poultices to use in bioterrorism in Union territory. London-born shoemaker and spy Godfrey Joseph Hyams, a former rebel agent stationed in Montreal and Toronto, was to take charge of the trunks when they arrived on the steamer *Alphia* on July 12. He earned $100 for transporting and selling the containers at auction in New York City, Philadelphia, and Washington, D.C. He also offloaded trunks filled with soiled goods at troop headquarters at Norfolk, Virginia, and New Bern, North Carolina, both of which were under Union control.

The scheme caught the attention of Charles M. Allen, U.S. ambassador to Bermuda. A plot and counterplot ended with the arrest of Hyams. The clothing did reach New Bern in summer during an epidemic of yellow fever that killed 2,000, only 280 of whom were Union soldiers. Supporters of Blackburn erroneously assumed that his plot was successful.

On September 4, Blackburn returned to Bermuda to collect more infected goods. During a subsequent yellow fever outbreak, a notice in the *Bermuda Royal Gazette* summoned island health workers to confer with him at the Hamilton Hotel, which officials had converted into a fever hospital. During Blackburn's service to the sick, he covered patients with blankets to induce sweating. He collected another three trunks of contaminated bedclothes and patient gowns, which he stored on Saint George Island until he could ship them to New York in spring 1865.

The plot emerged at the May 1865 trial of conspirators in the Lincoln assassination. Testifying to the Canadian-based Confederate league, the disgruntled and underpaid Hyams told how biological terrorists were using yellow fever "directed against the masses of Northern people solely to create death" (Steers 2000, p. 60). The *Montreal Gazette* castigated Blackburn for committing "an

act [that] cannot be held to belong to civilized war. It is an outrage against humanity" (ibid., p. 67).

Blackburn went on trial in Toronto for infringement of the Neutrality Act. During extensive testimony, prosecutors presented evidence of the outbreak in New Bern and of Southern agents attempting to pass fever-infected dress shirts to Abraham Lincoln. Although investigators found the second stash of trunks on Saint George Island, Blackburn, the "Confederacy's Angel of Death," was acquitted of attempted mass murder for lack of evidence (ibid., p. 69).

The U.S. media pursued the matter of bioterrorism. The Philadelphia *Enquirer* and the Washington *Intelligencer* reported the 50-day trial transcript to readers. A New Bern paper exclaimed:

> This hideous and long studied plan to deliberately murder innocent men, women, and children, who had never wronged Dr. Blackburn in any manner, is regarded here as an act of cruelty without a parallel—a crime which can only be estimated and punished in the presence of his victims in another world. [ibid., pp. 67–68].

When New York newspapers reported Blackburn's bioterrorism, they labeled him the "fever fiend." Corroborating the story were the writings of privateer Thomas E. Taylor, author of *Running the Blockade: A Personal Narrative of Adventures, Risks, and Escapes during the Civil War* (1896), who learned of additional plans to collect infected goods to send north. Unknown to the conspirators was the fact that yellow fever is transmitted solely by the bite of a disease-bearing mosquito. Still respected as a humanitarian and healer, in 1879, Blackburn served one term as governor of Kentucky. *See also* **January 1878.**

July–August 1864 During General William Tecumseh Sherman's assault on Atlanta, Georgia, epidemic dysentery, enteritis, and malaria (called typhomalarial fever), along with measles, mumps, pneumonia, scarlet fever, scurvy, smallpox, venereal disease, and yellow fever, produced huge losses for both Union and Confederate soldiers. In the weeks preceding capitulation of the Southern army in September, the major outbreaks resulted from climatic change and sleeping in close quarters under wet bedding. The most destructive disease to manpower was dysentery, which struck 21,094 Yanks and 8,353 Rebs. Sherman's forces, who lacked immunity to malaria, suffered nearly three times the losses to the Army of Tennessee. Scurvy worsened in late July after the Tennesseans withdrew into Atlanta, where they had no source of fruits and vegetables. Adding to the death toll was catastrophic erysipelas, gangrene, pyaemia, and tetanus from septic wounds and unclean hospitals and kitchens.

July 6, 1864 At Fort Larned, Kansas, an army outpost on Pawnee Fork eight miles from the Arkansas River, venereal disease was endemic among soldiers. Cholera broke out in July after the 38th Infantry stopped on their way to New Mexico territory. Although the surgeon general warned of contagion, the fort's commander admitted the troopers. After the first victim fell ill on July 6 and died within ten hours, the disease spread four days later, killing more soldiers.

September 17, 1864 Confederate authorities erected a prison stockade at Florence, South Carolina, to house 6,000 Union prisoners of war. Guarded by men too old or too young to fight for the Confederacy, the inmates suffered malnutrition, battle fatigue, smallpox, and yellow fever. Unsanitary conditions and overcrowding led to 2,802 deaths. Among the sick was a transdresser, Florena Budwin, who had followed her husband into the Northern army. She became the first female veteran buried in a national cemetery.

1865 In the Marquesas Islands, many native children died from an epidemic of "Chinese scabies," caused by an infestation of mites. Hastening the victims' deaths were secondary infections produced by scratching at the lesions.

1865 Among creoles on the French island of Guadaloupe, cholera produced a misguided response from government doctors, who at first diagnosed the disease as a pernicious fever.

1865 Louisville, Kentucky, fought epidemic cholera by declaring public health the city's social responsibility. Ordinances ended the dumping of refuse on city streets and required the penning of animals. The creation of a board of health and street cleaning and garbage collection services helped to keep the water-borne pathogen out of drinking water.

1865 Following the American Civil War, the Caddo and Hasinai, who had been removed to Texas, Kansas, and Colorado, fought back against U.S. agents. Returned to Indian Territory under army escort, the Caddo, headed by Show-e-tat (or Little Boy), were traveling with the Delaware and Shawnee when cholera struck. As the company approached the Washita River valley in west central Oklahoma, some 60 deaths decreased their number, along with another 47 when they arrived at Fort Washita.

May 1865 When smallpox infected Menominee living in Keshena, Wisconsin, enforcers of health regulations insisted on immediate interment of the dead. A disobedient Belgian priest, Joseph Mazeaud, conducted regular funeral rites and the disease infected all attendees. Of the 79 victims, most were Catholic converts, notably Chief Pegah Kenah. Authorities arrested the priest for violation of health regulations.

May 1865 Carried by sea from Singapore through Bombay to Arabia, the fourth Asiatic cholera pandemic struck the Middle East, targeting an Islamic hadj to Mecca and claiming one-third of some 90,000 pilgrims. When the faithful returned home, they carried contagion throughout the Muslim world and over Europe. In Ancona, Italy, Count Girolamo Orsi, a philanthropist who established a foundling hospital for orphans, interpreted the outbreak of disease as the result of moral failings. By fall, infection arrived in New York harbor. During an outbreak in Brooklyn, 19-year-old Susan McKinney Steward was so moved by suffering that she entered medical training and became the first black female physician in New York State. In the late 1860s, the disease invaded South America and Russia.

During the first half of the 1870s, the disease afflicted the Balkans and Germany, moving into the low countries and Scandinavia as well as France. From Marseilles, the pathogen traveled to Valencia, killing 120,000 as it spread over Spain. During an early winter visit to Saxony, German hygienist and physiological chemist Max von Pettenkofer, author of *Handbuch der Hygiene und der Gewerbekrankheiten* (*Handbook of Hygiene and Occupational Disease*) (1883), was able to observe one facet of the global epidemic and to assist local health workers. Cholera spread to Southeast Asia and Indonesia in 1873, but did its greatest damage to Japan from 1877 to 1879.

summer 1865 The Welsh of Swansea suffered the only recorded outbreak of yellow fever in the British Isles. Transported from Cuba by the vessel *Hecla*, the disease was not noted in the captain's report to harbor authorities. The media, lulled by past experience, falsely assured readers that the disease was not viable in a temperate climate.

June 1865 During the Civil War, with one in every 29 Union soldiers dying in the field, the U.S. Army had to combat enteritis, dysentery, and typhoid caused by filth, infectious latrines, and shallow wells. The distribution of quinine slowed malaria outbreaks. At the signing of the Confederate surrender at Appomattox, for a total of 806,755 troops, the death total was 359,528. Of these, 110,070 or nearly 31 percent died from combat; 224,586 or over 68 percent died from the multiple diseases that produced 6,000,000 cases of illness. In addition, there were 24,877 killed by accidents and other injuries.

A state-by-state tabulation of the number of soldiers killed by disease reveals the hardships that epidemics placed on the Confederate army.

After John Wilkes Booth assassinated Abraham Lincoln, disgruntled Northerners brought charges against Confederate Army doctor Henry H. Wirz, on staff at the notorious Andersonville Prison in Georgia. They accused him of engineering mass murder of Union inmates by pretending to vaccinate them against smallpox. According to eyewitness reports from September 1864 made by Joseph Jones, a Georgia-born medical investigator and professor of medicine:

> The haggard, distressed countenances of these miserable, complaining, dejected, living skeletons, crying for medical aid and food, and cursing their Government for its refusal to exchange prisoners, and the ghastly corpses, with their glazed eyeballs staring up into vacant space, with flies swarming down open mouths formed a picture of helpless, hopeless misery which it would be impossible to portray by word or brush [J. Jones, 1864].

In weakened condition from ill health, untreated wounds, lack of water for bathing, and poor camp sanitation, many died quickly of gangrene.

July 31, 1865 In Spain's last attempt to seize the Dominican Republic, Captain-General José de la Gandara met opposition to his occupation of the island after his troops faced both guerrilla uprisings and yellow fever. In all, 10,888 of Gandara's forces fell in combat against mulatto guerrilla leader Gregorio Luperón. More devastating was disease, which claimed 30,000 Spaniards.

The Dominicans lost approximately the same number as the Spanish to war and epidemic.

September 1865 Scarlet fever followed agents of the Hudson Bay Company up the Mackenzie River Valley in Canada, killing up to half the Yukon's native population.

October 1865 When Muslims returning from a hadj at Mecca spread the fourth great cholera pandemic, infection arrived in New York by ship from France, killing thousands. Surprisingly, the disease did not immediately spread north of France into the British Isles. A successful harbor inspection by Port Sanitary Authorities identified symptoms of disease before it could invade England.

When the disease made an appearance in London in 1866, it was the last serious outbreak of cholera in Great Britain. Some 4,000 died in London's East End. Against traditional attribution of disease to miasmas, William Farr, Edward Frankland, and Netten Radcliffe blamed contaminated water piped from the East London Water Company. The virulence of the epidemic and repeated accusations of unclean water influenced England's water treatment policy, bringing the utility into a more hygienic mode of assessing water quality.

That same year, to quell infection, the French promoted the Conseil de Santé Maritime et Quarantennaire (Council of Coastal Health and Quarantine), isolation stations set up in Crete, Egypt, Syria, and other heavily traveled commercial routes in the Ottoman Empire. The Conseil was successful and remained in operation for 62 years. However, marching military transported the disease to Italy during the Austro-Prussian War. Carried by pilgrims, tourists, and soldiers, the bacteria reached the Piedmont in fall and spread to the north and south.

Because of protracted conflict in central Europe, Bohemian civilians suffered a military occupation that contributed to the scourge. Military authorities made insufficient efforts to compensate residents for their losses from harboring soldiers and from resultant deaths. In recognition of peasant losses, an initiative from Prague launched public works projects that bolstered the economy.

Along waterways, cholera traveled into Persia from India through Afghanistan and Kurdistan. Into late winter 1868, the disease claimed a Persian prince and a governor and raged into north-western Europe. A recrudescence in 1869 continued the wave of deaths as far north as Kiev, Russia, and east into Central Asia.

early 1866 Malaria struck the British crown colony of Mauritius, claiming around 75,000 in a three-year assault. During an increase in global traffic, the disease reached the harbor at Port Louis from India or Madagascar and proliferated in the rain-soaked lands of the Albion sugar plantation. At the height of the epidemic, around 32,000 succumbed to the fever and swamped medical services with the demand for care. Citizen panic was so intense that many residents of Port Louis fled to higher ground in Curepipe, turning a small village into a major island city. Increasing island chaos were three cyclones. Multiple catastrophes frightened off colonial Franco-Mauritian investors, forcing the owners of factories to downsize their holdings from 259 to 124 sites.

winter 1866–1867 Cholera proved devastating for the plains tribes, particularly the Wichita, Caddo, and Pima. From depots on Governor's Island, New York, and Carlisle Barracks, Pennsylvania, the scourge swept into eastern Kansas, killing 1,499. Some epidemiologists attribute the contagion to the arrival of buffalo soldiers in the 38th Infantry Regiment from Jefferson Barracks, Missouri. As described by Joseph Janvier Woodward's *Report on Epidemic Cholera and Yellow Fever in the Army of the United States, during the Year 1867* (1868), the disease moved west with passengers aboard the Union Pacific Railroad, striking rail crews and military outposts on the frontier. A study of rail lines and frontier routes helped to explain the path of the outbreak.

When contagion advanced northward and threatened Canada, government officials convened medical experts at Ottawa to summarize causes and prevention methods. Because of lack of an understanding of the water-borne disease, participants squabbled over the type of warning to issue. Joseph-Charles Taché, the deputy minister of agriculture, completed *Memorandum on Cholera* (1866), which he based more on Catholic ethics than on scientific knowledge.

1867 An epidemic of puerperal, or childbed, fever in London's Charing Cross Hospital spurred nurse Mary Agnes Jones to write to Florence Nightingale about improper hygiene among doctors. Jones reported that hospital medical staff moved directly from the post-mortem room to the birthing area and refused to cleanse their hands and instruments before examining parturient women. As a result, infection spread in the lying-in ward, endangering mothers and infants. From close work with contagious patients, Jones died of typhoid.

1867 Yellow fever assaulted Lima and Callao, Peru, so virulently that Manuel Pardo, director of charities, reported that more than 52 percent of 8,478 cemetery burials in Lima were yellow fever victims.

1867 Cholera once more hit central Europe with deadly force. The severe economic disaster in Warsaw, Poland, brought down the city's first Yiddish weekly newspaper, the *Varshoer Yidishe Tsaytung* (*Warsaw Yiddish Newspaper*). When contagion assailed Zurich, Switzerland, the populace, lacking faith in the traditional social order, demanded reform of water and sewage service. The emergent Swiss Democratic Movement precipitated a constitutional reform two years later.

1867 Sergeant John Spring reported on an unusual scourge, an outbreak of malaria at Camp Wallen in Cochise County near Huachuca City, Arizona. The disease sapped the men's strength, reducing the number of details guarding wagon trains, U.S. mail, and frontier families.

early 1867 During the Paraguayan War, following the invasion of Mato Grosso, Brazil, and the retreat of Brazilian troops from Laguna, soldier health declined seriously from cholera, enteritis, and salmonella. Losses to both sides resulted in the worst carnage in international war in the Western Hemisphere. In the third year of the conflict, an outbreak of cholera followed by malaria felled some 280 per day among the allied Argentinians, Brazilians, and Uruguayans, led by Marshal Luis Alves de Lima e Silva, Duke of Caxias. Sickness temporarily halted hostilities until July. At the end of the fighting in 1870, total deaths approached 100,000.

February 6, 1867 When measles once more infected Australians, Sydney suffered its highest mortality rate. Into late spring, 780 or 6 percent of the 13,000 infected toddlers and preschoolers succumbed to the epidemic or to its sequelae, usually enteritis or pneumonia. Worst hit were residents of urban tenements.

April 1867 The re-emergence of cholera in India struck the Hardwar fair, an annual gathering that coincided with a Hindu ritual involving submergence of the faithful in the holy Ganges River and the sprinkling of cremated remains of family members on the surface. Of the three million devout, some died on the river bank, others en route to their homes. Pedestrian survivors spread contagion into the north and west to Pakistan, Kashmir, Persia, and Afghanistan; those traveling by rail bore the disease into the heartland and urban centers. The disease claimed around 117,200, killing merchants, British occupation forces, and the shawl weavers of Amritsar. *See also* **1891**.

May 4, 1867 As described in Elizabeth Bacon "Libbie" Custer's autobiographical *Boots and Saddles* (1885), the horse soldiers of her husband, George Armstrong Custer, battled scurvy and an outbreak of cholera at Fort Hays, Kansas. The regiment lived on bacon and hardtack with nothing more than subsistence stores that were stale, moldy, and unfit for consumption. As soldiers contended with sickness, hunger, cold, mud, and soggy tenting grounds, General Winfield S. Hancock ordered the immediate delivery of potatoes and onions as antiscorbutics to halt desertions.

July 1867 During Galveston's "year of crucifixion," yellow fever killed 1,100 during a four-month outbreak worsened by loss and damage from an October hurricane. Texans battled the human toll as well as the commercial loss caused by the diminished labor force and battered property and equipment.

late 1860s A cholera pandemic in the Senegal Valley killed thousands. The Fulbe, a Muslim colony formed of settlers from Bundu, Futa Toro, and the lower Senegal Valley, mounted a *hijra* (migration) to Karta, which Umar Tal had conquered the previous decade in *jihad* (holy war) against insurgent Europeans in Africa. Away from pestilence, the Fulbe summoned friends and relatives to join them at the new enclave.

1868 A military surgeon, Jean Antoine Villemin of Prey, France, established that tuberculosis is contagious rather than hereditary. To prove his theory, he exposed rabbits and human volunteers by mouth and nose to body fluids of infected patients. His findings, issued in *Études sur la Phthisie* (*Studies in Tuberculosis*) (1868), directed health workers toward a systematic defeat of rampant TB by hindering transmission of bacilli.

1868 During a seven-year pandemic of smallpox in Philadelphia and New York City, some 3,100 were stricken; 805 died. The victims were typically uninoculated children. Because of an increase in immigrants following the Franco-Prussian War, many new residents were not vaccinated. By May 1869, the New York City board of health had launched a door-to-door campaign of inoculations among 700,000 residents. Some parents refused, accusing doctors of further endangering their children and of profiteering from the pestilence. The ignorant concealed sick family members to spare them suffering in isolation hospitals.

1868 At Fort Jefferson, Florida, on Dry Tortugas in the Gulf of Mexico, Samuel Alexander Mudd served a life sentence at hard labor for aiding and abetting John Wilkes Booth's assassination plot against President Abraham Lincoln. For Mudd's assistance during a yellow fever epidemic, he obtained release through President Andrew Johnson in February 1869. Mudd returned to his Maryland farm, where he died of pneumonia in 1883.

1868 At the port of London, Harry Leach, the Medical Officer of Health, fought rampant scurvy in the British navy and merchant marine by circulating information on prevention and treatment. He compiled *The Ship Captain's Medical Guide* (1868), a handbook for officers who had no shipboard doctor. The popular text went through seven reprints. In 1967, Leach's original document entered its 20th edition.

mid–February 1868 Unsanitary conditions and starvation worsened epidemic typhus at Tunis, where 50,000 died of infection. Colonial Europeans forced local authorities to remove offal from streets, to cleanse jails and military posts, and to sequester the sick.

summer 1868 In the months preceding the opening of the Suez Canal, the British raj enacted strict sanitation measures to control outbreaks of cholera. To protect English investment in the venture, Indian health officials supported policies that denied the nature of contagion or the necessity for sanitary cordons and quarantine of incoming vessels. As a result, cholera moved more freely about the Middle East, Africa, and Asia.

July 1868 While working as an interpreter and scout for the cavalry at Camp McDermit, Nevada, activist Sarah Winnemucca of the Northern Paiute learned that her starving people were dying of measles and smallpox at Pyramid Lake Reservation. The Indian agent lied to his superiors that the people on reservation land were content and well provisioned. In actuality, Southwestern Indians were encountering tuberculosis, trachoma, smallpox, venereal diseases, and spinal meningitis. At the Cheyenne and Arapaho Reservation in Indian Territory, malaria struck a virgin-soil population, contributing to high death rates among women and children. The same scourge, called "ague" or "intermittent fever," appears to have thrived among Indians in the Columbia and Willamette valleys from British Columbia through Washington State and Oregon. All these diseases were new to the populace and presented medicine men with symptoms unknown to them and untreatable by native pharmacopoeia. *See also* **August 1830; 1910.**

1869 In Arizona, the feuding citizens of Gila and Graham blamed each other for a diphtheria epidemic and charged that peddlers carried the sickness. Escalating rhetoric included threats of separating the two counties by a wall or moat.

1869 The enslavement of Chinese coolies in Chile came to a rapid end during a yellow fever epidemic. The workers, who contracted themselves into virtual bondage until they paid their debts, emigrated from China to work Chilean mines. Because of dangerous employment and the wretched hovels in which they lived, they did not establish a thriving ghetto society.

1869 Elizabeth Matthews Harris, wife of Major General Webber Harris, earned the Victoria Cross for aiding sick soldiers after cholera overran her husband's British regiment, the 104th Bengal Fusiliers in Peshawar, Pakistan. When the men evacuated their barracks and took to the hills for three months, she was the only woman accompanying them to treat the sick.

July 1869 At Novgorod, Russians organized a trade fair that assembled 200,000 traders from Persia, Afghanistan, and central Asia. The outsiders brought Asiatic cholera, a pandemic that, over a three-year span, also assaulted Turkmenistan and Bukhara, Uzbekistan, and outlying Russian defenses along trade routes. By August 26, Tashkent, Uzbekistan, suffered an infection rate of over 8 percent and a death rate of over 5.5 percent.

fall 1869 Thousands of plains Indians and métis died of smallpox from contact with trading posts along the Missouri River. In northern Montana, the Assiniboin and Gros Ventre or Atsina lost half of their number. Many survivors suffered maiming and blinding. Historians blame natives for disinterring the corpse of a crewman of the steamer *Utah* to claim his burial clothes. The disease migrated from Fort Belknap to white scavengers who rifled native corpses packed in skins and secured in trees.

November 1869 Cholera advanced from Somaliland (modern Somalia) and Abyssinia (modern Ethiopia) through Kenya and Tanganyika to Zanzibar, where 70,000 or 70 percent of 100,000 victims died of the epidemic. Into 1870, ships taking on water from contaminated sources carried the disease down the east coast of Africa as far as Cape Horn, South Africa.

1870 Endemic typhoid fever in Norway, which began in 1864, dropped by 67 percent and shifted from spring outbreaks to winter emergence. The alteration in patterns of contagion resulted from the halt of the spring herring runs, when fishermen contracted typhoid and returned home with the contagion, which tended to kill teens and young adults.

1870 Within four years, acute Oroya fever or *Bartonellosis bacilliformis* killed 7,000 railway laborers in the Andes Mountains of Peru, where they laid rails from Lima to Oroya. Also called cat scratch fever or Peruvian wart, the malady, which produced warty skin lesions and a lethal hemolytic anemia, appeared to follow weather-related cycles. Clarifying symptoms of the disease in 1885 was a 28-year-old medical student, Daniel Alcides Carrion of Quiulacocha, the father of Peruvian medicine, who became a martyr to medicine after his research caused his death. In his honor, the malady earned the name Carrion's disease.

1870 The migration of French Canadians to Lowell, Massachusetts, to work in textile factories produced a high death rate. Historians blame untenable living conditions in Little Canada or Pawtucketville for worker debility from brown lung, pneumonia, and tuberculosis. Contributing to respiratory illness among weavers was the inhaling of cotton fibers through mouth and nose, which filled the air around whirring looms.

July 1870 The Franco-Prussian War generated loss to communicable disease. In an all-out attempt to seize Naples, Italy, and Sicily, the red-shirted followers of Giuseppe Garibaldi, patriot leader of the Italian Risorgimento, weakened from smallpox. When the nation-builders returned to Italy to celebrate the retrieval of Italy from papal power and extol unification under King Victor Emmanuel, they spread disease. Milan and Rome suffered 1,000 deaths each. At Cesena in northern Italy in 1871, public health officer Robusto Mori lessened the infection rate by inaugurating a general vaccination program.

In August, the Prussians' three-month siege of Metz successfully isolated the French amid epidemic typhoid and typhus, smallpox outbreaks, rampant syphilis, starvation, and a battle of nerves. The toll on the trapped forces created a crisis of leadership and eventual capitulation. To halt the spread of smallpox in the Prussian army, in 1871, 800,000 men underwent vaccination. Out of 8,360 cases of the disease, only 300 German soldiers died, as compared to 23,000 unvaccinated French.

1870s Christian relief agents rescued Chinese immigrants infected with leprosy after families abandoned them in a San Francisco ghetto. Asians suffered discriminatory laws and racist stereotyping, which limited their access to medical care. As tensions mounted among racists, in 1877, Denis Kearney led agitants who blamed the Chinese for an economic downturn, for spreading leprosy and smallpox, and for luring whites into opium dens. White workers launched a three-day riot, slew several residents of Chinatown, destroyed Chinese laundries, and raided the piers of the Pacific Mail Steamship Company, the main transporter of immigrant laborers from China.

1870s During the decline of the Ottoman Empire, Armenians immigrated in large numbers to the North American colonies. Because of endemic trachoma among newcomers, the U.S. Immigration and Naturalization Service opposed their entry into the country. Port authority agents held some patients in quarantine for months and checked their eyes with rough jabs. Because staff moved from bed to bed without washing their hands, they spread the disease to otherwise healthy people.

1870s Endemic malaria in Bengal (modern Bangladesh) placed constraints on the British raj and investors. To alleviate quarantines and inspections of ports, railroads, and fresh produce and meats, epidemiologists studied causes of recrudescences of the disease.

1871 The sick poor of Glasgow, Scotland, placed demands on hospital care during an outbreak of relapsing fever, which is caused by the bite of an infected tick. At the Glasgow Royal Infirmary, the only public facility open to the indigent, admissions were more common among people who had an in with the staff, primarily blood kinship.

January 22, 1871 At Carlton House, Saskatchewan, the board of the Hudson Bay Company reported high death rates among Indians and métis from smallpox. Cree chiefs consulted with William Joseph Christie, the agent in charge of the Saskatchewan district, concerning the possible spread of disease to Manitoba, squatters on their lands, and the depletion of the buffalo, their chief source of meat. The Cree pressed the Canadian government to negotiate immediate treaties to protect native rights.

1872 At Provincetown, Massachusetts, a year-long outbreak of smallpox produced 27 victims. The one-room smallpox hospital had declined since its construction in 1848. Because of limited facilities, nurses refused to staff the city's ramshackle pesthouse.

1872 At the end of a twelve-year war against English settlers, the indigenous Maori of New Zealand had lost land and power. During the struggle of the regular British army against local guerrilla warriors, 2,000 Maoris died in combat. Another 25,000 perished from hunger and widespread typhoid, diphtheria, scarlet fever, and measles. Poor data collection obscured exact causes.

May 13, 1872 Anders Daae, a physician at Bamble in southern Norway, observed his first case of viral Bornholm disease, causing myalgia or muscle ache. He reported torso pain and labored breathing in patient zero, from whom the infection spread to endemic proportions of 290 cases. The disease recurred in summer 1873 with

36 new infections. Complications involved inflamed genitals, pleurisy, and pericarditis, all resulting from infection by *Coxsackie B* pathogens.

July 1872 The panic generated by rumors of advancing cholera provoked a strike by one-third of the construction laborers of Kreenholm, Estonia. An example of early proletarian protest, the successful work stoppage spread to spinners and weavers, who complained to management about poor working and living conditions.

1873 The outbreak of Sweden's last smallpox epidemic began at Stockholm. Lessening fatalities was the work of Eberhard Munck of Rosenschold who introduced the smallpox vaccine in 1801. Pockets of resistance to mass inoculation derived from religious fundamentalism, protection of civil rights, fear of contracting the pox, and annoyance at revaccination.

The nomadic Saami or Lapps of northern Sweden at Jokkmokk, Gällivare, and Enontekis avoided contagion by retreating from contact with urban populations and their produce and trade goods. As the epidemic worsened, the Saami, fearful that the epidemic was a divine punishment, avoided all infected areas. Thus, their isolationism spared them from mass loss to smallpox.

1873 Serious pertussis hit New Zealand, targeting Auckland, Wellington, Nelson, and Marlborough. Of 356 deaths, over 95 percent were children under school age.

June 12, 1873 When cholera struck Birmingham, Alabama, as described by Mortimer H. Jordan, secretary of the Jefferson County Medical Society, citizens panicked at the rapidity of the appearance of symptoms and subsequent deaths. Poor sanitation and marshy ground spread the disease through cesspools and dirty streets and alleys. Attempts at burning mattresses and bed linens failed to treat the most serious problems— damp basements and contaminated sources of drinking water.

August 20, 1873 When yellow fever threatened Shreveport, Louisiana, it apparently derived from upriver boat traffic. Citizens evacuated the city in an orderly fashion. Glimpses of the controlled effort appeared in the letters of First Lieutenant Eugene Augustus Woodruff of the Army Corps of Engineers, a volunteer who died during the mission to save citizens. Remaining behind to treat the sick were Catholic priests assigned to the Caddo and DeSoto parishes. The 760 deaths out of 2,600 cases exceeded the fatalities reported in New Orleans and halted an undercurrent supporting secession from Louisiana to form the town of Shreveport, Texas. The city's historic Oakland Cemetery was so deluged with burials that city authorities had to make other arrangements for interring the dead.

August 21, 1873 Newcomers Fannie and William Davis and their baby from Needmore, Indiana, imported Asian cholera to New Elizabeth, Indiana. Infection killed a dozen citizens by early September, including visitors from the Christian Church and the town physician, John A. Dicks. In another week, coma, convulsions, and fatalities doubled, terrorizing families who fled the town, which was permanently affected by loss of citizens.

1874 When smallpox returned to Europe, it constituted 10 percent of deaths in Ljubljana, Slovenia. After contagion threatened Germany, a government directive forced the vaccination of children twice before their teens. Each patient was inoculated free of charge and received a certificate displaying a doctor's signature and naming date and place of birth, father, and date and place of inoculation. Because of thorough enforcement of the law, the empire suppressed the pathogen and gained the lowest incidence of smallpox on the continent.

1875 Of 17,600 Japanese soldiers, 26 percent bore symptoms of beriberi, a debilitating disease caused by a lack of thiamine in the diet. Three years later, data on citizens of Kyoto found that 1,093 of 230,000 citizens suffered the same nutritional deficiency.

1875 According to *Dreadnought*, the Seamen's Hospital Society journal, with improvements in diet and antisepsis, crew in the British Mercantile Marine suffered fewer cases of scurvy, erysipelas, pyaemia, and cholera. However, the men incurred greater infection from gonorrhea, syphilis, and tuberculosis, all of which plagued much of Europe and the Western Hemisphere.

1875 Because of high incidence of pellagra throughout China, Japan, India, and the Philippines and particularly in the Japanese military, Kanehiro Takaki, surgeon-general of the Imperial Japanese Navy, set out to determine the cause.

During a naval training voyage in 1872, he had witnessed the illness of 169 sailors and the deaths of 25 out of a crew of 276 and recalled his father's memories of an outbreak of pellagra among the Imperial Palace Guard in 1862. While other medical staff searched for a pathogen, in 1884, Takaki sent out two test ships on restricted diets for nine months. Because the rice-heavy diet caused disease, he concluded that pellagra was nutritional in source. To prevent subsequent occurrences, he recommended changes in shipboard rations to more vegetables, milk, fish, meat, barley, brown rice, and wheat bread. *See also* **1904**.

January 6, 1875 When Chief Cakobau and his extended family of around 100 traveled aboard the H.M.S. *Dido* from the capital city of Levuki, Fiji, for a state visit and colonial treaty-signing in Sydney, Australia, his oldest son, Ratu Timoci, and chaplain Mesako became ill with measles. While in port, the captain quarantined the men in a hut on deck and hoisted the yellow quarantine flag. Ratu ignored the restriction and crept out each night to local brothels, where he also contracted gonorrhea. When the family returned home 19 days later, he assumed that the two-week contagion period had passed with four days to spare.

On January 22, Fijians gathering to celebrate the historic occasion inadvertently spread the measles virus simultaneously throughout the archipelago to nearly all native inhabitants, killing Savanaca, the king's brother. The islanders signaled the outbreak by pounding the *lali* (death drum). With no one to aid the sick, people died of exhaustion, dysentery, dehydration, and hunger. Others triggered fatalities by plunging into the sea to cleanse themselves of contagion. Within four months, nearly 40,000 Fijians died of measles and dysentery, leaving the island population at 60 percent of its original strength.

Recovering islanders raised an armed rebellion against the British, whom they accused of attempted genocide. Queen Victoria demanded an investigation of so great a loss among her colonial subjects. Eugenicists claimed that natives died because they were genetically inferior. Rebuttal from Fijian eyewitness Josefa Sokvagone charged the white chiefs. He declared, "They are blighting us, the natives, and we are withering away.... They are great and we are insignificant" (Garnett 1999). Colonizers used the devastating plague as an excuse to seize prime island property and, four

years later, to import laborers from India. Fortunately for the Fijians, as a result of the epidemic, the British founded the island chain's first medical school.

late February 1875 In India, Bengalis died in large numbers from a pair of cholera strains that killed a total of 365,000. Pilgrims emerging from sacred grounds at Allahabad and Nasik carried disease south to Malabar and Ceylon and into the uplands and foothills of the Himalayas.

late September 1875 Australians suffered a severe strain of scarlet fever. The outbreak in Sydney sickened 10,000 and killed 584, forcing the closure of public schools. Nearly 33 percent of the fatalities resulted from pneumonia and other complications. As the disease made its way over New South Wales, South Australia, Tasmania, and Victoria, it killed 5,000 more. Government response to the infection was the beginning of the nation's public health initiatives. When the disease spread to New Zealand, it inundated the Otago district. By the following year, 195 died of the fever. In Westland and Otago, high fatality rates persisted.

1876 An American, Edward Livingston Trudeau, helped bring outbreaks of tuberculosis under control by treating his own lung disease at a mountain retreat in New York state. After founding the Adirondack Cottage Sanitarium at Saranac Lake, he set a standard of isolating patients and improving their wellbeing through fresh air, rest, and a healthful environment. Over a half-century, some 600 similar institutions opened in the United States, becoming a cottage industry in mountain communities like Asheville, North Carolina.

1876 When low income and the eruption of Mount Askja worsened rural conditions for fishers and farmers in Iceland, 1,200 took the suggestion of author Sigtryggur Jónasson's *Nya Ísland Í Kanada* (*New Iceland in Canada*) (ca. 1874) and emigrated to the Gimli and Riverton districts of Manitoba to found New Iceland. The "westerfarers" carried with them smallpox contagion that, in September, led to an epidemic felling 500. One hundred never recovered their strength.

The Manitoba government dispatched three doctors, but delayed quarantining New Iceland until November 27, 1876, because the disease was first diagnosed as chicken pox. Government workers turned a storehouse at Gimli into a hospital and set up a quarantine station and guard post at Netley Creek. At a Saulteaux village on Sandy River around Lake Winnipeg, two white doctors battled contagion among natives, whom smallpox quickly annihilated. The settlers bound thirteen Indian corpses on sleighs and transported them to a burial pit.

The quarantine lasted until June 20, 1877. Because of long sequestration, the Icelanders faced a famine from scarcity of seeds and tools for gardening. When John Taylor boarded the steamer *Mary Ellen* on a buying expedition for seed potatoes, he had to return to the colony. One Saulteaux, John Ramsay, lost his wife and two boys to the disease, which also scarred his baby daughter, Maria. He bore no ill will toward the settlers and helped them survive the terrible winter.

To evade quarantine patrols at Netley Creek, another pioneer, Fridjón Fridriksson, set out on foot and coerced two Indian women to ferry him over the creek. He purchased a sow and 300 bushels of seeds at Winnipeg. When the patrol lit smudge pots to kill mosquitoes, Fridriksson crept through the smoke and reunited with a jubilant colony. Subsequent epidemics of diphtheria, measles, scarlet fever, and scurvy further tested the mettle of Icelanders, many of whose children died. They relocated in Winnipeg and the Dakotas and adopted a constitution similar to the Icelandic Althing. Jónasson became the first Icelandic-Canadian elected to the legislature.

1876 In Rutland, England, the Uppingham School evacuated 300 boys and 30 staff members and their families to Borth, Wales, to escape an outbreak of typhoid. Health official traced the contagion to polluted drains and water contaminated with sewage. The 14-month relocation quelled a serious breach in purity standards.

July 12, 1876 Settlers aboard the S.S. *Collingwood* continued New Zealand's problems with imported disease, particularly measles, mumps, and scarlet fever among children. The New Zealand *Times* reported, "It is our painful duty to have to chronicle the arrival of the above ship with so much sickness on board, and we regret to add that death has been very busy amongst the immigrants during the voyage from London"

("Arrival"). Passengers incurred stays of up to three months in harbor quarantine.

August 6, 1876 Typhus in Ethiopia and Sudan resulted from an Egyptian invasion from Massawa on the Red Sea shores to Hamasien Province in modern Eritrea. Within three days, 160 of the soldiers of Egyptian commander Rateb Pasha required hospitalization. Within two months, 2,000 Arab, Egyptian, and Sudanese troops died along with one quarter of the town of Tegray, Ethiopia, and two-thirds of Adwa. Foreigners concluded that the contagion began as a cattle zoonosis.

1877 Timothy Lewis, a British surgeon with the Royal Army Medical Corps, identified the trypanosome parasite in a rat's blood in his laboratory at Bombay, where the microbe caused pandemic sleeping sickness. The organism acquired the name *Trypanosoma lewisi.*

1877 Tahiti, already reeling from decades of epidemic European diseases, battled typhoid fever imported by the crew of the armored cruiser H.M.S. *La Magicienne.* Similar contagion killed half of residents and animal life on the Cook Islands, Fiji, Hawaii, Samoa, and Tonga. Judgmental outsiders tended to blame depletion of native Polynesians on infanticide, violence, opium, alcohol, and sexual depravity.

1877 The Bengali suffering from smallpox perpetuated the worship of the goddess Mata Sitala (Mother Cool One), which originated in the 12th century B.C.E. and re-emerged in the 1200s. By identifying infectious disease as a form of divine grace or *mayer daya* (mother's mercy), the sick had reason to celebrate the honor of bearing the goddess's mark. They processed with her image, held ritual baths at the river, and wrote and recited *mangal* (sacred verses) explaining the complex human truth of blessing through suffering.

1878 During an epidemic of smallpox, at the gold rush town of Deadwood, South Dakota, Seth Bullock became the pioneering community's treasurer of the board of health, its first elected official. One of the few females in town, 26-year-old Martha Jane Canary (or Cannary), a transdressing teamster and Indian scout commonly known as Calamity Jane, nursed the local miners and later cared for families stricken with diphtheria and mountain fever. According to the many legends connected with her life, when she died from alcoholism and pneumonia in 1903, a huge send-off was the community's thanks for her kindness. One man whom she had saved had the honor of closing her casket.

1878 The reduction of cases of typhoid fever in Stockholm resulted from improved sanitation, more effective food inspection and control of substandard products. By making surprise visits to producers and retailers and by confiscating suspect produce, health enforcers prohibited the sale of contaminated meat and other table goods to Swedish consumers.

January 1878 The emergence of the Bronze John or Yellow Jack from Cuba and the Caribbean burgeoned into the worst yellow fever epidemic in Mississippi River history. Whole families died; others fled the state to take up residence permanently in safer climes. The economy was devastated; Beechland turned into a ghost town. Within the year, 3,227 people perished.

By the end of the catastrophe, the state's toll was significant. Because Mississippi officials were slow to respond, the lack of quarantine allowed infection to claim 350 out of 1,050 cases at Grenada and 84 out of 490 cases at Jackson. Seriously hit were Meridian, Port Gibson, and Vicksburg. At Holly Springs, six nuns of the Sisters of Charity offered skillful care until all six died of the disease. A relief fund collected in various states totaled $75,472.

At Water Valley, Mississippi, physician H. A. Gant praised the assistance of Bob Reed, a black volunteer nurse's aide who was immune to yellow fever after surviving it in Natchez years earlier. Reed carried a red flag, rang the contagion bell on city streets, and supervised care of the sick, both white and black. His daily chores involved feeding patients and burying the dead.

In New Orleans, after the steamer *Emily B. Souder* arrived in late May with contagion from Havana, 27,000 came down with yellow fever, which killed 4,046. Of 211,000 inhabitants, 40,000 fled to healthier climes. Strictly quarantined, the city became known as the "wet grave." Citizens on both sides of Louisiana rejected boats and trains carrying passengers who might be infected. At Crescent City, officials burned carbolic acid and sulfur as disinfectants of the noxious atmosphere. Homeopaths produced a greater cure rate than traditional doctors dispensing calomel, but none of the measures halted the sweep of Yellow Jack.

As fever-carriers aboard the *John D. Porter* churned upriver, disease blanketed the Mississippi Valley at the port of Vicksburg, Mississippi, and beyond to Memphis, Tennessee, and Cairo, Illinois. In Hickman, Kentucky, local physician Luke Pryor Blackburn attended the sick in the far west as yellow fever killed 200. His dedication won him election to the state's governorship in 1879 and a tombstone marked "The Good Samaritan." Farther south at Mobile, Alabama, the "Can't Get Away Club" consisted of distinguished men who stayed behind to treat fever victims. To stem the transfer of pathogens from ships to states, President Rutherford B. Hayes enacted the Quarantine Act, which placed the Marine Hospital Service in charge of quelling epidemic conditions. *See also* **August 9, 1878.**

spring 1878 Famine and heavy rains in the Punjab preceded heavy losses to malaria, which killed around 180,500 residents of northern India. By epidemic's end, the region lost 440,492 people to endemic fever. Like the ancient Romans, Punjabi officials placed blame on "bad air" from fens, waterlogged railroads, and flooded land.

August 1878 At Fortaleza, Brazil, black smallpox, a deadly hemorrhagic strain, swept rapidly over the citizenry, claiming half who caught the disease. Spreading to Arcati, Baturité, and Pacatuba, the disease cut down columns of hungry, weakened natives fleeing drought in the interior. In all, the epidemic killed two-thirds or 100,000 of 150,000 refugees. At Christmas, families loaded the dying in slings and bore them to church for confession. City laborers, bribed with cash and liquor, worked steadily to cart bodies to plague pits.

August 9, 1878 D. R. Brown, secretary of the Memphis board of health, surveyed a yellow fever epidemic in the town of Grenada, Tennessee, where 367 of 2,000 citizens died. Four days later, Memphis suffered the first epidemic-related death; within three days, panic ensued as the death toll rose to 200 per day. Contagion targeted Irish immigrants, but left the black citizenry largely unscathed. In all, the fever struck 17,000 people, killing 5,000 or 29.4 percent.

As 39 local physicians and 72 volunteer doctors treated the sick, citizens deserted Memphis city streets. A backlog of coffins lined a sidewalk outside the mortician's office. Heavily represented among the dead were Irish immigrants, who had developed no immunity to the disease. The city's health officer, John Erskine, a veteran of the Civil War, also succumbed to fever. Refugees from the city carried infection to Saint Louis, Missouri; Cincinnati, Ohio; Louisville, Kentucky; and Little Rock, Arkansas.

At the Memphis *Commercial Appeal*, staff depleted to two workers issuing the paper, which came to be called "Old Reliable." City land values plummeted, crops went unharvested, and the city sank into bankruptcy. The media printed advice that Memphians burn the city and start over elsewhere, but officials chose to sell bonds to cover upgrades to sanitation and drainage and the paving of streets.

Humanitarians of the New Orleans Howard Association traveled by train with nurses and medication to aid the stricken. Upon finding Memphis Mayor John Flippin dead, civic chaos increasing, and more new cases than they expected, the volunteers telegraphed for additional doctors and nurses. Still primly Southern, conservative elements criticized female nurses for working outside the home. Only nurse-nuns avoided censure for violating gendered proprieties.

By the epidemic's wane in October, some of the volunteers were added to the list of the dead, whom a cadre of city gravediggers buried in the town's Yellow Fever Cemetery. The heroes of the moment included two Episcopal priests, Charles C. Parsons and Louis S. Schuyler, and the Mary Magdalene of Memphis, Annie Cook, a madam who converted her brothel, Mansion House, into a contagion ward and superintended nurse care by her employees until her death from the disease on September 11. The following year, a Howard Association volunteer, John McLeod Keating, editor of the Memphis *Daily Appeal,* issued an account, *The Yellow Fever Epidemic of 1878 in Memphis, Tennessee* (1879). He included among the victims Jefferson Davis, Jr., a ne'er-do-well and only surviving son of the former president of the Confederacy, who had accompanied his sister, Margaret David Hayes, and her husband, Addison Hayes, to nearby Buntyn Station, Tennessee.

As a result of the yellow fever epidemic, Memphis residents, only a year out of the Civil War and Reconstruction, reconsidered the hiring of black police officers to bolster the force during the epidemic and quickly stripped them of their badges. Officials restructured city government through municipal reform. To bolster the devastated region, they instigated the Memphis Plan by placing a weak mayor, John R. Flippin, over a powerful centralized commission of health. Until the end of the century, an elite coterie ruled over the underclass, comprised largely of Irish immigrants.

In Jackson, Tennessee, quick action by authorities held down panic. Having learned the need for calm and reason the previous year during a smallpox outbreak, they appointed a board of health, quarantine the city, and set a watch on the roads and rail stations to halt incoming visitors showing symptoms of yellow fever. The board also countered rumors with informative public meetings. Because their action spared the city the expense of serious disease, they could collect money, bedding, medicines, and food to send to hard-hit areas.

1879 Microbiologist Louis Pasteur, professor of chemistry and dean of the science department at the University of Lille, France, cultured fowl cholera bacteria to create the first vaccine. His injections of hens with full-strength microbes killed them all. He discovered by accident that leaving the microbial mix to weaken for a day resulted in the right strength for injection. The hens survived and didn't contract the disease.

1879 The Cuban government appointed epidemiologist Carlos Juan Finlay of Camagüey to join an American mission studying yellow fever prevention. He began searching for the vector that spread contagion. In February 1881, he outlined theories on mosquito transmission at the International Sanitary Conference in Washington, D.C. He surmised that the *Aedes aegypti* mosquito relayed the fever to humans. After the Spanish-American War, U.S. Army doctor Major Walter Reed investigated the theory for the Yellow Fever Board. In honor of Finlay's innovative work, the Cuban government founded the Finlay Institute for Investigations in Tropical Medicine.

1880 An epidemic of ankylostomiasis (or ancylostomiasis), a tropical disease common to Ceylon, Egypt, and India, caught the attention of Europeans after Italian laborers in the Saint Gothard tunnel, a massive highway engineering project under the Swiss Alps, contracted anemia from the infestation of hookworms, a blood-sucking parasite, in their intestines. Impervious to the danger from landslide, cave-in, and explosions of gas and dynamite, the men worked from 1872 to 1880 for around fifty cents per day and survived primarily on polenta, a congealed corn-meal mush. Weakness and vertigo from infestation of the nematode compromised their safety. As a result, ankylostomiasis acquired the folk names miner's or brickmaker's anemia or Goth-ard tunnel disease. The 1st International Congress on Occupational Diseases saluted the memory of some 10,000 men who died digging the tunnel linking Italy and Switzerland.

1880 Health officials founded the Iowa State Board of Health to monitor an epidemic of small-pox and to improve health statewide. Beginning with campaigns to prevent and control the out-break over a mostly rural state, staff fought con-tagion in Keokuk and in Worth and Fremont counties. They quarantined the sick, dispatched doctors, and issued prevention information. To halt new influxes, members petitioned the U.S. government for screening and vaccination of immigrants taking trains west from port cities.

1880 New Zealand became the home of choice for British tuberculosis victims seeking a more salubrious climate than the cold and damp of England. One faction of New Zealanders wel-comed immigration for the quality of newcom-ers, who included the wealthy and educated, and encouraged the investment of their money in the islands. A xenophobic faction urged restricting the immigrants carrying infectious diseases.

1881 Upon accepting a mission to Rotuma in the Society Islands of the South Pacific, the Rev. William Allen of the Australasian Methodist Church observed the misery of elephantiasis, the worst of endemic maladies:

> The scourge of the place is elephantiasis and ter-rible ulcerous sores, which not only eat the flesh but the bone. Large numbers of the natives—men and women—are suffering from elephantia-sis, and some cases are pitiable in the extreme. It attacks the arms, legs, heart, and generative organs. The last form of the disease is the worst [Wood 1978].

He noted that whites sometimes contracted the disease. Those who remained in the area could die of elephantiasis, but those who returned to a cold climate regained their health. He also remarked, "European diseases have been intro-duced which the natives do not understand; there being no resident doctor, the people quickly succumb" (ibid.).

In 1979, anthropologist Alan Howard remarked that the year 1881 marked the end of freedom for islanders, who came under European rule. As a result, epidemics followed in rapid order: dysen-tery in 1882 and 1901, pertussis in 1884, dengue fever in 1885, influenza in 1891 and 1896, and fish poisoning between 1885 and 1887. He estimated that, over two decades, Rotuma lost 46 per thou-sand to infectious disease. He also noted suffer-ing from endemic scourges—scrofula, yaws, trachoma, and elephantiasis. *See also* **1885; Jan-uary 11, 1932.**

1881 A fifth Asiatic cholera pandemic, diag-nosed by epidemiologist Robert Koch, author of *Investigations into the Etiology of Traumatic Infec-tive Diseases* (1880), incubated in Bengal and the Punjab before moving on to Korea, Siam, China, and Japan. In the Philippines, the disease claimed economics scholar Suganuma Teifu. In its third year, the scourge struck France, Italy, and Iberia, but did less damage than previous outbreaks because of improvements to sanitation, diagno-sis, and quarantine. Cholera bypassed Britain and North America, but exacted a heavy toll in Ar-gentina, Brazil, Chile, and Uruguay. Still virulent into the mid–1890s, it made a final appearance in northern Africa.

After Koch identified the cholera bacterium, the disease became a political issue. British sci-entists disputed the need for quarantine and claimed that the disease derived solely from filth. As a means of fostering trade through the newly completed Suez Canal, politicians supported the opinions of scientists over those of Koch. In 1885, Edward Emanuel Klein and Heneage Gibbes compiled a report rejecting the theory of bacterial transmission of cholera.

1881 Sir Patrick Manson, educated at the Uni-versity of Aberdeen, Scotland, became the father of tropical medicine for his work for the Royal Customs Service at Formosa and the island of Amoy off the southeastern coast of China. Dur-ing his work at Amoy, he isolated the mosquito vector of filariasis, an endemic disease marked by disfiguring edema. His research, issued in the classic text *Tropical Disease* (1898), was the first to demonstrate a vector-borne disease. In 1899, he established the London School of Tropical Medicine; in 1905, he lectured in San Francisco at Cooper Medical College. *See also* **1894**.

1881 As thousands of Jews emigrated from Eastern Europe to the Northeastern United States, they lived in dark, crowded urban tenements that spread tuberculosis, called the "Jewish disease." Over four decades among Jewish immigrants on New York's Lower East Side, sweatshop workers in the garment industry suffered a high rate of respiratory illness lumped under the heading "consumption." Tubercular workers blamed workplace environments rather than physical weakness. The International Ladies' Garment Workers' Union championed the cause of seamstresses and cloth cutters as a means of protecting laborers from infection.

1881 A resurgence of scarlet fever in New Zealand killed 104 in Auckland. The death toll rose to 153 the following year, but spared Christchurch the heavy fatality rate experienced in Auckland, Nelson, and Otago. To lessen infection, harbor authorities inspected and quarantined all ships that reported fever on board.

May 25, 1881 A concurrent bout of chicken pox and smallpox created confusion and hysteria among the inner-city residents at "The Rocks," the harbor area of Sydney, Australia. After disease killed a Chinese baby, smallpox worsened over the rest of the year and into 1882, targeting mostly children. In response, New South Wales empaneled its first six-member board of health. Police and ambulance crews forcibly removed from private residences all victims and people exposed to infection. Municipal health authorities sequestered the detainees in tents and barracks at a quarantine station at North Head and established the Coast Hospital at a distance from residential areas to spare citizens additional contagion.

As reported in the Sydney *Morning Herald*, Anglo-Australian suspicion of "the colored races" placed additional hardship on local Asians, whom enforcers forcibly vaccinated. Because of rampant racism, the Chinese lost their businesses to boycotts and looters; health squads roughly disinfected or burned homes in the Asian sector. Local whites accosted many Asians publicly and charged them with filthy lifestyles. The provincial assembly passed racist laws limiting entry of Asians to the country and quarantining vessels carrying Chinese immigrants.

June 1881 In Panama on February 1, 1881, Ferdinand de Lesseps and La Compagnie Universelle du Canal Interoceanique began cutting timber and slicing through mountains to make way for the digging of a canal. The mammoth task required 200 technicians and 800 laborers. By June, around 200 workers had died of infectious disease at a rate rising to 40 per day and extending into the next three years. In a cartoon for *Harper's Weekly*, Thomas Nast posed the question, "Is M. de Lesseps a Canal Digger or a Grave Digger?" ("T.R.'s Legacy").

One cause of the alarming mortality was the use of cauldrons and earthen bowls of water beneath bed legs to trap crawling ants. Meanwhile, the stagnant water bred mosquitoes carrying malaria and yellow fever. To maintain high interest in engineers and laborers, the French media suppressed news of the epidemic, but failed to offset the financial crisis that struck in December 1888.

1882 Jean Antoine Villemin and Robert Koch isolated the tuberculin bacillus, *Mycobacterium tuberculosis*, which they found to be endemic in people and livestock. To stop the spread of the disease and rehabilitate patients, Koch summarized in *Die Aetiologie der Tuberculose* (*On the Nature of Tuberculosis*) (1882) a regimen of hygiene and cleanliness, pure air and water, and sufficient rest. Carrying his precepts to crowded New York City, tenements, foul with overflowing privies and contaminated water supplies, were public health nurses, who educated patients one family at a time.

1882 In Alexandria and Cairo, Egypt, epidemic diphtheria felled 3,500, striking large numbers of children over a three-year span of contagion. Paralleling the loss of life in northern Africa was a three-year outbreak of smallpox in Cape Colony and Cape Town, South Africa. The emergence in May of deadly infections in the Boshof district threatened the survival of the Khoikhoi or Hottentot. Health workers implemented fumigation and border patrols to prevent infection; elite whites used the epidemic as justification for corralling poor blacks in ghettos. To protect employees of Kimberley diamond mines, the company hired Hans Sauer, a Boer who helped Cecil Rhodes to buy investment properties, examine patients, vaccinate those lacking immunization, and quarantine any who rejected protection. Sauer's observation netted 14 cases of the pox.

Sauer discovered that company doctors were falsifying screening of patients to conceal from the Bantu the peril of contagion in the area. Without those workers, the mines would close. As a result of Sauer's courageous stand on professional ethics, the government enacted a public health statute requiring inoculations and records of infection. Concerning his pioneering days during South Africa's emergence as a world leader in gem mining, he compiled *Ex Africa* (1937), later added to the First Gold Series of Rhodesia (modern Zimbabwe).

1882 When smallpox swept Ontario, health care workers focused less on prevention than on treatment. Officials were lax in enforcing compulsory vaccination and delegated responsibility for public wellness to municipalities. Edward Playter, editor and publisher of the bimonthly *Toronto Sanitary Journal* and author of *Elementary Anatomy, Physiology and Hygiene for the Use of Schools and Families* (1882), influenced medical opinions during establishment of the Provincial Board of Health in 1882 and a subsequent epidemic in 1883.

August 10, 1882 A five-month upsurge in cholera in Manila killed 30,000 Filipinos. As the number of deaths each day outpaced the city's ability to collect and bury victims, corpses littered the approaches to San Lazaro Hospital and to graveyards. A lazaretto at Mariveles offered additional space for the sick.

Contributing to civic chaos was the large number of homeless children. Bands of orphans roamed the streets, picked pockets, and panhandled pedestrians for coins to buy food. A group of benevolent women petitioned Father Salvador Font to intercede with the governor to build Asilo de Nuestra Senora de la Consolacion orphanage for boys at Paco. Officials transferred homeless girls to Mandaluyong.

1883 The spread of syphilis worsened over Europe. In Italy, as a means of improving the well being of poor women and their families, government commissioner Agostino Bertani crusaded to end legalized brothels. He campaigned for civil laws against government-regulated prostitution and supported women's rights.

In France, venereologist Jean Alfred Fournier,

author of *Traité de la Syphilis* (*Treatise on Syphilis*) (1899), differentiated between syphilis and gonorrhoea. As chief of staff at Hôpital Saint-Louis, where he taught dermatology and syphilitic diseases, he correctly identified syphilis as the source of numerous degenerative ailments, notably blindness, paralysis, and insanity. In 1883, three years after he organized the Société Française de Prophylaxie Sanitaire et Morale (*French Society of Sanitary and Moral Prevention*), he characterized rampant congenital syphilis in newborns.

A greater understanding of the disease coincided with public hysteria from rampant infection. Because law forbade doctors from warning prospective brides that their future husbands were infected with syphilis, women often suffered needless pain and social stigma out of adherence to patriarchy. In an effort to protect husbands' reputations, doctors tended to offer infected wives substandard treatment.

1883 Christiaan Eijkman, a physician in the Dutch military, along with Cornelis Winkler, a neurologist from Utrecht University, and Cornelis Adrianus Pekelharing, a professor of pathology, traveled to the Dutch East Indies to observe epidemic beriberi. At Atjeh in the jungles of Java, Eijkman saw natives as well as soldiers sidelined by the dietary deficiency, then called polyneuritis endemica perniciosa. As the nation's medical officer, he directed a government laboratory to determine the cause of the disease. By studying monkeys and chickens afflicted with the paralysis of beriberi, he determined that a diet of cooked white rice resulted in sickness, weakness, and death in the animals, just as it did in humans.

1883 When measles overwhelmed Perth, Australia, the media reported that bronchopneumonia was prevalent among measles victims, most of whom were children. The telegraph office languished for lack of delivery boys. The agent posted one wire into the hands of a child standing near the office door. The boy delivered the message to Lady Barker, who remarked in a letter to her son on the severity of the epidemic and on her gratitude to the child, who earned a tip of sugarplums. The disease remained active for the next three decades.

July 15, 1883 When cholera traveled by ship from Bombay to Alexandria, Egypt, it moved rapidly toward Cairo, forcing the evacuation of

English military to the Suez. Tewfik (or Tufik) Pasha, the khedive of Egypt, set an example for his staff by joining his wife, Bazmi Khanum, in visiting contagion wards to speak to patients. Bazmi also instructed poor women on sanitary procedures to reduce infection.

During the year-long siege in Egypt, European epidemiologists studied the disease for Germany's Cholera Commission. As a result of autopsies performed on victims, bacteriologist Robert Koch, head of the team, located a bacillus *Vibrio comma* in the intestinal mucosa. When the disease subsided, he journeyed east to Calcutta, where cholera remained active, and announced his conclusions to health officials on January 7, 1884. Based on his research, city planners determined that hygienic disposal of sewage and purification of water rid cities of recurrent cholera.

August 25, 1883 The end of the first phase of the Tonkin War produced sobering figures that dismayed the French imperialists seeking a toehold in the rich Mekong Delta of Cochin China (modern Vietnam). After the collapse of the Vietnamese resistance of Liu Yung-fu's Black Flag Guerrillas in August 1883, the French, led by Captain Henri Riviere, controlled the area, but lost 4,500 insurgents, most to epidemic disease. A new offensive in 1885 pitted the French against the combined forces of the Vietnamese Black Flag and the Chinese army and fleet. The Treaty of Tientsin, signed on June 9, 1885, attested to the futility of a war that cost the French colonials and regulars 4,222 combat fatalities and 5,223 deaths from sickness.

1884 Cholera once more scavenged Europe with a fatality rate of 50 percent. When the bacteria advanced on western Italy from traders arriving from Alexandria, Egypt, enforcers of health regulations battled the disease with extreme measures, but failed to identify raw mussels as a source of contagion. In Naples, authorities demolished or burned many buildings to rid the area of contagion, leaving the city looking haggard. When the disease overran Garfagnana, Italy, as a means of calming panic, the government dispatched Italian soldiers to establish a sanitary cordon to isolate the valley.

In Spain, cholera arrived at Alicante on a vessel from Oran, Algeria. In response to widespread misery and death, bacteriologist Jaime Ferran y

Clua resolved to create an anti-cholera vaccine in a period when 7.7 percent of the population acquired the disease. He cultured the pathogen in a special broth and vaccinated thousands of healthy citizens against infection. His inoculations inflicted cholera on only 1.3 percent. When he refused to cooperate with the Pasteur Institute the following year, epidemiologists negated his contribution to immunization. A street and statue in the town square of Corbera honored Ferran as a native son.

1884 At Valetta Hospital in Malta, Australian bacteriologist Sir David Bruce, a surgeon general in the Royal Army Corps, and his wife and partner, artist and medical technologist Lady Mary Elizabeth Steele-Bruce, set up a makeshift lab to investigate Malta fever or brucellosis. The disease, which infested three densely populated islands, felled 100 British soldiers per year. The Pasteur Institute in Paris corroborated the Bruces' identification of the *melitensis* microbe. *See also* **1887**.

1884 Among Australians, pulmonary tuberculosis, a lethal respiratory infection, peaked as the leading cause of death. The epidemic, which began in 1788, produced 135 fatalities per 100,000 residents. During a heroic age of curative treatment, sanitoria replaced healing societies or private clinical care of outpatients.

Contributing to the number of illnesses was the erroneous European image of Australia, especially Victoria, as an environment conducive to recovery from pulmonary tuberculosis. Patients often took recuperative voyages to the South Seas to seek long or permanent residence as a form of therapeutic emigration. The stereotype remained viable until late in the 19th century, when doctors promoted high-altitude sanitoria, such as those springing up in the Swiss Alps. *See also* **1890**.

1885 At Prairie du Chien, Wisconsin, diphtheria quickly claimed around half the population. Most of the victims were children. Families swabbed sore throats with a carbolic solution. Because of the danger of contagion, families held no public funerals.

1885 The severity of dengue fever in the Leeward Islands of the South Pacific impressed historians, who noted the infection of a European missionary. Imported from New Caledonia, the disease appears to have advanced to Huahine.

1885 After the introduction of underground mining in Kimberley, South Africa, the DeBeers Consolidated Company instituted closed compounds for laborers, ostensibly as a form of social welfare. However, pneumonia ran rampant among African miners, casting doubt on the altruism of white overlords and their model communities.

1885 In Chandag, India, Mary Reed, a Methodist missionary-nurse from Lowell, Ohio, operated a leper's colony for over 50 years to rescue the diseased from homelessness. She happened on the area by accident and made the sick her life's work. After contracting leprosy in 1890, she remained involved with treatment of lepers until her death in 1943.

1885 Medical giant Louis Pasteur, upon observing Joseph Meister dying of rabies, inoculated the boy 13 times in ten days, a regimen that Pasteur had tried on dogs. After a 17-week observation, he declared Meister cured of a formerly fatal disease.

1885 Over an eight-year period, Savill's disease, a fatal eruptive dermatitis named for English dermatologist Thomas Dixon Savill, struck 163 of the elderly poor in London's Paddington Infirmary and Workhouse. Because the disease was most common in Jewish males, doctors surmised that it was psychogenic. He reported his findings in an article, "On an Epidemic Skin Disease," in the 1891 issue of the *British Medical Journal*.

1885 The spread of syphilis in Cape Town, South Africa, forced authorities to reexamine the First Contagious Diseases Act of 1868, which was largely unenforced, and to enact the Second Contagious Diseases Act of 1885. At four port cities, mandatory internment and physical examination of prostitutes resulted in a peasant riot. In northern and central Africa, the battle against syphilis as well as endemic malaria, smallpox, and typhus received the notice of German explorer and medical writer Gustav Nachtigal. Over a decade, he recorded folk medical practices in his three-volume *Sahara und Sudan* (1889).

1885 The increase in tuberculosis among the working poor of Buenos Aires, Argentina, continued for two decades. Worsened by inadequate sanitation and crowded tenements, the disease failed to respond to prevention procedures and treatments imported from Europe and the United States. The poor, whom elitists blamed for provoking disease through immoral lifestyles, avoided the hostility and bigotry common to clinics, hospitals, and sanatoriums and resorted to self-medication or folk cures.

early spring 1885 In Montreal, an urban stench derived from a smallpox epidemic that killed 3,164, mostly French-Canadian children. Citizens rioted against vaccination programs, yet decried such unhealthful conditions as malnutrition, dank residences in the poor neighborhoods, impure water, and adulterated milk, all of which boosted infant mortality and fostered enteritis, tuberculosis, and typhoid. The smell from the epidemic forced citizens to search for methods of cleansing the city to purify the air. However, their primitive efforts fought the results of disease, but not the causes.

Meanwhile, the economy faltered because buyers refused goods shipped from Montreal. Vendors and passengers from the area who could not produce proof of vaccination were ejected from trains. The spread of disease caused U.S. border enforcers to reject emigrants fleeing the area. As a result of the isolation of French-Canadians within Canada, the city of Montreal enforced inoculation programs.

October 6, 1885 The influx of cholera over Iberia exacted a death toll of 800,000 Spaniards. At Tunis, the spread of contagion from southwestern Europe outpaced Tunisian efforts to prevent and contain the disease. Until the end of the epidemic in mid–December, the contagion slew 2 percent of Muslim Tunisians, 4 percent of Afro-Europeans, and 94 percent of Jews living in the winding alleyways of the ghetto. Because of the ancient architecture and dense occupancy, city officials were unable to mount urban renewal of aqueducts and sewage systems.

1886 After the defeat of the Apache and the capture of Geronimo, the 500 Chiricahua sent from the Arizona Territory to prisons in Florida died at a rapid rate, despite the care of U.S. Army surgeons. At Fort Marion, Saint Augustine, overcrowding, poor sanitation, malnutrition, and malaria killed 22 in six months. Those moved to the Mount Vernon Barracks in Alabama suffered malaria, which killed 10 percent. Of the 112 children sent to the Carlisle Indian School in Pennsylvania, 30 died, primarily from malaria.

1886 The onset of smallpox in Ethiopia targeted children, killing 300 of the 7,000 residents of Odowa. On the coast of the Red Sea, the Mansa Bet Abraha of Eritrea lost 700 tribe members. At Tigre, people blamed God's anger for their losses.

1886 The failure of the farmers' institute movement, mobile agricultural schools that advanced before the Civil War from New England to the Midwest, occurred in part because of social and economic disorder caused by yellow fever. Set up each summer as one-day tutorials by promoter Jordan G. Lee, they declined in Louisiana as farmers and legislators turned from farming to matters of public health.

1886 In South Africa, miners' phthisis, a form of pulmonary tuberculosis, concerned colonial managers of the gold mines at Witwatersrand. Worsening the outbreaks of this industrial anomaly was the increase in dust from mechanical rock drills, which caused silicosis.

1887 During an epidemic of paralytic polio in Sweden, pediatrician Oskar Karl Medin, a professor at Karolinska Institutet and head physician at Stockholm's Children's Hospital, made the first official diagnosis of the disease, which emerged outside Umeå in northern Sweden in 1881. He also described two phases of infection. In his honor and that of German orthopedist and medical writer Jakob von Heine, the scientific community began calling polio the Heine-Medin disease.

1887 English epidemiologist David Bruce identified the source of endemic fever on Malta as brucellosis, a pathogen that lodged in the spleen. The pasteurization of milk and cooking of meat halted the zoonosis, which caused muscle and joint pain, undulant fever, anemia, headache, and profuse sweating in humans and spontaneous abortion in parturient women. Complications from endocarditis caused death.

January 1887 At Tucumán, Argentina, a peasant revolt against "gringos" and Masons resulted from government health strictures during a serious cholera epidemic. Frustrated by high food prices and the displacement of individual farms by huge sugar plantations, smallholders protested the destruction of their harvests and the ban on eating fruit to halt disease.

1888 Civil War heroine Clara Barton headed the American Red Cross in Jacksonville, Florida, during a bout of yellow fever. Contagion broke out in the Mayflower Hotel. Citizens demanded that the building be burned. Writer Ellen K. Ingram proposed application of concussion to kill pathogens. On her suggestion, city officials imported four cannon from Saint Augustine to add to the cannon at the Wilson Battery. Cannoneers began discharging the weapons to purge the downtown area each night of germs. The continual firing broke store windows, but did nothing to eradicate fever. Officials added to the noise the burning of tar barrels, which killed mosquitoes. Homeowners disinfected rooms with copperas, sulphur, and lime. Rumors spread about the natural immunity of most blacks.

Panic sent 10,000 or 40 percent of the populace to packed government trains that removed the healthy from the city to the mountains. Armed guards warned incoming passengers to stay on board rather than disembark in Jacksonville. According to eyewitness William F. Hawley:

> [The trains] were packed to the limit, even the roofs of the cars [were] crowded with terrified citizens…. Some people in their haste left their homes with fires burning, food in preparation for the noonday meal, and doors wide open [1999].

To circumvent the disease, the postmaster general inspected baggage. To fumigate mail, workers in the mail car perforated letters with spiked paddles, loosened packets, and scattered individual pieces on wire netting, under which they lighted sulfur in iron kettles.

On the south bank of the Saint Mary's River, Florida's governor Edward Aylesworth Perry set up a Marine hospital contagion ward named Camp E. A. Perry in his honor. People exposed to disease remained within the perimeter under guard for up to ten days. Because 400 of the 5,000 victims died of yellow fever, Floridians established a state board of health headed by Joseph Yates Porter, who upgraded and modernized methods of containing disease, notably, by becoming first president of the Florida Anti-Mosquito Association.

1888 As beriberi contributed to death tolls in the Malay Federated States, W. Leonard Braddon, an English physician, studied the role of dietary deficiency in the deaths of thousands of Chinese immigrant laborers. He determined that no beriberi troubled the Malays who home-milled their rice, Tamils who parboiled their rice, and Europeans who did not eat rice. Only the Chinese who ate imported rice suffered the brunt of the disease. Braddon deduced that a toxin formed in white rice during storage. He concluded that a diet rich in white rice caused the nutritional deficiency, but he was incorrect in establishing a cause. From his findings, he compiled *The Cause and Prevention of Beri-beri* (1907).

1888 Epidemic cholera at Rizal province in the Philippines forced officials to take sanitary measures to counter infection. Manila recorded 2,403 deaths or 13.9 percent of a total 17,280 fatalities. Reports of victims buried alive resulted in a required waiting period before corpses could be interred.

summer 1888 Susan "Bright Eyes" La Flesche Picotte, the nation's first female Indian physician, returned home to eastern Nebraska for the summer after her second year at the Woman's Medical College of Pennsylvania to tend members of the Omaha nation suffering a measles epidemic.

1889 The most devastating plague to strike Africa was epizootic rinderpest, a cattle plague that killed up to 95 percent of herds across the continent. The animal virus created madness, debility, and death in herds from Ethiopia to South Africa. According to observer Lord Frederick John D. Lugard, first British governor of Nigeria and author of *The Dual Mandate in British Tropical Africa* (1922), the people "are half-starved-looking, most of them, and covered with itch—a most filthy looking disease which is most contagious, and the body is covered with open sores like smallpox" (Reader 1998, p. 590).

Human responses to the epizootic were disastrous. Devoid of their wealth, some herders lost their sense of self, retreated into hallucination, and killed themselves; about three-quarters of the Kenyan Maasai contracted smallpox and influenza and died. The Ndebele of Rhodesia (modern Zimbabwe) raised a doomed revolt against British settlers. The Tlharo and Tlhaping of British Bechuanaland and the South Sotho of Lesotho and Cape Colony reached the same impasse of confusion and discontent. In an era of social upheaval, European exploiters in southern Africa divided Herero leadership and extended German control over ancestral lands.

One tribe that survived loss of herds was the Mpondo of Pondoland. Labor recruiters offered them opportunities to abandon failing herds and join the migrant population finding jobs in South African mines and homes. The Mpondo maintained their faith in an animal-based economy and demanded cattle rather than monetary wages. Additional details of tribal coping skills derived from chronicler Samuel Blackbeard, author of *Khama's Country* (1930), which he compiled from a wagon journey through Tswana towns Shoshong, Palapye, and Serowe in Cape Colony.

1889 Ponca healer Susan La Flesche Picotte, the first female Indian trained in the white world, became a reservation doctor and medical missionary at the Omaha reservation, treating some 1,300 cases of cholera, conjunctivitis, dysentery, influenza, tuberculosis, and typhoid. In 1913, she opened a hospital in Walthill, Nebraska.

1889 At Santos, Brazil, Oswaldo Goncalves Cruz, a sanitarian and bacteriologist from Sao Paolo, began amassing data on an epidemic of bubonic plague that threatened Rio de Janeiro. He worked toward serum and anti-plague vaccines and eradicated Rio's yellow fever pandemic. The Institute he directed bears the name Instituto Oswaldo Cruz.

1889 In *Drusenfieber* (*Glandular Fever*) (1889), Emil Pfeiffer, a German pediatrician and balneologist from Wiesbaden, characterized outbreaks of infectious mononucleosis, a chronic disease common in children that caused fever, tenderness in the neck, and swollen lymph glands.

1889 Following war, drought, herd epizootic, locusts, failed harvest, and a famine named *Ya Kifu Qan* (the Cruel Days), Ethiopia and Somalia experienced a 13-year epidemic of Asiatic cholera. Allegedly imported from pilgrims on a hadj to Mecca, Arabia, the disease was so virulent that people of Asmara, Eritrea, were forced to cremate corpses. Adding to the death toll were outbreaks of dysentery, influenza, smallpox, and typhus, all flourishing in depleted bodies, killing 80 percent of the peasantry.

December 1, 1889 Raj Kumar Sen described the path of a deadly plague of Asiatic or H1N1 influenza from Europe to Bokhara, Turkistan. Called "Russian flu" in England, the pandemic began in St. Petersburg and caused high mortality. Doctors puzzled over the force of the outbreak. A quarter million Europeans died, impressing the Rev. Daniel Bell Hankin of East London, the vicar of Mildmay and keeper of a daybook from 1875 until 1904 on events at Stoke Newington. He delivered a chilling pulpit announcement to parishioners about the disease as contagion hit Paris. The flu spread over the South Pacific before striking North America, Latin America, and China in February 1890, with a total mortality count of one million.

1890 In Queensland, Australia, doctors J. Lockhart Gibson and A. Jefferis Turner of the Brisbane Hospital of Sick Children reported widespread lead poisoning among the very young. In 1904, Gibson concluded that the source of the disease was the eating of lead-based paint fragments, an observation that he and Turner wrote up for the *British Medical Journal*.

1890 On 11 of the Ellice and Gilbert islands, measles overran the British colony, killing 1,000. Complicating the health outlook was a parallel outbreak of dysentery. On Tabiteuea, Kiribati, contagion killed one-ninth of the 4,500 islanders.

1890 Medical researcher William Boog Leishman, while serving with the Royal Army for eight years, observed an epidemic of fever at Dum-Dum outside Calcutta. He studied symptoms by making slides of tissues from the spleen, lungs, and livers of victims of kala azar, called Dum-Dum fever. By 1903, he was able to publish in the *British Medical Journal* an analysis of the trypanosomes that caused the disease, named leishmaniasis in his honor. He later taught pathology at the British Army Medical School and became director general of British Army Medical Services.

1890 In the Alps Mountains at Leysin, Switzerland, innkeepers turned the mounting tuberculosis epidemic in Europe and the United States into a profitable business by providing live-in care centers for the Société Climatérique. Adding to the reputation of Swiss doctors for successful treatment was the work of Auguste Rollier, creator of heliotherapy for bone and joint disease.

1890 Beriberi afflicted 22,670 Thai after an upsurge at the Bangkok jail. The incidence was the result of institutional provisioning with white milled rice.

1890s When cases of leprosy increased in number in Iceland, the Danish Odd Fellows funded the building of the Leprosy Hospital at Laugarnes, a suburb of Reykjavík. Contributing to the containment of infection was the passage of a law requiring that doctors register leprous patients and isolate them from the public. The successful enforcement of Iceland's anti-leprosy campaign rapidly reduced the outbreak. As of July 1979, after the death of the last patient at the Landspítalinn hospital in Reykjavík, the disease was officially exterminated.

1890s In South Africa, European residents of the lowlands of the eastern Transvaal hesitated to invest in clearing land for farming and livestock breeding because of endemic malaria. In contrast to poor white immigrants, the more informed newcomers suffered lower infection rates because they screened their buildings, drained pools, poured oil on standing water, and took quinine.

early 1890s A scourge of typhoid fever struck Chicago, producing a higher fatality rate than in other cities in Europe and the United States. To assure fair-goers' health at the World's Columbian Exposition, Chicago officials established an on-site purification plant, trucked in water from Wisconsin, dug a sanitary canal, and drew water from offshore sites on the Chicago River.

1891 Among the 180 Gaelic-speaking residents on Saint Kilda Island in the Outer Hebrides, neonatal tetanus or tetanus infantum, which islanders called the "eight-day sickness," endangered the very young. Around 80 percent died within days of birth after contracting the disease from bacterial spores that infected the severing and dressing of the umbilical cord. Local midwives blamed neonatal deaths on stale air in huts, chill rooms, oil-rich food, improper birthing techniques, and cousin intermarriage. The last island minister, Angus Fiddes, investigated the mysterious deaths, caused by use of fulmar oil stored in the stomachs of gannets. After he sent to the mainland in July for Nurse Chisnhall to institute more sanitary practices, islanders lost no more babies to tetanus.

1891 When 60,000 devout Indians performed a rare Hindu water ritual, cholera spread among their number along departure routes, killing 580,000. The disease migrated west to Pakistan, Afghanistan, Persia, and Russia. To pre-empt additional outbreaks in central India, the provincial government curtailed the Hardwar fair, halted the sale of rail tickets to the area, and ousted 200,000 celebrants. In this era of disease, Rudyard Kipling, an Anglo-Indian journalist, composed one of his most poignant romances, "Without Benefit of Clergy" (1891), a short story about a British bureaucrat whose unacknowledged native wife and infant die in a cholera epidemic. *See also* **1892**.

September 1891 A shift in the Asiatic influenza virus produced a pandemic, which originated from a single focus and advanced along heavily traveled routes with high incidence, but low mortality rates. The spread of the disease to one-quarter of Sydney's populace followed a year of infections in Australia, New Zealand, and Tasmania. The recrudescence claimed 234 of some 130,000 reported cases in New South Wales.

1892 When cholera moved north from Astrakhan on the Kazakhstan border into the Russian heartland, sickness followed a year after a major famine. In May, the emerging epidemic of cholera in Russian Turkistan followed rail lines and waterways. Disgruntled serfs spread rumors that nobles conspired with doctors to wipe out the underclass. In the port cities of Astrakhan, Ekaterinoslav, Samarkand, Tashkent, and Volga, rioting expressed the despair, fear, and frustration of laborers that overlords considered peasants expendable. As contagion invaded Iuzovka in the Ukraine, squalid conditions among miners living along the Don River contributed to infection and death. A pioneer of social medicine, Anton Chekhov, who studied medicine before making a survey of the penal colony on Sakhalin Island, took an interest in epidemiology. When cholera broke out in Moscow, he treated patients and attempted to counter contagion.

The harsh reaction of civic leaders to the epidemic led to open rebellion against mandatory hospitalization and quarantine in barracks. Rioting produced more deaths, particularly among medical staff, and precipitated the destruction of the commercial center. Violence prefaced the eventual uprising of the peasantry in 1917 against tsarist brutality and neglect. Because cholera failed to yield to standard medical treatment, the epidemic produced bizarre gossip. Russian writer Avdotia Panaeva, author of *Vospominaniia* (*Memoirs*) (1927), recorded rumors that the Poles or doctors were poisoning Ukrainians or that assassins accepted bribes to exterminate the under-class. *See also* **June 18, 1892; August 1892; 1893; fall 1893.**

1892 A two-year typhus epidemic in Guanajuato, Mexico, arrived with an influx of immigrants and infected 432. The city, which was spared a high death rate, used the threat as an impetus to investigate contagion within families and to upgrade community sanitation rather than invest in isolation of patients.

summer 1892 Passing through Odessa, Ukraine; Constantinople and Smyrna, Turkey; and Naples, Italy, 268 Russian Jews and 470 Italians arriving in New York harbor aboard the S. S. *Massilia* from Marseilles, France, carried typhus. Health officers admitted cabin class and Italian emigrants and sequestered only Jewish passengers and their local contacts. Housed in tents, the Jews remained in quarantine near the Riker's Island garbage dump on North Brother Island, where 50 people developed the disease. Because the detainees lacked soap and water and washed in the freezing East River, six people died of exposure and complications. Officials either cremated or buried victims in sealed metal cans. New York Jews protested the inhumane treatment, hunger from lack of kosher meals, and unorthodox burial procedures.

By boosting the cost of steerage immigration for steamship companies, President Benjamin Harrison hoped to slow the flood of undesirables covertly. At the New York port, strict immigration regulation was the work of Tammany Hall appointee William T. Jenkins. The 21-day detention of ships in quarantine targeted only steerage passengers. Jenkins's xenophobia worsened rumors about Russian Jews, whom anti–Semites denigrated for their poverty and illness.

June 18, 1892 Cholera in Asia and Europe produced serious civic concerns. On June 18 at Baku, Azerbaijan, an inadequate and substandard water supply had been a long-standing civic problem. Because officials scrapped the idea of piping water from the Kur River, they chose a less expensive project, the building of a desalination plant to process water from the Caspian Sea. The meager plant produced 30,000 buckets per day of rusty, foul-smelling water.

In desperation for water, peasants turned to their old wells. Of the 800 in use, only 100 were untainted. As a result of the change in water sources, a cholera epidemic struck the town.

Authorities, realizing that impure water was sickening the citizenry, closed suspect wells and instituted disinfectant programs.

August 1892 In Hamburg, Germany, 8,594 residents died of cholera within four weeks, with a fatality rate of over 50 percent of citizens. Because contagion proliferated among the poorest, officials had to face up to blatant social irresponsibility and elitism that precipitated epidemics. The aristocracy, enriched by foreign trade, preferred investing in a warehouse complex and extension of the harbor to combating tenements and providing clean water to the city. The disease polarized the populace, placing the laboring class of Social Democrats in direct confrontation with Polish immigrants, a caste reviled as *classes dangereuses* (Kakimoto 1988). The proletariat accused the newcomers of spreading disease through unclean, unhealthful lifestyles.

Epidemiologist Robert Koch, author of *Deutscher Bakteriologe* (*German Bacteriology*) (1890), whom the government appointed as health inspector, found that the bacteria traveled through contaminated water supplied by the Elbe River. On August 25, he remarked: "In no other city have I encountered such unhealthy living quarters, such dens of pestilence, such breeding grounds," and added to a company of officials, "Gentlemen, I forget that I am in Europe" (Bastian 1994). To save face, Hamburg's medical authorities rejected his findings and downplayed evidence that the Altona suburb suffered fewer infections because the favored residents drank purified water. Koch aided Hamburg's poorer areas by distributing public service brochures, organizing disinfection teams, establishing stations to boil water, and offering imported stocks of potable water.

1893 Robbed of their civil rights as the rest of the islanders obtained their freedom, native Hawaiians exiled to a leper colony on the Kalaupapa Peninsula on Molokai evolved a microcosm marked by rage and rebellion. One patient, Koolau (or Kaluaiko'olau) of the Kalahau Valley of Kauai, fought off health authorities because he didn't want to leave his wife, Pi'ilani, and their child. Koolau survived until 1897 outside the captivity reserved for island lepers.

1893 At Samarkand in Turkistan, cholera swept two army regiments camped on a stream. A third body of soldiers avoided infection entirely because the commanding officer insisted that the men drink and wash only with boiled water. At Ashkhabad, cholera was on the wane when a governor honored Tsar Alexander III with a banquet. Within a single day, half the guests died, as did half a regiment and 80 percent of the military band. By the end of the second day, the area had lost one-tenth of its citizenry.

1893 North America's first major outbreak of polio on record began in Boston and spread over New England. The next summer, Charles Solomon Caverly of Burlington, Vermont, head of the state board of health, investigated infections at Rutland and Wallingford and concluded that the disease was impartial to class and economic status. The first public health officer to study polio, he was also the first to report nonparalytic cases.

Caverly observed the distribution of cases in valley communities along the sewage-contaminated Otter Creek. He noted in an article for the *Yale Medical Journal* a total of 132 cases, 30 instances of paralysis, and 18 fatalities. Concerning his epidemiological study of the nation's first large polio epidemic, he published *Infantile Paralysis in Vermont: 1894–1922* (1924). The naming of the Caverly Preventorium at Pittsford honored his work.

1893 Imported from New Zealand, measles afflicted South Pacific islanders at Samoa and Tonga. The latter island, which suffered the most hysteria, reported 1,000 deaths and widespread hunger. Following a civil uprising, Samoa, which experienced a concurrent outbreak of dysentery, noted only 300 deaths from measles.

June 22, 1893 After Vice Admiral Sir George Tryon hoisted a signal flag, the H.M.S. *Camperdown*, an armor-plated barbette captained by Admiral A. H. Markham, accidentally collided with and scuttled the H.M.S. *Victoria*, the flagship of the British Mediterranean fleet. The sinking off Tripoli, Libya, in North Africa drowned 358, including Tryon. An inquiry attributed the tactical error and Britain's worst peacetime naval disaster to crew suffering the effects of Malta fever, later identified as brucellosis.

August 1893 A smallpox epidemic in southern Muncie, Indiana, produced confusion among health experts. Because some identified the symptoms as chicken pox, officials delayed for weeks before issuing a quarantine order. The outbreak halted commerce, education, church activities, and amusements and cost a serious outlay of municipal funds, requiring a rise in taxes. Roving patrols to arrest violators of quarantine, the posting of guards at individual houses, and removal of the sick to a special hospital angered citizens, who sported guns to defend citizens' rights. Some health enforcers were shot in the melee.

fall 1893 A sudden and virulent outbreak of Asiatic cholera at the Nietleben insane asylum outside Halle, Belgium, demonstrated the elements of true malignant cholera. Confined within the facility, the disease allowed epidemiologists a glimpse of a controlled population. Arndt of Greifswald investigated the jettisoning of in-house sewage and examined the water source for the institution. He determined that a cycle of effluent passing from a farm to the river Saale and into the water intake circulated the cholera pathogen perennially through inmates. The conclusions of the study aided epidemiologists to pinpoint sources of the water-borne disease.

1894 At Carville, the state of Louisiana operated the Louisiana Leper Home, the nation's only leprosy hospital at the time. Confined like prisoners, patients endured treatment like pariahs and isolation from the public. For 46 years, doctors at the institution evolved treatment with chaulmoogra oil until chief medical officer Guy Paget introduced sulfones, which altered thinking about contagion and allowed victims to live normally.

1894 Hong Kong, a major port for global trade, suffered the world's last bubonic plague pandemic, which originated in Yunnan and Canton two years earlier and killed 100,000. The virulence of disease among tenement dwellers caused authorities to raze the Chinatown neighborhood at Tai Ping Shan. One of the observers of the Hong Kong outbreak, Japanese internist Aoyama Tanemichi of Edo, later applied his experience while on staff at Tokyo University and as director of the Institute for Infectious Diseases.

Contagion spread to ports of call in Taiwan, Japan, India, Portugal, Scotland, Australia, and San Francisco. The disease yielded to antibiotics and strict public health containment measures. Almost simultaneously, the laboratory research of Louis Pasteur's student Alexandre Yersin in France and of Robert Koch's Japanese pupil Kitasato Shibasaburo cultured and identified *Pasteurella pestis* as the bacillus causing bubonic plague.

1894 When Australian bacteriologist Sir David Bruce, a surgeon general in the Royal Army Corps, witnessed the extent of sleeping sickness among humans and nagana in cattle in Zululand, South Africa; Natal, Mozambique; and Uganda, he determined that the vector was the tsetse fly. In 1902, the medical adviser to the British Empire, Patrick Manson, who had been publicizing the dangers of sleeping sickness since 1891, convinced the Royal Society to send a research team to Africa to study it. The effort failed. Nonetheless, Manson was elected to the Royal Society in 1900 and knighted in 1903; his colleague, Ronald Ross, was also knighted and won the 1902 Nobel Prize for physiology or medicine for isolating the vector of malaria in 1898. *See also* **1897.**

1894 During a 200-member Austro-Hungarian Congo expedition, Viennese explorer Oscar Baumann, a naturalist and agent of the German Anti-Slavery Committee in Tanzania, commented on the emergence of jiggers or sandfleas (*Sarcopsylla penetrans*) across the African continent. They first arrived at Ambriz, Angola, in ballast from Rio de Janeiro aboard the British vessel *Thomas Michell* in 1872. Over caravan routes, jiggers spread as far east as Zanzibar.

The insects threatened all who did not constantly inspect their feet. Baumann observed that the best way to prevent infection was to extract the jiggers before they spread disease. If left to grow, they produced sores, blood poisoning, and death.

In *Dutch Massailand zur Nilquelle* (*Through Maasai Country to the Source of the Nile*) (1894), Baumann noted the results of jigger infestation in areas lacking medical skills: "Its impact can be devastating. We saw people in Uzinza whose limbs had disintegrated. Whole villages had died out on account of this vexation" (Reader 1998, p. 588). Hastening death to sufferers was their inability to harvest crops and feed their families.

summer 1894 Milwaukee's smallpox epidemic damaged Wisconsin's public health program. Health Commissioner Walter Kempster began a widespread vaccination campaign, quarantined infected patients at home, had the sick forcibly removed by ambulance to the Isolation Hospital, banned public funerals, and distributed information on wellness. German and Polish immigrants spurned his efforts as infringements on citizen rights and ethnic cultures. On August 5, mobs armed themselves with knives and clubs and attacked the health department. Based on the virulence of opposition, in February 1895, the Common Council voted him out.

1895 At Lille, France, endemic tuberculosis caused Catholic social volunteers and the followers of Désiré Verhaeghe to perpetuate the stereotype of victims as drunken, vice-ridden degenerates who deserved sickness as a punishment for their sins.

1895 Endemic sleeping sickness, dubbed "negro lethargy," killed 5,000 Congolese living in the Congo River basin. As of 1903, only 100 natives survived in the city of Kinshasa. The upshot of the virtual annihilation of the black labor force was a scathingly anti-imperial report issued in 1904 to the governor-general of the Congo Free State by British ambassador Sir Roger David Casement of Kingstown, Ireland. Casement recorded the filthy, inhumane treatment of workers on rubber plantations at Lukolela of the Upper Congo.

In his diaries, Casement commented specifically on sleeping sickness and the appalling loss of life:

> The cataract region, through which the railway passes ... is ... the home, or birthplace of the sleeping sickness—a terrible disease, which is, all too rapidly, eating its way into the heart of Africa.... The population of the Lower Congo has been gradually reduced by the unchecked ravages of this, as yet undiagnosed and incurable disease, and as one cause of the seemingly wholesale diminution of human life, which I everywhere observed in the regions revisited, a prominent place must be assigned to this malady.... Communities I had formerly known as large and flourishing centers of population are today entirely gone [Casement 1904].

To Casement's charges, Belgian overseers riposted that blacks expired solely from sleeping sickness, not from hunger or abuse. After an 11-year pandemic of sleeping sickness in the Congo River basin, by 1896, over a half million victims had died of the disease. Paralleling the misery of central Africa was a surge in infections in southern Sudan and in northern Uganda in the Nile basin, where 200,000 died of the disease.

1895 During the year-long Sino-Japanese War, which began on July 25, 1894, China and Japan fought over claims to Korea and mobilized iron-clad battleships, cruisers, and gunboats. After the Chinese surrendered on April 17, 1895, the Japanese reported 1,177 killed in combat and 15,860 from epidemic disease. The hardships of the winter campaign contributed 56,138 disabilities from infections. The Chinese kept no records, but appear to have lost twice what Japan incurred.

Epidemic cholera spread from Manchuria into Korea as far as south Seoul. At its height, the disease killed 300 per day. The scourge overran the city pesthouse, killing 95 of its 135 patients. In all, 300,000 died of cholera, including 5,000 at Seoul.

1896 After a flu epidemic in Jaffa, Israel, northwest of Jerusalem, two Zionist philanthropists, the Baron Edmond de Rothschild and Baroness Clara Hirsch, widow of banker Baron Maurice de Hirsch, donated the money to found an Israeli hospital. Working from her Paris office, Clara Hirsch contributed a total of $25 million to humanitarian projects. Supporting her effort was a bequest from the will of Wolf Segal, a member of the First Aliyah (exodus) of 40,000 European immigrants to Israel.

1896 The German Red Cross, the source of South Africa's first professional nursecare, opened at Windhoek, where German Catholic sisters arrived nine years later to staff wards. As the center branched out with mission stations, one pioneer nurse, Sister Winkelman, a victim of typhus, became the first to die in country. Other nurses faced constant contagion during daily exposure to bubonic plague, diphtheria, leprosy, malaria, measles, syphilis, and tuberculosis, the white plague that had paralleled the growth of industrialization and urbanism in Europe. Crowded ghettoes, poor sanitation, low wages, and insubstantial nutrition spread tuberculosis, causing fever, chest pain, bloody sputum, malaise, and collapse. Up to sixteen per thousand perished each year. Close, unventilated quarters encouraged the spread of the pathogen to whole families.

1896 In Oslo, Norway, the *Norske Kvinners Sanitetsforening* (*Women's Public Health Association*), an offshoot of the women's rights movement, offered personal care for victims of endemic tuberculosis.

1896 Spain's last-ditch effort to hold onto Cuba cost General Valeriano Weyler heavy losses— 9,413 dead in combat and 13,313 lost to yellow fever. Another 40,127 died of other diseases. In Spanish concentration camps, epidemic disease killed tens of thousands of Cuban civilians.

1896 During a period of danger to citizens, the quality of public service in Duluth, Minnesota, impacted the political career of Caspar Henry "Typhoid" Truelsen, a Danish-German immigrant who ran for mayor. Opposing Truelsen were the conservative candidate, Charles Allen, and corporate interests, backed by the major newspapers. Because of recurring typhoid fever, the issue of clean water became the core of campaign debate. The privately owned water company refused to disinfect its supply or to reduce its price to the city. After Truelsen's election to the first of four consecutive terms, he maneuvered a low price for the utility and enacted statutes to protect the city in future discussion of public services.

August 1896 A devastating reemergence of bubonic plague caused Austrian, British, Dutch, Ceylonese, Egyptian, German, Italian, and Russian bacteriologists to gather to study links between the bacillus and rats. When the disease struck Bombay and Calcutta, only 20 percent of victims survived. After the scourge killed six million Indian citizens, survivors attributed sickness to Hindu and Islamic deities. Additional contagion in Hong Kong brought death to the poorest citizens. Within a year, Japanese bacteriologist Ogata Masanori, author of *Über die Pestepidemie in Formosa* (*Concerning the Plague Epidemic in Formosa*) (1897), isolated the bacillus in Formosa. At Bombay, French epidemiologist Paul Louis Simond, author of *La Propagation de la Peste* (*The Spread of Plague*) (1898), accomplished the same breakthrough.

In September at Bombay, colonial and municipal authorities floundered over what action to take. The All India Act to Provide for the Better Prevention of the Spread of Dangerous Epidemic Disease, of February 1897, empowered enforcers of health regulations to isolate suspected victims, hospitalize the sick, destroy contagious property, relocate citizens, ban street vending and pilgrimages, and hold and examine travelers on streets and in cars, ships, and railway coaches. In March 1897, the Bombay Plague Committee decontaminated drains and cesspools with carbolic acid and seawater, disinfected alleys, and cleansed or burned slums.

The upshot of panic and reaction to the committee's hard-handedness was the flight of 425,000 citizens. Some, believing rumors, feared vaccination with the anti-plague serum that Russian-Jewish bacteriologist Walter (or Waldemar) Mordechai Haffkine of Odessa, a Louis Pasteur trainee, compounded from sterilized cultures of plague bacillus at his Plague Research Laboratory in Bombay and tested in the Calcutta slums. The ignorant spread outlandish claims that the vaccine would kill or sterilize them. The Parsi, India's devout Zoroastrians, doubted that the procedure was acceptable according to scripture in the *Vendidad*. To demonstrate the safety of inoculation, the Aga Khan III submitted publicly to an injection of vaccine.

Offsetting folk qualms were demographic studies that proved the serum valuable. At Dharwar, 5.2 percent of uninoculated locals died of plague; at Poona, the number of deaths neared 8 percent. After safely inoculating 45,000 citizens of India and reducing the mortality rate 90 percent, Haffkine was lionized as a savior of humankind. His laboratory was renamed the Haffkine Institute.

The Epidemic Diseases Act of 1897, which affected all of British India, gave the government the right to inspect ships and passengers, to isolate and detain anyone suspected of plague, and to destroy infected goods and personal property. License for colonial agents to search, disinfect, evacuate, ventilate, or raze any building suspected of harboring contagion rapidly eroded native rights. The British government could also suspend fairs and pilgrimages, examine travelers on the road and in rail cars, and perpetrate whatever indignities and inconveniences necessary to ward off plague.

1897 The squalor of British plantations in Ceylon thrust immigrant workers into a perpetual state of hookworm infestation, a disease worsened by walking barefoot in areas contaminated by nightsoil. The parasites caused fever, debility, anemia, diarrhea, dysentery, and dropsy, the edema caused by congestive heart failure. Of the 262 reported deaths from anchylostomiasis (hookworm disease), 80 percent were Sinhalese laborers and Malabar coolies. By 1904, hospitals received 1,755 infested patients. Five years later, infection compromised the health of 90 percent of workers in British colonies. While entrepreneurs enriched themselves on cheap labor and saved money by housing crews in squalid shacks, racism shamed the Asian coolie for an inability to live clean, sanitary lives. A pilot program in Matale altered the self-defeating attitude by diagnosing infestations in 10,000 laborers and treating the sick with chenopodium oil in sugar syrup. By 1917, the regimen cured some 78 percent of 40,000 patients.

1897 In British East Africa during the building of the Uganda Railroad, East Africa's first rail line, smallpox overwhelmed those natives who lacked immunity. Worsening conditions for the poor at Kitui was a drought that killed cattle and caused mass starvation. Within two years, the Maasai were reduced to one-quarter their former number and the pastoral Kikuyu to around 30 percent. The Rendile, camel herders of the uplands, were virtually depopulated. Traditional herders migrated to Swahili towns, refugee centers that introduced newcomers to wage labor. Because foreigners suffered few infections, the underclass grew suspicious as colonists enriched themselves by seizing abandoned lands and villages.

1897 At the São João de Deus asylum in Bahía, Brazil, epidemiologist Raimundo Nina Rodrigues of the Bahía School of Medicine studied the source of beriberi. She was puzzled why the disease ravaged inmates over a period of eight years, but not their keepers. According to mortality data at the institution, beriberi was the cause of 66 percent of deaths. The disease was also endemic among slave infants and toddlers.

August 20, 1897 To end endemic malaria, Sir Ronald Ross, an Anglo-Indian epidemiologist and surgeon in the Indian Medical Service at Secunderabad, identified a parasite in the gut of an anopheline mosquito that transmits the disease by biting humans. His work coincided with that of Giovanni Battista Grassi, recipient of the Darwin Medal, who issued his findings on persistent malaria in Rome in 1898. In 1899, Ross established a long career as lecturer at the Liverpool School of Tropical Medicine. He composed a poem acknowledging the misery caused by malaria:

> With tears and toiling breath,
> I find thy cunning seeds,
> O million-murdering Death ("Ross").

October 1897 During the Maidstone outbreak of typhoid fever in Kent, England, 132 out of 1,900 patients died of the contagion, which targeted youth. The scourge of *Salmonella typhi* began in Farleigh Spring from foul water supplies that the city refused to have analyzed. Authorities implicated migrant hops pickers as scapegoats for a local pestilence. The town received the first phone-tree emergency communication and the first coordinated response to an epidemic, beginning with sterilization of water mains with chloride of lime. Citizens boosted sale of bottled beer and mineral water. Hospitals and domestic services set up laundries to disinfect 62,000 items of bedding and garments with high-pressure steam. Sulfur fumigation, carbolic scrubs, wallpaper removal, and fresh whitewash purified patient rooms. At a psychiatric hospital, 84 staff and their aides received the first prophylactic inoculations in England. All were protected from contagion. Of the unvaccinated, 16 contracted the fever, which raged for five months. One of the nurses, World War I martyr Edith Cavell, received the Maidstone Epidemic Medal. *See also* 1937.

October 23, 1897 When yellow fever beset Bay Saint Louis, Mississippi, in mid–October 1897, editors of the *Seacoast Echo*, the town's newspaper, kept the disease secret from readers. The vow of silence followed the paper's castigation of physicians at nearby Ocean Springs for concealing the disease the previous September. On October 23, the staff had to acknowledge 42 infections and two deaths to a consortium of independent physicians. Simultaneous flare-ups struck four Southern states:

City	cases	deaths
New Orleans, Louisiana	266	54
Mobile, Alabama	74	5
Wagar, Alabama	45	3
Scranton, Mississippi	35	3
Montgomery, Alabama	25	3
Whistler, Alabama	25	2
Memphis, Tennessee	19	10

1898 Príncipe, a Portuguese island in the Gulf of Guinea, suffered a 15-year epidemic of sleeping sickness that killed 3,500. Transmitted from 600 laborers imported from Portuguese West Africa (modern Angola), Gabon, and the Gold Coast to work cacao plantations, the disease spread from unsanitary conditions at the rate of 70 infections per month. By 1907, the labor force had diminished by 88.3 percent from 3,000 to 350. The loss created an economic catastrophe.

To exterminate sleeping sickness, health authorities resorted to spreading sticky "rat varnish" to trap flies and distributing hoods and neck drapes for workers to wear in the field while they felled trees in virgin forests, cut brush, and drained bogs, vastly altering the terrain. Swine-killing squads eradicated wild pigs, which were reservoirs for the disease. The only immunizations involved injections of atoxyl, a form of arsenic recommended by Portuguese epidemiologists even though it caused blindness and insanity. Through over a half century of disease control, residents finally enjoyed freedom from epidemic by 1956.

1898 During epidemic smallpox among the Navaho and Jicarilla Apache in Arizona and the Zuñi, Hopi, and Pueblo of New Mexico, the last major Native American smallpox outbreak killed 300 of the surviving 1,798 Zuñi. Of 900 Moqui at the Pueblo Reservation, Arizona, 590 contracted the pox; 184 died. The scourge reached epidemic proportions among Hopi of First and Second Mesa, Arizona. In anger at whites, Hopi elders added catastrophic disease to their list of complaints about forced enrollment of children at boarding schools and dislocation of natives from traditional homelands. At Oraibi, Chief Lomahongyoma and Tawahongnewa welcomed resistant families and organized a revolt against fumigation of residences, washing in acid solution, vaccination, and the burning of corpses.

Indian agent George Hayzlett and the Navaho police arrested and imprisoned rebellious Hopi, who watched helplessly as 632 sickened and 10 percent of their people died. Of 220 infected Hopi who refused Western medicine, 163 died. Only 26 percent survived. Of the 412 who accepted new medical treatments, only 24 or 6 percent died. The result of hard-handed government intervention was a revival of traditional ways among Hopi survivors. By April 1899 at First and Second Mesa, 600 more Hopi suffered infection and 187 deaths from smallpox and famine.

Among Pueblo Indians of Laguna Reservation, New Mexico, federal health services failed to meet the demands of so great an upsurge in infection. Lacking a chief medical officer to apply the latest advances in epidemiology, the people continued visiting from site to site to celebrate ritual occasions. They died in record number, with villagers at Paguate incurring a 40 percent mortality rate. Most white U.S. citizens were unaware of the tragedy.

1898 When journalist, writer, and explorer Félix Dubois, author of *Timbuctoo the Mysterious* (1897), attempted to introduce automobile travel to upgrade the African economy, he mobilized a transport company in French Sudan (modern Mali) between Kayes and Bamako. The 15-year project failed because of inadequate roads, engine problems, and endemic yellow fever, which robbed investors of local laborers.

1898 While Europe battled tuberculosis, statisticians at Lódz in central Poland considered the infection a social disease of Jewish ghettoes. Doctor Seweryn Sterling and other theorists developed a reputation for lessening the suffering of tubercular patients at an up-to-date hospital TB ward through hygiene.

1898 Upon visiting Rotuma in the Society Islands of the South Pacific, naturalist and travel writer John Stanley Gardiner remarked on endemic disease, particularly yaws, which natives called *coko* or *tona*. He noted that the malady had emerged recently, leaving large scars from ulcerations on the mouth and nose as well as blindness and crippling. Families urged their children to contract the disease from sick playmates to make them immune to further outbreaks.

Out of interest in islanders' health, Gardiner commented on the source of the infection:

> I cannot resist the idea that really these ulcerations and yaws are of a syphilitic nature and give immunity from this disease, which is absolutely unknown on the island; other diseases of a venereal nature too are very rare, owing to the extreme cleanliness of the women [Gardiner 1898].

He also noted the outbreak of pneumonia, tuberculosis, and other respiratory ailments in the 19th century. He observed the frequent incidence of ringworm, malarial fever, and mild typhoid fever. He described a two-day operation for elephantiasis, a common cause of scrotal enlargement. He witnessed clouded corneas and blindness from epidemic trachoma.

Concerning epidemics, Gardiner estimated the loss to the island population from European diseases:

I estimate that the population in 1850 cannot have been short of 4,000, and that at the beginning of the century there were nearly 1,000 more…. In this last period of ten years there were four epidemics, viz., dysentery in 1882, whooping-cough in 1884, dengue in 1885, and influenza [1898].

Concerning historical records, he concluded, "Many epidemics are remembered, though few details are known" (ibid.).

April 22, 1898 After American naval forces blockaded Cuba's harbors at Daiquiri, Havana, Cienfuegos, Cardenas, Bahia Honda, and Santiago, some 200,000 military volunteers went to war. More than 10 percent contracted dysentery, malaria, and typhoid. The worst was typhoid, which felled 20,926 or 10.5 percent of the men. To accommodate the ingathering of patients, the medical corps opened emergency tent cities. To add to American Red Cross emergency measures, Anita Newcomb McGee formed a D.A.R.–led Hospital Corps Committee and enlisted 8,000 nurses, who began setting out for the medical command post at Key West, Florida, on May 7. The deployment of 723 male orderlies and nurses two months later added to a staff of 6,000 untrained corpsmen, who dismayed superiors by sloshing bedpans and handling food and sick men with unclean hands.

The poor performance of amateurs inspired military authorities to hire more female nurses to manage the outrage of the war—the epidemics that felled 49,500 or 30 percent of the 165,000 volunteers who remained stateside until needed. One nurse at Chickamauga Park, Georgia, Lucy Shook Huxtable of Kansas, battled foul latrines and flies. A New Yorker, nurse Anna C. Maxwell, bought chloride of lime and carbolic acid in Chattanooga to disinfect surfaces and laundry to rid them of typhoid bacilli. In 1904, majors Walter Reed, Edward O. Shakespeare, and Victor C. Vaughan produced an historic monograph, *Report on the Origin and Spread of Typhoid Fever in U.S. Military Camps during the Spanish War of 1898*, which compiled details of the 19,000 cases in 89 volunteer army regiments in the eastern United States.

July 1898 According to data collected in the 1950s by Henry Oliver Lancaster, author of "The Epidemiology of Deafness due to Maternal Rubella" in *Acta Genetica* (1954), the high incidence of deaf-mutes and blindness at Darlington, New South Wales, were the result of epidemic rubella that lasted until March 1899. The disease passed from the mother through blood in the placenta to fetuses in the early formative stages of the first trimester of pregnancy.

July 3, 1898 Esther Voorhees Hasson of Baltimore, the first chief of the U.S. Navy Nurse Corps, served as surgical nurse aboard the hospital steamer U.S.S. *Relief*, which was commissioned on May 18 to evacuate the sick and wounded from Cuba during the Spanish-American War. After the battle of Santiago Bay, she and five nurses received casualties and victims of dysentery, malaria, typhoid, and yellow fever at Siboney, Cuba. Within two months of service, her staff treated 1,485 patients, of whom 95.7 percent survived.

In mid–July, as the war came to a close, 729 U.S. nurses transferred to Santiago, Cuba, to fight a yellow fever epidemic. Distressing nurse Anna Turner as she recorded patients' fluid intake and urine output were symptoms of intestinal pain and nausea, streams of malodorous black vomit and diarrhea, jaundice, and bloody gums and skin. She participated in medical research on yellow fever, which established that the disease did not pass between human carriers but rather from the bite of the *Aedes Aegyptus* mosquito. *See also* **August 3, 1898; September 10, 1898.**

July 11, 1898 Near the Klondike gold-mining fields at Dawson in the Canadian Yukon, Mary of the Cross, Mary Joseph Calasanctius, and Mary John Damascene of the Sisters of Saint Anne staffed Saint Mary's Hospital along with three untrained aides—nuns Mary Zephyrin, Pauline, and Prudentienne—during a typhoid epidemic. With little equipment and sawdust sacks for mattresses, they tended 140 victims in a two-story log clinic.

August 3, 1898 Theodore Roosevelt, who was assistant secretary of the U.S. Navy and organizer of the Rough Riders, feared that his men would perish in Cuba if not evacuated to a healthier climate. The army moved 1,000 sick soldiers to isolation wards at Camp Wikoff at Montauk Point, Long Island, where patients overran barracks. Sleeping on the ground in tents, 2 percent died of exposure and inadequate rations.

Two weeks later, American Red Cross nurse Anne Williamson of New York, author of *50 Years in Starch* (1948), was overwhelmed by the extent of typhoid infection. She found up to six feverish men piled into a single ambulance and pleading for water to moisten swollen throats. Their bodies were wracked by fever and crusty sores and contaminated with dead insects. There were so many sick in wards that nurses hosed them down as a group rather than sponge them individually. After taking a break, Williamson returned to duty and found a co-worker dead at her desk.

Nurse-author Jean S. Edmunds added details in *Leaves from a Nurse's Life's History, 1906.* Under the command of Chief Nurse Anna C. Maxwell, Edmunds struggled with the regimen of broth feedings, ice packs, and constant stripping of sweaty sheets and airing of mattresses. The day's work kept her busy for 14-hour stints with only two 20-minute breaks for lunch and dinner. So many nurses caught dysentery from treating diarrhea that they taxed facilities in staff latrines. Nurses Helen B. Schuler and Florence M. Kelly remarked that some of the veteran nurses of the Spanish-American War suffered chronic war-related ills all their lives.

September 10, 1898 After the Spanish-American War, Sisters of Charity and the American Red Cross aided the local nursing staff in transferring 1,000 sick soldiers from Montauk Point, Long Island, to civilian hospitals in Boston, New York, Philadelphia, and Providence, Rhode Island. The crisis passed by the end of October, but public outrage humiliated the U.S. military for its inept handling of typhoid and yellow fever cases. Investigators concluded that 90 percent of the 5,000 who died from disease could have been saved. To prevent future fiascos from inept male staff, Congress passed the Army Reorganization Bill and initiated the all-female Army Nurse Corps.

1899 After treating the poor in Berlin, Germany, Rosalie Slaughter Morton, of Lynchburg, Virginia, the first female on staff at New York Polyclinic Hospital and at Columbia University's College of Physicians and Surgeons, conducted postgraduate work in Asia. She studied the management of outbreaks of bubonic plague during her six-month stay in India.

1899 At Bangalore, Lucknow, Pindi, and Rawal in India, bacteriologist and immunologist Sir Almroth Edward Wright, head of the Institute of Pathology at St. Mary's Hospital, London, vaccinated over 3,000 of the Indian army with his

typhoid fever vaccine made from heat-killed bacilli preserved in phenol. Assisting him was Sir William Boog Leishman of the Army Medical School, who later directed military medical services. Their experiment began with volunteer officers from the Indian Medical Corps and successfully prevented disease in 63 percent. The British government deemed the two experimenters' procedures lifesaving for soldiers and paid for clinical expenses from the treasury. In subsequent outbreaks of disease, the inoculated soldiers were protected from contagion.

1899 In a pattern of contagion, the sixth in the great Asiatic cholera pandemics began in India, raising the death toll for the country to 805,698. In a repeat of earlier outbreaks, the disease blanketed the Muslim world when it spread from Mecca, Arabia, via pilgrims on an annual hadj. Most of the Middle East and Asia suffered the disease, which died out in Russia in 1907. Echoes of the pandemic continued hammering Asia into the 1910s and 1920s, producing the greatest loss of life in India.

1899 When yellow fever invaded the newly founded city of Miami, Florida, citizens fumigated infected homes and isolated the sick on vessels in Biscayne Bay and at a camp north of the city. Through the epidemiological control established by four local doctors, authorities held the loss to 14 deaths and lifted the quarantine in January 1900.

1899 When yellow fever invaded Grand Bassam, the capital of the Ivory Coast, it nearly annihilated European residents. To protect the health of whites from the fever and from endemic malaria, French colonial authorities chose Bingerville as a new capital. The location offered protection from disease-rich harbors on the Atlantic Coast and from too close contact with the native Ebrie. They rejected the shift of their capital and fomented a rebellion that burst into a three-year insurrection in 1903.

1899 In Valencia, Spain, Francisco Moliner y Nicolás initiated the nation's first government-sponsored campaign against epidemic tuberculosis. Based on the era's concepts of contagion, the effort influenced a nationwide crusade established in 1906.

April 1899 Bubonic plague made its way from Asia to Argentina and Paraguay and by sea to

South Africa. As a result of the successful anti-plague campaign mounted by Oswaldo Goncalves Cruz of Brazil, a researcher and parasitologist educated at the Pasteur Institute, the nation's congress named the Oswaldo Cruz Institute of Rio de Janeiro for him in 1908.

June 20, 1899 A decade before Carlos Chagas identified the protozoa causing Chagas disease, U.S. victims reported an epidemic of swollen eyes and fever from the bite of the "kissing bug." For the next months, 100 cases and a few fatalities from "American trypanosomiasis" cropped up in major newspapers from Atlanta, Boston, Chicago, New York, Texas, and Washington, D.C., *see* 1909. The silent killer remained undetected until the protozoa compromised the heart muscle, esophagus, and colon, causing fibrosis, sensory collapse, confusion, dementia, and death.

October 1899 The arrival of plague on a flotilla from Hong Kong to Honolulu, Hawaii, precipitated disease throughout the island chain. The scourge entered the wharf area after the docking of the freighter *Nippon Maru* from Hong Kong at the Pacific Mail pier. While the cargo remained unclaimed, workers noticed hundreds of rat corpses littering the area and swept them into the bay.

Sickness first struck Asian islanders and fanned out to other parts of the island cluster. In November, the death of 22-year-old Malaoa Momona signaled the onset of pestilence. That month, Queen's Hospital admitted twice its usual number of patients. One issue of the *Pacific Commercial Advertiser* observed that Hawaiians were three times more likely to die of plague than were whites and predicted that indigenous islanders would soon die out completely.

Henry Cooper, head of the board of health, initiated a practice of sprinkling lime, killing rats, and slitting mail or clipping the corners of envelopes to allow fumigation of contents to stem contagion. A colleague, Charles Allen Peterson, volunteered to treat plague victims. To spare his family, he exiled himself to quarantine with health professionals at a building behind Iolani Palace. When an experimental plague serum became available, he volunteered to test it on himself.

Because the disease roused suspicions of Chinese residents, the 6th Artillery enforced a quarantine in Chinatown and torched suspect refuse,

mattresses, residences, even entire city blocks. In January 1, 1900, as the world welcomed a new millennium, the burning of infected property got out of control, destroyed much of Honolulu, and dispossessed 5,000 Chinese. The harbor ceased to receive steamers. Supply boats and troop carriers were rerouted to Hilo, which remained plague-free through some controversial public health measures. Honolulu's stringent rules remained in effect until April 30, 1900.

October 4, 1899 A smallpox epidemic struck Laredo, Texas, causing a face-off between Mexican-Americans and Texas Rangers. Mayor Louis J. Christen appointed a committee of inquiry. By January 31, 1899, the disease had caused 100 deaths. William Thomas Blunt, a state health officer from Austin, advised quarantining, a general vaccination of barrio residents, and incineration of suspect clothing and bedding. To Hispanics resisting the measures, he dispatched Texas Rangers to invade homes and enforce immunization and isolation of the sick. Locals pelted the rangers with stones; rangers retaliated with buckshot. When a shootout between officers and one Mexican family provoked a riot, on March 21, the rangers called in Captain Charles G. Ayers and the Tenth Cavalry from Fort McIntosh. By April, the epidemic eased.

October 12, 1899 In South Africa during the Boer War, when the British fought Dutch Calvinist farmers, disease became far deadlier than violence. Stymying health agents was widespread contamination of food and water with human feces. As a result, 13,139 locals died of disease, primarily enteritis. Epidemic typhoid fever infected 42,741 troops, killing 26 percent, nearly five times more than died in combat. In one instance, on the march to Pretoria, the 2,000 men of the Second Battalion Royal Canadian Regiment fell below half strength from typhoid.

The fever quickly became the most lethal threat to life. One serious incident in April 1900 involved the halt of British forces at Bloemfontein to drink at the Modder River, which was heavily contaminated. After ingesting fast-acting pathogens, 10,000 men fell sick with typhoid. The outbreak delayed the deployment of Lord Frederick Roberts's forces for six weeks. At the siege of Ladysmith, Australian bacteriologist Sir David Bruce, a surgeon general in the Royal Army Corps, and his wife and partner, artist and medical technologist Lady Mary Elizabeth Steele-Bruce, battled an outbreak of typhoid fever that killed 393 of the 563 wartime fatalities. The courageous 180-day scourge earned Mary Bruce the Royal Red Cross Medal. David Bruce was promoted to Colonel.

A dark element of the Boer War was the concentration camp system set up by Lord Horatio Herbert Kitchener, the commander in chief, as an expedient means to contain refugees and encourage genocide of Boers. Cornish activist Emily Hobhouse, founder of the Relief Fund for South African Women and Children, protested that most detainees were mothers, their children, and elderly male "hands-uppers," those who had surrendered willingly. All fled dynamited outbuildings, harvests burned to the ground, and livestock bayonetted as part of the English scorched-earth policy against 30,000 Boer farms. The English rounded up survivors in open-top cattle trucks and flatbed trains or forced them to walk to the camps carrying their children.

To punish the Boers, Kitchener set up 50 *laagers* (camps). As featured in *Diary of a Nurse in South Africa* (1901), compiled by Alice Bron, a Belgian nurse and social activist, the concentration camps had limited washing, disinfectant, and toilet facilities, with Aliwal North supplying only one toilet per 170 detainees. Notable for its wretchedness was Standerton on the mosquito-infested shores of the Vaal River, where, according to Maria Fischer's *Tant Miem Fischer se Kampdagboek* (1912), 40 inmates shared each tent. Similarly miserable were camps at Brandfort, Orange River, and Springfontein, where 19 people occupied a single tent and slept on bare ground. Merebank, located in a swamp south of Durban, was even more despicable. Out of 8,000 inmates, 453 died mainly from measles, enteritis, and dysentery.

In all, 150,000 inmates survived on a half-pound of meat and a pound of maggoty flour or meal per day. The families of active commandos received only meal until the men surrendered. Babies and children received no milk or vegetables. Many developed scurvy, an ailment described in Alida Badenhorst's *Tant Alie of Transvaal: Her Diary 1880–1902* (1923). To comfort as many as possible, 1,700 nurses and volunteers from England staffed infirmaries.

Few inmates left camp hospitals alive. At Kroonstad compound, the death rate reached 878 per thousand. Of the 28,000 total who died, 81 percent were children. William Thomas Stead, a British newspaperman, commented on the deaths of the young:

> Every one of these children who died as a result of the halving of their rations, thereby exerting pressure onto their family still on the battlefield, was purposefully murdered. The system of half rations stands exposed and stark and unshameful as a cold-blooded deed of state policy employed with the purpose of ensuring the surrender of people whom we were not able to defeat on the battlefield [Barnard, "The Concentration Camps"].

Most of the casualties suffered from pneumonia, typhus, and typhoid fever compounded by starvation, exhaustion, and despair.

Measles rapidly depleted the weak. The inmates at Irene Camp in Pretoria crammed 14 victims in one isolation tent. At Barberton and Nylstroom, two concentration camps in Transvaal, an upsurge in measles in a virgin-soil population increased adult fatalities by nearly half and children's deaths by two-thirds. Increased migration to populated areas carried infection from farmlands to cities. In an act of biological terrorism, commanders transferred Afrikaners and Africans to zones of contagion, where they infected minority populations.

December 1899 Within months of the U.S. takeover of the Philippines, plague invaded Manila via Chinese steamers from Hong Kong. The infection flourished in harbors and spread along roadways to the poorest and most populated areas. Isolation of the sick received little understanding or respect among the ignorant and ill-fed peasants. Health officials contained the disease by disinfection facilities at Cebu, Iloilo, and Mariveles. U.S. military health initiatives, headed by Richard Pearson Strong, president of the Board for Investigation of Tropical Diseases in the Philippines, quarantined shores from foreigners bearing contagion.

1900 Scots physician Andrew Balfour of Edinburgh, an expert on sleeping sickness and author of *Health Problems of the Empire* (1924), gained experience in tropical disease in Pretoria, South Africa, at a typhoid camp. Of the typhoid epidemic, he said:

> There one saw the disease at its worst, witnessed wretched, stuporous patients in stinking khaki

taken from trains and ambulance wagons, heard the droning buzz of accompanying cohorts of filthy flies, saw peeling and crusting lips, teeth coated with sores, and tongues dry as those of parrots [Balfour 1921, p. 227].

His year's service won him the South African Medal and the directorship of anti-toxins at the research labs built by an American manufacturer, Sir Henry Solomon Wellcome, who partnered with Silas Burroughs in a pharmaceutical business in London. Wellcome chose the site for the Gordon Memorial College in Khartoum, Sudan, because of the abysmal health conditions among the Sudanese.

Simultaneous with the typhoid fever epidemic in Pretoria was the devastating pneumonia epidemic among migrant workers at the Crown gold mines. Over 40 percent of the 111,500 laborers succumbed to infection, which killed 6,500. Worsening the outbreak were hardships of travel, change in climate, hunger, and unhygienic living conditions. An undercurrent of criticism of colonial entrepreneurs sparked protests at the Rand operations. For a decade, investors lost one-tenth of their workers to pneumonia.

1900 North of Lake Victoria, 100,000 Ugandans and Tanganyikans died during a decade of lethal sleeping sickness. Spreading disease were water-borne vendors navigating new trade routes in the Congo River basin. The disease killed 11,000 at Busoga and 43,000 at the Buvuma Islands. Natives of south Busogo disseminated a legend, "The Bishop's Head," explaining the sleeping sickness epidemic as a curse caused by the murder of Bishop James Hannington, whom natives speared to death on October 29, 1885.

British commissioner Sir Hesketh Bell evacuated people from hot zones to less infested areas as work on combatting the disease progressed. A pair of English doctors, Albert Cook and his brother Jack Cook, identified the Gambian strain of parasite in victims' blood. At Tanga to the west, German bacteriologist Robert Koch formulated atoxyl, a cure made from arsenic, which made some patients blind or insane. By 1902, Patrick Manson and the Royal Society organized the first sleeping sickness research expedition to Africa. Distribution of atoxyl and other methods of containment combined with medical breakthroughs to end the epidemic by 1909.

1900 In Havana, Cuba, sanitary chief William Crawford Gorgas of Mobile, Alabama, rid the city of filth, but did not prevent 1,400 new yellow fever infections. Cuban epidemiologist Carlos Juan Finlay, author of *Estudios sobre la Fiebre Amarilla* (*Studies on Yellow Fever*) (1945), contended that the *Aedes aegypti* mosquito spread the disease. A U.S. army surgeon, Walter Reed, in collaboration with doctors Aristides Agramonte, James Carroll, and Jesse W. Lazear, tested the theory on a human volunteer while disproving beliefs that humans were infected through contact with diseased body fluids and soiled clothing. During the intense study, both Carroll and Lazear caught the disease after exposing themselves to mosquitoes. Lazear died. Playwright Sidney Coe Howard captured their heroism in 1934 in the drama *Yellowjack*, which MGM filmed two years later, starring Robert Montgomery and Charles Coburn.

Gorgas reduced the yellow fever epidemic citywide by controlling mosquitoes. The regimen required upgrading general sanitation, screening windows, draining ditches and wet spots, and covering remaining pools with a layer of oil to inhibit insect breeding. At Las Animas Hospital, John Guiteras inoculated volunteers to build immunity. Within the year, reports of new infections fell to 37. Oswaldo (or Osvaldo) Goncalves Cruz duplicated Gorgas's methods in Rio de Janeiro, Brazil, with parallel results. To rid Santos and Belém of the disease as well as of plague, he set up his own laboratory and crusaded for city sanitation over a nine-year period. His methods helped end outbreaks of disease. *See also* **March 1901; 1904; 1906.**

1900 The return of cholera to Bengal, India, targeted foreign mill workers and laborers at Bombay and infected southeastern Asia and Oceania as far east as the Philippines. In worker *chawls* (boarding houses), some 270 died per week from plague, contributing to the toll from cholera and tuberculosis. The deaths of 806,000 Indians produced the highest number of fatalities ever recorded during epidemics of the disease. Under British orders, workers cleaned Bombay and built Sydenham Road and Princess Street to the coast to convey ocean breezes into the fetid slums.

1900 During an upsurge in malaria in Croatia, German epidemiologist Robert Koch spent two years on the island of Brijuni near the Istrian peninsula eradicating the sources of disease. The island atmosphere was so insect-ridden that it had been deserted of human habitation for two centuries. After Viennese industrialist Paul Kupelwieser bought Brijuni, he began developing the land as a vacation spot, using convict labor to produce roads, paths, habitats for wild animals, and vineyards that produced Brioni wine.

To rid the area of mosquitoes, Kupelwieser hired the Nobel Prize–winning biologist to supervise improvements. Koch instructed civil authorities on the drainage of swamps and cutting of underbrush at Istria and Brijuni. The island immediately became a tourist attraction to notables, including Emperor Wilhelm II, Archduke Franz Ferdinand, and Josip Broz Tito, the president of Yugoslavia, who established a summer residence at Brijuni. Stone monuments on the site preserve Koch's image.

1900 After the transmission of bubonic plague from Chinese trading vessels to Rhodesia (modern Zimbabwe) and South Africa, the pestilence figured in the professional observations and correspondence of Australian bacteriologist Sir David Bruce, a surgeon general in the Royal Army Corps, and his wife and partner, artist and medical technologist Lady Mary Elizabeth Steele-Bruce. The threat of both plague and dysentery created a social and political climate rife with racism. Colonial authorities attempted to limit contagion by segregating blacks from whites. The concept inspired apartheid, an extreme separation of the races.

1900 As with Uganda, Zululand, and South Africa, recurrent epidemics of malaria at Swaziland inundated struggling farmers and herdsman, making them more dependent on Europeans. At a low economic point in the area's history, drought, famine, malnutrition, and heavy rains decreased residents' ability to fight mounting infestations by mosquitoes, which increased the incidence of sickness and death from fever. Colonial developers exploited the tenuous position of local tribes and gradually stripped them of self-sufficiency and rights to land.

1900 In political exile in the Marzuq Desert of Libya, Turkish physician 'Abd al-Karim Abu-Shwairb made use of his skills by observing and treating epidemic syphilis and tuberculosis in the city of Fezzan, a major nexus of the caravan trade.

ca. 1900 In Scotland, a prevalent myth that intercourse with a virgin would cure males of venereal disease encouraged incidents of rape and sexual abuse of young girls. As a result of exploitation by adults, the children incurred a disproportionate infection with gonorrhea and syphilis. Although the court system prosecuted rapists, doctors tended to misdiagnose the symptoms of genital infection in little girls as filth or parasites.

ca. 1900 Compelled by numerous deaths from epidemic tuberculosis, Mefodiy Romanowski, a Polish doctor, developed medicines to treat pulmonary disease in the Ukraine. After he was drafted into the military during the Russo-Japanese War, he halted his research temporarily. Later, he evolved Phosphacid to repair cell damage. After his death from typhus in 1911, his wife continued his work.

January 19, 1900 A plague epidemic in China spread to the Tonkin region of Vietnam and advanced by ship southeast to Sydney, Australia. The disease first claimed Arthur Payne, a carter employed by the Central Wharf Company at "The Rocks," the area's harbor. Panicked by media predictions, islanders were already anticipating that China's epidemic would move in their direction via Hong Kong and Noumea, New Caledonia.

The pestilence turned into a racial issue. The Sydney *Morning Herald* and other papers linked the disease to Asia and encouraged racist notions among Anglo-Australians. Among outlandish rumors were claims that local "Syrians and Hindoos" were public menaces. To carry solace and treatment to those least likely to receive it, members of the Sydney Home Nursing Service packed hampers with supplies and went door to door. By late spring, after 303 of 1,200 mostly young adult victims had died, only ten were Asian. Nonetheless, the national assembly began examining the need to limit immigration of Asians to Australia. In late April, the *Northern Miner* noted the arrival of a plague victim aboard the S.S. *Cintra* out of Cairns and Brisbane. Authorities quarantined the patient in Townsville. Rumors shifted blame from Chinese ghettoes to the rodent-infested Charters Towers gold mines in Queensland for fostering contagion.

In response to a need for a general clean-up, health officials isolated 2,000 sick citizens at North Head and dispatched workers to kill 45,000 rats, cover them with boiling water, and remove their remains with tongs. Under command of the Sydney Harbour Trust, the squads also cleared rubbish from under piers at "The Rocks," demolished ramshackle structures, and cleansed, purified, and whitewashed 4,000 infected residences. To prevent future accumulations of garbage and silt, the city erected a permanent municipal incinerator.

March 5, 1900 When the transoceanic ship *Nippon Maru* brought bubonic plague to North America from China, infection struck first in San Francisco, where staff at the Globe Hotel discovered the corpse of Wing Chut King. To protect commerce, Governor Henry T. Gage chose to deny the beginnings of an epidemic. By 1904, 118 or 97.5 percent of 121 cases proved fatal. Although ground squirrels transmitted the disease, Asian immigrants suffered isolation in Chinatown and other indignities and endured the blame for the spread of infection beyond the city. It was three more years before the concept of an animal vector carrying the bacillus was generally accepted. By that time, 89 deaths among 167 cases turned attention permanently from racial or cultural causes to environmental factors.

Hygienist Joseph James Kinyoun of North Carolina operated the U.S. Quarantine Station and promulgated a theory of microbial transmission. Until he was discharged from his office in 1901, he perfected the Kinyoun-Francis sterilizer, a portable steamer that disinfected isolation wards and ships. He also furthered knowledge of diphtheria, identified the cause of cholera, and promoted tetanus vaccinations. At the Marine Hospital at Stapleton on Staten Island, New York, he founded the Hygienic Laboratory, a one-room forerunner of the U.S. National Institutes of Health.

June 1900 Alaska's Great Sickness generated a high number of fatalities from epidemic influenza and measles. Because the region lacked medical care and adequate shelter and nutrition, native Alaskans, a virgin-soil population devoid of natural immunity, died at an alarming rate. The disease wiped out the Ingalik village of Chuathbaluk and the Dena'ina Athabascan village of Tyonek, and affected residents of the village of Meshik at Port Heiden. At Nulato, the epidemic coincided with starvation and measles which reduced the Koyukon Athabascans by one-third.

In this same period, endemic tuberculosis was the prime killer of Alaska's Indians, Eskimos, and Aleuts. In combination—flu or measles with tuberculosis—the lethal pathogens reduced the anticipated life span well below that of white Alaskans. Even with state-of-the-art care, the survival rate was barely 60 percent. Some native survivors converged at Saint Michael's, the northernmost Russian settlement in Alaska, to create a new commune.

Among the Kutchin of Yukon Territory, Canada, influenza arrived along with expeditioners, colonists, and missionaries. Because of the obvious connection between outsiders and illness, the virus acquired the name "white man's disease" (Cohen 2001, p. 384). At the first sign of winter weather, illness exacerbated conditions in Indian camps, where natives were already suffering widespread dysentery, measles, and pneumonia. By December, the pestilence migrated along the Pacific coast to British Columbia. As a result of adult deaths, in 1903, Jesuit missionaries set up a mission down-river from Saint Mary's, Alaska, at Akulurak to house and educate the Yup'ik children orphaned by flu.

July 1900 A puzzling epidemic of peripheral neuritis, skin thickening and discoloration, and impairment of reflexes and motor skills leading to paralysis in Manchester, Liverpool, and Salford, England, derived from arsenic poisoning in beer drinkers as young as age two. Of 6,000 victims, 70 died. The deaths of two women resulted in lawsuits.

Health officials traced the heavy metal to contaminated sugar and malt in the processing of beer at 67 breweries. The accession of Edward VII to the throne preceded his attention to the epidemic and to polluted foodstuffs. Brewers recalled tens of thousands of gallons of beer and poured the quaff into sewers.

August 1900 In Siam, widespread substitution of white grain for hand-milled rice in urban areas elevated the number of hospital admissions for Thais suffering beriberi. Thiamine deficiency also afflicted patients at a mental institution in Bangkok. In contrast, rural folk who still prepared their meals in traditional fashion avoided nutritional deficiency.

November 1900 The arrival in New York City of a touring minstrel show from the South initiated a two-year smallpox epidemic that afflicted 2,100, claiming around 400, more than 19 percent. In March 1902, health workers immunized 810,000 citizens at the rate of 10,000 per day. The disease advanced to Albany, Gloversville, and Schenectady, New York, and to Boston, Cleveland, and Philadelphia. *See also* **May 1901**.

early 1900s As pellagra produced a major epidemic in the American South, revelation that the disease resulted from nutritional deficiency of the B vitamin niacin humiliated and embarrassed local people. After New York physician Joseph Goldberger published his findings in *The Prevention of Pellagra* (1915), Southern legislators and medical leaders castigated, then ignored the obvious cure for the disease for two decades.

1901 In response to endemic leprosy in the Philippines, the U.S. military removed patients from the Saint Lazarus Hospital and settled them at Culion Island, a small island cluster 150 miles southeast of Manila. Modeled on the colony at Molokai, Hawaii, the facility segregated victims by race and offered new treatments to rid them of disease. In 1909, beriberi proved lethal for 329 inmates at the Culion leprosarium because of government purchases of polished rice for use at prisons, lighthouses, and welfare facilities. The institution, which grew into the world's largest leper colony, adorned its own aluminum coins along with the caduceus, emblem of the medical profession.

1901 During a smallpox epidemic, a Scottish physician, Elizabeth Beckett Scott Matheson, of Onion Lake, Saskatchewan, established credibility as a female doctor by her handling of contagion. She confined her work to a mission school that she founded and managed with her husband, John Richard Grace Matheson, a missionary to the Cree. The outbreak subsided because of her visits to each family, strict patient isolation, enhanced diet, and the replacement of bedding and clothing burned to kill the pathogen. As a result of her skill and competence, the Department of Indian Affairs chose her as a government physician.

1901 During the Moro wars, Captain John Joseph "Black Jack" Pershing of Camp Vicars on Mindanao, the Philippines, established peace and order by soothing the egos of Moro chiefs and by aiding ethnic Muslim islanders during a cholera epidemic. Despite quarantines, the scourge killed 100,000 civilians and soldiers and suppressed trade with island markets and with

Singapore. Natives retaliated with a fake coffin cortege carrying children pretending to be corpses. The "mothers" leading the procession turned out to be knife-bearing warriors in women's dress. For pacifying the Moro uprising, Pershing earned the rank of brigadier general. In 1939, Gary Cooper played a heroic army doctor aided by David Niven and Broderick Crawford in *The Real Glory*, an MGM version of the cholera outbreak.

1901 At Sierra Leone, reputedly the most malaria-ridden sector of the British Empire, officials opted to separate whites from blacks. They cited as their reasons two disease control expeditions of the two previous years, when British epidemiologists studied methods of preventing infection. City planners mapped out a hill station, Freetown, complete in 1904, where whites could live apart from the sickness that dominated black residential areas. Anglo-Indian epidemiologist Sir Ronald Ross led mosquito brigades to identify breeding grounds and drains and quell insect propagation.

1901 A winter outbreak of pertussis struck a quarter-million preschool children in Florence, Genoa, Milan, Rome, and Venice, Italy, killing 75,000 in a three-year epidemic. Lessening chances of survival for urban Italians were hunger, crowded housing, and parallel infections with enteritis and pneumonia.

January 5, 1901 After the incarceration of William Hysnick in an unlighted basement-level Seattle, Washington, jail, typhoid fever spread from filthy conditions that even the guards criticized. The dark cells contained only 6" X 12" exterior windows. Inmates ate and slept on the floor. Worsening the situation was an order that Hysnick join chain gangs working in snow on city streets. Guards complained of stench; prisoners panicked and revolted because contagion felled a dozen men. As reported in the *Seattle Star* and the *Seattle Post-Intelligencer*, after the attending physician sent Hysnick to the county hospital, other inmates and guards rebelled against the staff's neglect and cruelty.

March 1901 The heroine of the yellow fever epidemic that ravaged the military during the Spanish-American War was nurse Clara Louise Maass of East Orange, New Jersey, who had served in Florida, Cuba, and the Philippines. After volunteering at Havana's Las Animas Hospital under William Crawford Gorgas, author of *Sanitation in Panama* (1915), she and 19 male volunteers submitted their hands to mosquitoes. On August 24, she became the only one of the human guinea pigs to die of the bites.

After Maass's burial in Newark, nurse Leopoldine Guinther collected money for a suitable memorial—a pink granite headstone. On the 50th anniversary of the experiment, a Cuban postage stamp commemorated Maass's courage. At Wayne, New Jersey, congregants at the United Methodist Church honored her with a stained-glass window. In 1952, the Lutheran Memorial Hospital was renamed the Clara Maass Memorial Hospital, currently known as the Clara Maass Medical Center. *See also* **December 9–15, 1901**.

May 1901 Boston suffered its last outbreak of smallpox, an urban phenomenon that spread from New York and infected people in numerous neighborhoods. Officials blamed the homeless for spreading the disease. The city's board of health, headed by Samuel Holmes Durgin, a faculty member at Harvard Medical School, set up voluntary vaccination clinics, virus squads to inoculate vagrants, an isolation center at the detention hospital on Southampton Street, and a branch quarantine unit at Gallop's Island. Health care workers obtained serum from heifers at vaccine farms and introduced it subcutaneously in glycerinated form through a sharpened bone or ivory point. Over a 22-month period, 270 people or 16.9 percent died among 1,596 diagnosed cases, bringing the death rate to three per 1,000 citizens. As a result of a landmark court case challenging the right of the state to enforce vaccination on anti-vaccinationists, in 1905, the Supreme Court ruled in the state's favor.

December 1901 An outbreak of beriberi in the Philippines limited the quality of surveillance that police and scout units provided the U.S. Army. At Bilibid Prison, the disease engulfed captured guerrillas with cardiac problems, edema, nerve pain, wobbly gait, unclear vision, weakness, mental fog, and paralysis. Because of insufficient thiamin in the diet, the disease sickened 5,448 and killed 229. The incidence didn't ease off until a year later. Paralleling the epidemic was a less serious occurrence of beriberi at Lingayen Prison, killing one-fourth of all victims.

December 9–15, 1901 While trying to eradicate endemic yellow fever, Walter Reed, leader of a team of medical researchers in Havana, Cuba, attempted to protect the U.S. garrison from sickness. His theory based contagion on the bite of an *Aedes aegypti* mosquito. He successfully produced symptoms in four human subjects, who volunteered to be bitten by contaminated insects. To halt the scourge, he prevented mosquitoes from feeding on infected patients.

1902 Social worker Katherine Pettit of Lexington, Kentucky, co-founded Hindman Settlement School to relieve the poverty and misery of rural Appalachia. With the aid of teachers May Stone and Linda Neville, Pettit provided vocational instruction and crafts as well as wellness projects intended to eradicate endemic blindness from trachoma. Neville, called the "angel of the blind," established a fund to help students seeking eye treatment. From the Hindman clinic grew a national crusade to end trachoma in the U.S.

1902 The hard-fought rebellion of Filipinos against the United States army produced three years of fighting as José Rizal and Emilio Aguinaldo led limited forces against the superior fire power of American troops. U.S. forces lost 1,073 to combat and 2,572 to widespread disease. Indigenous people suffered 16,000 battlefield deaths and lost 200,000 civilians to starvation and sickness.

Successive epidemics of malaria, plague, and cholera produced a serious demographic decline of Filipinos in Batangas, which reported 34,000 deaths. Evidence of an 11-month spread of cholera over the Philippine Islands following the Spanish-American War derived from reports compiled by the Chief Quarantine Officer for the Philippine Islands in over 440 hamlets. His detailed mapping reconstructed the routes by which contagion enveloped the archipelago and sickened the nation, killing around 200,000 from cholera, including 50 U.S. soldiers.

Epidemiologists connected the spread of infection from Canton to Manila, where Chinese truck gardeners fertilized crops with human feces. Lieutenant Colonel Louis M. Maus, chief surgeon in the Philippines and president of the board of health, prohibited sale of contaminated vegetables and secured the purity of city water from the Mariquina River by banning swimming, washing of livestock, and sewage disposal. The adaptation of a detention facility at San Lazaro into a plague hospital for 2,500 as well as a morgue and crematory proved a poor choice for natives already panicked by the epidemic and doubtful of government containment measures.

Because of Filipino inefficiency, the American military posted Army surgeons to municipal boards of health. These outsiders took charge of public sanitation and screened food and water for impurities. Under martial law, islanders were forced to boil drinking water, dispose of garbage, direct sewage away from streams, quarantine the sick at detention camps, and bury their dead according to regulations. The island did not quell cholera until spring 1904.

1902 For a year, Emílio Ribas, sanitarian and administrator of the Serviço Sanitário de São Paulo, Brazil, tried to locate the cause of yellow fever while treating an influx of patients at the city's Hospital de Isolamento. Over a period of three years, he battled the mosquito, the vector of the disease, at Sorocaba, São Simão, and Ribeirão Preto. Political factors favored the eradication of the contagion that afflicted mainly European immigrants and delayed profits from the expanding railroad and plantations. For his work, Ribas's name survives in the Institute of Infectious Diseases Emílio Ribas. *See also* **1910**.

February 1902 Seven cases of smallpox at Reno, Nevada, forced health officials to close public schools and quarantine the University of Nevada campus. The situation worsened with the diagnosis of five more patients. George Springmeyer, editor of the *Student Record*, the campus newspaper, led a student revolt against restriction to the grounds. As a result, university officials expelled one-third of the rebels. The staff relented and reinstated most of the activists except for Springmeyer, who remained critical of staff policy.

summer 1902 Volunteers were appalled that a small Italian-American community of run-down tenements was responsible for one-sixth of Chicago's deaths from infectious disease. Bacteriologist and reformer Alice Hamilton, a specialist in occupational medicine and the first female faculty member at Harvard Medical School, joined Chicago's Memorial Institute for Infectious Diseases and studied the problem. In the environs of Hull House, a settlement house founded by Jane Addams, Hamilton identified flies as the source of a typhoid epidemic and warned that insects carried infection from open privies to families. Addams described Hamilton's crusade for ghetto hygiene in Chapter 13 of *Twenty Years at Hull House* (1910). While Hamilton worked at Hull House, she established medical education classes and a well-baby clinic and reorganized the city health department.

August 1902 During an epidemic of cholera, Naguib Mahfouz, the father of obstetrics and gynecology in Egypt and the Middle East and founder of the region's first midwifery school, volunteered to treat patients at Asyiut on the upper Nile. He traced contagion to a polluted public well. According to folksay, an Egyptian Muslim returning to Moucha from a hadj to Mecca, Arabia, brought a container of holy water, which he emptied into the well to bless it. The water, contaminated with cholera, caused a deadly surge in infections, killing 26,554 over a three-month period. The epidemic postponed the opening of Victoria College, an English public school that offered education to the children of Europeans posted in the Near East.

fall 1902 In the Canadian Arctic, an outbreak of typhus (also identified as typhoid fever, dysentery, or influenza) on Southampton Island in Hudson Bay erupted from Inuit contact with a sick whaler aboard the *Active*, which anchored off Cape Low northwest of Coats Island. The Sadlermiut Eskimo had already suffered from a series of epidemics, which whittled down their numbers to under 60. Racing through a virgin-soil population, the pestilence claimed 51 of the 56 residents. By winter, the Sadlermiut were extinct. According to an article for *National Geographic* by Henry B. Collins:

> [Microbes] infected the last remnant of the Sadlermiuts. Visitors to Native Point the following winter found a scene of death. Not an Eskimo had survived. Some lay in their houses on sleeping platforms; others on the ground outside, where their dogs still ran about [1956, p. 674].

According to an eyewitness, an infant skeleton lay in a corner of stone ruins. With the loss of the last Sadlermiut, a culture, language, and lifestyle disappeared.

October 30, 1902 The vaccination of 19 villagers in Malkowal, India, against plague killed them all. A colonial commission, headed by medical and epidemiological experts, deduced that a single serum bottle was contaminated with tetanus before it was sealed. The members charged a technician with faulty sterilization of instruments during the filling procedure. A media pillorying of vaccine-designer Walter (or Waldemar) Mordechai Haffkine, a Russian Jew from Odessa, Ukraine, resulted in his vindication against professional critics and anti–Semites.

November 21, 1902 At the end of the War of the Thousand Days, Colombians lost Panama to the U.S. in a conflict that cost them 50,000 in the field and an equal number to epidemic disease. After the battle at Palonegro, Colombia's bloodiest engagement, hundreds of human and animal corpses lay rotting. The runoff of decay contaminated water and food supplies with dysentery, amoebiasis, and gastroenteritis.

1903 In Maryland, the first paid tuberculosis nurse, Rieba Thelin, investigated burgeoning lung infection among Baltimore's poor, who incurred 1,206 infections in the calendar year. Thelin conducted tutorials on diet and hygiene and dispatched relief workers to assist poor families. When she signed on with Lillian Wald's Henry Street Settlement at the end of the year, Nora Holman replaced her as visiting nurse to consumptives and increased the campaign's effectiveness through funds collected by Grace Linzee Osler, the wife of Canadian histologist and physiologist William Osler of Johns Hopkins Medical School.

1903 Demographic disaster resulted after a priest introduced smallpox to the Cayapo, an agrarian people living in South America between the Triangulo Mineiro, western Sao Paulo, Goias, and southern Mato Grosso, Brazil. Initial infections with measles and tuberculosis continued to lower the headcount. From a population of 8,000, the tribe shrank to 500 survivors in 1918 and to 27 in 1927. By 1950, there were only three native Cayapo left alive.

1903 South African mining industrialists recruited laborers from the heartland, but found Central Africans susceptible to pneumonia, tuberculosis, scurvy, enteritis, and spinal meningitis. The working conditions of the shafts were so debilitating to miners that the disease killed 5,022, too many to make the program lucrative. Life in compounds in concrete barracks stuffed up to 50 men into an unfurnished, dirt-floored environment heated by smoky coal fires. Diet limited workers to moldy meal and meat. To increase survival rates, miners searched for an anti-pneumonia vaccine. After a decade of epidemic respiratory infections, the South African Union government curtailed the hiring of applicants from Congo, Tanzania, and Zambia and replaced them with Chinese men.

1903 The Ithaca, New York, typhoid epidemic created a rift between town and gown. At first, physicians diagnosed symptoms as elements of influenza. Of 13,156 citizens, 1,350 sickened with the fever, which killed 82. The public health service quarantined 522 residences and engaged sanitarian George A. Soper, author of *The Air and Ventilation of Subways* (1908) to quell the outbreak. He tested 1,300 wells and ordered the cleaning of 1,300 privies.

The concentration of infection in the Cascadilla section caused Ithacans to avoid ghetto dwellers, who also suffered the highest incidence of tuberculosis. After specialists at Cornell University reported that sewage from the area was contaminating the water supply, the middle class refused to pay for sewer systems or filtration. Instead of taking action, anti-immigrant, anti–Catholic forces blamed neighborhoods of newcomers in the marshland on the Cascadilla Creek inlet for harboring disease. Opposing forces applied settlement house models to the area and began classes to teach hygiene and vocational skills.

May 29, 1903 During a pandemic in Zululand, which killed 330,000 Ugandans by 1922, Australian bacteriologist Sir David Bruce, a surgeon general in the Royal Army Corps, headed the Sleeping Sickness Commission in Uganda. He worked in collaboration with his wife and partner, artist and medical technologist Lady Mary Elizabeth Steele-Bruce, who drew laboratory images of microbes, and with Italian pathologist Aldo Castellani, who identified the protozoa in the blood and spinal fluid of the sick. In support of medical research, Ugandan bishops posted local "fly boys" to stand in infested areas and collect flies that landed on their naked torsos. Bruce instructed endangered natives of ways to avoid the tsetse fly and to discourage its propagation. On May 29, 1903, the Royal Society received the team's findings and published them in the *British Medical Journal*. Identification alone did not stop the disease, which spread to the Ivory Coast in 1904.

1904 Clara Barton headed the American Red Cross in Butler, Pennsylvania, during a typhoid fever epidemic that began in December 1903. The onset of *Salmonella typhi* and secondary infections killed 20 percent of 1,277 patients. A young idealist, who admired Barton's tender care, recalled:

> And we pictured the light [of her lantern] going on and on through the night until it should stop over the stricken town of Butler, and the suffering people there would look upon it as the light of a great soul that had come to them out of the darkness, bringing comfort and healing and the calm spirit that banishes all fear ["Clara Barton"].

In her last field effort, she completed distribution of medicines and supplies, lectured citizens on relief efforts, and left to local health professionals the sterilization of water polluted by sewage.

1904 William Crawford Gorgas, author of *Sanitation in Panama* (1915), began battling yellow fever and malaria in Panama. In Panama City, home to 20,000, he supervised sanitation efforts to fumigate every building, install plumbing, drain swamps, cleanse drainage lines, spray for mosquitoes, dispose of garbage, and pave roads. During two of his clean-up sweeps of the city, yellow fever outbreaks led people to question his methods. He persisted, launching a third city-wide sweep. He prohibited one of the local cemetery customs involving exhumation of remains after two years to free burial space for rental to another family.

During Gorgas's campaign, in 1905, Samuel Taylor Darling, head of the Board of Health Laboratories, led an investigation of malaria among canal laborers. From his data, the team conceived a control program that included application of Darling's larvicide, an insect preventive that paralleled Gorgas's drive to eradicate mosquitoes. By December 1905, yellow fever was officially eradicated from the isthmus unless it arrived by ship. Any newcomers entered isolation to halt new infections.

1904 When cholera migrated from Mesopotamia, the annual hadj to Mecca, Arabia, spread contagion throughout Persia. In Baghdad, Iraq, 1,100 or 73.3 percent of 1,500 patients died. The Ottoman Empire reported 10,466 infections and 9,192 fatalities. Azerbaijan recorded 30,000 deaths; Tehran, Iran, incurred 26,000. At Qatar,

Kuwait, Bahrain, and Muscat, peasants faced an increase in bread prices and displaced their anger and frustration on Belgian border guards and American and British physicians. Citing the Koran, Muslims expressed resentment of foreign intervention in their homeland and complained that officials were lax in following the dictates of Islam.

1904 When measles again emerged in northwestern Iceland, it arrived in summer by a Norwegian whaler from Bergen. Infection followed the standard route of epidemic invasion of an island dependent on goods and foodstuffs imported by sea from Europe. Most of the 2,000 patients were teenagers. Rural social functions such as church confirmations and receptions were the typical source of dissemination.

1904 Plague returned to India in full force. As refugees fled contagion over a four-year pandemic, infection spread from Bombay to central India, slaying some two million. Health officials immediately relieved panic by disinfecting streets and sewers, cleansing residences with carbolic acid, and removing the dead and dying. In 1905, the Lister Institute of Preventive Medicine established the Commission for the Investigation of Plague in India to analyze recent bacteriological and epidemiological information about fleas on *Rattus rattus* and their link to the spread of pestilence.

1904 Nearly 98,000 of the Japanese troops fighting the Russo-Japanese War developed beriberi, a nutritional imbalance resulting from lack of thiamine or vitamin B1. Of the upsurge, around 4,000 cases were fatal. During this same period, the navy reported no disease. Historians concluded that army dietitians failed to follow the directive of Kanehiro Takaki, surgeon-general of the Imperial Japanese Naval Medical Services, who, in 1884, restructured shipboard rations to the European regimen to prevent inadequate nutrition. Primarily, he advised a diet that preserved the nutrients in brown rice over the standard Japanese diet of polished white rice. His contribution to wellness helped Japan overcome the Russians.

spring 1904 An onslaught of endemic smallpox in Rio de Janeiro killed hundreds. Into the summer, a nationwide inoculation program, advised by Brazilian epidemiologist Oswaldo (or

Osvaldo) Goncalves Cruz, gradually enticed more volunteers to seek protection in spite of the *Revolta da Vacina* (*Revolt against Vaccination*), a vigorous protest by an anti-vaccination faction. By fall, a new law required vaccination, yet could not assure compliance by migrant workers who moved freely in and out of the city. On November 14, a renewed protest against inoculation emerged from armed rebel cadets at a military school. A recrudescence of the disease in April 1908 reinforced the need for vaccinations.

April 1904 In camps at Onjatu in the German colony of South West Africa (modern Namibia), 439 German troops died of typhoid fever during their war on the Herero (or Bantu) and the Nama (also Hottentot or Khoikhoi). Unsanitary quarters, exhaustion, inclement weather, and lack of typhoid fever inoculation contributed to the death rate of nearly 54 percent of the troops of Kaiser Wilhelm II. *See also* **1908**.

August 27, 1904 America's most notorious disease carrier, "Typhoid Mary" Mallon (also Malone), an Irish immigrant from County Tyrone, first spread *Salmonella typhi* in ice cream and peaches at the summer home of Charles Henry Warren at Long Island's Oyster Bay. When doctors found no contaminants in water or milk supplies and began questioning kitchen staff, she fled to Manhattan and continued taking short-term domestic jobs. *See also* **March 1907; 1915**.

1905 The threat of syphilis in South Africa spawned a mission station founded by the Berlin Mission Society and staffed by Helene and Robert Franz, who offered free treatment. The government-funded syphilis hospital at Bochum placed Helene Franz in charge of the venereal disease clinic, which evolved into a treatment center for leprous and tubercular patients as well.

1905 An upsurge in smallpox at Recife and Rio de Janeiro, Brazil, precipitated citywide vaccination and a public health cleanup of *cortiços* (slum housing) and *favelas* (shantytowns). Some citizens resisted with open rebellion. When wealthy entrepreneurs examined the pros and cons of the costs, they anticipated the loss to coffee plantations if contagion scared off future immigrant laborers from Italy, Portugal, Spain, Germany, and Japan. Most investors believed that inoculation was worth the outlay in vaccine and clinic costs.

1905 Increased incidence of yellow fever in New Orleans required traditional quarantine along the Mississippi Valley. The question of power escalated rivalries between Louisiana Governor Newton C. Blanchard and Mississippi Governor James K. Vardaman. The regional set-to required federal intervention by militia and navy patrol of borders and waterways. Federal health authorities took control of the epidemic situation, the last yellow fever outbreak experienced by the American South. The event was a turning point in Gulf Coast public health policy because of city reliance on modern science and bureaucratic supervision.

June 14, 1905 Australian bacteriologist Sir David Bruce, a surgeon general in the Royal Army Corps, along with Themistocles Zammit, a Maltese bacteriologist, and ten other investigators of Malta fever isolated goat's milk as the source of the microbe *Brucella melitensis*. The germ caused brucellosis, previously known as Malta fever, Mediterranean fever, and undulant fever, which affected hundreds of British soldiers, killing a few, and requiring three months of recovery time in survivors.

summer 1905 After an outbreak of polio in rural Sweden afflicted thousands of children, paralyzing hundreds, Ivar Wickman, a pupil of Swedish physician Oskar Karl Medin, went house to house studying the nature and transmission of the disease. Wickman compiled his findings in *Akute Poliomyelitis* (1911). He concluded that the disease was infectious in nature and passed from person to person. He at first suspected milk as the carrier after a dairying family at Ukla experienced one case of polio on October 6, 1905. The outbreak infected another son and four neighbor children as the microbe migrated to a total of six families, all of whom depended on the dairyman for milk products.

Four years later, Austro-American immunologist and pathologist Karl Landsteiner of the Royal-Imperial Wilhelminen Hospital in Vienna joined German pathologist Erwin Popper in collecting samples from the brain and spinal cord of a child killed by polio. They used the tissue to induce poliomyelitis paralysis in a rhesus monkey. The experiment proved that the disease was viral. *See also* **summer 1911**.

July 1905 Against the outcries of land developer Horace S. Crowell of Boston and other petitioners, the Massachusetts State Board of Charity opened a 60-acre leprosarium, Penikese Hospital, on Penikese Island in Buzzards Bay. The first four male and one female patients rode a special rail car to a dock for transport by the facility's sloop *Keepsake* to the island, where they would occupy private dwellings. Superintended first by Louis Edmonds of Harwich, the leprosarium passed to the management of physician Frank H. Parker and his wife, Marian Parker, who encouraged patients to organize a self-sufficient community. Father Stanislaus of the Order of the Sacred Heart ministered monthly to spiritual needs.

1906 At a fearful upsurge in infections in Toronto, where 445 died of tuberculosis, Christina Mitchell, an experienced nurse from the City Missions of New York, concentrated on treating TB patients. By collaborating with staff at the area's Nursing Mission and City Mission and with volunteers from the YMCA, YWCA, and Anglican, Methodist, and Presbyterian churches, she visited homes and received the sick at a string of outpatient clinics. Toronto expanded the outreach by hiring Elizabeth "Lilly" Lindsay, the first "city nurse."

1906 During a welter of conflict pitting the Ontong Javanese, residents of the British Solomons off New South Wales, against missionaries, traders, and the colonial government, the islanders of the atoll lost up to 40 patients or 1 percent of their 3,000 residents each day during a lethal influenza outbreak in a virgin-soil population. As a result of depopulation, the citizenry of Luanguia and Pelau decreased; other settlements became ghost towns from the combined onslaught of yaws, dysentery, pertussis, measles, chicken pox, and hookworm.

1906 During the digging of the Panama Canal, 21,000 or 81 percent of the 26,000 workers sought treatment at hospitals for malaria. The concerted effort of William Crawford Gorgas to wipe out malaria on the isthmus of Panama eradicated the disease. To do the job, he hired 1,200 workers organized into divisions to cover 25 sanitary districts. They set about draining swamps, burning insecticide, clearing weeds, and spreading carbolic acid and caustic soda. To protect individuals from contagion, Gorgas distributed screening and bed nets and posted men on rail cars to swat and chloroform mosquitoes. Methods of defeating the vector, the *Anopheles* mosquito, included an oil-drip barrel to layer still water with a coating that prevented hatching. Staff traveling in horse-drawn wagons sprayed ditches. Other workers moved into infested areas on foot to eradicate larvae with knapsack sprayers, screen rain barrels, and enclose verandahs.

The improvement to public health dramatically ended years of outbreaks. Fifty cases were recorded in 1903, 40 in 1904, and 210 in 1905. In 1906, there was one case, followed by zero recurrences. For his success, Gorgas earned respect for achieving the world's greatest sanitary project and for enabling engineers and workers to complete the canal project. Gorgas was named to the Panama Canal Commission and knighted by King George V. Gorgas later served as surgeon general of the U.S. Army and chaired the American Medical Association. Greatly admired worldwide for his contributions to epidemiology, he was eulogized at Saint Paul's Cathedral and buried in Arlington National Cemetery.

1906 In South Africa, epidemic tuberculosis continued to ravage the population. Disease targeted port cities and felled underfed slum dwellers living in unhygienic conditions. Of a total of 5,000 deaths, 1,129 occurred in Johannesburg. Mortality rates for blacks rose from 5.5 to 8 per 1,000.

1906 In Vietnam, plague struck the southern half of the country. At Saigon, the market center incurred heavy rodent populations transported in shipping cartons from Hong Kong and Canton, China. Indian merchants of the Cholon sector suffered the highest incidence of disease.

1907 An epidemic of polio hit New York City and claimed 2,500 victims, predominantly young males, with fatalities largely among older children and teens. The first instance of the crippling disease as a major health hazard in the United States, the epidemic persisted for nine years and pushed the death count to 7,000 or 28 percent. Health professionals knew little about the source of contagion or treatment.

Spread in clusters following human traffic patterns, the disease appeared to be infectious but not contagious. Municipal officials closed swimming pools, gyms, theaters, and religious classes for children and banned team sports. Street cleaners sprinkled chlorinated lime as a disinfectant and battled flies with insecticide. Pathologist

Simon Flexner, editor of the *Journal of Experimental Medicine* and author of *Natural Resistance to Infectious Disease and its Reinforcement* (1910), deduced that the pathogen was a filterable virus rather than a bacterium.

1907 A yearlong outbreak of bubonic plague at San Francisco followed within months of the great earthquake of 1906. Of 205 cases, half died of the disease. The epidemic, which began at Oakland to the north, received an appropriate response from the Citizens' Health Committee and from expert epidemiologist Rupert Blue, bacteriologist of the U.S. Public Health Department and surgeon at the Marine Hospital Service. Out of respect for Blue's expertise, President Theodore Roosevelt personally appointed him to combat the epidemic.

Blue coordinated efforts citywide to establish sanitary streets, tear down 1,700 abandoned structures, seal dirt floors in barns and basements, and inspect each residence and all food industries. To rid the area of rats, he ordered volunteers to mop floors with carbolic acid, distribute flea powder, clean chicken coops, and poison rodents. He rewarded volunteers with a bounty of five cents per dead rat. Through a joint effort, workers lessened the number of fatalities that might have resulted. The next year, the citizen committee toasted Blue at a public dinner.

1907 According to a rough estimate of Long Island physician Allen S. Busby, half of the cases of blindness in Poland resulted from trachoma, which European Jews called "Egyptian ophthalmia." Busby based his figure on examination of Polish and Russian Jews emigrating to the United States, which incoming Poles called the *Goldene Medene* (*Golden Land*). At Kiev and Lodz, up to 250 patients per thousand suffered the infection.

Because of the high concentration of disease among poor Jews, according to Kiev ophthalmologist Max Mandelstamin, Russian gentile medical journals scapegoated Jews as the source of the disease. He refuted their racist notions with data showing that "in New York City where the Jewish population consists of 900,000 people, there are [no] trachoma epidemics" (Markel, 2000, p. 539). Corroborating his assertion was the writing of Julius Boldt, a German ophthalmologist and author of *Trachoma* (1904), who attested to the role of poor sanitation rather than

ethnicity in spreading the infection. To assist Jews in arriving healthy at their new homes, before they embarked for America, the National Jewish Immigration Committee aided Eastern Europeans with medical inspections, which involved everting the eyelids and examining the surface for granulation and inflammation.

1907 New Zealand experienced a new epidemic of pertussis, which assailed Auckland and Otago, but not Marlborough or Westland. Of the 307 who died, 99 percent were preschoolers.

March 1907 In Manhattan, where "Typhoid Mary" Malone (or Mallon) worked as a cook, she spread typhoid to 22 people. One died. At this point in her notorious career, engineer and epidemiologist George A. Soper of New York City's Department of Health began the hunt for Mallon, whom he traced through employment agencies.

early summer 1907 Imported to Reykjavik by crew of a Danish ship from Copenhagen, measles infected Iceland, where stringent quarantine kept susceptible children indoors. According to the state document *Heilbrigdisskyslur* (*Public Health in Iceland*) (1908), one child was released too soon at Stykkishólmur. Upon her return to Reykjavik, she sustained a mild case of measles that quickly infected the city's children. A severe variety of the scourge gained speed from October until February 1908 and killed 354 out of 7,398 patients.

1908 For five years during the building of the Canadian Northern Railway, the increase in immigration of Irish navvies to British Columbia produced epidemic typhoid. Increasing contagion were inadequate water and sewage services and too many shanties crowded with the poor. As a result, citizens contracted sickness from contaminated fish, milk, and water.

One respected physician at Mission City, British Columbia, Alexander James Stuart of Scotland, had practiced medicine in the area for a decade, performing surgery, operating a dispensary, and mixing his own drugs. During the epidemic, he served as the health officer, coroner, and doctor to the Whonnock tribe, whom he visited by horse, bicycle, and rail to battle typhoid fever as well as endemic smallpox, diphtheria, tuberculosis, pneumonia, and spinal meningitis. To protect school children, he mailed out vaccination notices.

1908 Some 8,000 Ecuadorians contracted bubonic plague after the docking of a ship from Paita, Peru, at Guayaquil harbor. A wise community action plan ordered citywide immunization, the burning of infected houses, and a rodent squad to suppress the vector of the infection. The disease targeted more port entry points at El Oro, Guayas, Los Ríos, and Manabí and traveled into the South American uplands along the main rail lines on rats in freight cars. As a result of the epidemic, Ecuador formulated a comprehensive public health initiative.

1908 Natives of the Marquesas Islands blamed the crew of the French gunboat *La Zélée* for spreading epidemic influenza at their port. The population was already fragile following epidemics in 1791 to 1863, which slew approximately 80 percent of native Polynesians. Over the next three years, more died from leprosy, tuberculosis, and pertussis which arrived by sea on European vessels.

1908 Rampant tuberculosis, especially among children, compelled the League for the Suppression of Tuberculosis in Providence, Rhode Island, to collaborate with public school officials to establish the first special school for tubercular or at-risk students. To assure their health, the staff offered a regimen of fresh air, exercise, moderate classroom load, and nutritious meals. Other U.S. school districts emulated the Providence model.

1908 During the four-year Herero Uprising, at the forced labor camps built by German imperialists in southwestern Africa, 9,000 of 17,000 civilian Herero (or Bantu) and Nama (also Hottentot or Khoikhoi) inmates died of disease and bayoneting. Under orders of General Lothar von Trotha, staff fenced black inmates into cramped areas with barbed wire or thorn-bush barriers. Prisoners made do on minimal rations—a half-cup of rice, salt, and water. The filth, insufficient bedding and clothing, and lack of medical care allowed typhoid, typhus, and smallpox to proliferate.

Contributing to deaths were mistreatment and floggings along with forced labor for starving men, women, and children. The combined loss from war and epidemics constituted the most expensive tribal revolt in the European partition of sub–Saharan Africa. From the racial experiments of German eugenicist Eugen Fischer came the book *Menschliche Auslese und Rassenhygiene* (*The Principles of Human Heredity and Race Hygiene*) (1931), an influence on Adolf Hitler's concepts of a master race in the 1930s. At the University of Berlin, Fischer taught his principles of racial superiority and genocide to Josef Mengele, a despicable tormentor in the death camps of World War II. At the end of the war in South Africa, at Shark Island, the most notorious camp, the German government raised a monument to the Germans who died during the war. The plaque made no mention of dead camp inmates.

January 5, 1908 Shortly after a severe rat and livestock epizootic, residents of the Gold Coast (modern Ghana) died of plague with a fatality rate of over 87 percent of those stricken with either the bubonic or pneumonic form of the disease. Hardest hit were the tribal Ga, whose *mantse* (chiefs) and shamans resisted modern methods of disease suppression. The next month, a British epidemiologist, Sir William John Ritchie Simpson, professor of hygiene at Kings College, London, arrived with the latest epidemic-fighting technology available at the Liverpool School of Tropical Medicine. Experienced at fighting cholera in Calcutta, he poisoned rats in drains, fumigated with sulfur in Jamestown and Usshertown, burned grass huts in infected hamlets, and placed stringent controls on travel that required proof of vaccination of visitors to the area.

To explain the health regulations to natives, staff nailed up posters and issued leaflets in Ga, Twi, and Arabic. The sick entered isolation at the Contagious Diseases Hospital; those exposed to plague were quarantined in pest huts across the lagoon from Accra at the Korle Gonno Camp. Aided by Ga Chief Kojo Ababio and some elders, local officials distributed blankets and sleeping mats. Keeping order in the sanitary cordon at Accra were 25 watchmen from the Gold Coast regiment.

Inoculation was more appealing than isolation or restricted travel. In the words of Acting Governor Major H. Bryan:

> I am glad to be able to report that the natives at Accra have not only offered no objection to inoculation, but that they are presenting themselves in such large numbers as to throw a heavy strain on the medical staff. I was inoculated on Thursday last and several other officials were publicly inoculated by Professor Simpson. The courtyard, entrances, and passages of the build-

ing were packed with natives of all classes who were so eager to present themselves that the police on duty had some difficulty in clearing a passage [Roberts 2001].

Altogether, 16,000 received the vaccine. Contributing to native trust in prophylactic measures were the deaths of several tribal herbalists and shamans from plague, notably Kuao, a medicine man and the brother of Chief Tago of Labadi.

fall 1908 Epidemic malaria returned to the Punjab, striking Lahore, India, and interrupting rail lines from serious loss of manpower. Of some half million victims, a disproportionate number were infants and young children, who had developed no immunity to the infection.

1909 To stem tuberculosis in the United States, Jane Arminda Delano, supervisor of the U.S. Army Nurse Corps, founded the American Red Cross nursing program. The corps started the Town and Country Nursing Service in farm country to identify and treat the sick and to teach families about sanitation and cleanliness. Expanding the outreach was the work of Joseph W. Schereschewsky of the Division of Industrial Hygiene, who investigated unsanitary conditions that produced an excessive rate of tuberculosis among garment workers.

As the disease swept over Indian tribes, it inflicted severe loss among indigenous people of the Southwest. The U.S. government waged a campaign against tuberculosis that included erection of the Phoenix East Farm Sanatorium, adjacent to the Phoenix Indian School in Arizona. Within two years, the TB eradication program extended to nonstudent Indians.

1909 The onset of a five-year epidemic of smallpox forced citizens of Calgary, Alberta, to rethink their lax attitude toward prostitution. Instead of light sentences and fines for streetwalkers, police blanketed the red light district, forcing sex workers out of business.

1909 While studying an epidemic erroneously identified as malaria during the building of the Central Railroad Brasil at Lassance in northern Brazil, bacteriologist Carlos Chagas of Oliveira, an expert on epidemiology from his study of yellow fever in Rio de Janeiro, examined immigrant laborers. Microscopic study of blood samples identified the cause of their illness as flagellates that he named *Trypanosoma cruzi* after a mentor,

Oswaldo (or Osvaldo) Goncalves Cruz, the nation's leading parasitologist and expert on vaccines and serums at the Serum Therapy Institute in Manguinhos. Similar to the cause of African sleeping sickness, the pathogen, spread by the triatomine or kissing bug, causes a potentially fatal coronary malady known as Chagas disease, an endemic sickness among the poor that afflicts 18 million in 21 countries extending from Mexico south to Argentina.

1909 A catastrophic fire and an epidemic of typhoid fever were the results of poor city planning in the northern mining boomtown of Cobalt, Ontario. Because Canadian city planners tended to react to outbreaks of disease rather than prevent them, the area grew rapidly, filling available acreage with buildings erected on inadequate lots and crowding people into an area lacking sanitation and public services. The fever epidemic derived from too many people, too much untreated sewage, and a suspect water supply.

1909 A specialist in endemic leprosy, Mitsuda Kensuke of Yamaguchi, Japan, established a leprosy ward at the Tokyo Shi Yoikuin before founding Tokyo's Zensei Hospital, the nation's first public leprosarium. For his work as president of the National Leprosarium and for discovering the Mitsuda reaction, a test differentiating types of leprosy, in 1951, he received the Order of Culture.

1909 Influenced by the *New York Sun*'s publication of speeches of Charles Wardell Stiles of the U.S. Agriculture Department, the Rockefeller Sanitary Commission launched a seven-year campaign to eradicate hookworm as a source of anemia, low IQ, and low energy levels in the people of rural North Carolina and ten other Southern states. To the detriment of the crusade, the reformers maintained a superior attitude toward the public, who identified the disease as "ground itch," "dew poison," or "toe itch" (Boccaccio, 1972). The outsiders' attempts to alter unsanitary conditions humiliated farm families and violated rural standards of independence, creating ill will that hindered progress. To enlighten families to the dangers of privies and unsanitary disposal of waste and to the need for handwashing, the commission tackled the restructuring of school curricula, thus threatening communities and parents with loss of control of schools.

1910 The Philippines and Thailand suffered endemic problems with thiamine deficiency disease. In Manila, around 40 percent of newborns died of beriberi between the ages of eight and 20 months. Simultaneously, Thai residents of Bangkok reported 22,670 cases and 1,063 deaths from endemic beriberi. However, the malady did not harm European travelers and government agents, who did not rely on the traditional rice-centered diet.

1910 During an era of pandemic syphilis, bacteriologist Hata Sahachiro, who trained at Tokyo's Institute for Infectious Diseases and at the Koch Institute in Berlin, aided researcher Paul Ehrlich at the National Institute for Experimental Therapeutics in Frankfurt am Main, Germany. The team synthesized arsphenamine, sold under the trade name Salvarsan, an arsenic compound that controlled syphilis, but at great risk to the patient's sanity. One famous Danish patient, memoirist and fiction writer Karen Blixen, known by her pen name Isak Dinesen, survived the lengthy treatment without loss of mental faculties.

As explained in Judith Thurman's biography *Isak Dinesen: The Life of a Storyteller* (1982), Dinesen contracted the disease from her husband, Bror Blixen, who had been intimate with a native woman during World War I: "Bror appeared at her camp having spent the night in a Maasai *manyatta*. Syphilis was almost epidemic among the Maasai, the cause of widespread sterility among Maasai women" (p. 150). Dinesen traveled to Zurich, Switzerland, to be treated with "Salvarsan, Dr. Erlich's 'magic bullet'" (p. 152). The treatment required ten to twelve weeks of Salvarsan alternating with twelve weeks of intramuscular bismuth. Unfortunately for Dinesen, the disease was arrested in the infectious secondary stage, but was neither cured nor controlled. *See also* **1916**.

1910 Australian Bush nurse Elizabeth Kenny of Warialda, New South Wales, treated her first encounter with endemic infantile paralysis in a toddler in North Queensland. Instead of splinting painful legs, Sister Kenny developed an unheard-of regimen of hourly application of blanket strips soaked in boiling water followed by massage of wry limbs. The second stage of her regimen involved stretching muscles to ease pain and relax spasms. Three years later, Kenny opened a cottage polio clinic in Clifton. After service in World War I, she tried introducing her treatment at Brisbane, but did not convince dubious male Australian doctors of her success until 1939. *See also* **1940**.

1910 According to reports from the Bureau of Indian Affairs, in the American Southwest, 40 percent of some tribes were suffering scarring, diminished vision, and blinding from trachoma. On the Crow Indian Reservation in Wyoming, photographer Richard Throssel compiled a picture portfolio illustrating the lifestyle and discomforts of native people during a trachoma epidemic. At the Fallon Indian School in Stillwater, Oklahoma, the first symptoms appeared in March 1910 and spread rapidly. The malady began with pain and bloodshot eyes, contracted pupils, and diminished sight. Baptist missionaries screened adults as well as children. Administrators hired only people certified free of infection.

The disease remained a major scourge for the next three decades. In this same period, a 1911 health survey in the hills of Kentucky disclosed that 13 percent of residents suffered trachoma. Many were already blind or only partially sighted. By comparison, Native Americans living on reservations suffered from 65 to 95 percent infection with the disease.

1910 At Duala, Cameroon, German imperialists seized more land from the Witbooi Nama. Colonial officials also used epidemic malaria as a reason for deporting native people from the harbor town. Citing European doctors, the German minority insisted on segregation of whites from blacks as a form of racial hygiene to quell disease.

1910 In Mozambique, Basutoland, and Nyasaland, sources of cheap African labor for colonial mines at Witwatersrand, tuberculosis and pneumonia became major causes of death for black workers. Miners contracted lung disease in epidemic numbers because of the sharp variance in atmosphere between the tunnels and their residences. *See also* **1912**.

1910 The Ila of Namwala in Northern Rhodesia (modern Zimbabwe) acquired an undeserved reputation for sexual license after colonial health workers misdiagnosed pandemic yaws as syphilis. Erroneous reports were the impetus to health policies and regimens to annihilate venereal disease.

1910 According to city documents, maps, and health surveys, a cholera epidemic in Barletta, Italy, sparked social disorder, violence, and rioting. At the heart of dissension lay the despair of peasants that the town's medical authority and sanitation system ignored the working class. Adding to their frustration was the superior upper class, which blamed the poor for the outbreak.

1910 When an unidentified pestilence beset São Paulo, Brazil, for four years, two physicians—an Italian, Antonio Carini, director of the Instituto Pasteur de São Paulo for 35 years, and Emílio Ribas, sanitarian and administrator of the Serviço Sanitário de São Paulo—argued over a diagnosis. While Carini insisted the disease was smallpox, Ribas sought to prove it was alastrim or milk pox.

1910 In Barnsley, England, the opening of a special school for students suffering pulmonary tuberculosis helped to control the disease, which was a common cause of death in young citizens. Designed by Robert Philip, the school dispensary offered standard instruction and tutorials for parents on care of patients and protection of family members from contagion. In London, Kensal House, an urban open-air TB school, operated more like a sanitorium.

June 1910 When cholera struck the Russian Ukraine in the coal-mining center, infection of 230,000 killed nearly 48 percent. Terrified that widespread illness would rob the area of laborers, mining magnates begged for strong containment measures by the government health agency and the Red Cross. The dispatch of pioneer health care reformer G. E. Rein, head of the Medical Council, to the Donets coal basin resulted in a joint effort that successfully combatted the disease. In October, he petitioned Tsar Nicholas II to form the Ministry of Public Health. The initiative did not reach fruition because of the outbreak of World War I.

September 1910 From a yearlong outbreak of pneumonic plague in Manchuria and northeastern China during the late empire, 60,000 died as the pathogen moved over 2,000 miles from Siberia southwest to Peking. The disease overran enclaves of migrant workers and spread along rail lines. A tyrannic health program based on German and Japanese models served Shenyang. Police initiated a militaristic quarantine; mer-chants set up isolation wards that reduced infection. As a result of the parallel systems, the public realized that the police-run health program was more thorough than programs instituted by other agencies.

The disease ravaged victims until the spring of 1911 and produced a mortality rate approaching 100 percent. That winter, Wu Lien-teh, a Cambridge-educated doctor and personal physician to the Dowager Empress Cixi of China, earned international fame for stemming the contagion. As a result of his success, in 1912, he founded the Manchurian Plague Prevention Service, the beginning of China's modern medical service. He spent the next three decades of his career performing humanitarian service and published an epidemiological handbook, *Plague: A Manual for Medical and Public Health Workers* (1936). *See also* **1928**.

In anticipation of other epidemics, the Imperial Chinese government summoned a global plague conference in Manchuria that drew representatives of 11 countries. Richard Pearson Strong, the U.S. delegate, studied the pathology, transmission, and treatment of the disease for which his drugs proved useless. He learned that droplets of saliva spread the pathogens, which derived from the popular tarbagan (marmot) fur coat that replaced sable. Three years later, while chairing Harvard's department of tropical medicine, he established a reputation for epidemiological expertise that survived over a quarter century.

November 1910 Over three years, tens of thousands died of the plague in Java, which was virgin-soil territory for the disease. It struck Surabaja in the east. Borne along with a shipment of Burmese rice, the pathogen afflicted the area village by village, killing 16,406 islanders by 1915. The Dutch administration's plague detail detained and disinfected travelers, exterminated rats, destroyed roof thatch to limit rat breeding grounds, quarantined the sick, and ordered examination of corpses to determine cause of death. Eradication required a quarter century of health care and improvement to public sanitation. *See also* **1932**.

December 18, 1910 Across the U.S. South, hookworm disease infested the bodies of the poorest people. The most affected displayed anemia and were frequently judged retarded for their lack of energy and mental acuity. Health workers identified the sickness in over 81 percent of 884 Southern counties, where visitors ridiculed residents as lazy half-wits. A one million dollar gift from the Rockefeller Sanitation Commission for the Eradication of Hookworm Disease enabled Dr. Waller S. Leathers to launch a campaign in Mississippi, beginning with the Jackson State Fair and the Harrison County Fair. Leaflets introduced parents to the nature of infestation, the need for shoes, and methods of sanitizing privies. Diagnoses in 166,623 people in 78 counties found more than one-third of adults infected and 36.7 percent of children weakened and compromised by the disease. Doses of tetracholorethylene killed the parasites.

1911 As reported in an annual assessment from the Imperial Japanese Central Sanitary Bureau, Manchuria, a Japanese protectorate since the 1904 Russo-Japanese War, encountered simultaneous epidemics. Japanese colonial officials determined that Chinese medicine failed to alleviate outbreaks of cholera, syphilis, typhoid fever, and typhus. Through the activism of statesman Shimpei Goto, the Japanese government founded the South Manchuria Medical College and 13 Red Cross hospitals.

1911 Among Fijians, a severe outbreak of measles derived from the crew of the ships *Mutlah* and *Sutlej* from India. When the disease invaded Rotuma to the northwest in spring, gastrointestinal complications and tuberculosis worsened mortality figures, which reached 12.8 percent. The Polynesian population dropped to 1,983, its lowest point before rebounding.

1911 Australian bacteriologist Sir David Bruce, Director of the Royal Societies Third Commission on Sleeping Sickness, along with his wife and partner, artist and medical technologist Lady Mary Elizabeth Steele-Bruce, investigated an epidemic at Nyasaland at Kaviondo on Lake Nyasa. He hired local "fly boys" to collect up to 300 tsetse flies each day for testing dogs, goats, and monkeys for transmission of sleeping sickness. Bites of the flies infected the animals at the rate of one in 500. Bruce examined wildlife and discovered the epidemic in the duiker, hartebeest, reedbuck and waterbuck.

1911 Back-to-back typhoid fever epidemics in Ottawa, Ontario, generated public support for reform of urban health measures. Within eight weeks, 83 or 8.4 percent of victims out of 987 cases expired from this disease. In July 1912, a recrudescence proved even worse, afflicting 1,878 citizens and killing 91. Typically, disease targeted the French-Canadians of Lowertown, where poor drainage generated deadly pathogens that claimed more infants and toddlers than in Anglo-Canadian neighborhoods. Against the opposition of business leaders and conservative politicians, reformers insisted that city officials examine and eradicate causes of the crisis.

1911 In the Marquesas Islands, the crew of the French ship *La Gauloise* brought pertussis to the central island of Mangareva. The disease afflicted toddlers and young children at an alarming rate. Because of high mortality, the scourge worsened chances that native islanders would prosper in Oceania.

1911 Pneumonic plague felled hundreds in Vietnam, where the disease followed trade along the Mekong River. In moist conditions, the pathogen flourished in rodent populations at river towns, killing 1,119.

1911 At Lausanne, Switzerland, where health officials classified tuberculosis as a social disease, pioneering hygienist Charlotte Olivier directed the Lausanne antituberculous dispensary. Her method began with the treatment of infection and progressed to containment of its spread within the social environment of the lower class. A plaque honors her 15-year commitment to health in Vaud canton and her mobilization of Swiss women to supervise family wellbeing. In 1996, the establishment of La Fondation Charlotte Olivier at the Université de Fribourg perpetuated her concepts of public health.

spring 1911 An outbreak of encephalitis in Skagway, Alaska, killed physician Emil Pohl, who had worked out of a log cabin and carried treatment to distant patients by dogsled. His widow and former partner, obstetrician Esther Pohl Lovejoy, the first woman to practice medicine in the Klondike, returned to the United States. At Portland, Oregon, she became the first female director of a board of health in a large U.S. city.

May 1911 Plague killed around 10,000 Moroccans living in the flea-ridden slums of Casablanca and Rabat and threatened French soldiers posted on the coast of North Africa during the threat of civil war. When Morocco became a French protectorate, French doctors allied to upgrade sanitation and to protect residents from endemic malaria, syphilis, trachoma, and tuberculosis. During this period of social betterment, the medical community faced epidemics of plague, smallpox, and typhus.

summer 1911 During Sweden's worst eruption of polio, 3,840 suffered infection. The size of the epidemic made it the world's largest to date. Leading the investigation of infection was medical writer Carl Kling, who isolated the virus from living patients. At Stockholm's State Bacteriological Institute, more thorough study of the pathogen and its transmission pinpointed the migration of the virus through the human body.

summer 1911 During the Italo-Turkish War, cholera traveled west to Libya from Arabia and Egypt. The pathogen engulfed Tripoli's homeless beggars and spread to Italian forces, killing 3,000 or 30 percent of 10,000 victims. When the disease advanced north to Marianopoli, Sicily, the town council appointed Giuseppe LoDico, a pharmacist, as one lone agent to check passengers on trains from Palermo, disinfect them and their luggage, and aid physician Vincenzo Ferrara in containing the disease before it overran the island. By the end of the two-year war, on the Libyan front, Italians suffered 6,000 deaths. Turkey lost 14,000 men, some 2,000 of them from epidemic disease.

summer 1911 Epidemic polio at Devonport, Plymouth, Stoke, and Stonehouse, England, alarmed citizens. A. Bertram Soltan reported in the *British Medical Journal* that the emergence of 73 cases among 250,000 people terrorized the community. He puzzled over the fact that the poor of the Barbican section of Plymouth were scarcely hit by infection as contrasted with the wealthy of Plymouth. The data refuted common conceptions of polio as a disease of squalor.

1912 Over nearly three decades, a pandemic of sleeping sickness reduced the Sara of southern French Equatorial Africa (modern Gabon, Cameroon, Chad, and Central African Republic) to two-thirds their original number. The disease arrived over traditional caravan routes and sickened villagers living along rivers as early as 1901. In poor, unsanitary homes at Fort-Lamy, Gorée, Moundou, and Moissala, 55 percent died within

the first decade of contagion. Treating the needy after 1916 were physicians Albert Schweitzer and Eugene Janot. *See also* **August 15, 1913.**

In northern Rhodesia (modern Zimbabwe), overly cautious officials saddled natives of Mweru-Luapula with harsh restrictions. The prohibitions halted fishing, hampered trade and religious ceremonies, inhibited missions, and limited workers commuting to jobs in Katanga. Into 1922, resettlement efforts caused idling of workers, malnutrition, and disgruntlement with colonial bureaucracy.

1912 Rampant tuberculosis among the gold miners of Witwatersrand, South Africa, spawned racist notions among colonial labor leaders that the Kaffirs of rural areas and black Africans in general were physically vulnerable to the disease. For this reason, management did little to upgrade the living and working conditions that spread respiratory disease among mine laborers. Instead, administrators set up stringent screening procedures before hiring new workers.

1912 In Wales, the incorporation of the Welsh National Memorial Association to fight tuberculosis operated on a one-to-one basis. Staff applied medical treatment to individual patients rather than broad public programs to improve the standard of living for the poor.

1912 On the Kenya-Tanganyika border, plague assailed Gassenia, a minor village near Mount Kilimanjaro. Communal care of the sick and dying spread the infection, which killed most of its victims. The following year, German army officer and medical writer Richard Lurz published his account of the pestilence.

November 14, 1912 When a typhoid epidemic raged out of control in Moose Jaw, Saskatchewan, the disease devastated Hungarian immigrants and local métis. Father Francis Woodcutter, pastor of Saint Joseph's Church, requested a contingent from the Sisters of Providence of Saint Vincent de Paul in Kingston, Ontario, to institute public health measures. Mother Superior Mary Angel Guardian and Sister Mary Camillus immediately left Daysland, Alberta, by train. They bought a building for quarantining the sick, opened a clinic, and, within two weeks, began admitting patients.

1913 The empaneling of Britain's Royal Commission on Venereal Disease and the founding of treatment centers resulted from an outcry of feminists and social hygienists against endemic syphilis. In an era of rampant prostitution, the disease afflicted 10 percent of England's urbanites. Heads of government, burdened by a disastrous economic and demographic threat, proposed treatment centers in London hospitals. The success of these 190 centers helped lower the spread of infection.

1913 An upsurge in disease in New Zealand at first appeared to be chicken pox. Later identified as smallpox introduced by a Mormon missionary, infection engulfed the indigenous Maori, a virgin-soil population. Government agents were slow to vaccinate natives, whose travel laws restrict movement. Marginalized by racism and poverty, the Maori incurred a high incidence of disease because of the failure of health officials to quarantine the sick.

May 13, 1913 As the Ottoman Empire crumbled, the First Balkan War resulted from the Turks' last-ditch effort to retain a coastal empire extending from the Adriatic to the Black Sea. As Greek forces broke the Ottoman hold on southeastern Europe, protracted fighting cost Turkey 30,000 deaths in combat and 20,000 from cholera and dysentery epidemics. The Balkan allies from Bulgaria, Serbia, and Montenegro lost 110,000, equally divided between deaths from battle wounds and from disease.

August 15, 1913 Nurse Hélène Bresslau Schweitzer and her husband, physician Albert Schweitzer, journeyed to Gabon and opened Lambaréné, originally a chicken coop. To seek treatment, patients arrived by canoe to the clinic dock on the Ogowe River. The service, which developed into a four-room hospital, offered first aid, surgery, and medicines to natives suffering from leprosy, malaria, and sleeping sickness. In November 1917, Hélène Schweitzer, suffering from tuberculosis, returned home to Strasbourg, France, for care. *See also* **February 1918.**

1914 At a time when polio infected 30,000 in the United States and killed 5,000, pathologist Simon Flexner, chief of the laboratories of the Rockefeller Institute of Medical Research in Manhattan, and Hideyo Noguchi, a Japanese bacteriologist, identified the poliomyelitis virus. Flexner's optimistic reports to the nation gave Americans hope that a cure for polio was possible in the near future.

1914 To curtail endemic trachoma in Israel, then under Ottoman control, Jewish doctors convened a three-day medical conference to mount the nation's first campaign of social and environmental prevention among Jewish and Arab residents. A German epidemiologist, Aron Sandler of Posen, Prussia, joined a consortium of European physicians to improve sanitation and hygiene in Palestine.

1914 During World War I, disease became a significant issue to all nations involved in the conflict. In September 1914, Australian bacteriologist Sir David Bruce, Commandant of the Royal Army Medical College at Netley, Southampton, studied the effects of anti-tetanus inoculation during outbreaks of tetanus, primarily caused by puncture wounds. His efforts at immunization reduced mortality by 27 percent.

In November 1914 at Trentham Camp, New Zealand, 7,000 soldiers packed into accommodations intended for 2,000. An outbreak of measles proved impossible for military medical staff to contain. By July 1915, the number reached 1,035 victims, who overran the Wellington Hospital. Officials were forced to adapt a building as a quarantine hospital. In all, 190 or 18.4 percent died of the epidemic or of secondary pneumonia.

Glanders appears to have been the first biologic weapon of the 20th century. As a means of in-country sabotage, German undercover agents exposed animals of several countries to the lethal equine disease *B. mallei*. Throughout Canada, Great Britain, and the United States, a control program eradicated the disease in livestock by slaughtering infected animals. However, scientists did not develop a vaccine or effective treatment.

A German-American surgeon, Anton Dilger, a wound specialist at Johns Hopkins, obtained anthrax and glanders pathogens from German bioterrorism labs to culture at his home laboratory in Washington, D.C. In league with sabotage specialist Frederick Hinsch, Dilger dispatched a liter of the lethal materials to longshoremen in Baltimore to inject into 3,000 cattle, horses, and pack mules being shipped to the Allies in Europe. The diseases may have infected hundreds of soldiers as well.

In France, the rise in venereal disease at the beginning of the conflict forced French officials to fight syphilis by suppressing professional and amateur prostitution. After the first months of the war, physicians blamed middle-class Frenchwomen for selling their favors to earn enough money to maintain their pre-war lifestyles. By war's end, to assure a healthy military, officers alerted soldiers to the threat of casual sex with local women.

Beset by fears of widespread pestilence, French leadership began responding to the agitation of the neoroyalist Action Française, which declared that widespread syphilis and tuberculosis were evidence that democracy and republicanism were basically flawed philosophies. Literary imagery of Marianne, the French symbol of the commoner's ideal of liberty, began picturing her as a debauched bearer of degeneracy and national disorder. The more extreme visions depicted France as feminized and sapped of moral fervor. *See also* **1919.**

As demonstrated by the correspondence of American and French commanders, the two nations varied widely in attitudes toward celibacy and sexual hygiene. Because American soldiers believed the French to be immoral, upon their arrival at Nantes and Saint-Nazaire, they patronized brothels in expectation of exceptionally licentious pleasures. American officers, fearing widespread sickness, pressed for the closure of government-inspected brothels. Because the French had not ended their system of government-supervised prostitution, American commanders declared houses of prostitution off-limits to troops and distributed condoms to soldiers. The French adapted the better of two systems—they kept supervised prostitution and copied the system of preventing disease with free prophylactics.

Infection posed near disaster for the Carrier Corps of East Africa. Military laborers in the local campaign suffered horrendous loss to disease. Some 40,000 died, more than the losses sustained by all other units taken together. Stricken by dysentery, hookworm, and malaria, they ceased to tote supplies, medicines, and ammunition for British Forces as they forged a way through German East Africa (modern Rwanda, Burundi, and Tanzania).

The documentation of disease included not only governmental data, but also eyewitness anecdotes and journals. A *krestovaya sestra* (Red Cross Sister) and diarist, Florence Farmborough of England, was stationed in a flying column on the Russo-Austrian front in the Carpathian Mountains. During her service, she compiled experiences with infection in *With the Armies of the Tsar: A Nurse at the Russian Front, 1914–1918* (1974). Under artillery shelling, she treated the wounded and isolated patients suffering cholera, smallpox, typhoid, and typhus. In 1981, dedicated military service won her membership in the Royal Geographical Society.

In the Middle East, a rise in schistosomiasis cases at Basra, Iraq, struck Indian forces, who encountered river snails while they swam in local waters. Officials battled the parasite by emptying irrigation ditches, ridding water supplies of *schistosoma*, and pressurizing and agitating water supplies to discourage additional infestations. In this same period, Australians posted in Egypt felt more like vacationers than soldiers. In an alluring, sensual environment, they encountered a rise in syphilis from contact with native women. *See also* **November 1914; spring 1915; July 1915; September 1915; April 16, 1917; 1918; early summer 1918.**

1914 When bubonic plague struck urban Dakar, Senegal, 3,653 died. The natives protested European authoritarianism, which violated individual liberties and commerce. The discriminatory, obviously anti-black, anti–Muslim segregation, quarantining, and sanitary measures in city ghettos violated native culture and customs of living in straw huts and cohabiting with farm animals. Peasants rebelled by refusing to sell foodstuffs to privileged French imperialists and their house staffs.

1914 During the Atjeh (or Aceh) War in Sumatra, Dutch colonials fought natives in disease-ridden jungle combat. Of the total Dutch levied for service, 2,317 died in battle; 23,000 suffered tropical leprosy and malaria. The heavy loss on both sides reached a stalemate, forcing the Dutch to retreat to coastal areas, leaving the interior to the indigenous Atjeh.

1914 Among the 5,618 immigrants refused entry to the United States because of dangerous or contagious diseases, 3,051 were diagnosed with trachoma. As lines of new arrivals filed through the immigration facility on Ellis Island, examiners spread the disease with contaminated fingers and instruments. Secretary of Labor and Commerce Victor H. Metcalf reported that

Doctors made the examinations with dirty hands and with no pretense to clean their instruments, so that it would seem to me that these examinations as conducted would themselves be a fruitful source of carrying infection from diseased to healthy people [Markel 2000, p. 550].

Although physicians employed by the U.S. Public Health Service denied the charge, photographs of the reception area and observation of newcomers supported Metcalf's suspicion.

When native-born residents suffered the disease, the U.S. Public Health Service blamed contagion from "trachomatous aliens" (ibid., p. 536). The stigma of filth and disease clung to Asians, Eastern European Jews, Greeks, Italians, and Syrians. Distributed by the National Council of Jewish Women in European Jewry were copies of Cecilia Razovsky's *What Every Emigrant Should Know* (1922), a booklet warning potential emigrants of the seriousness of eye infection during inspection of newcomers to Ellis Island.

The numbers fell significantly over the following ten years (ibid., p. 535):

Year	Examined	Diagnosed	Debarred
1915	562,263	3,137	1,842
1916	481,270	2,5939	20
1917	528,618	3,799	1,116
1918	278,736	3,556	1,319
1919	339,375	5,057	1,166
1920	762,217	5,216	1,028
1921	1,137,682	7,165	1,331
1922	551,454	1,243	425
1923	745,515	1,710	669
1924	874,962	2,372	154

summer 1914 While investigating episodes of pellagra at state institutions, Joseph Goldberger of the U.S. Health Department witnessed the demise of orphans at a pair of Jackson, Mississippi, facilities. The previously healthy children sickened on a diet high in grits, biscuits, syrup, molasses, and cornbread. The dirty, overcrowded orphanages, which had produced 204 cases of pellagra, instituted Goldberger's diet rich in beans, eggs, meat, milk, and oatmeal, all sources of niacin. As a result, the children thrived. However, because staff did not accept his regimen as a permanent change in operating procedure, the deficiency disease returned.

June 1914 Bubonic plague in New Orleans, Louisiana, required the teamwork of rat-proofing crews organized by Public Health Service officer Charles V. Aiken. Rat-catchers baited traps around the city and checked them twice a day. Agents labeled up to 5,000 trapped rats a day with the address, then secured them in galvanized buckets or fabric bags and dispatched them to a laboratory for examination. These measures remained in use for two years until no evidence of the disease existed.

November 1914 During a war between the Balkans and Turks, typhus struck the Serbian troops, who were still recovering battle strength after a cholera outbreak during the Turko-Bulgarian wars. Historians deduce that the disease spread from Albania and inundated 60,000 Austrians held at a prisoner-of-war camp at Valjevo, Serbia. To spare the army, Serbian officials isolated enemy prisoners away from the combat zone.

Among Serbian soldiers, up to 2,500 per day entered the infirmary, while civilian populations suffered three times that many cases. In April 1915, some 9,000 new cases per day contributed to the total count. By May 1915, 150,000 died of the disease. Adding to the total were accounts of typhoid fever and relapsing fever.

Paralleling the Serbian epidemic, typhus produced disastrous infection in Russia, striking soldiers and refugees in huge numbers. Imported from the Volga, Romania, and Petrograd (modern St. Petersburg), the three-pronged attack remained active for four years, infecting over five million people. Massive disinfection efforts decontaminated rail lines and towns.

1915 The U.S. Public Health Service fought tuberculosis in Maryland by dispatching specialist nurses as field workers, especially among poor immigrants. Baltimore's tuberculosis nurse, Ellen N. La Motte, a medical writer for *The Masses*, *Harper's* and *Atlantic Monthly*, summarized the role of the nurse and the regimens that brought results in *The Tuberculosis Nurse* (1915). Later in her career, she described infectious disease during World War I in *The Backwash of War: The Human Wreckage of the Battlefield as Witnessed by an American Nurse* (1916) and joined the League of Nations' fight against the opium trade.

1915 Tuberculosis shortened the lives of reservation Indians in North America. At the Yakima County Health Department in Washington, staff blamed bacteria, diet, inadequate housing, and seasonal work for the heavy toll among youth and young adults of 14 tribes. The Quecha (or Yuma) of the Fort Yuma Indian Reservation in southeastern California suffered increasing losses from starvation and declining health. Contributing to their deaths over a decade in addition to tuberculosis were epidemic enteritis, pneumonia, and syphilis.

1915 The first incidence of Rift Valley fever, an acute viral zoonosis caused by disease-bearing mosquitoes, occurred in Kenya, striking domestic herds. It was a common cause of death among lambs and cattle during the wet season and a source of infection to shepherds and cattlemen. The virus, which passed directly from animals to humans and caused hemorrhagic fever, remained unidentified until 1931.

1915 When over 10,000 pellagra patients in the United States died of the disease, Joseph Goldberger of the U.S. Public Health Department studied the epidemic, which killed at the rate of five percent. The affliction tended to strike poor, rural farm folk and destroy their mental health. First to die were children and those adults already weakened by tuberculosis. While on staff at a 45-bed hospital in Spartanburg, South Carolina, Goldberger began analyzing diet after learning that 190 patients in the state asylum at Milledgeville developed pellagra, but the staff remained healthy.

Refuting notions that the disease resulted from the sting of the stablefly or that Jews were immune, Goldberger, himself a Carpathian Jew educated at Bellevue Hospital Medical College in New York City, perused case histories derived from the city's winos, prostitutes, and outcast immigrants. After traveling farms and mill villages in Alabama, Florida, Georgia, Kentucky, Mississippi, South Carolina, and Virginia, he concluded that a preponderance of cornmeal, molasses, rice, and fatback in the diet caused the disease, which results from niacin deficiency. The laboratory that sampled the Southern diet evolved into the National Institutes of Health.

1915 Health officials investigating typhoid outbreaks at sanitariums in Newfoundland and New Jersey, and at Sloane Maternity Hospital in Manhattan discovered that the "Mrs. Brown" working as a cook was really "Typhoid Mary" Mallon. Without understanding how her body harbored and spread the fever, she had infected 25 staff members, two of whom died.

1915 Austria, France, and Romania suffered *encephalitis lethargica*, a mysterious 12-year pandemic that moved to the Western Hemisphere, Africa, and Australia, infecting around five million people before disappearing abruptly. Its symptoms included drowsiness, inattention, fixed stare, drooling, double vision, palsy, and difficulty swallowing. Identified two years into its advance in Vienna by Austrian neurologist Constantin von Economo, the virus acquired the names Von Economo disease, sleeping sickness, and second-stage Parkinson's disease. Some theorists incorrectly connected the disease with the Spanish Influenza epidemic. Robin Williams publicized the peculiar anomaly with the film *Awakenings* (1990), in which victims revive briefly from chronic invalidism. Directed by Penny Marshall, the film received Oscar nominations for best picture and best actor for Robert De Niro, who played the first patient to emerge from the malady's effects. *See also* **1919**.

1915 The Consorzio Antitubercolare Provinciale in Milan, Italy, attempted to curtail tuberculosis through prevention and treatment. Stymying the effort was the high cost of improvement to peasant housing, where families crowded into unsanitary rooms and shared beds and drinking cups.

spring 1915 In Serbia, the devastation from infectious disease drew volunteers from the American medical community and some 130 American humanitarian organizations, notably, the International Red Cross, Committee of Mercy, Rockefeller War Relief Commission, Serbian Relief Committee, and Serbian Hospital Fund. Epidemic typhus required the aid of French military medical workers to implement preventive and sanitary control. During the Austro-Bulgarian invasion in September, French physicians treated ailing Serbs on the retreat through Albania, on Corfu, in Greece, and during the 1918 liberation of Serbia.

After the armistice ended World War I, typhus advanced through the Balkans, Poland, and Russia. The U.S. Army Medical Department, directed by Colonel Harry L. Gilchrist, took charge of battling disease in Poland. As a result, the American-Polish relief mission updated the nation's health service.

July 1915 Cholera beset Russia in an eight-year epidemic, forcing a military retreat from Poland the next month. As panicky soldiers and peasants fled contagion, they spread sickness across the nation as far east as the Ural Mountains. Over 66,000 contracted the disease.

September 1915 The outbreak of pandemic back and head pain, vertigo, stiff joints, and fever in the trenches of the Western front along Belgium, France, and Germany during World War I appeared to be a relapsing fever of unknown origin. By war's end, trench fever claimed 800,000 victims, affecting the Allies more frequently because of the boggy ground they occupied and the mucky, poorly ventilated trenches and pits in which they worked, ate, and slept. Lice, the vectors of the fever, infested straw, blankets, and uniforms, requiring delousing, naphthalene insecticide, and steam disinfection. Compounding the problem was inadequate drainage on land contaminated by privies and human and animal corpses. In this same period, 6,000 people contracted typhus daily on the Eastern front of Belgium and France. To prevent the spread of trench fever to England, military stations at Boulogne, Calais, Dieppe, Dunkirk, Le Havre, and Rouen sanitized demobilized men and their kits.

1916 The first report of tegumentary leishmaniasis occurred in Argentina. Carried by female sandflies, the disease traveled via rodents and dogs in vegetation growing along waterways in remote areas. Periods of high rainfall boosted the number of sandflies and increased outbreaks of the malady.

1916 To stem pandemic syphilis, the British Royal Commission opted to offer free treatment with salvarsan or arsphenamine, a mercury-based drug formulated by Nobel laureate Paul Ehrlich. Despite its side effects, which included irreversible insanity, English officials preferred the drugs rather than confront thorny social issues of promiscuity and prostitution. By the end of World War I, Scottish health officials relaxed their concern as cases of syphilis declined. However, because gonorrhea remained a medical threat, doctors continued to press for compulsory treatment and reporting of cases.

1916 Nurse-author Mary Sewall Gardner, a specialist in rural public health issues, compiled *Public Health Nursing* (1916) from her work in Genoa, Italy, for the American Red Cross Commission for Tuberculosis. Dispatched by Army Nurse Corps head Jane Arminda Delano, compiler of *The American Red Cross Textbook on Home Hygiene and Care of the Sick* (1918), Gardner and 14 others set up an international study of public health and well being. They taught Italian medical personnel the American method of ending perennial outbreaks of tuberculosis and the deployment of public health workers to treat the sick. In 1919, Gardner's work won her a Genoese bronze medal. On a second mission in 1921, she gathered information about disease containment in the Balkans, Czechoslovakia, France, and Poland.

1916 Following defeat in World War I, Romania countered an outbreak of typhus by rationing food to halt famine and instituting a successful cleanup campaign to reduce lice infestations. Artists and reporters applied their talents to boosting the national self-image.

1916 The battles of World War I produced huge loss from sickness. At Gallipoli, Turkey, on the Dardanelles, an engagement lasted from February 19, 1915, to January 9, 1916. Poor organization, monotonous diet, and inadequate medical facilities forced constant evacuation to hospital ships and base hospitals in Egypt, Lemnos, and Malta. At the height of summer, worsening sanitation spread flies and pathogens, felling 120,000 British soldiers and 20,000 French from dysentery, enteritis, trench fever, and typhus. The youngest casualty, 14-year-old Private James Martin from Australia, expired from disease.

At sea, the main killers were measles and pneumonia. Among Australians posted to Egypt, easy access to women resulted in rampant venereal disease. Schistosomiasis infected those who swam in the Nile or who showered with impure water pumped from the canal. As allied troops advanced on Damascus, Syria, malaria struck in epidemic numbers. In Macedonia, malaria overran British, Bulgarian, and French armies, forcing a hiatus of hostilities. The British hospitalized 162,512 or over 37 percent of their soldiers.

January 1916 Paralytic polio, which had plagued New Zealand since 1914, reached serious proportions among young children. In all, 1,018 Anglos contracted the disease and 123 died. By 1961, when immunization ended the outbreaks, 10,000 islanders had been stricken, of whom 70 percent incurred spinal paralysis. Meningitic polio resulted in head and back pain, vomiting, and distortion of spine and neck. Worsening health conditions, onsolaughts of enteritis, influenza, and measles afflicted young adults.

summer 1916 An outbreak of polio along the Atlantic Coast killed mostly American children. In New Jersey and New York, epidemic polio terrified citizens, who dispatched 50,000 children to the countryside. The adults who remained behind shut themselves indoors and sealed entrance and window crevices to ward off infection. Officials closed public dumps and delayed the opening of the school term. The first city to evolve a plan of rehabilitation was Newark, New Jersey, where viral infections flourished.

In New York, health commissioner Haven Emerson pinpointed each block reporting a new case of polio and dispatched canvassers to go door to door to search for unreported illness. He believed rumors that Italian immigrants spread the disease and reported to the *New York Times* that he had asked the Italian consulate to determine if any hometowns of new arrivals reported incidence of polio. To spare children from exposure to the sick, officials closed schools and playgrounds. Some 50,000 wealthy people decamped to summer homes in the Catskills and along the Atlantic shores. In Hoboken, New Jersey, police guarded against outsiders under age 16 at train, subway, and ferry stations and on major roads. Less well-to-do families shuttered their windows and kept their children indoors.

Parents of sick children resented having to send their loved ones to isolation hospitals. Immigrant families resisted ghetto quarantines and cowered from the approach of public health nurses dispatched by the Department of Social Betterment of the Brooklyn Bureau of Charities. Medical workers for the board of health convinced some parents to enter their ailing children in isolation wards, where caregivers saved lives by immuring patients in long rows of cylindrical iron lungs. Called the Drinker-Collins tank respirator, the iron lung, which the Emerson Company manufactured, was the invention of physiologist Philip Drinker, physiologist Louis Shaw, and Warren Collins. These tanks plus constant attention, heat treatment, exercise, and rehabilitation lessened the extent of paralysis. The regimen earned the approval of the expert of the era, Robert W. Lovett, a Harvard professor and author of *The Treatment of Infantile Paralysis* (1916). Lovett also treated Franklin Delano Roosevelt.

1917 Pneumonic plague struck Mongolia and Kiangsu, China, infecting 16,000. To halt contagion, two years later, the government erected three pesthouses in Hangchow, alerted police to report infection, and set up the Central Epidemic Prevention Bureau in Peking.

1917 During the rise in tuberculosis infections that accompanied World War I, Sir Pendrill Varrier-Jones established a sanatorium outside Cambridge, England. Called Papworth Village Settlement, it combined a tuberculosis hospital and the convenience of a sanatorium along with industrial employment to provide patients with a means to earn a living. To prevent patient despair, his medical village admitted the sick along with their families. Within a decade, the facility was self-supporting. The pleasures and reassurance of a normal community helped patients to feel less isolated and more productive.

January 1917 A surprise epidemic of plague infected Northern Rhodesia (modern Zimbabwe) with a virulent form of the disease, which produced a death rate over 96 percent. Spread from British East Africa along the Luangwa River, the outbreak struck seven communities before moving east to Nyasaland.

spring 1917 After a year of heavy rains and increased disease-bearing mosquito populations, Murray Valley encephalitis struck Australians along the Darling River with a severe attack on the nervous system. The rare strain, a mosquito-borne flavovirus found only in Australia and New Guinea, invaded New South Wales, Queensland, and Victoria, felling young children with swelling and bleeding in the cranium, brain damage, and death.

April 16, 1917 During stateside preparation for World War I, the U.S. military mobilized 3,704,630 men in crowded camps, which experienced epidemics of influenza, measles, mumps, meningitis, and pneumonia. On May 20, 1918, Surgeon General William Crawford Gorgas persuaded the War Department to create the Pneumonia Board to investigate methods of curbing outbreaks.

1918 To stem endemic malnutrition, trachoma, and tuberculosis among the Polish poor, the Society for the Protection of the Health of the Jewish Population established 54 branches throughout Poland to examine students in public schools and to offer summer play centers for children.

1918 In the last stage of World War I, disease infiltrated the fighting forces at the Salonika front in Greece. There were 9,714 soldiers killed in combat and 481,262 sickened by rampant disease. Of that number, 3,744 died. On the Romanian front, over half the Allied casualties or around 8,000 suffered from exhaustion and disease. The total for Serbia was 45,000 killed in battle and 82,000 dead from frostbite, influenza, malnutrition, and typhus.

Ironically, influenza helped bring an end to World War I. Among General John Joseph "Black Jack" Pershing's American Expeditionary Force, 35,000 died of battle-related injury and 9,000 from flu. Pershing himself came down with flu, as did the three negotiators at the peace conference at Versailles—President Woodrow Wilson, England's Prime Minister David Lloyd-George, and Premier Georges Clemenceau of France. Clemenceau, a former advocate of a free press, had reluctantly opted to expunge mention of flu in 16,000 newspaper articles until October 1919. To prepare for influenza in the United States, nursing heroine Lillian Wald chaired the home nursing commission of the Council of National Defense and the American Red Cross Nurses' Emergency Council in Atlanta, where she recruited volunteers and set up health-care agencies.

early 1918 Imported from Uganda by porters mustered into the German East African Army, a lethal cerebrospinal form of meningitis struck Mongalla, Sudan, with a kill rate of 80 percent. Worst hit by the bacteria were the Dinka and Nuer tribes living in substandard housing.

February 1918 Influenza turned into an epidemic in the village of San Sebastian, Spain. Local people tried to conceal the outbreak to protect tourism. When Alfonso XIII came down with the same ailment that sickened a third of Spain, eight million in Madrid alone, the disease acquired the name "Spanish influenza" and "the Spanish lady" (Kohn 1995, p. 319). If not exposed to sunlight, the airborne virus survived and flourished in adults. It was deadly because of the muscle aches, headache, and joint pain that disabled victims, making them vulnerable to cyanosis and pneumonia. People who went back to work before a full two weeks of recuperation were likely to collapse and die from an accumulation of black fluid in the lungs.

The disease migrated directly to Sierra Leone from England after 200 sailors disembarked from the ship *Mantua* during a coaling stop at Freetown. Spread from stevedores inland to two-thirds of the populace, the disease infected additional passengers and crew of the H.M.S. *Africa* and the H.M.S. *Crepstow Castle* from New Zealand. From the initial outbreak, the disease progressed to Gambia, Ghana, Nigeria, Senegal, and South Africa. When influenza struck Kumasi, Ghana, on September 8, 1918, it claimed 9,000 of the Ashanti people. For six weeks, up to 20 a day died of the pestilence. The high mortality rate forced the closing of schools, military installations, gold mines, and roads.

Considered South Africa's greatest demographic disaster, flu moved rapidly. Within six weeks, it killed 300,000 or around five percent of the populace, altering the age and gender ratios considerably. In Cape Town, South Africa, even the assistant medical office of health, Frederick Willmot, admitted to panic. At Windhoek, South Africa, nurse Albertina M. Walker from Ireland supervised an emergency flu ward for three weeks without assistance. At Bloemfontein in the Orange Free State, 1,300 deaths in the month of October proved a brutal truth—that the death rate for black Africans was double that of Europeans. Folk analyses of the pestilence blamed God's anger against sinners, complicity of the British to kill black Africans, or poison gas released during World War I.

In Rhodesia (modern Zimbabwe) the loss of the labor force from disease affected colonial investments. The stoppage of work in cities and mines in the south over a four-year period provided natives with leverage against management to force demands for higher wages. Assisting the unemployed and families in need were mutual aid societies established by migrant workers. Along the Gold Coast (modern Ghana), authorities were virtually helpless to control the contagion, which killed 100,000. Because of a vigorous shipping trade, port cities were the first to suffer illness. The virus spread into the interior along the colonial transportation network.

The scourge entered Nigeria through Lagos, the chief harbor, where troop ships and tourists aboard ocean liners introduced the virus. In all, 512,000 died. Among animists around Abaja,

Benin, panicky bush dwellers brandished Muslim amulets and used camphor to combat a contagion that afflicted 90 percent of the Igbo. At Aladura, Nigeria, influenza introduced doubt among newly won Christians. The Christ Apostolic Church, the largest and most influential outreach, stressed prayer and, at Ijebu Ode, organized a prayer network, the Precious Stone Society. Essential to faith was the renunciation of medicine in favor of trust in the divine. When the concept moved to Lagos, the prayer society merged with the American Faith Tabernacle.

West of South Africa, the disease savaged British Bechuanaland (modern Botswana) in October and November. The Kgatla applied folk preventives and cures, but were unable to end the sickness that cost the tribe many deaths. Enabling their survival as a culture, members stressed traditions, kinship bonds, and political unity.

In Gabon, Lambaréné hospital shifted its outreach from tropical medicine to focus on victims of Spanish influenza in 1918–1919. In 1925, two nurses and two doctors returned to Gabon with physician Albert Schweitzer to treat the sick and to train native health-care providers and medical technologists. For Schweitzer's selfless work, in 1952, he received a Nobel Peace Prize. Six years later, his daughter, Rhena Eckert, joined the staff to work as a medical technologist.

Influenza made its way to the Caribbean and Central and South America. The disease struck Jamaica, followed by advance into Belize and Guyana. Less serious was the infection rate in the eastern Caribbean. In October 1918, around 100,000 Caribbean islanders died of flu, including a disproportionate number of Native Americans and East Indian immigrant workers, whose under-nutrition and low standard of living lessened chances of survival.

At San Marcos, Guatemala, indigenous people suffered higher death rates than North Americans because the San Marcans constituted a virgin-soil population. Without natural immunity, they lacked defenses against the virus. On coffee plantations, management protected their labor forces through quarantine. Other areas received little government aid and sustained pervasive loss and poverty into the 1920s.

In Bogotá, Colombia, influenza thrived among the poor in urban slums. As a result, city planners, with help from church leaders, modernized the inner city by eradicating working-class housing and by relocating laborers to worker villages such as Quiroga south of Bogotá. The shift, paralleled at Medellín, modernized Colombian metropolises.

March 4, 1918 The emergence of influenza in the United States began with the military. At Fort Riley, Kansas, sickness made an alarmingly abrupt appearance. On March 11, army cook Albert Mitchell contracted the virus. Within hours, 107 soldiers came down with symptoms, which appeared in 414 others by the second day. Because the epidemic preceded the invention of antibiotics, pneumonia threatened their lives. In all, 48 soldiers died of the scourge, which drowned their lungs in fluid and blood.

Ultimately, the hero of the Kansas battle against influenza was Samuel J. Crumbine, secretary of the state board of health. Experienced at public health campaigns, he organized treatment for the sick and issued orders in compliance with federal dictates. By closing public institutions and urging municipalities to ban gatherings at offices and businesses, schools, churches, and theaters and amusements, he reduced the impact of contagion on citizens. Despite his noble intentions, the arbitrary selection of gathering places annoyed Kansans. The *Philadelphia Inquirer* muttered, "What are they trying to do, scare everybody to death?" (Iezzoni 1999, p. 127).

At Camp Taylor in Jackson Purchase outside Louisville, Kentucky, unsanitary conditions encouraged the spread of influenza, which claimed 1,500 soldiers. When crowds gathered for war bond rallies, local people sickened with the disease, which killed 15,000 Kentuckians by spring 1919. A pervasive desperation boosted the sale of folk remedies concocted by itinerant quacks. The economic quandary, worsened by layoffs and lost business, forced families to wash and prepare their own dead for burial and to shop for homemade coffins at general stores.

In San Diego, California, influenza infected 5,000 residents, generating 368 fatalities, a mortality rate of 7 percent. The outbreak refuted the area's reputation as a health haven. Medical officials banned gatherings in schools and churches, distributed nasal spray and facemasks, and recommended higher altitudes for those fleeing contagion.

April 1918 In France, the beginning of the flu epidemic struck Paris. In all, 11,500 died of *la* *grippe*, many at the height of infection in July. Because World War I required a preponderance of medical supplies and personnel, civilians suffered poor care in crowded clinics that lacked enough cots for all victims. In death, they also lacked coffins and the civility of daylight funerals as burial crews worked into the night to keep up with the onslaught. City officials were slow to close schools and entertainment centers and hesitated to impose disinfection of residences.

Sergeant John C. Acker of the American Expeditionary Force described widespread influenza, which brought spiking fevers, flushed skin, and aching body and head. Physicians evolved the principles of bacteriology and immunology by focusing on chemical stimulants, vaccines, and immune serums. Despite advanced care, after victims sweated out the virus, they remained weak and melancholic for months.

April 1918 At Addis Ababa, Ethiopians witnessed the onset of a serious influenza pandemic, which overran the city's only hospital and threatened the life of Emperor Haile Selassie I.

early summer 1918 Following the infection of King George V with influenza and the inundation of the British royal fleet with 10,313 sick sailors in May, the onset of the pandemic in London began suddenly and spread to catastrophic proportions. Dubbed the "Flanders grippe," it quickly reduced strapping young adults to bedfast invalids and proceeded to inflict coma and death (Kohn 1995, p. 319). On June 30, the British army postponed dispatching the 29th Division into La Becque near the Nieppe Forest, Flanders, because of lost manpower from illness.

July 1918 When influenza invaded Denmark, it proved difficult to contain. Fortunately, it did not strike the survivors of the 1890 outbreak. Medical officials identified the cause of widespread death as pneumococcal infection, for which there was no treatment. The disease panicked Danes until September 1920.

mid–July 1918 Spanish influenza did not spare Asia, where the disease was called three-day fever or wrestler's fever. According to the *Chinese Journal of Medical History*, the virus killed 20 million people in China. Some 1.5 million died in Indonesia, where ships' passengers introduced the pathogen. The area around Sulawesi lost 20 percent of its citizenry in 21 days; at Lombok,

36,000 islanders perished. By January 1919, when contagion subsided, folktellers had added flu lore to their stock of oral tales.

Around 1,500 flu victims perished at Bombay, where the virus had raged for six weeks. The advance of infection halted mail delivery and railway service, disrupted law courts and schools, filled gutters and alleys with corpses, and seriously impeded factory output of cotton and jute. Because of a scarcity of woodcutters, Hindu families released uncremated bodies into the sacred Ganges River. Survivors sickened and died from famine and pneumonia. In Assam, Raj Kumar Sen guessed at a death toll of 15 million, a higher percentage than the Chinese incurred.

mid–July 1918 The arrival of the vessel *Somali* in Quebec and the *Nagoya* at Montreal brought influenza from outsiders. The aggressive nature of the pathogen at first caused health officials to misdiagnose it as pneumonic plague. People retreated into their homes, deserting churches, schools, theaters, and auditoriums. The province of Quebec suffered a total of 530,704 illnesses. Contagion advanced more critically to the east into Labrador and less at New Brunswick. In central Ontario, 300,000 illnesses swamped hospitals and caregivers. Nationwide, the military suffered a loss of over one-sixth of its personnel.

August 27, 1918 When influenza reached New England, it spread rapidly via military movement. On August 27, the disease emerged in Boston at the Commonwealth Pier, where three sailors collapsed in crowded barracks. Within five days, the virus prostrated 266. The disease produced high fever and secondary pneumonia as it advanced to 2,000 sailors. By the first week in September, Boston's civilian population began entering hospitals for treatment. So many coffins piled up at Dorchester's New Calvary cemetery that the owner pitched an army tent beside the chapel to conceal the backlog.

Outside Boston, soldiers at Fort Devens began to collapse with influenza. Two weeks later, a team of physicians headed by William Henry Welch, a pathologist at Johns Hopkins, investigated the outbreak, which infected 12,600 or 28 percent of a total of 45,000 men. Welch found the lungs so oxygen-deprived that victims turned huckleberry blue. At this stage, there was no treatment to save them from death.

One doctor at Fort Devens in Middlesex

County, Massachusetts, worked 16-hour days to treat 50,000 patients. He wrote, "The camp is demoralized and all ordinary work is held up till it has passed. All assemblages of soldiers are taboo" (Kolata 1999, p. 13). He reported an average of 100 deaths per day and the need for special trains to transport the dead for burial. One barracks was adapted for a morgue.

From Fort Devens, soldiers from Maine carried contagion during furloughs home. At New Haven, Connecticut, staff separated beds in each ward with sheets to contain contagion. After disease invaded Waterbury, Connecticut, local people turned to Chief Two Moons Meridas, the Indian name of herbalist Chico Colon Meridan. A peddler who sold his own medicines in Philadelphia, he developed a successful nostrum called Bitter Oil, which eased influenza patients ostensibly by cleansing poisons from the intestines. Analysis of his formula turned up nothing out of the ordinary. In New York, the gullible bought an onion-based nostrum hawked by financier Hetty Green, dubbed the "witch of Wall Street."

August 31, 1918 The onset of Spanish influenza surprised Chief Navy Nurse Grace MacIntyre, who served aboard the steamer R.M.S. *Briton* off Whitepoint, Ireland. Within two days, her patient load ballooned from six to 160. As rough seas dampened men lying on deck, she removed patients to the indoor smoking saloon.

At a naval hospital in Portsmouth, Virginia, nurse Ada McGrath met throngs of evacuees from France and housed those suffering diphtheria, measles, meningitis, and scarlet fever at temporary bungalows and contagion wards. One troop carrier transported 700 medical cases requiring quarantine. At the Great Lakes Hospital, staff worked double shifts to manage 2,800 patients and sent an overflow of 1,000 to a nearby camp. One overtaxed medical worker, Maude Coleman, succumbed to flu on September 22, 1918. She became the first navy nurse interred at Arlington National Cemetery. The navy accorded military honors to two other nurses who died of flu at the Brooklyn Naval Hospital.

Among civilians at Puget Sound, a train carrying Navy draftees from Philadelphia brought influenza to Washington State, where many victims succumbed to fluid in the lungs and bacterial pneumonia. Among the military population, officials banned gatherings, church and funeral serv-

ices, dancing, and spitting in public. Medical staff received patients at a women's dormitory at the University of Washington and at the former city hall. The state Department of Health and Sanitation and local doctors successfully protected 10,000 shipyard laborers with vaccination. On November 11, 1918, when celebrants of the armistice dropped their gauze masks and frolicked in the streets, mortality rates climbed for the next month and did not cease until March 1919.

fall 1918 Germany faced famine caused by loss of war stocks and by an Allied blockade that impeded new shipments. Influenza, which they called the Blitzkatarrh, advanced rapidly among the starving, demoralized people, claiming 186,000 (Iezzoni 1999, p. 67). At Nuremberg, where four of every 1,000 citizens perished, some 40,000 survivors demanded that the armistice be completed and signed. In the United States, the misinformed ignored Germany's sufferings and spread rumors of a Hun plot. The most inventive assured the naive that German sympathizers were manufacturing the pathogen to pour into American drinking water, spray over the populace, or implant in cigarettes. Evangelist Billy Sunday exhorted his audience with empty rhetoric: "There's nothing short of hell they haven't stooped to do since the war began. Darn their hides!" (ibid., p. 68).

September 1918 The spread of influenza in Australia began in Sydney. J.H.L. Cumpston, national director of the Australian Department of Health, deflected the pandemic by quarantining people arriving by ship. Three months later, the virus reached New South Wales, but in weakened form. The virus continued to reach out into Melbourne and across Victoria into mid–1919 and worsened in New South Wales. Less hard hit were Queensland, the west, and Tasmania.

In New Zealand, 1,130 Maori and 5,559 Europeans perished. At Auckland, authorities shortsightedly failed to quarantine a vessel, the S.S. *Talune*, which bore the virus, then charged islanders with so poor a diet, hygiene, and lifestyle that they contributed to their own deaths. The ship's next ports of call were Suva and Levuka, Fiji, where lax harbor control allowed infected passengers to mingle with unsuspecting locals. When the ship moved on to Western Samoa, 22 percent of islanders perished after colonial administration did nothing to stop contagion.

The Sydney, Australia, *Daily Telegraph* reported on a relief effort to islanders:

> Troopers with their motor-trucks are doing wonderful service day after day gathering up the dead, who are simply lifted out of their houses as they lie on their sleeping-mats. The mats are wrapped around them, and they are deposited in one great pit at Vaimea [Iezzoni 1999, p. 93].

Out of their hatred of New Zealand's arbitrary rule and its overt racism, Samoans blamed outsiders for the infection and resented the resulting economic inflation. The disease accompanied the *Talune* to Tonga, where the pandemic claimed 1,595 lives.

In Japan, a quarter million people died of the flu. As compared to the rest of the world, the unusually low mortality rate reflected national immunity acquired from a previous mild outbreak. Local response to the danger increased chances of survival. A traditional approach involved chauvinistic propaganda. Patriotic Japanese lauded the nation's protective "divine wind" and claimed that it was equal to invasive "devil wind" derived from alien sources (Palmer and Rice "Divine Wind," 1992).

To bolster national resistance to infection, Japanese health officials launched a propaganda campaign of slogans stating how to avoid contagion. Health workers vaccinated the citizenry and advised on widespread use of masks, disinfectant gargles, Chinese medical treatment, and herbal home remedies. One innovative physician, Gomibuchi Ijiro of Yaita-Cho, injected 124 patients with diphtheria serum. All survived the flu epidemic.

September 1918 In Philadelphia during the great influenza epidemic, the know-nothings prevailed. Wilmer Krusen, chief of the Department of Public Health and Charities, made little effort to stay the spread of influenza. Although forewarned, he left relief work to volunteer groups and failed to reduce unhygienic conditions by cleaning the streets. As a result of oversights, in 1920, C. Lincoln Forbush replaced Krusen as head of public health.

September 1918 In Berkeley County, West Virginia, a three-month surge of influenza reduced the community severely. Of the 6,000 afflicted, one-twelfth died. Schools and businesses closed as 80 percent of the work force sickened with the virus.

September 27, 1918 As Vermont infantrymen decamped from the Hudson River area bound for France, they began collapsing with influenza. Of those who advanced to New York to board the U.S.S. *Leviathan,* a troop ship in the American convoy to Brest, France, 120 of some 12,000 aboard had to be evacuated for medical care. After embarking on September 29, the ship housed some 2,000 flu-stricken men, including Franklin Delano Roosevelt, then assistant secretary of the Navy. Staff created isolation areas, forcing healthy men into cramped quarters.

The *Leviathan*'s influenza patients crammed the sick bay, littered the decks, and drooped in their bunks. When one-sixth fell ill with the virus, staff stopped trying to quarantine them. Adding to the confusion were the illnesses of the chief surgeon, physicians, and nurses, who left medics to carry out the heavy work.

Because the dead overran the embalming facilities, officers revived burial at sea at the rate of over 15 per day. The ship arrived at Brest with more dead and some 1,300 sick with flu. On the *Leviathan*'s docking, 969 military personnel had to be transferred to city hospitals. Of those who marched through a storm to camp, 600 fell by the way for volunteers to collect.

October 1918 Norway battled the Spanish influenza on land and sea. The virus struck half the 100 miners in Svalbard. At Longyearbyen, in one week, seven crewmen aboard the fishing trawler *Forsete* died of influenza. So many local people caught the flu that the dead fishermen remained unburied for 20 days. A Norwegian liner, the *Bergensfjord,* arrived in New York harbor carrying 200 sick passengers. Another four died en route and were jettisoned over the side into the Atlantic.

October 1918 In Persia, where peasants were recovering from famine, flu spread along highways, striking Baghdad, Iraq, and Tehran, Iran, killing outlanders, and threatening British occupational troops at Ahwaz, Iran. Arrivals from Qazvin, Persia's ancient capital, carried infection to Tehran, where it claimed over 2,000. At Bushire, one-tenth of the populace succumbed to the scourge; Hamadan lost 1,000. A milder form of the virus afflicted 100,000 at Tabriz, Azerbaijan. In Palestine, a more dire situation confronted the Egyptian cavalry, which encountered both malaria and influenza during a two-weeks campaign against the Turkish army. The Egyptian military lost a total of 1,767 men to the diseases or resultant pneumonia.

October 3, 1918 The Spanish Influenza pandemic reached the Pacific Coast. At Seattle, Washington, doctors at the University of Washington Naval Training Station reported one death among 700 cases. Within six months, 1,600 persons died. Letters and journals such as *The Memoirs of Eleanor Castellan: The Years in the Pacific Northwest, 1910–1919* (1999) recorded personal suffering and loss.

To combat infection, Seattle officials closed theaters and schools, banned public gatherings, and distributed gauze masks. The American Red Cross harangued the public with a jingle:

> Obey the laws
> And wear the gauze.
> Protect your jaws
> From septic paws [Iezzoni 1999, p. 84].

Streetcar conductors rejected passengers for not covering their faces with surgical masks. Soldiers of the 39th Regiment wearing gauze masks marched through town on their way to war in France.

Heavily hit was Montana, especially the Butte area, where influenza killed 5,000 citizens, or 1 percent of the citizenry. The state board of health closed the University of Montana in Missoula. According to the *Ninth Biennial Report of the State Board of Health for 1917–1918*, staff ranged over the state, but were not numerous enough to cover remote mountain homes, where victims received no care. Farther south, the Ute suffered a death rate four times that of white Americans. Amid fears of the Spanish influenza in Iowa, in early October, an epizootic of swine flu struck the Cedar Rapids Swine Show and infected thousands of pigs. The virulent disease raised concerns for the spread of the virus to humans.

October 11, 1918 Reports of influenza among civilians and military draftees were so disturbing that the U.S. Congress allotted one million dollars to battle the epidemic. Epidemiologist Rupert Blue of North Carolina led the U.S. Public Health Service in compiling statistics and tracking the spread of the virus. He realized that infection moved too fast for the establishment of quarantines, but he chose to ban major gatherings.

Because many doctors and nurses were involved in the war effort, Blue summoned help from the Volunteer Medical Service Corps. One valuable colleague, Joseph Goldberger of the U.S. Public Health Service, tried to prove his surmise that Pfeiffer's bacillus was the cause of the scourge. To test his theory, he secured 68 volunteers among inmates at a naval prison in Massachusetts to inject with the microbe. When none of the guinea pigs sickened, Goldberger's theory crumbled. By the end of the month, the U.S. 31-day death toll from flu stood at 200,000.

November 1918 Influenza moved up the Pacific Coast to Alaska, killing around 3,000. The disease surged in indigenous people and the young, especially those living along the coast. Some native Alaskans attempted to escape the disease by settling with the Aleut, Eskimo, and Yup'ik at Egegik. Near the mouth of the Yukon River, oral histories of disease in the village of Pastuliarraq described how survivors abandoned the site and resettled at Caniliaq. Among the Eskimo and Athabascan of Dillingham, no more than 500 survived; the Dena'ina Athabascan village of Tyonek was nearly wiped out. At Wales, Alaska, disease damaged the economy by claiming some of the region's finest whalers.

At Teller Lutheran Mission on Seward Peninsula, missionaries reported on the stacking of frozen bodies in an igloo until they could be buried in spring and of marauding dogs ripping the corpses apart. The decimation of adults left 46 children orphaned. At the Ingalik village of Anvik, so many children were orphaned that they became wards of a mission. *See also* **August 20, 1997.**

To the east, the emergence of "la grippe Espagnole" in Canada sickened 300,000 in Ontario and killed 8,700. The numbers were higher in Quebec where 14,000 or 2.8 percent of a half million victims died. Valiant for their efforts to preserve life, *les Soeurs Grises* (Grey Sisters) staffed the hospital in Nicolet, Quebec. Hampering Cecil S. Mahood's efforts at Calgary, Alberta, was the absence of doctors and nurses still assigned to the European front.

Other Canadians felt overwhelmed by the demands of so much sickness. So many died in Montreal that the city pressed street cars into service as hearses. In Saskatchewan, over 5,000 died. Hardest hit were the young and Native Americans, whose death rate stood at 33.7 per thousand as compared to 6.5 per thousand for whites. Among the Swampy Cree and metís of the Norway House, Oxford House, and God's Lake communities of Keewatin, Manitoba, pathogens traveled water and overland routes, along which outsiders brought furs and imported mail and trade items. According to the Norway House Anglican parish registers near Lake Winnipeg, within six weeks of the arrival of sickness in December 1918, the pandemic may have killed 18 percent of local people, predominantly indigenous people of the fur-trading communities. Following the epidemic, provincial public health and hospital care improved for Canadians.

December 1918 The Spanish influenza continued to devastate the plains and western United States. When influenza struck the Menominee reservation in Kenesha, Wisconsin, it killed the youngest and oldest. The Ojibwa surviving the disease believed their land cursed and moved to their traditional home at Grassy Narrows, Ontario. The Wisconsin State Board of Health totaled the damage: within four months, 8,459 died and 103,000 were disabled as life expectancy dropped from 51 to 29.

According to Army doctor Major Curtis Atkinson of Wichita Falls, Texas, the emergence of influenza among military personnel required the hiring of civilian nurses. On a Sunday afternoon at Call Field, football players began to sink with flu. Atkinson explained the quarantine procedure:

> The commanding officer immediately ordered the game stopped and sentinels posted at the gate of the field with orders that no one was to be admitted. It was very hard for the citizens of Wichita Falls to learn that a military quarantine could not be evaded. Within an hour the two ambulances were very busy taking men from the different parts of the camp to the hospital, and by the next day the hospital was filled to its capacity [Atkinson "Interview," 1999].

As routine halted, a triage system placed the sickest men in the main hospital. Atkinson took pride in the camp's lowest fatality rate of any U.S. military installation. In contrast, Amarillo and Canyon, Texas, suffered a combined loss of 143 residents.

1919 After founding the American Zionist Medical Unit, social reformer and health worker Henrietta Szold returned to Palestine to improve the health of homeland Jews in Eretz, Israel. Appalled by epidemic trachoma, rickets, and high infant and maternal mortality rates, she dedicated the next 14 years to improving conditions among the poor. With other Zionist women, she formed Hadassah, a relief organization named for the Hebrew for Esther, a biblical hero. Szold's activism survives in the Henrietta Szold Hadassah-Hebrew University School of Nursing in Baltimore. *See also* **mid–1940s.**

1919 A 12-year epidemic of encephalitis in Great Britain caused health officials to record incidence of the disease, which claimed 11,619. Stymying physicians was a lack of clinical experience with the virus, which Austrian pathologist Constantin von Economo had identified as encephalitis lethargica in April 1917 and written about in *Die Encephalitis Lethargica* (1918) as a post–World War I anomaly.

1919 When yellow fever advanced to epidemic proportions along the northern coast of Peru, American doctor Henry Hanson, head of the Rockefeller Foundation intervention team, applied authoritarian methods of eradicating disease. During this period, Japanese bacteriologist Noguchi Hideyo, a member of the Rockefeller Institute of Medical Research and author of *Experimental Studies of Yellow Fever in Northern Brazil* (1924), observed the outbreak while perfecting a yellow fever vaccine. Over nearly four years, the disease dogged illiterate people who had no folk remedies or methods of treatment. Although Hanson generated hostility for his arrogance, Peruvian health officials duplicated his technological method in subsequent outbreaks.

1919 The post–World War I spread of tuberculosis in France was the result of a virtually untreated epidemic that began at the dawning of the century. Public demand for containment and cure spawned numerous medical campaigns and anti-tubercular facilities. John D. Rockefeller, a philanthropist who set up the Rockefeller Foundation, aided the National Committee for Defense against Tuberculosis in funding treatment programs in 58 departments and in establishing anti–TB committees throughout France.

The creation of student health centers in France into the 1920s produced a front-line defense against early death from tuberculosis. Because of post-war health consciousness, the National Union of Students made treatment available and built facilities that allowed youths to receive care while continuing their education.

To broaden the value of the centers and make them more appealing, staff offered politics, movies, theater, and other cultural activities. Within three decades, the rate of tuberculosis decreased.

May 1919 Ten percent of the citizens of Mauritius died of flu during a five-month siege that arrived after most of the globe had experienced the Spanish influenza. Spread from the harbor at Port Louis, the scourge struck Asians, Europeans, and natives alike, afflicting 80,000. Most of the reported fatalities resulted from pneumonia or cardiac complications. Overworked health officials mustered troops of Boy Scouts as aides.

July 9, 1919 Epidemic cholera transmitted from India through southern China assailed Manchuria, infecting 45,251 with a mortality rate of over 60 percent. Hardest hit was Harbin in the northeast, where 4,500 caught the water-borne sickness. Because disease traveled by rail, medical authorities isolated junctions at Dairen, Mukden, Newchwang, and Port Arthur.

1920 The final accounting of the influenza pandemic passed through a number of recalculations, which gradually upped the totals to give a more accurate estimate of the huge number of deceased. Within 18 months, the disease ran its course and disappeared without being fully identified. Health officials later noted that more people succumbed to the epidemic than the total people killed in three wars—World War I, the Korean War, and the Vietnam Conflict. Estimates of underreporting in underdeveloped countries tripled the total to 60 million. Overall, of the two billion infected with the virus, 800,000 died in the United States and 25 million worldwide. Subsequent estimates claim that 5 percent of the world population succumbed to the virus.

1920 The Workers' Society of Friends of Children in Lodz, Poland, fought endemic tuberculosis with a crusade to prevent and treat illness among the working class. The activists set up outpatient clinics and a nursing home as benevolent care centers for young consumptives. Four years later, a health kitchen for the poor set up the previous year at Otwock, Poland, developed into a sanatorium for tubercular Jews. For the next 14 years, the facility offered hope of cure to some 5,000 needy patients.

March 1920 In Hawaii, the global influenza epidemic of 1918 claimed over 2,500 islanders, who had already suffered three lethal pan-island epidemics. After deaths peaked across the United States in 1918, Hawaiian mortality continued to rise, reaching 602 in March 1920.

August 1920 A resurgence of pneumonic plague in Manchuria spread during the early winter, reaching the Russian port of Vladivostok in January 1921 and killing 9,300. To stem the infection, the Manchurian Plague Prevention Service conferred with health officials from China, Japan, and Russia on set-up of plague housing and use of rail lines and enforced harbor controls, including the rat-proofing of vessels.

September 1920 An influx of alastrim (or Kaffir pox) from Cuba or Brazil to Hampstead, Jamaica, in April struck four months later, sickening some 6,000 islanders with high fever, headache, malaise, and numerous papules. Kingston alone reported 700 cases. The pustulation, scarring, streptococcal infection, and mortality rate of this mild form of smallpox reached below 2 percent. Health workers boosted immunization with a vaccination program.

The infection rate over the next year reached 5,283 in April 1921. The Isolation Hospital admitted 2,912 victims. The virulence dropped steadily, generating 15 cases in November 1921. Port authorities issued anti-rodent regulations for vessels from other countries. Nonetheless, the epidemic resurged in June 1922, with 234 new infections in seven parishes.

November 1920 The conclusion of the four-year Russian Revolution left heavy carnage—the Red Army suffered around 702,000 deaths, 283,079 from epidemic disease, notably, cholera, influenza, typhoid, and typhus. The Whites lost 175,000 to combat and 150,000 to sickness. Of the total cost of nine million Russian civilian deaths, an unofficial list of fatalities resulted from reprisals and executions, famine, and disease.

1920s In colonial Kenya, a national campaign reduced infection in rural villages by applying Western medical principles, as with the outreach at Chogoria and the Tumutumu Yaws Camp. Health officials launched a rodent reduction program and taught villagers to dispose of sewage safely. Uneducated Kenyans came to respect inoculation as a defense against such diseases as tertiary yaws, smallpox, plague, leishmaniasis, pleural pneumonia, tuberculosis, and rheumatic fever, a cause of cardiac weakness.

1920s To improve the worthiness of military recruits in Newfoundland, government commissioners and health officials attempted to end beriberi by supplying brown flour to welfare recipients and by teaching islanders to select a varied diet.

1920s During a lengthy bout of malaria, epidemiologist Luis Figueras Ballester, the first medical officer of the Spanish Naval Air Service, fought the disease from a command post at the naval air station of Prat de Llobregat outside Barcelona. By limiting military assignments to Morocco and isolating sick soldiers, the service controlled outbreaks.

1920s At Hainan, an island off the southern shores of Guangzhou, China, a 60-year campaign battled malaria. Health workers gradually controlled protozoa-bearing *Anopheles* mosquitoes to drop rates of infection, which reached 10.3 percent.

1920s Among Australian aborigines, the prevention and treatment of leprosy was primitive, mainly because of racist attitudes of whites. In lieu of more progressive care, leprosaria remained the first line of defense until the 1980s. Immurement automatically conferred a life sentence to the sick, who remained institutionalized out of sight and contact with normal society. For the average patient, doctors tended not to intervene until the final and most debilitative stage of the disease.

1921 A Seventh-Day Adventist couple, Leo and Jessie Halliwell, staffed the *Luzeiro*, a floating clinic along the Amazon River in Brazil, to care for some 2,000 natives ill with elephantiasis, leprosy, malaria, trachoma, and yaws. After moving on in 1929 to Bahía, Manaus, and Belém, they remained in service until 1958, when the nation presented them the Brazilian Cross.

1921 An outbreak of lethal pneumonic plague in Manchuria killed victims in less then 48 hours. At Nanking, a Japanese research experiment with plague, typhus, cholera, and anthrax under General Ishii Shiro tested the effectiveness of infection as a weapon of mass extermination. Plague spread to Korea, where colonial officials blamed Chinese laborers as the carriers.

1921 Demographic movement at Saratov, Russia, the largest city on the Volga River, derived from famine and epidemic cholera and typhus. Hardships on refugees caused increased infant mortality, spread of disease, and decreased population.

In the former Kazan province of eastern Russia, the Tatars experienced a yearlong drought, crop failure, starvation, and epidemic typhoid fever. The population shrank after 218,000 died and a half million more emigrated, leaving 75 percent of the original headcount. Industry shriveled from lack of capital, fuel, and raw materials. To revive the area economy, the Soviet government advanced Kazan 7.5 million rubles by 1925.

1921 Soldiers from French Guinea returning from World War I posts in Algiers and Morocco carried relapsing fever, which killed 15,000. The pathogen spread to Burkina Faso, Chad, Gold Coast, Mali, Niger, Nigeria, and Sudan, claiming 21,000 more victims in a four-year span.

July 1921 Just as Joseph Goldberger of the U.S. Health Department predicted, the fall of cotton and tobacco prices in the American South and a boll weevil infestation returned a previously prosperous area to its former state of poverty. Pellagra was so rampant that President Warren G. Harding studied the problem personally. The Southern media denied Goldberger's charge that insufficient diet caused the outbreak and rejected relief shipments of free condensed milk from the Borden Company and free meat from the Institute of American Meat Packers.

1922 A decade of smallpox in Chad killed 20 percent of the 40,000 victims. The disease appeared to enter the borders when traders and nomads arrived from Cameroon and Nigeria, where suffering generated the Mbula prophet movement at Adamawa. Hardest hit among Chadians were the indigenous Sara, who suffered deaths, spontaneous abortions, disfigurement, and blindness.

1922 In Southern California, the youths and young adults of the Cahuilla, Chemehuevi, Cupeño, Kumeyaay, Luiseño, and Serrano at the Mission Indian Agency experienced 180 fatalities from tuberculosis. Their death rates outnumbered those for the rest of the United States. By 1926, the loss of life to infectious disease jumped to 661, its peak during a 24-year period. Contributing to a subsequent decline was the work of village elders and field nurses dispatched by public health agencies to teach native people about the cause and prevention of respiratory infection. *See also* **1928**.

1922 Malaria overran Russia in the worst assault in modern times. Thousands fell ill along the Volga River following heavy flooding. The infection of 72.4 million people over a two-year span produced 60,000 deaths. A lack of quinine and medical supplies forced the Russian government to bolster its laboratory, pharmacy, and hospital readiness.

January 1924 Epidemic bubonic plague beset Western and Central Africa. Contagion invaded Madagascar, raising the death toll to 2,000. When a yearlong bout of plague struck Kumasi, Ghana, it killed 90 of 166 victims. Percy Selwyn Selwyn-Clarke, compiler of *Report on the Outbreak of Plague in Kumasi, Ashanti* (1926), initiated the "Kill rats and stop the Plague" campaign to halt infestations of vermin in railroad centers.

summer 1924 Japanese B encephalitis, spread by mosquitoes that infested coastal Asia from India to Singapore, killed around 3,800. Only 38 percent survived the aggressive viral inflammation of the brain; those who lived faced a possibility of partial or total paralysis. An eyewitness account from Shiro Ishii, a doctor with the Imperial Japanese Army and proponent of bioterrorism, reported on patient misery on the island of Shikoku. A recrudescence in 1927 lowered the survival rate to 29 percent.

October 29, 1924 When pneumonic or pulmonary plague emerged in the Mexican barrio of Los Angeles, California, it produced the worst epidemic in U.S. history. Of 33 cases, all but two victims died. Because county health workers were slow to diagnose plague, the press carried reports of a malignant pneumonia. Within two months, the city had mounted a portside battle against rodents and quarantined the harbor. The contagion ended early in 1925.

December 1924 Polio in Wellington, New Zealand, infected 1,185 European settlers. Over half were under school age.

1925 During a diphtheria outbreak at Nome, Alaska, a series of 20 dogsledders transported the needed serum from the Nenana depot to the sick nearly 700 miles away. The public praised the medical mission and lionized two dogs, Balto and Togo, and mushers Leonard Seppala, Charlie Olson, and Gunnar Kaasen, who covered a total of 165 miles from Shaktolik in a blizzard. The annual Iditarod commemorates the humanitarian run with a competitive sled race in sub-zero temperatures over 1,180 miles of rough, late-season ice and snow. The route takes competitors from Eagle River outside Anchorage to Wasilla, Knik, Big Lake, Skwentna, Finger Lake, Rainy Pass, Rohn Roadhouse, Nikolai, McGrath, Takotna, Ophir, Cripple, Sulatna Crossing, Ruby, Galena, Nulato, Kaltag, Unalakleet, Shaktoolik, Koyuk, Ellim, Golovin, White Mountain, Safety, and Nome.

1925 Endemic leprosy in the Belgian Congo received more interest after the Belgian Red Cross began an anti-leprosy program. To contain contagion, medical workers isolated the sick in government-run colonies. Because of inadequate treatments, lepers received mainly medical observation and palliative care until sulfa drugs offered a cure in 1946.

July 1925 Epidemic influenza began in Russia, producing a high incidence in Moscow. Over a six-year period, the disease remained stubbornly concentrated in urban areas.

1926 The first black registered nurse in Birmingham, Alabama, Pauline Bray Fletcher, treated consumption in young campers at Kelly Ingram Park. During the civil rights movement, the area became the organization point for demonstrations and sit-ins.

1926 Dengue fever infected 50,000 at Durban, South Africa, after progressing from Stanger and Pinetown and continuing to Kelso Junction. Epidemiologists surmised that travelers returning from India imported the disease.

1926 Because of rampant tuberculosis in Mendoza, Argentina, agencies and individuals fought the disease through a number of innovative methods. The Liga Argentina contra la Tuberculosis campaigned for improved sanitation and advanced epidemiological research. Despite these efforts, the disease proved difficult to control until the development of a vaccine and the improvement of drugs in the 1940s.

1926 Ontong Java, part of the British Solomons off New South Wales, experienced an epidemic of influenza, which recurred in 24 months. Most vulnerable were infants, who tended to die of pneumonia.

1926 Pneumonia in the Union of South Africa increased mortality among miners and infants. Hard hit were the Bantu employed at the gold mines of Witwatersrand. Worsening their chances for survival were influenza outbreaks, alcoholism, and the hardship of their employment. Because over eleven percent of the 41,394 victims died, managers had difficulty recruiting replacement laborers. Insufferable conditions extended the epidemic into 1940.

May 1926 Yellow fever made its first large-scale appearance in West Africa at Asamankese 50 miles north of Accra, Gold Coast (modern Ghana), putting to rest false assumptions that the black race was immune to the disease. To maintain political control, native shamans concealed the outbreak and rejected European methods of disease control. The village chief used human sacrifice to propitiate angry gods and hired a juju expert from Togo to dispel the curse of sickness. The people put no stock in the fetishist after his son died of yellow fever. Worsening the credibility gap was the death of the chief's sister. The people of Asamankese allowed white health officials to spray for mosquitoes and to quarantine those villagers bearing contagion. Within four months, the epidemic ended. Ironically, while Japanese bacteriologist Hideyo Noguchi, on staff at the Rockefeller Institute of Medical Research, was studying the pathogen at Accra, on May 21, 1928, he died of the fever.

June 1926 Among Ugandans, a six-year outbreak of lethal bubonic plague overran the southeastern area, producing around 1,850 infections with a death rate of over 85 percent. Increased health initiatives brought government workers in conflict with religious dissident Semei Lulaklenzi Kakungulu of the Koki kingdom, a district chief in Buganda province during religious wars among Muslims, Catholics, and Protestants that preceded establishment of British rule over much of Uganda. After abandoning politics and the military for religion, Kakungulu turned to strict Jewish practice based on Old Testament law and opposed doctors, medicine, and immunizations. Because he and his followers of Abayudaya Judaism violently rejected vaccinations against plague, British authorities exiled Malaki, one of

his lieutenants, who died during a protracted hunger strike.

The plague refused to cede to increased vaccination. Infection figures increased the next year, when the disease claimed 1,863 for a two-year total of 3,437 deaths. The numbers doubled in 1929 as cotton farm laborers suffered seasonal infections of pneumonic plague. By 1931, the death rate for 15,000 cases neared 90 percent.

September 1926 At Darfur, Sudan, 200,000 people died of relapsing fever during an 18-month epidemic. Staff identified the louse as the vector and set about spraying delousing agents, boiling garments, shaving body hair, and isolating patients. The epidemic killed 25 percent of the indigenous Fur.

late 1926 An upsurge of cerebrospinal meningitis in Mongalla, Sudan, killed 1,243 in a six-year period. Hardest hit were the Dinka and Nuer people, who battled insect infestation, dust storms, and famine and hovered in small huts against cold and misery. In close quarters, they spread the infection.

1927 After a severe flooding of the Mississippi River, 112 counties in 12 states experienced disaster and a revival of endemic pellagra from inadequate diet. Joseph Goldberger of the U.S. Public Health Department was testing a theory that brewer's yeast provided the appropriate nutrient to prevent disease. In answer to needs in the South, he instructed the American Red Cross to distribute brewer's yeast to residents. In a matter of weeks, the disease abated.

The battle against pellagra prevailed in the rural South over the next 20 years. Additional aid came from diversification of crops, refrigeration, and the instruction of Southern families in vegetable gardening and home canning, a project of 4-H and Home Demonstration clubs. To legitimize the value of domestic science in agricultural reform, experts produced such teaching films as *The Happier Way* (1920) and *Helping Negroes to Become Better Farmers and Homemakers* (1921). Fieldworkers Mary Engle Pennington and Margaret H. Kingsley of the Household Refrigeration Bureau taught rural women Cornell University's Home Bureau curriculum, which included instruction in nutrition.

1928 Physician Charles McKenzie recorded his medical service to the people of Ituna Bon

Accord, Saskatchewan, noting the diphtheria epidemic of 1928. Multiple roles required him to examine school children, treat sickness, and serve as county coroner.

1928 To combat endemic trachoma and tuberculosis, the U.S. Indian Service (U.S.I.S.) stopped sending field matrons to Indian reservations and ended the routine performance of tarsectomies, a surgical procedure on patients' eyes to rid them of the curling lids caused by trachoma. Instead, the U.S.I.S. began dispatching public health nurses. Originally called "tuberculosis nurses," these predominantly middle-class, unmarried, and white women conducted a program evolved at Johns Hopkins University in 1903. Into the outback, the women traveled by car to patients and fought the native disdain for "white" medicine by educating families on sanitation, nutrition, and wellness.

1928 A three-year onslaught of plague struck Manchuria and Mongolia. Attesting to 268 deaths was Wu Lien-teh, a Cambridge-trained Chinese physician, hero of the plague of 1910. He created the Manchurian Plague Prevention Service and co-authored *History of Chinese Medicine* (1936). After modernizing health service and medical training, he founded the Chinese Medical Association and directed the building of 20 hospitals and research labs. At Chengchiatun, his staff set up temporary headquarters and a laboratory and recovered bodies of plague victims where survivors abandoned them. He described personal involvement in *Plague Fighter: The Autobiography of a Modern Chinese Physician* (1959).

June 1928 Schistosomiasis recurred at Basra, Iraq, after children and laborers swam in trematode-infested water during the months of summer and early fall. Children completed the life cycle of the parasite by excreting the eggs in urine.

1929 The first of a series of malaria outbreaks in the Union of South Africa began on Natal's sugar cane plantations, where managers housed non-immune laborers in despicably dirty, cramped barracks. Over seven years, the disease centered on immigrant workers, who incurred the highest mortality. At the epicenter, George Park Ross, a philanthropical English surgeon, distributed quinine and organized an anti-mosquito program, which saved worried investors the massive loss of laborers to disease.

The Zulu spread rumors that whites caused malaria and that quinine aborted black babies and caused sterility. A hostile mob ejected agent Nicholas Bhengu when he offered them quinine and advice on how to avoid infection. By the time the disease was contained in 1935, 21,000 had died of it. Some 7,700 were residents of Zululand.

summer 1929 A virulent form of tuberculosis and "ship's illness" threatened the lives of the Coppermine (or Copper Inuit) Indians of Coronation Gulf north of Nunavut in Canada's central Arctic region. The scourge, combatted by physician Russell D. Martin, tested the government's commitment to indigenous Inuit. Some of the survivors moved to Dettah in the early 1930s to work and live with the Dogrib and Chipewyan.

The Coppermine experience survived on paper and in art and film. Residents of Holman in the Northwest Territory value stories of the epidemic as a part of their history. William Kagyut, one native victim of tuberculosis, wrote a song about spending a decade at an Edmonton hospital. He dreamed of an elder who confided that, if he sang of his illness, he would be healed. The singing produced a cure. In 1991, author Walter J. Vanast worked with the National Film Board of Canada in chronicling the epidemic with the story of Jennie Kanajuq, who died of TB in Coppermine near the end of the contagion in 1931. Her experience attested to the callousness of health care professionals and the neglect of religious and political leaders.

summer 1929 Along the Cuyuni River, 1,000 Venezuelans, mostly forest laborers, contracted yellow fever. The jungle variety of the pathogen, carried by mosquitoes and monkeys, killed only .05 percent, a death rate much lower than for other incidences of the disease.

mid–July 1929 Psittacosis or parrot fever, considered rare, reached pandemic proportions after it spread from Córdoba, Argentina, among hobbyists and pet wholesalers at an auction of wild Brazilian parrots. The disease advanced to the Atlantic coast to humans and non-psittacine birds in Buenos Aires. By water, the infection traveled to Paris, Switzerland, Germany, Poland, and Austria. By 1931, it had made its way over Czechoslovakia, Denmark, Holland, Iceland, Italy, Spain, Sweden, Algeria, Egypt, the Middle East, and Canada. When it reached the United States in parrots imported for Christmas sale, the disease afflicted people in Philadelphia; Providence, Rhode Island; and Warren, Ohio, infecting 1,000 and killing around 20 percent. Internationally, strict laws involving importation of psittacine budgerigars and parrots from South America stemmed future outbreaks.

fall 1929 Following heavy rains throughout the summer, around 40,000 Pakistani died in the Sind district west of India from a virulent form of malaria. The disease struck those too young to have developed immunity and caused complications in the spleen.

1930 A mysterious paralysis of the legs called Jake paralysis or the "jake leg" originated in the Southern and Midwestern United States and grew into a puzzling epidemic across the land. Health officials connected the disorder to Jake, an extract of Jamaican ginger root used as medicine. Consumers added it to soft drinks to cloak its unpleasant flavor. Although 70 percent alcohol, Jake was exempt from Prohibition laws because it was too strong-tasting to be a recreational drink. Health officials analyzed its makeup and identified a toxin, tri-ortho-cresyl phosphate, which weakens the limbs from a form of polyneuritis. The U.S. Food and Drug Administration traced the outbreak of paralysis to customers purchasing Jake from Hub Products of Boston.

1930 At Lübeck, Germany, to stem endemic tuberculosis, health authorities mobilized an oral vaccination campaign for newborns. Practitioners misused the vaccines developed in 1921 by microbiologist Leon Albert Calmette of Nice, France, founder and director of the Institute of Bacteriology of Saigon, and his partner, Jean Camille Guérin, inventor of a method of immunization against TB. As a result, 71 of 252 babies sickened and died from tuberculosis. A public outcry targeted the designers of the vaccine, but subsequent investigation disclosed that the directors of the German laboratory carelessly contaminated the culture with tuberculin bacilli. Both men were imprisoned.

July 1930 Deriving in India, a lethal form of cholera emerged in Kabul, Afghanistan, and spread to Charikar, Ghazni, Jalalabad, Kandahar, and Makur Kalat. Health officials warded off catastrophe by monitoring borders and importing vaccine from India and Russia. For a decade, the disease remained virulent, striking again in

1936, 1938, and 1939, when it moved southwest, killing 849.

November 1930 Some 45,400 Nigerians contracted smallpox, which killed over 23 percent of victims within an eight-month siege. Over a five-year period, the disease prevailed, leaving survivors disfigured or blind. Hard hit were the Igbo and Yoruba. The latter rejected inoculation and propitiated Shapona (also Sopona or Babilu or Obaluwaye), a smallpox god and warrior who supposedly spread contagion with medicinal brooms. A vengeful spirit whom the colonial British tried to suppress, Shapona forced victims to resign themselves to the disease without sadness or regret. Pious families turned over property of the dead to Shapona's priests. The deity was so important to Nigerians that he survived in Brazil, Trinidad, and Cuba as Omolu among blacks whose ancestors transported their beliefs on slave ships. In the Western Hemisphere, Shapone, the father of worldly materialism, found new converts among voodoo worshippers and Santerians. His powers increased to the god of AIDS.

1930s Methodist missionaries from England attempted to relieve the ostracism and misery of lepers in Nigeria. Because the Igbo stigmatized the sick, colonial missionaries encouraged self-esteem and contribution to society at the Uzuakoli Institute at Owerre Province in eastern Nigeria.

1931 Borne by *Rattus norvegicus*, plague in Shansi and Shensi, China, afflicted 20,000 with both bubonic and pneumonic forms of the disease. Prompt preventive measures contained the epidemic in the two provinces.

1931 In the Islamic emirates of northern Nigeria, a rescinding of bans on Christian missionaries allowed proselytizers to influence residents through medical care. In Uganda, Anglicans from England collaborated with government health initiatives to aid lepers. The missionaries converted some Muslims by opening a medical clinic to treat lepers, but made few lasting conversions after patients returned to their homes.

summer 1931 When polio invaded the United States, it produced a mortality rate of over 12 percent or 4,000 deaths. Hardest hit were New York City and New Haven, Connecticut. George Draper, author of *Infantile Paralysis* (1935), compared the siege to the London plague of 1665. Of the militaristic nature of disease control on Long Island, he said:

> Deputy sheriffs, hastily appointed and armed with shot-guns, patrolled the roads leading in and out of towns, grimly turning back all vehicles in which were found children under sixteen years of age. Railways refused tickets to these selected youngsters, the innocent victims of ignorance and despair [Gould 1995, p. 3].

The cruelties of policing the epidemic created widespread skepticism and suspicion toward government health officials.

In June, John Rodman Paul, a professor of preventive medicine, and James Trask, a pediatrician and researcher at the Rockefeller Institute of Medical Research, established the Yale Poliomyelitis Study Unit, which laid the groundwork for inoculation programs of the 1950s. Their campaign involved ringing doorbells, canvassing neighborhoods of polio patients, and isolating the virus. Extensive paperwork required patient histories, physical exams, and collection of throat cultures for injection into laboratory monkeys. For contributions to epidemiology, Paul was named to the National Academy of Science and the Polio Hall of Fame.

1932 For the third time in its history, West Java suffered a staggeringly lethal form of plague that thrived in mountainous Priangan province. Unlike the situation the island administrators faced in November 1910, this outbreak demanded more mobilization than the plague service could manage. Islanders refused to cooperate with the puncturing of the spleen of corpses to diagnose the cause of death. By 1935, the health service initiated a plague inoculation program among two million citizens that halted recrudescence of the disease.

January 11, 1932 When anthropologist Gordon MacGregor of Hawaii's Bishop Museum journeyed to Rotuma in the Society Islands of the South Pacific, he took notes on the treatment of endemic elephantiasis, a disease that 19th-century travelers had observed. To be cured, island men went to Vaitupu in the Ellice Islands for testicular surgery. Others performed self-surgery by stretching the scrotum and piercing it with an ax to drain fluids and rid their bodies of shameful swelling.

August 1932 After cholera moved from Canton and Shanghai, China, to Hankow, Hunan, Shensi, Suiyuan, and Nanking, contagion reached Manchuria during heavy floods. Of 100,000 cases, nearly a third died. The misery spread to Japan, but did less damage than it had on the mainland.

1933 Recurrent yellow fever in Brazil and Colombia derived from infection of monkeys by *Haemagogus leucocelaenus* mosquitoes.

1933 In the Ingham district of North Queensland, Australia, the rise in illness and deaths from leptospirosis, a hemorrhagic fever also known as Weil's disease or coastal fever, occurred from rats spreading the disease among sugar cane workers. Increasing the rate of contagion, unusually heavy rains thickened vermin nesting grounds. A local practitioner, Gordon Morrissey, the superintendent of the Ingham District Hospital, first diagnosed the fever. Raphael Cilanto crusaded to destroy the rat vectors but, out of concern for profits, opted not to burn cane fields to end rodent infestations.

1933 At Saint-Hilaire-du-Touvet, France, the need for treatment centers for tuberculosis patients created a demand in 1918 that required 15 years to satisfy. The building of a 646-bed facility, one of the nation's largest, provided the populace with a sanatorium sufficiently large to serve infected patients.

1933 A field trial of the Calmette-Guérin vaccine for tuberculosis among 609 Assiniboin, Cree, and Saulteaux infants born in the Qu'Appelle reserves in southern Saskatchewan proved the value of the preventative. Robert George Ferguson, medical superintendent of Qu'Appelle Sanatorium and medical director of the Saskatchewan Anti-Tuberculosis League, chose the natives because their mortality rate from TB was ten times higher than the rest of Canada. To his dismay, he encountered a 20 percent death rate among the subjects from enteritis and pneumonia, typical diseases for Native Americans in squalid living conditions.

1933 At Madagascar, epidemic plague struck around 14,000 people with a fatality rate of nearly 86 percent. Moving from the harbors inland, the disease afflicted families living in huts overrun by rats. Officials failed to contain the disease until 1937, when incidence declined through government disinfection of homes and bedding, immunization programs, and vermin control.

mid–1930s Malaria recurred in the United States during the Great Depression. Because poverty lowered living standards, families ceased replacing window and door screens, draining marshy ground, applying insecticides, and taking quinine to ward off disease.

1934 When the fall of agricultural prices idled workers in Ceylon, the public health initiative grew lax in suppressing mosquitoes. The emergence of a yearlong malaria epidemic was devastating. When the fever felled three million out of 5.5 million islanders, only 3 percent died because of the immunity built up from years of exposure to the parasite. The disease tested the efficacy of home rule prescribed by the Donoughmore constitution after years of colonial control by Britain. Medical historians considered the outbreak a turning point in island health service.

1934 After a period of drought, poor harvests, and hunger in South Africa, louse-borne typhus killed 1,662 or 13 percent of 12,782 patients. Most victims lived in Orange Free State and Cape Province. When the disease resurged the next year, the death rate rose to nearly 15 percent.

May 1934 Some 1,300 Californians entered the Los Angeles County General Hospital for treatment of polio. Dr. John Rodman Paul investigated the outbreak and discovered a pervasive dread over the populace: "It was as if a plague had invaded the city, and the place where cases were assembled and cared for was to be shunned as a veritable pest house" (Gould 1995, p. 62). An epidemiological team discovered around 1,200 cases of hysteria, which masked chronic fatigue syndrome as polio. Those encountering the lookalike symptoms included 198 medical workers at the hospital.

1935 United States troops in Panama contracted malaria at such high rates that the high command abandoned field maneuvers. At the recommendation of General James Stevens Simmons, the military tested the effectiveness of atabrine, a synthetic quinine made from coal tar. The German firm I. G. Farben developed the drug as a preventive and evolved preparedness and medical intervention methods for future engagements in the tropics. Simmons's foresight

was prophetic of needs of the Pacific theater during World War II.

1935 For two years in Brisbane, Queensland, Australia, Edward Holbrook Derrick, director of the Queensland State Health Department Laboratory of Microbiology and Pathology, battled epidemic pneumonia and hepatitis in abattoir workers. He named an unknown virus Query fever, a common malady among cows, goats, sheep, and other domestic livestock. When it infected sheep shearers, ranchers, and meatcutters, it sometimes resulted in endocarditis. In 1935, Sir Frank MacFarlane Burnet, author of *Virus as Organism* (1945), confirmed the zoonotic disease, which acquired the name Q fever.

summer 1935 An epizootic of sylvatic plague among rodents of the Union of South Africa generated an epidemic, which afflicted 543 with a death rate of around 65 percent. Most susceptible were Asians, with the Bantu suffering a 59 percent rate and whites, 30 percent. A recrudescence the next year produced 253 infections with a mortality rate of 66 percent.

summer 1935 Under Marxist Communism, the Soviet government prepared for war against the capitalist world by setting up labs to produce toxins for biological warfare. After the People's Health Commissariat ceded their endeavors to the Red Army Biochemical Institute, experimenters on the Mongolian-Siberian border field-tested anthrax, cholera, and plague on political dissidents and Japanese prisoners of war. At the Volga field station and in Ostashkov near Lake Seliger, staff extended the number of usable deadly sera to include dysentery, paratyphoid, tetanus, typhoid, and tularemia. On March 17, 1937, Britain's Sub-Committee on Bacteriological Warfare reported to the Committee of Imperial Defense that the Japanese had enlisted Soviet agents to spread bacteria on troop trains. The report warned that the English were in danger of biological warfare directed at crops, animals, or the populace.

1936 At Hawk's Nest, West Virginia, 764 tunnel workers died of acute silicosis, an occupational disease caused by inhaling mineral fragments and dust arising from drilling and blasting silica rock. Of the 2,500 laborers, 60 percent or 1,500 developed the disease, which results from fibrosis in the lungs. The project, which provided water from the New River for the Gauley Bridge power generator, brought congressional attention to the importance of ventilation, respirators, and wetting down surfaces to suppress dust. The U.S. Bureau of Mines issued alerts to the dangers of multiple jobs involving high particulate in the air, such as farming, marble and ceramic tile, pottery, pigment grinding, quarrying, cutting, oil refining, and sandblasting.

1936 Basing its attack strategy on the germ warfare research conducted by General Shiro Ishii, professor of immunology at Tokyo Army Medical College, in 1932, Japan's Imperial Army Unit 731 began testing biological and chemical weapons 40 miles south of Harbin, Manchuria, at the village of Beiyinhe. Concealing the monstrous lab as a water purifier, a staff of 3,000 erected a compound and methodically executed 9,000 Chinese and Russian prisoners of war. *See also* **June 30, 1938.**

Aizawa Yoshi, on staff at the Harbin High School for Japanese Women, concluded that the typhoid epidemic that struck students in 1940 derived from the biological warfare unit. She blamed Unit 731 by name, but had difficulty convincing the Japanese that military laboratories were the source of contagion. *See also* 1942.

1936 A six-year outbreak of sleeping sickness in the Gold Coast infected 34,651 Ghanians, sickening the Bawku, Gambaga, Nakpanduri, Tamale, and Walewale. The disease, which emerged in the Congo, had moved overland to Cameroon, Dahomey, Nigeria, and Togo, arriving at the Gold Coast by 1930 and reaching a height of virulence along riverbanks. At Gambaga in 1937, health officials waged the "Trypanosomiasis Campaign" by destroying the breeding grounds of tsetse flies, trapping adult insects, and unearthing their larvae.

1936 South Sea Islanders encountered new outbreaks of European diseases, which invaded some virgin-soil populations. Simultaneous with an outbreak of measles in the United Kingdom, in the Ellice and Gilbert Islands, the spread of the disease from Fiji assailed the British colony, felling 53 percent and killing around 500 of the 27,000 islanders. At the atoll of Ontong Java, part of the British Solomons off New South Wales, a resurgence of influenza arrived aboard the trader *Southern Cross*. The infection continued into the next year, proving more lethal in Luanguia than

in Pelau. Over the rest of the decade, local officials applied the same medical restrictions to ports and emigration that protected European nations.

1936 In Spain, the regime of dictator Francisco Franco attempted to end endemic tuberculosis with a rigorous program of prevention and treatment and a form of national anti–TB insurance. *See also* **1945.**

early 1936 At Italian Somaliland, the incursion of uninocculated Ethiopian soldiers during the Italo-Ethiopian War introduced smallpox. A resultant epidemic claimed about 42 percent of the 1,142 victims, mostly children who lacked immunity.

January 4, 1936 An outbreak of influenza at Angmagssalik Island in eastern Greenland cost the nation 26 of its best seal and bear hunters.

February 26, 1936 Epidemic influenza felled citizens of Moscow and Leningrad as western Russia came under viral attack. Doctors and ambulance crews struggled to cope with thousands of new cases reported each day. The disease favored adults living in crowded, poorly heated tenements.

August 28, 1936 At the Wallington public hall, health officials opened discussions of epidemic enteritis in Surrey, England. Suspecting the underground River Bourne, residents ceased using the Cheam well, which may have picked up pathogens from a feeder stream that flowed near two garbage dumps.

November 1936 Recurrent polio in New Zealand hit Dunedin and spread over North Island for nine months, felling both Europeans and Maori and favoring preschoolers. Of 896 infections, nearly 95 percent survived with some paralysis.

1937 More than any other disease, the impact of polio against Canada's middle class resulted in a unique national response—state medical programs. At Ontario, the second of four major upsurges in the disease forced officials to set up disease control, treatment centers, polio hospitals, and post-polio rehabilitation. Because of the national dedication to victims, survivors could look forward to more mobility as they recovered from the virus.

1937 After Robert Skidmore Ecke of Eliot, Maine, ended his medical training at Johns Hopkins and joined the staff of Notre Dame Hospital in Newfoundland, for the next four years, he battled disease brought on by poverty, illiteracy, malnutrition, and limited hygiene. As described in *Showshoe & Lancet: Memoirs of a Frontier Newfoundland Doctor* (2000), chief among endemic ills were beriberi and tuberculosis, the region's major killer.

1937 Botulism killed over half the populace of a village on Baffin Island, Canada. Victims sickened after eating meat from a beached whale that was frozen after its death months before. The disease was endemic in Alaska, Canada, and Greenland because of reliance on natural cold to preserve foods.

1937 The second largest waterborne epidemic of typhoid fever in England occurred in Croydon. Following a failure to purify recently repaired wells with chlorination of water supply, 43 of 341 patients died of the outbreak.

early 1937 In the dry atmosphere and chilly nights of French Equatorial Africa (modern Chad, Congo, Gabon, and Central African Republic), cerebrospinal meningitis killed at a rate of 65 percent. Sleeping in cramped, unventilated mud homes increased the native's inhalation of bacteria. Worst hit the following year were the Sara, particularly residents of Doba, where 1,008 died of the disease. Only 30 percent survived infection.

January 4, 1937 A mild influenza epidemic struck London the last five days of December 1936, emptying offices of staff and depriving factories in the Midlands and mills to the north of workers and the Chelsea and Birmingham football clubs of players. By the first week of January, health workers at Guy's Hospital and at regional nursing homes were infected, leaving their posts unfilled and forcing healthy staff to perform extra duty. Within the first four weeks, there were 3,160 deaths, mostly in London; to the north, Glasgow, Scotland, reported 120. All regions except London recorded low incidence of pneumonia.

May 1937 Cholera swept over China from Kwangtung to Kwangsi and into the Yuan River and Yangtze River basins. The outbreak of the Sino-Japanese War on July 7 spread contagion over the land into Hunan and Manchuria, felling over 50,000 into the next year. The disease slowed the Shanghai Expeditionary Force at Gosho, at the mouth of the Yangtze River. Hardest hit were war refugees, who also suffered dysentery and typhoid fever. At Shanghai, swarms of flies carried contagion from unburied bodies. City officials converted the 400-bed Chungshan Hospital into a cholera pesthouse. Special water filtration devices halted the disease in urban Shanghai by September. For the military, heavy incidence of disease lessened the efficiency of 14 million male draftees. By war's end, 1.4 million were dead of infection and starvation.

November 1937 When measles invaded North Auckland, New Zealand, of the 4,000 cases, 76 died, 60 of whom were Maori. Increasing the virulence of the attack were complications with hemorrhaging, pleurisy and pneumonia, heart and kidney impairment, and tuberculosis. To the east, half the indigenous children came down with measles, forcing the Maori to accept European medicine and concepts of personal cleanliness and community sanitation. The total deaths reached 375, most of them young Maori.

1938 In Australia, a widespread bout of rubella lasting three years struck children and adults. As a result of observing numerous cataracts and 15 infant deaths following the spread of measles, in October 1941, an ophthalmologist and surgeon, Sir Norman McAlister Gregg of Sydney, addressed the Ophthalmological Society of Australia at Melbourne on the connection between rubella and first trimester pregnancies. His summary established that the fetus can acquire a congenital syndrome producing anomalies to hearing, speech, vision, and cardiac function. To spare families handicapped children, the Australian government made available vaccine to prevent maternal rubella infection.

1938 A lethal epidemic of diphtheria afflicted 19,000 residents of South Africa, claiming over five percent. At greatest risk were the children of poor blacks and Asians, the ones least likely to obtain inoculation and medical care. Milk and milk products from diseased cows and hand-to-mouth infection in classrooms spread contagion.

1938 An abundance of *Anopheles gambiae* mosquitoes bred during the rainy season in Brazil produced epidemic malaria among dwellers along the Jaguaribe River. The scourge claimed 26,000 or 9 percent, many of whom were children and babies. Fred Lowe Soper of Hutchinson, Kansas, educated at Johns Hopkins and Regional Director of the International Health Division, launched an extermination project involving oiling and draining the surface of stagnant pools and spreading paris green, the vermin-suppressant of choice before the invention of DDT. In 1942, he applied his knowledge in Cairo to a malaria outbreak monitored by the U.S. Typhus Commission by spraying selected regions with insecticide. For his work and the publication of *The Organization of Permanent Nation-Wide Anti-Aedes Aegypti Measures in Brazil* (1943), he earned the first annual Lasker Award of the American Public Health Association. In Soper's honor, the Pan American Health Organization funds the Fred L. Soper Lecture and issues the Fred L. Soper Award for Excellence in Health Literature.

January 1938 President Franklin Delano Roosevelt, a victim of polio at age 39, conceived the National Foundation for Infantile Paralysis (NFIP), a forerunner to the March of Dimes. To fight the virus six years before he became a victim of paralytic polio, he founded the Warm Springs Foundation. Singer-comedian Eddie Cantor originated the term March of Dimes and suggested that donors send coins directly to the White House. Comic actor Danny Kaye and singer Bing Crosby popularized the fund drive during its early campaigns. Yale University was the first research facility to use the funds in a drive to prevent outbreaks.

June 30, 1938 Major-General Rensuke Isogai established the Pingfan Special Military Zone at Harbin, Manchuria. His staff of 3,000 operated an airfield, railroad siding, incinerator, powerhouse, prison, insectarium, disease research center, and animal house. The Japanese Secret Service stocked research labs with Russian citizens, servicemen, escapees, political agitators, intellectuals, and spies. Used as guinea pigs, 3,000 victims called *marutus* (logs of wood) died during the study of epidemics, infectious dosages, biological warfare, and immunization. Among the pathogens unleashed on the unsuspecting were anthrax, bubonic plague, cholera, dysentery, paratyphoid, syphilis, tuberculosis, and typhus.

The experiments attested to a monstrous disregard of humanity. Some victims were forced to stand in water bearing typhoid germs or to eat melons injected with the bacteria. Pregnant women and infants were infected with syphilis and later dissected for study. No inmate survived the experiments; all were consigned to the incinerator after death.

1939 The amassing of Southern Chinese refugees into Hong Kong during the two-year Sino-Japanese War taxed the colony with simultaneous outbreaks of infectious diseases. According to the China National Relief Commission, a total of 150 million people left their homes. Officials slowed the immigration by requiring border passes, entry and residence certification, and cash.

Within the city, where homeless families sheltered in handcarts, police rounded up vagrants and the unemployed. Without hope, the poor and homeless died of malnutrition, dysentery, leprosy, smallpox, and typhoid. Tuberculosis alone killed 4,000. At Hong Kong and Shanghai, cholera was widespread; malaria rocked Central and Southern China at Hunan, Kiangsu, and Kiangsi. Over all areas where armies marched, syphilis was epidemic.

1939 During a typhus pandemic in Spain at the end of the Spanish Civil War, the emerging dictator, Francisco Franco, manipulated health policies to validate his political policies and consolidate power. On February 10, British journalist Nancy Cunard wrote about the epidemic in the *Manchester Guardian*. After observing thousands of Spanish refugees at the central camp at Le Boulou, France, she reported on sores caused by typhus, which was rampant in fascist Spain.

1939 Prefatory to the atrocities committed during World War II, German tropical medicine researchers and the drug industry evolved inhumane experimentation and quarantine practices that guided professional ethics through 1945. To regain global prestige and to develop footholds in Africa, German lab technologists gradually relaxed humanitarian standards through involuntary experimentation on humans during malaria research. After seizing Western Poland, Hitler falsely claimed that Poles of the Warta region were suffering a tuberculosis epidemic that threatened the health of Germany. To contain the bogus contagion, the Reich set up isolation wards that were the forerunners of death camps.

Subsequent studies of health data proved that the incidence of the disease among Poles was lower than that for Germans.

In this same period, at Lódz, Poland, ghetto dwellers fought starvation, dysentery, tuberculosis, and typhus for six years, incurring a death toll of 43,000 or 37 percent of residents. In Warsaw, government staff treated a typhus outbreak by ghettoizing Jews. An internal program of secret training in sanitation and epidemiology at two Jewish hospitals under Juliusz Zweibaum and Ludwik Hirszfeld produced 500 ghetto experts. When the disease recurred and beset the populace into 1942, the Jewish Health Department and the Warsaw Health Council took charge of containing the epidemic.

1939 In Singapore, English pediatrician Cicely Williams presented the speech "Milk and Murder," an eye-opening castigation of the infant formula market begun by Henri Nestlé and industrialist Jules Monnerat, inventor of canned condensed milk. She charged that a multi-million dollar business in packaged foods freed women from breast-feeding and encouraged them to enter the job market at the expense of their children. Malnourished infants and toddlers bore the brunt of product misrepresentation in kwashiorkor, skeletal malformation, and other nutritional deficiencies. From a concerted humanitarian campaign against unconscionable consumer fraud, a global boycott of Nestlé products ensued.

January 1939 In West Africa, people of the French colony of Upper Volta suffered a catastrophic four-month epidemic of cerebrospinal meningitis, which killed nearly 46 percent of 6,800 cases. Imported by a visitor from Niger, contagion advanced from Kiembara and invaded Fada N'Gourma, Ouagaduogou, Tenkodogo, and Tougan. By spring, new outbreaks occurred in the French Sudan and Gold Coast (modern Ghana).

1940 Eleanor Josephine Macdonald, an epidemiologist for the Connecticut State Health Department, initiated the first population-based cancer record registry and follow-up program for survivors. Until 1948, she compiled hospital records of 1,800 cancer survivors and of the most effective treatments to prevent a recurrence of the disease. *See also* **1944.**

1940 The first major epidemic of polio in an army hit New Zealand's army in the eastern Mediterranean. Milder than the form that strikes and paralyzes children, the virus infected 40 adult men and killed only four. That same year, an Australian nurse, Sister Elizabeth Kenny, demonstrated hot wet-pack treatment for the spasms of paralytic polio at the Minneapolis General Hospital. The supercilious male medical establishment denigrated her lack of formal training, ridiculed her creativity, and refused to implement the revolutionary compress method of treating paralytic polio.

Kenny left the rehabilitation clinic for polio victims in Australia because her unorthodox regimen to relieve spasms contradicted accepted medical treatment calling for immobilization of paralyzed limbs. Against a deluge of negative articles in learned journals, the results spoke for themselves. In 1942, acceptance of the method in the United States preceded the establishment of the Sister Kenny Institute of Minneapolis, a training center for care of polio victims.

1940 In Japan, the Imperial Diet enacted the National Eugenics Law, the nation's first sterilization legislation, which framers intended to rid the country of endemic social and medical ills. Provisions of the first stage of the plan called for the sterilization of alcoholics, the insane, cripples, and criminals along with victims of inherited disease, leprosy, syphilis, and tuberculosis.

1940 A six-year epidemic of louse-borne typhus assailed Egypt, striking 110,000 with a fatality rate of over 18 percent. Spread by the clothing and bedding of poor migrant workers of the Nile Delta region, the disease was most virulent in urban areas around Alexandria and Cairo.

1940 At Papua New Guinea, American forces fought endemic malaria, which contributed to extensive illness at Guadalcanal in the Solomon Islands and precipitated the Japanese victory over the allies at Bataan, Philippines. The disease infected half of American troops and 74 percent of Australian soldiers in the South Pacific. At Bataan, the army lost 20,000 American and Philippine fighting men to the fever. To contain the disease, health squads fought mosquitoes with DDT, insect repellant, long-sleeved clothing, and mosquito netting. The medical corps distributed atabrine and chloroquine, two synthetic drugs that replaced quinine after the Japanese cut off supplies of cinchona from Dutch reserves in the Javanese uplands.

1940 During the second Franciscan mission to Portuguese Guinea (modern Brazil), under the aegis of the Vatican and the government of Portugal, agents erected a seminary and organized a cadre of hospital nuns. Among the good works of the outreach were sanitary measures, orphanages, and leprosy clinics.

September 1940 Some 14,378 people visited thrice-weekly clinics in Halifax, Nova Scotia, to avoid the virulent strain of diphtheria that felled patients. Swelling and membranous coverings of pharynx, glands, uvula, and soft palate caused severe bleeding from the mouth and nose in adults, who surpassed children in number of infections. Four patients died before tracheotomies could relieve the obstructions to their airways; 22 children succumbed to cardiac complications. Doctors saved some hospitalized patients with tonsillectomies and with a painting of tincture of metaphen and glycerine.

October 1940 To the northeast in Anglo-Egyptian Sudan's Nuba Mountains, the native population suffered its last great epidemic of yellow fever, which began in the spring rainy season and reached full strength in autumn, killing 15,000. Victims required attentive nurse care to recover. To protect British, Indian, and Sudanese forces posted on the eastern border to forestall an Italian invasion of Abyssinia, officials inspected residences and quarantined the mountain area to prevent traffic from spreading the disease. By the time that residents were vaccinated in the early winter, the contagion was already waning with the annual depletion of the mosquito population.

October 27, 1940 The Japanese war ministry, which had long plotted bioterrorism, dispatched planes over Ningpo, China. The crew scattered infected grain and possibly fleas. Two days later, local people collapsed with bubonic plague, which killed hundreds.

On November 12, 1940, a Japanese plane released infected fleas and grains of rice and wheat over the Chinese village of Chuhsien south of Shanghai. On October 4, 1940, bubonic plague struck the populace, killing 21. It was the first plague outbreak for the area. Another fly-over on November 28, 1940, released tiny granules carrying plague bacilli over Kinhwa, but the attempt at deliberate infection failed.

The biological terrorism continued in November 4, 1941, at Changteh in Hunan, China. Kiyoshi Ota, chief of Japan's biological warfare division, sent one plane to spray plague-carrying fleas over the 50,000 citizens of Changteh in Hunan province. Within a week, people began contracting the disease. Numbered among the dead were 10,000 Chinese and 1,700 Japanese who were unintentionally infected.

December 1940 A four-year epidemic of a lethal form of sleeping sickness killed Ugandans in Busoga at the rate of 10 percent of 2,500 cases. The outbreak arrived among laborers from Burundi and began with the death of a child from the bite of a tsetse fly. Health officials failed to contain infection, which traveled to lepers in Buluba and on to Ikulwe, Kityerera, and Kyemeire. In January 1942, the scourge reached Kenya. The use of sequestering at a camp in Bugiri and the natural suppression of fly propagation during heavy rains ended the outbreak in 1943.

1940s Kuru, the first documented prion disease and a form of bovine spongiform encephalopathy, began attacking the Fore, a tribe living at Okapa in Papua New Guinea's eastern highlands. Natives contracted the fatal brain sickness from cannibalistic rites involving the dismembering, oven-baking, and eating of corpses as a form of honor and respect to dead relatives and as a means of preserving within the family the deceased person's powers. The widow received the honor of preparing the meal and distributed portions to her children and relatives. Women and children were most susceptible to kuru because they ate the brain and other organ meats; men ate only muscle tissue.

The consumption of the brain's infectious proteins set in motion up to 30 years of incubation of an anomaly the Fore called "laughing disease," which slowly reduced brain tissue to sponge. The Fore people attributed to sorcery the horrible brain-wasting malady, which is similar to mad cow disease. After they realized the true source, they abandoned the practice of devouring their beloved dead in the 1950s. In 1959, a law officially banned funerary cannibalism.

Still active in the bodies of older tribe members, the disease ran a three to twelve-month course, attacking the central nervous system and causing headache, joint pain, a tottering gait, tremor, a masklike smile, uncontrollable laughter, and dementia. By the time that microbiologist Daniel Carleton Gajdusek, author of *Melanesian Journal* (1968), chief of the Laboratory for Central Nervous System Studies at the National Institute of Neurological Disorders and Stroke in Bethesda, Maryland, and winner of the 1976 Nobel Prize for medicine, investigated New Guineans with wasting brain disease, kuru had reached epidemic proportions. As of April 1997, 2,500 had died horribly. In a model of biological adaptation, survivors developed genetic resistance to kuru. *See also* **February 6, 2001**.

1941 Omsk hemorrhagic fever received its first notice in western Siberia around Kurgan, Novosibirsk, Omsk, and Tyumen. The encephalitic virus apparently passed to some 1,500 humans from ticks on muskrats and small mammals during late spring and summer. Unpasteurized goat milk also presented possible sources of transmission. Symptoms—rash, low white cell count, and bleeding from nose, mouth, skin, and uterus, increased over the next five years as epidemiologists struggled to differentiate the outbreak from tularemia and typhus. Proof of the link to muskrats appeared in high incidence of the virus among muskrat hunters and preparers of muskrat skins. Some 97 percent of victims survived the virus.

1941 When World War II increased the incidence of syphilis, Great Britain experienced a 70 percent rise in infections. With the passage of Regulation 33 B, the national health policy shifted from voluntary to obligatory treatment. Officials, still unsatisfied with their anti–VD program, pondered Scandinavia's more stringent measures against a deadly scourge. By June 30, 1943, when the number of new cases reached double the pre-war figures, Ernest Brown, the Minister of Health, blamed promiscuity and the wartime strain on the family. In Scotland, the rise of casual sex between local women and soldiers produced a worrisome boost to syphilis cases. Against the advice of the Association of Moral and Social Hygiene, Glasgow's municipal officials acceded to public demand for compulsory reporting and treatment, but failed to implement its get-tough policy.

1941 In the Jewish ghetto of Lodz, Poland, tuberculosis produced pasty white skin and listlessness. Among Jews in Vilna, Lithuania, doctors and health workers initiated a program to stem initial outbreaks of contagion in the cramped community. Working from the Jewish Hospital, they set up a clinic and a sanitary-epidemiological control unit to superintend a highly successful standard of cleanliness. The policy was effective against influenza, scabies, tuberculosis, and typhus.

1941 The East African campaign during World War II produced horrendous loss as the Allies cleared Italian fascists from the Horn of Africa. The British reported 10,000 downed by dysentery and an equal number stricken by malaria. The death toll from illness was 744. In Kenya, epidemic plague, emerging from rat-infested grain depots, struck the British colony in January.

March 1941 After a rainy season bolstered the ranks of *Anopheles* mosquitoes, the onrush of malaria from Morocco to Constantine, Algeria, engulfed the city for eight months, infecting 198,000 and killing 7,725. By December, the disease had gained a hold on the Usambara Mountains of northeastern Tanzania, where migrant laborers introduced the pathogen. The outbreak produced serious complications in the circulatory system, kidneys, and liver.

April 4, 1941 Because of the privations of the civil war, which cost the Spanish adequate nutrition and soap, a typhus epidemic emerged in Madrid. City officials rounded up vagrants for inspection and delousing and "[ejected] dirty looking passengers" from city buses, trams, and subways ("Typhus Epidemic," 1941, p. 3). Up to 30 cases a day entered reports. Of the first 300 cases, 27 victims died, including two physicians. The disease spread to Almeria, Granada, Murcia, and Seville. Herald R. Cox of Hamilton, Montana, author of *Ultrafiltration of the Virus of Vesicular Stomatitis* (1934), formulated a vaccine that the Rockefeller Foundation dispatched to Spain aboard the *Atlantic Clipper*. Madrid lab technicians began making batches according to Cox's instructions.

fall 1941 The interrelated horrors of World War II echoed in the spread of disease. After the emptying of the ghetto at Bialystok, Poland, Nazis moved inmates by freight trains to work stations at a labor camp at Blizyn. Inmates repairing lice-infested German uniforms contracted typhus. The rounding up of Gypsies or Rom from Austria resulted in resettlement of 2,500 detainees in Lodz, Poland. In November all were either gassed in mobile extermination vans or died of epidemic typhus.

In Russia, the Polish army, which relied on inmates from labor camps for recruits, lived in unhygienic conditions rife with pathogens. German public health officers connected the rise in disease to Jewish ritual, a source of cultural contention that led to anti–Semitism. Weakened by malnutrition and contagion, Polish soldiers transported the disease to Kazakhstan. Into June 1942, the epidemic killed 1,290 or 18.4 percent of over 7,000 cases. When the men were reposted in Iran, they faced a new threat from epidemic malaria.

winter 1941 A two-year epidemic of meningitis in Santiago, Chile, arrived from Valparaiso. The scourge afflicted 5,885 people and killed 928, primarily children, young teens, and soldiers. Cold, crowded housing and hunger worsened conditions, as did the scarcity of sulfa drugs.

December 1941 The German high command excused as unavoidable a typhus epidemic among two million POWs at the Molodecno camp near Minsk, Belarus, on the Polish border. The truth of the situation was clearly one of neglect of sanitation and delousing and an absence of soap plus the debility of health caused by slow starvation. In secret, the German Commissar ordered guards to shoot the sick immediately to halt contagion, but the men apparently ignored the order. Testifying to the heinous situation were depleted inmates who survived a transfer to other locations, where they collapsed and died after their first meal.

December 8, 1941 After the bombing of Pearl Harbor, Hawaii, the U.S. military entered the war the next day. The surgeon-general's office was already prepared for the possibility of epidemics and of biological warfare involving dissemination of contagious material on military and civilian populations. To circumvent disease, by May 15, 1942, the government set up a comprehensive medical department. Under the executive officer, a preventive medicine division comprised five divisions: epidemiology, venereal disease control, sanitation and hygiene, sanitary engineering, and medical intelligence and tropical medicine.

Facilities were soon swamped with potentially disastrous pathogens. Military nurse Laura Cobb and volunteer Filipino and British nurses at the Japanese internment camp at Los Baños, Luzon, the main island of the Philippines, used whatever supplies and medicines they could scrounge to treat beriberi, dysentery, fever, kidney infection, scurvy, and tuberculosis. Along the Burma Road, refugees streaming in from China traveled through the Manipur Valley in India to Tamu, Burma. They had no food and bore evidence of dysentery, malaria, and smallpox. To the southeast, despite the watchfulness of the wartime Australian New Guinea Administrative Unit, dysentery found its way to the New Guinea highland. Off the southern coast of India, the Ceylonese, who suffered endemic dysentery, experienced a serious upsurge that afflicted over 59,000 islanders, killing 2,275.

At the barracks and jail of Changi Prison on the northeastern tip of Singapore, 30,000 Australian, British, Eurasian, and New Zealand inmates—men, women, and children—began experiencing serious beriberi by February 1942. Contributing to Allied prisoner ill health were cholera, diphtheria, dysentery, malaria, pellagra, and tropical skin ulcers. An Australian nurse-social worker, Mary Seah, the "angel of Changi Prison" and winner of the Order of Australia, heroically risked her safety by disguising herself as a food vendor to gain entry to prisoners. Ignoring threats of beheading and torture by guards who forced her to stand in the sun and rain, she smuggled in food, medicine, and herbs as well as supplies and radio parts.

1942 General Shiro Ishii, Japan's genius of bioterrorism, carried germ warfare outside the test confines at Harbin, Manchuria, to Chinese citizens, killing tens of thousand with aerosolized anthrax, bubonic plague, cholera, dysentery, and typhoid. At Ningpo and Changde, China, planes blanketed the area with plague-carrying fleas. Ground troops contaminated lakes and wells with cholera and typhoid cultures. While they exposed Zhejiang Province to cholera and typhoid, the operation backfired and killed 1,700 Japanese soldiers. *See* **fall 1945.**

1942 Tropical disease complicated the work of the military during World War II. In the South Pacific, soldiers suffered thousands of cases of elephantiasis and schistosomiasis, which infected marines wading up the beach at Leyte Gulf in the Philippines. Malaria, the most serious health

hazard to Allied troops, felled five times more soldiers than combat injuries for a total of 100,000
cases. At Bataan, Philippines, American soldiers
contracted malaria at the rate of 85 percent. At
Efate, New Hebrides, two-thirds of advance
troops incurred malaria. The outbreak was so
severe on Guadalcanal, Solomon Islands, that
100,000 were sickened by the parasite. Tertiary
fever threatened military effectiveness and, ultimately, the success of the military campaign. *See
also* **fall 1944.**

Lesser incidents struck soldiers on Espíritu
Santo in New Hebrides, Tulagi-Florida and the
Russell Islands in the Solomons, and Munda,
New Georgia. Australians fleeing New Britain fell
victim to tertian fever after they ran through their
supplies of quinine. Those soldiers who took
treatment and suffered relapses developed immunity that ended their infection, but the loss of
manpower seriously compromised troop strength.

When American occupation forces landed at
Efate in the New Hebrides in March 1942 to
build an airfield at Vila, they encountered *Anopheles* mosquitoes at their bivouac outside a native
labor camp. The lack of bed nets and required
night work exposed them to endemic malaria,
which they battled with quinine. When the disease proved resistant to the drug, medical personnel distributed atabrine weekly. On July 28,
1942, mosquito-control teams went to work suppressing insects, spraying huts with pyrethrum,
and screening quarters to reduce the number of
infections. *See also* **May 1943.**

Among the Japanese and their prisoners of
war, the malaria epidemic was worse. Fever along
with beriberi, dysentery, and malnutrition killed
9,000 Japanese of the 18,000 men stationed on
Guadalcanal. In addition to these outbreaks,
insects spread roundworms causing filariasis and
tsutsugamushi disease, which is borne by mites.
The most serious affliction to the war effort was
dengue fever, a virus that was more crippling than
malaria. Introduced to port cities from ships sailing from Malaya, Shanghai, and Singapore, the
disease spread from mosquito bites. It produced
two million cases by war's end.

South America also produced its share of malaria. During the dry season at Amapá, Brazil,
endemic malaria developed into an epidemic. In
addition to the 300 natives stricken with chills
and fever, seven American service personnel collapsed. The U.S. Army medical team lacked qui-

nine tablets and atabrine and had to inject the
soldiers with liquid quinine or administer it intra-
venously. Prevention teams evacuated sick laborers, offered blood screening to natives, and
studied means of controlling the mosquito population.

In June 1943, an epidemic of malaria assailed
villagers in Belém, Recife, and Natal, Brazil.
Along the Pirangi Valley, U.S. forces stationed at
Parnamirim Field remained healthy, but those
leaving the base to service the warning lights near
Rio Cajuiranga and the radio range five miles
from the base incurred disease. Those visiting
Natal against orders also suffered bites of the
Anopheles mosquito.

In Ecuador, where epidemics of plague, smallpox, and typhus were common events, as of May
1943, 7.8 percent of 62,183 illnesses were caused
by malaria, the fourth most invasive disease.
Investigation by U.S. Army personnel found that
drainage, ditching, and oiling aided in mosquito
control in the valleys, but the coastal swamps
defeated local efforts to drain stagnant water.

1942 Among allied troops on Iceland, prompt
inoculation prevented an epidemic of mumps
that struck the civilian population. Only 128 soldiers caught the disease. More serious to the war
effort were emotional disorders resulting from
isolation, boredom, inactivity, adjustment to the
climate, and a lack of participation in combat.

1942 Typhus seriously impacted the health of
combatants and civilians during World War II,
particularly among 18,000 Americans and Australian troops in Burma, New Guinea, the Philippines, and Rarotonga, the main island in the
Cook cluster. To prevent infection, soldiers cleared
brush around billets. Patrols and combat soldiers
relied on short-term disinfectant spray for bedding and clothing.

Hard hit by numerous diseases, detainees languished in filthy German internment camps. Vice
Admiral Sheldon F. Dudley, Medical Director
General of the British Navy and author of *Human
Adaptation to the Parasitic Environment* (1929),
commented that "[Though] Montgomery said
the Eighth Army won, Rommel claimed the victory for dysentery" (Moser 1965, p. 80). In January, Professor Lemierre and Major Sohier
conducted a mission to Germany to immunize
French prisoners of war. As a protective measure
for men living in crowded barracks without ade-

quate sanitation and bathing facilities, the two Frenchmen inoculated inmates against typhus.

In Algeria, an epidemic of typhus, which killed over half of the 52,000 fever victims, held steady for two years, overlapping an outbreak of relapsing fever, which felled 2,000 of 31,847 victims. Typhus proved more fatal among whites, who had less immunity than Arabs and Berbers. The disease spread across Morocco, Tunisia, and Egypt. Moroccans at Casablanca, Fez, Marrakesh, Meknes, and Rabat suffered a similar outbreak, which struck 40,200 and killed 20 percent. A recrudescence the following year infected 8,200. The U.S. Army protected troops through vaccination and distribution of DDT powder to kill pathogen-bearing lice.

In Persia, the eruption of pandemic typhus arrived with starving Polish refugees and military from Russian camps. Because medical officials quickly stemmed the disease through a successful delousing campaign, doctors reported only 1,102 cases. However, a new outbreak the next year produced over 25,000 victims. Allied troops incurred 170 illnesses with a death rate over 21 percent.

1942 Among British occupational forces in India, young officers were most likely to suffer a pandemic of polio, which struck the natives. Inspectors blamed the infection on a greater number of contacts with diseased civilians and the lack of sanitation in army billets.

January 26, 1942 At Athens and Piraeus, Greece, International Red Cross workers noted that 40,000 died of a combination of starvation, childhood diseases, and endemic infection. The main killers were diphtheria and pertussis among children and malaria among adults. Officials at Chios west of Turkey and at Syra in the Cyclades telegraphed the Greek mainland, "Send wheat or coffins" ("Famine" 1942, p. 4).

June 1942 Epidemic plague in French West Africa (modern Mauritania, Senegal, Sudan, Guinea, Ivory Coast, Upper Volta, Dahomey, and Niger) began at Dakar, Senegal, and killed 907 within the first year. Crowded, unhygienic housing contributed to the burgeoning of rodent populations that spread the disease.

July 10, 1942 At Auschwitz, a concentration camp in Poland, German SS medical service began experimentation on methods of biological warfare. Under the tenets of the master race theory, administrators considered expendable any inmates who lacked the typical traits of Aryans. Staff arranged and observed inhumane exposure to gangrene bacilli and injections of anthrax and typhus on 100 undesirables—Jews, Gypsies, Freemasons, Jehovah's Witnesses, homosexuals, Seventh-Day Adventists, and political prisoners. *See also* **late 1944.**

fall 1942 In Africa's most virulent cerebrospinal meningitis epidemic in a quarter century, around 7,000 of 11,700 affected Tanganyikans died, most in the northwest. The disease returned the next year, infecting 8,800, but producing a lower death rate.

late 1942 The U.S. Army medical team in Egypt and Libya battled infectious hepatitis, smallpox, and bubonic plague, which assailed the people of Dakar and those living along the Suez Canal. Immunization and control measures shielded American troops from contagion. Hepatitis at El Alamein, Egypt, weakened New Zealanders of the Eighth Army before striking Australian and British units. Along with malaria, hepatitis returned the next year in the Middle East and Italy.

November 1942 An upsurge in paralytic polio at Malta and the nearby island of Gozo afflicted 483, mostly small children. Also suffering the outbreak were British soldiers. The constant state of siege worsened contagion by forcing islanders to crowd into bomb shelters.

1943 At the University of Michigan, epidemiologists Jonas Salk and Thomas Francis developed a formalin-killed-virus vaccine to control type A and B influenza viruses.

1943 The first emergence of Q fever outside Australia occurred in 1943, when it struck German soldiers in the Crimea, Ukraine, causing concern that Russian biochemists might be incubating the disease as a form of bioterrorism. Identified in 1935 by Victoria-based medical researcher Edward Holbrook Derrick, director of the Laboratory of Microbiology and Pathology in the Queensland Health Department, the tick-borne disease of sheep, cows, and goats had previously afflicted only Australian cattle ranchers, dairy farmers, sheep shearers, and slaughterers. In the eastern Mediterranean, Q fever struck

Allied and Axis forces during the last months of World War II.

1943 At Fort Bragg, North Carolina, 40 trainees came down with an unidentified fever that resembled hepatitis. Nine years after the staff froze serum samples of "Fort Bragg Fever," medical investigators identified the outbreak as leptospirosis. The men apparently swallowed spirochetes while swimming in a pond polluted with the urine of cows, pigs, and rats.

1943 In Ceylon, scrub typhus or tsutsugamushi disease caused by *Rickettsia tsutsugamushi* afflicted British forces, who carried infected mites with them on an advance to the border between Burma and India. The disease advanced on Chinese and U.S. soldiers at Assam, India, and into northern Burma. Of 1,100 cases, nearly 9 percent were fatal.

early 1943 As allied military forces proceeded down the Nile from Asyut, they encountered the Egyptian *Gambiae* mosquito, which spread epidemic malaria. Egyptian government forces and the U.S. military set up a cooperative eradication project that continued until August 1945. In addition to applying paris green larvicide and pyrethrum adulticide, distribution of atabrine to soldiers suppressed the epidemic by February 1945.

April 1943 In the Persian Gulf; Cairo, Egypt; Bone and Oran, Algeria; Sicily; Salerno, Italy; and Tunis and Bizerta, Tunisia the U.S. Army Epidemiological Board and hygiene specialists of the British First Army investigated multiple strains of viral pappataci (sandfly) fever, poliomyelitis, and infectious hepatitis, all prevalent in Mediterranean rim states. At the same time, the German Army reported 4,941 hospitalizations for pappataci, which produced swollen spleen, eye and head pain, fever, body ache and chills, and sensitivity to light. Findings ruled out diagnoses of dengue fever and malaria and added to global knowledge of infectious diseases endemic in Algeria, Corsica, Italy, Sardinia, Sicily, and Tunisia.

April 18, 1943 Among 2,070 weak and demoralized British and Dutch prisoners of war on the island of Harukoe in the Moluccas during the Japanese occupation of the Dutch East Indies, epidemic disease was unavoidable. Without adequate food and medicine, the men endured drinking water from a polluted river, sleep in unhygienic shelters, and beatings with bamboo sticks. The unhealthful atmosphere produced dysentery complicated by beriberi from inadequate diet.

To relieve the survivors of mud and muck, healthy prisoners began building huts in May and isolated those 550 suffering dysentery at the rate of 100 to 125 per hut. By May 12, the deaths became everyday occurrences as the number of sick reached 1,300 and deaths climbed to 415 or nearly one-third. To counter beriberi, the doctors boiled and soured rice to make yeast and made tempi from soybeans. Additional problems from anorexia, malaria, and skin ulcers required adaptation of their limited supplies. After the Japanese moved the surviving prisoners to Java on August 15, 1944, every prisoner suffered from beriberi and other nutritional deficiencies.

May 1943 Of the 500,000 troops that the U.S. army sidelined with malaria, only 301 died. Although the fatality rate was small, disease weakened survivors and lowered morale. General Douglas MacArthur, commander of the Southwest Pacific Theatre, commented, "This will be a long war if for every division I have facing the enemy, I must count on a second in the hospital with malaria and a third division convalescing from this debilitating disease" (Moser 1965, p. 80).

To supply the military with adequate antimalarial drugs, in 1943, the American Pharmaceutical Association pooled all available quinine by collecting full and open bottles of powder, tablet, and capsule forms of cinchonine, cinchonidine, quinine, and quininidine. By October, the effort had stockpiled 13 million doses. Aiding the military response to increased infection, graduates of a program instituted in the American South by sanitarian Colonel Joseph Augustin LePrince taught locals to drain swamps, dig landfills, and screen homes. With additional prophylaxis and prevention, American troops incurred only 62 per thousand cases, as compared with 251 per thousand before distribution of atabrine. For Australian troops, the drop was more dramatic, from 740 per thousand to only 26 per thousand.

May 24, 1943 In Germany, the death toll increased among civilians as health deteriorated and the birth rate plummeted. Fatalities for adults rose from suicide and rampant tuberculosis;

among children, death frequently resulted from outbreaks of diphtheria and scarlet fever.

summer 1943 Tropical sprue or celiac disease, an unidentified digestive ailment inhibited human digestion. In addition to causing flaccid muscles, diarrhea, hyperpigmentation, and lesions of the mucous membranes and tongue, sprue disrupted the absorption of nutrients and salts and afflicted 3,000 members of Allied forces in Burma and India. One-third of the victims were too anemic to perform their jobs. Additional reports of sprue emerged in Vietnam, Thailand, Malaya, Hong Kong, Singapore, Borneo, Indonesia and the Philippines.

July 1943 An epidemic of louse-borne typhus reached Naples, Italy, as Italian forces returned from the Balkans. According to U.S. Brigadier-General Leon A. Fox, the disease derived from German sources. The epidemic felled 2,009 locals, killing 21 percent. In foul, cramped prisoner-of-war compounds, the disease endangered captives from Tunisia and Yugoslavia. To protect the Fifth Army, the military re-inoculated the men and declared the city off-limits. By December, the disease raged among civilians in Naples, particularly young slum dwellers weakened by malnutrition. Sanitation workers dusted civilians with anti-lice powder and pumped insecticide against clothes and into air raid shelters, where people lived in close contact.

By mid–January, the infection rate was fifty times worse than in the previous month, in part because of the influx of refugees in northern Italy. A flying squadron of medical officers distributed vaccines flown in from the Middle East and North Africa and set up two delousing stations that treated up to 72,000 per day for a total of 1,300,000 civilians. Because of the outstanding performance of DDT, manufactured by J. R. Geigy of Basle, Switzerland, British troops began wearing standard issue DDT-impregnated shirts, which warded off lice.

The successful program resulted in the first typhus epidemic in history contained by epidemiologists. American doctors offered to assist the Russians in quelling their own problem with typhus, but a proud military command rebuffed the offer. The gesture of aid from the U.S. also raised outrage among the British.

November 20, 1943 British relief agencies struggled to meet the needs of Bengal, which suffered serious famine and rampant malaria along with lesser upsurges of cholera, dysentery, enteritis, and syphilis. The absence of rice harvesters threatened to intensify malnutrition. At Dacca, one-tenth of the population was destitute. To feed families, 13,000 sold their property; daily, women and children filled soup kitchens as gloom engulfed villages. Relief workers dispatched trucks to retrieve the sick from streets, achieving a 10 percent rescue rate for some 100,000 homeless people.

December 13, 1943 During extremely cold winter weather, epidemic influenza in the eastern United States struck 25,000 in Louisville, Kentucky, 30,000 in Toledo, Ohio, 100,000 in Washington, D.C., and twice that number in Philadelphia for a total of 355,000. Although few deaths resulted from the outbreak, factories, offices, schools, and shopping centers lost human traffic as people took to their beds.

1944 At the University of Texas, Eleanor Josephine Macdonald plotted data for an El Paso cancer registry and follow-up program. She surveyed a high incidence of the disease over a period of 23 years through information she collected from hospitals, clinics, laboratories, nursing homes, and physicians' offices. Her compilation was the first to target Hispanics and the first to connect exposure to the sun's rays to endemic melanoma and other skin cancers.

1944 During World War II, Ruth Council, an orthopedic specialist, set up a polio treatment center in piedmont North Carolina called the Miracle of Hickory. Superintendent Ethel M. Greathouse of Louisville, Kentucky, and other on-loan nurses converged from Florida, Illinois, Louisiana, Mississippi, Pennsylvania, South Carolina, Virginia, and Wisconsin learned how to monitor swallowing and breathing of victims in iron lungs. When electrical power failed, nurses cranked generators by hand. In 1998, Alice E. Sink summarized the work of the polio camp in *The Grit Behind the Miracle: A True Story of the Determination and Hard Work behind an Emergency Infantile Paralysis Hospital, 1944–1945, Hickory, North Carolina.* The text emphasized that, in nine months of operation, the camp lost only twelve or 2.6 percent of its 454 patients.

In Ogden, Utah, a cadre of Polio Emergency Volunteers aided hospital workers by changing linens, feeding patients, distributing comic books and toys, and reading aloud. Aides learned Sister Elizabeth Kenny's method of applying hot packs and massaging shriveled arms and legs, a concept she evolved in Australia. Because of sharp division of labor between male doctors and female nurses, she had to leave Australia and demonstrate her regimen in Minnesota to receive an unbiased hearing.

1944 After a lull in malarial outbreaks in Italy, the turmoil of World War II caused serious rises in numbers on the northwestern and southwestern coasts. Shelling, bombing, trenching, and flooding left craters of stagnant water where mosquitoes multiplied. A vulnerable civilian population experienced 20 times the usual incidence of infection. Fatalities increased in Cassino, Fondi, Gaeta, Terracina, and Veneto, the areas most disturbed by combat and troop movements.

1944 The spread of relapsing fever from Libya east to Egypt coincided with major outbreaks of malaria and typhus. For two years, the louse-borne spirochete sickened 128,541 with fever and claimed around 3,300. American and English epidemiologists provided insecticide to halt infection from insects; local volunteers deloused patients and treated them with arsenic-based Salvarsan.

spring 1944 For Operation Vegetarian, a late–World War II bioterrorism plot, British pathologist Paul Fildes, a naval surgeon during World War I and director of biology at Porton Down, England, undertook a project to bombard Germany with anthrax. Under the project name N bomb, the British involved the U.S. military, which planned to produce anthrax at a microbiology plant in Terre Haute, Indiana. Canadian allies agreed to load the bacteria into cluster bombs.

Fildes manufactured five million linseed cakes permeated with anthrax bacilli to drop from RAF bombers on Hanover and Oldenburg, Germany. Strategists sought to kill beef and dairy animals and infect civilians. After securing the grain from the Olympia Oil and Cake Company in Blackburn and the pathogens from Ministry of Agriculture and Fisheries in Surrey, in 1942, the military passed the cakes to J & E Atkinson of Bond Street in London, perfumers and toilet-soap manufacturers, for cutting to the most effective size and shape. Before a planned drop in summer, when lush pasture grass was scarce, scientists tested the cakes on Gruinard Island, Scotland. Because the launching of D–Day on June 6, 1944, ended the need for biological warfare, the military incinerated the cakes.

The British military faced a sticky situation in keeping secret their plot while explaining the need to control Gruinard. To the amazement of scientists, the anthrax thrived. In 1986, the government launched a mass cleanup of the island. Of a range of chemical agents, formaldehyde worked best. After pelting the island with 280 tons of formaldehyde diluted in seawater, researchers concluded that the island no longer harbored deadly anthrax.

July 1944 In 1944, Kentucky health figures listed 800 new cases of polio. In Louisville, the sick overran General Hospital and Kosair Crippled Children's Hospital, both of which espoused Sister Elizabeth Kenny's hot wet-pack treatment of atrophied muscles followed by massage and exercise. Because patients were isolated from parents for the labor-intensive treatment, volunteers augmented the staffing by the National Foundation of Infantile Paralysis and the American Red Cross to soothe and comfort sick children.

August 1944 The emergence of plague at Bizerte on the Tunisian shore resulted in a seven-month siege. The epidemic threatened the landing of allied forces in southern France and the coordinated sweep over pockets of fascist resistance that remained.

fall 1944 In the China-Burma-India theater of World War II, the most serious malaria epidemic occurred at Karachi, Pakistan, following heavy rains in July and August. The optimal conditions for breeding disease-bearing mosquitoes in stagnant water required increased malaria precautions to restore control of vectors.

late 1944 After the German high command transferred survivors from Auschwitz, Poland, to the concentration camp at Dachau, Germany, a typhus epidemic raged into early 1945 because of poor sanitation and the staff's failure to rid inmates of lice. Of the total death toll, over 41 percent occurred in the final four months. Allied liberators found 900 survivors critically ill and malnourished because they were unable to digest food. In May and June 1945, 2,422 succumbed to typhus. The situation required American application of DDT and inoculation of survivors to counter the contagion.

November 1944 U.S. soldiers encountered an outbreak of polio on Leyte in the Philippines. During a troop landing, which inflicted heavy losses on the Japanese fleet, the virus produced 47 cases, 79 percent of which paralyzed victims. On December 27, a contingent of U.S. Army nurses aboard the U.S. hospital ship *Emily H. M. Weder* reached Leyte to offer aid. Rampant schistosomiasis added to the difficulty of securing military health.

1945 A scourge of sleeping sickness and famine among the nomadic WaNgindo of southern Tanzania south of the Rufiji River ended their peripatetic life of farming, hunting, and gathering. Government officials resettled them at the Eastern Selous Game Reserve at Njinjo.

1945 Following the Spanish Civil War, the Children's Tuberculosis Prevention Center in Almería, Spain, developed a workable campaign against lung disease, which spread rapidly after World War II. Despite a limited budget, the center and children's camps accepted young needy patients, many from Córdoba and Madrid.

1945 Alaskan Indians continued to suffer endemic tuberculosis after most of the United States had lowered rates of infection. The situation in Alaska worsened because of a lack of sanitarium space, poorly ventilated residences, and impeded transportation of doctors and visiting nurses during the coldest months.

January 1945 A surge in spinal meningitis cases on the Gold Coast (modern Ghana) killed at the rate of 1,000 per day. The four-month siege hit hardest at Lawra and West Gonja. The fatality rate shot up to one-tenth of the 34,000 victims.

February 1945 Relapsing fever afflicted 2,000 Kenyans with a 40 percent death rate. After ravaging the war-weakened populations of Arabia, Libya, Morocco, and Tunisia, the lice-borne infection threatened military outposts and sickened crews of Arabian vessels that docked at Mombasa, Kenya. Quarantine and travel restrictions failed to slow the epidemic, which was flourishing by September along trade routes to the north and west. By 1946, the fever had felled 2,000 Algerians, a similar number of Moroccans, and over 125,000 Kenyans for a total of 129,000. The scourge required a concerted program of dis-

infecting residences, boiling laundry, isolating the sick in fever wards, and banning traditional funeral rituals.

March 1945 At Bergen-Belsen, a prison in north central Germany outside Hanover where Nazis deported 15-year-old diarist Anne Frank and her 19-year-old sister Margot in October 1944, the camp held six times its original capacity of 10,000. Inmates suffered disease from foul water supplies, starvation, and exposure from inadequate clothing. After Margot died from typhus, Anne perished from the same disease, which killed an unknown percentage of the camp's 35,000 fatalities. British troops liberated the complex on April 15, 1945. For multiple atrocities and the pitiable conditions that the Germans fostered, British officers hanged Josef Kramer, the camp commandant.

fall 1945 Japan's germ warfare stockpiles contained 400 kilograms of anthrax to be distributed through fragmentation bombs. Japan's high command intended to defeat the U.S. with a plot named "Cherry Blossoms at Night," which would deploy kamikazes on September 22, 1945, to spread plague through flea vectors dropped on San Diego or by balloons carrying anthrax and plague. Before the Manchurian site could be captured in the final days of World War II, Japanese troops covered their heinous plot by detonating and burning Unit 731. Upon U.S. capture of the facility, General Shiro Ishii brokered staff immunity from prosecution for crimes against humanity by sharing his data with the U.S. military. However, the retreating forces left behind enough plague-bearing animals to kill 30,000 citizens of Harbin by 1948.

early November 1945 An upsurge in typhus in Hokkaido, Japan, overwhelmed workers of the Yubari coal mines. By December, the disease advanced on Osaka, Tokyo, and Yamagata. One source of contagion was an inmate of an Osaka jail incarcerated for selling infected blankets he acquired in Korea and Manchuria. The scourge ballooned once more in March 1946 at Tokyo, again transported from Korea.

Simultaneous with epidemic typhus was the introduction of malaria brought home by 600,000 Japanese troops mustering out of service at the end of World War II. The disease proved lethal to Yaeyama islanders, half of whom fell ill with fever. Application of American methods of controlling insects with DDT suppressed infection.

1946 An advanced system of monitoring disease prevention and the outbreak of epidemics became the work of the World Health Organization (WHO), organized by the United Nations in Geneva under the leadership of nurse sanitarian Olive Baggallay. At the time, the UN was collecting data on epidemic cholera, smallpox, tuberculosis, and typhoid and sponsoring relief to refugees, concentration camp victims, and displaced persons in China, Czechoslovakia, Greece, Italy, Poland, Ukraine, and Yugoslavia. A parallel agency, the United Nations International Children's Emergency Fund (UNICEF) broadened the outreach to orphans and abandoned children.

1946 At Chungkai and Nakom Paton, Thailand, Jacob Markowitz, author of *Transfusion of Defibrinated Blood in P.O.W. Camps* (1946), observed how inmates died rapidly from the spread of debilitating diseases. He noted the importance of whole blood to their depleted bodies. A contemporary, humanitarian Ernest Edward Dunlop, wrote to the *British Medical Journal* on the dangers of rampant dysentery in Burma and Thailand during a scarcity of specific amoebicidal drugs. Under wretchedly unsanitary conditions, health workers had to rely on primitive substitutes for blood and medicines.

1946 Cholera swept a prison camp in Hankow, China. Prison officials relocated Japanese POWs to Nanking and Shanghai, spreading contagion to their urban populations. By fall, the disease gripped much of China into Manchuria and Mongolia.

1946 At Matto Grosso, Brazil, along the Juruena River, the Mangabal do Juruena Catholic Mission, established in 1935, ceased proselytizing the Xingú Indians following an epidemic of measles. So many people died or deserted that missionaries abandoned the outpost.

1946 African laborers migrating from Bechuanaland (modern Botswana) carried smallpox into southern Rhodesia (modern Zimbabwe). Despite inoculation of residents, the disease spread beyond the borders and remained virulent in 1948.

February 8, 1946 According to the *British Medical Journal*, influenza virus B infected England with the same result that had gripped the United States in 1936 and 1940. The nation seemed relieved that patients suffered fewer lung complications and survived at a much higher rate than they had from the great influenza pandemic after World War I.

June 8, 1946 Bornholm's disease or pleurodynia struck 125 British soldiers posted at the Singapore naval base. Contracted through fecal contamination of food, the disease required hospitalization of 52 patients. Four years later, microbiologist John Franklin Enders and Thomas H. Weller determined that the contagion resulted from a *Coxsackie B* virus. By August, the infection created new concern in Aden, Yemen, by producing back pain, diarrhea, fever, vertigo, pleurisy, and respiratory distress. Secondary infection inflamed the testicles of male patients.

late 1946 At military training camps, U.S. forces suffered acute streptococcal infections that were resistant to sulfa drugs. When 60,000 recruits shipped out for Japan and Korea, they sickened during the miserably cold, uncomfortable sea voyage. By the time they disembarked at Inchon on January 15, 1947, more than half carried hemolytic streptococcus. Some 9,000 were hospitalized. Adding to the loss of manpower among occupation troops were upsurges in scarlet fever and rheumatic fever.

December 14, 1946 A disturbing outbreak of lethal enteritis killed one-tenth of the infants born at Saint Joseph's Hospital in London. The virulent intestinal disease caused the facility's closure as well as that of the Leicester City General Hospital.

1947 A serious upsurge in poliomyelitis infections in England and Wales produced 7,500 cases, one-third of whom were under five years of age and another third between five and fifteen. Because the remaining one-third were adults, doctors proposed removing the term "infantile" from the name of the disease. Some 3,000 of the victims required extended hospital stays. Staff placed over 300 in iron lungs.

The outbreak caused the medical community to re-examine the findings of early researchers Thomas Buzzard and Frederick Batten, who characterized poliomyelitis as a viral infection and the source of epidemics as rural and coastal areas. They also studied the findings of James Collier, who predicted the widespread epidemic of 1947. By tackling extensive research, devising methods of immunization and treatment, and educating the public, British public health officials began reducing the threat to the nation.

August 1947 Following the five-month Paraguayan Civil War, an outbreak of alastrim, a mild form of smallpox, sickened 2,328, approximately .023 percent of the population. Medical staff fought the epidemic with vaccines from Rio de Janeiro, Brazil.

September 18, 1947 A resurgence of cholera in Alexandria and Cairo, Egypt, raged for three months, sickening 32,978 and generating a mortality rate above 62 percent. The outbreak began among servants at a British military installation at El Korein. Spread by a trade fair, the scourge traveled with merchants throughout the Nile delta. From his observation of the severe dehydration of patients, U.S. Navy Commander Robert Phillips pioneered methods of rehydrating the sick.

When news of suffering in Egypt reached Iraq on October 27, the epidemic captured the imagination of poet Nazik Al-Malaika. Before anthologizing her first collection of poetry, A'shiqat Al-Layl (*Lover of the Night*) (1947), she left home to sit at a construction site to compose. Her poem "Cholera" summarized the horror of stacks of corpses atop horse-drawn carts. Her spontaneity freed Arabic verse from rigidity and traditional forms and rhymes. In a second anthology, Shazaiya wa Ramad (*Shrapnel and Ash*) (1949), she defended free verse as a vehicle for raw emotion such as that generated by mourners of the epidemic.

October 18, 1947 In the Punjab (modern Pakistan), infections of cholera and dysentery claimed around 600 per day at Delhi, India, for a total of over 1,200 deaths. To immunize a highly mobile population in refugee camps, the British army joined with relief agents to inoculate citizens in Delhi and the East Punjab.

November 1947 Polio infiltrated the South Pacific, causing paralysis and death. For eight months, the virus assaulted New Zealand, striking 1,406, many in the Plymouth area and rural districts. The outbreak paralyzed over 58 percent of the victims and killed 71 Europeans and six Maori. On the Nicobar Islands off eastern India,

the reposting of British forces following World War II exposed a virgin-soil population of islanders to the polio virus. Over two months, the resulting epidemic struck 800 islanders, mostly children and youth, with a rate of paralysis reaching over 70 percent and a death rate of nearly 15 percent.

December 1947 During post–World War II personnel movement on Guam, a complex pattern of contagion generated illness and death. A double epidemic of Japanese encephalitis B and mumps exacted a heavy toll in the southern portion of the island. Encephalitis sickened 46 natives and eight military personnel. Highly contagious mumps struck around 15 percent of islanders, nearly half of whom did not report their illnesses.

The encephalitis epidemic continued into the first week of April 1948, about the same time that doctors reported no new cases of mumps. Medical experts surmised that hot, moist weather increased the population of *Culex Annulirostris* mosquitoes, which carried the virus. Some survivors succumbed late in the year to an outbreak of measles.

December 19, 1947 Cholera gripped Syria at Damascus and Dera and spread to the plain. The World Health Organization helped local authorities contain the epidemic, which killed over 23 percent of 77 victims. Within a month, the epidemic waned.

1948 According to Canadian x-ray technician Bruce Norton, who joined a treaty party crossing northern Manitoba, Native Americans on reservations suffered so high an infection rate from tuberculosis that the team made 7,038 x-rays and placed the sick at a treatment center at Clearwater Lake outside The Pas, a lumbering center in western Manitoba. The team stopped reading all the x-rays after the number of infected filled all available hospital beds.

August 16, 1948 Mother Teresa began nursing lepers, victims of cholera and malaria, and the squalid poor of the streets of Calcutta, India. From one-on-one treatment, her outreach grew to a two-room clinic at the pagan temple called Kalighat, which she named the Nirmal Hriday or Place of the Immaculate Heart. Gradually, Hindus came to accept help from a Catholic nun.

1949 A sudden outbreak of polio struck 420 in San Angelo, Texas. All community agencies coalesced for a unified relief effort to rehabilitate victims and assist their families.

1949 Vegetable growers, dairiers, and farm and irrigation workers in the Sharon plain reported around 1,500 cases of leptospirosis, which was endemic in Israel. Health investigators located the disease in voles and began eradicating rodent colonies. Veterinarians, sewer workers, slaughterers, and herders at risk of infection received inoculations against the spirochete.

1949 When endemic meningitis resurged in northern Nigeria, the pathogen claimed over 16 percent of its 98,500 victims. During dry weather, cooler temperatures promoted sleeping indoors in close quarters, where families shared pathogens. At Kano, Katsina, and Sokoto, the disease reached terrifying infection rates, particularly among the very young and the old and frail. Natives resented the intrusion of British health workers, who banned marketing and attendance of schools and funerals, closed roads and borders, and restricted troop movements.

winter 1949 In the year that the Chinese Communist Party founded the People's Republic of China, a mild form of bronchial pneumonia struck Peking and Shanghai, targeting the very young. Families had difficulty seeking care in Peking, which had too few doctors and only 139 beds in the city's nine maternity and child care centers. To keep soldiers in the field, the government promoted traditional Chinese medicine, which was cheap, available, and acceptable to the populace.

mid–1900s To combat trachoma in children, Rose Kaplan, a Russian-Jewish nurse and reformer from Petrograd, worked for Hadassah, a Jewish women's Zionist society. Along with Rachel Landy and Helena Kagan, Kaplan organized nurse-care in Jerusalem, Israel, and among Jewish refugees encamped at Alexandria, Egypt.

1950 Spread from Fiji by sailors, measles struck French Polynesia and moved on to the Marquesas Islands early in 1951. The disease flourished among the young, who had no immunity, and killed half the islanders.

1950 Australian health officials introduced the myxomatosis virus among rabbit populations, the main source of endemic tularemia. The virus began killing 99.8 percent of the animals. However, within a year, the rate dropped to 90 percent. In 1958, myxomatosis killed only a fourth of the rabbits, which developed an immunity to the pathogen.

1950 In New Guinea, victims of kuru, called "laughing death," intrigued Australian physician Vincent Zigas to learn why the progressive, fatal brain ailment gradually reduced dozens of Fore victims to powerlessness. A decade later, veterinary pathologist William Hadlow compared kuru to scrapie, a degenerative disease that destroys the nervous systems of livestock. The colonial government and missionaries stemmed the disease by discouraging cannibalistic funeral rites for relatives.

1950 In a huge swell of newcomers to Israel, diphtheria afflicted 1,660 during a terrifying polio epidemic. Worsening the upsurge was the cessation of diphtheria immunizations during a summer outbreak of paralytic polio at Haifa, Jaffa, and Tel Aviv. Not until 1952—a year after a recrudescence of diphtheria in fall 1951—did the Israeli government require nationwide inoculation and eventually eradicate diphtheria.

In this same period, areas north and south of Tel Aviv reported summer outbreaks of West Nile fever. At Pardes Hannah, 636 inmates of a military camp contracted the disease at an infection rate of 64 percent. The fever receded until the next summer, when it struck the Maayan Zvi kibbutz, afflicting 41 percent of the populace. At Hadera and Ramlah-Lydda in 1953, 242 people suffered the bird-borne fever.

1950 In Italy and Sicily, outbreaks of typhoid fever struck children, youth, and young adults, killing 2,614 within two years. The high incidence of disease topped reports in other European countries. Contributing to infection, particularly in Naples, was the contamination of water, fresh shellfish and salads, and milk and milk products from polluted waterways, garbage, and sewage.

March 1950 In Victoria, 19 Australian babies suffered from acrodynia or pink disease, which had produced 502 cases in Melbourne the previous year. Local doctors connected the disease to use of calomel-laced Steedman's Teething Powder. The contaminant caused pain, anorexia, sensitivity to light, convulsions, coma, and a bright pink color to nose, ears, hands, and feet.

As the disease gained hold of infants worldwide, particularly England, Australia, Canada, South and East Africa and other countries of the British Commonwealth, answers came to light from the work of Viennese physician Josef Warkany of Cincinnati, Ohio, who studied 120 patients over a four-year span. With the aid of Donald Hubbard, a chemist at Kettering Laboratory, who devised a method of locating mercury in urine samples, the men solved the problem and published their findings in *Lancet* in 1949 and in the *Journal of Pediatrics* in 1953. In children suffering pink disease, tests revealed toxic amounts of mercury. As a result, manufacturers of teething powder removed calomel from the formula.

June 1950 Viral hepatitis and dysentery threatened American troops upon their arrival to Korea at the beginning of the Korean War. By December, a concentration of forces beneath the 38th parallel decreased control of sanitation. Because a surprising upsurge in cases followed transfusion with whole blood and plasma, medical personnel concluded that irradiating blood supplies with ultraviolet rays was ineffective. By late winter 1951, the army had to transport 573 ailing men to the Army Hepatitis Center set up at Kyoto, Japan. Within two years of operation, the center received 2,000 cases.

The hantavirus, spread by brown rats, *Apodemus* mice, and voles, afflicted 3,000 American and Korean soldiers, generating a 10 percent death rate from a variety known as Korean hemorrhagic fever. The late stages produced shock, hemorrhaging, and kidney failure. A quarter century later, collaboration between virologists Ho Wang Lee of South Korea and Karl M. Johnson of New Mexico, an American specialist in tropical diseases, isolated the virus. *See also* May 14, 1993.

October 1950 A high incidence of sickness from malaria engulfed Saudi Arabia. Over four months, the illness felled 32,000. A consortium of Saudi and Egyptian health workers and members of the Arabian-American Oil Company combatted the *Anopheles gambiae* mosquito, the vector of infection.

November 1950 By the glow of ultraviolet light, school nurses in Sault Sainte Marie, Ontario, identified 1,366 cases of ringworm in scalps of local pupils. An itchy fungal infection typically found in the short hair of sweaty boys and skin folds of pudgy children, the disease required immediate action. To halt the drop of hair on playgrounds and in classrooms, parents shaved heads of the infected, scrubbed hair in detergent, and soaked scabby scalps in saline before applying alcohol-base fungicide. They placed white caps over healthy hair to prevent spores from spreading the circular eruptions.

1950s In the Central Kalahari Game Reserve of Botswana, Africans suffered recurrent smallpox. The San, nomadic hunter-gatherers, who had allied with the rural Kgalagadi from Maun and Molepolole since the 1890s, shared barter, goat herding, and melon gardening until epidemics and migrations to outside jobs weakened their ties.

1951 In an era of widespread panic over epidemic polio, Jonas Salk of the University of Pittsburgh School of Medicine differentiated three types of polio viruses and identified the most contagious. Backed by the National Foundation for Infantile Paralysis, he immediately began developing a vaccine to end polio outbreaks. The American war on infantile paralysis unified citizens when the disease reached its peak in 1952, infecting 58,000, paralyzing 21,000, and killing 3,000. Two years later, 38,000 U.S. citizens sickened with the virus.

1951 The incidence of leukemia peaked in Japan following the atomic blasts in Hiroshima and Nagasaki on August 6 and 8, 1945, which also precipitated high rates of breast cancer and thyroid carcinoma. Commenting at length on the connection between radiation exposure and cancer was John Hersey's *Hiroshima* (1946), which he updated in 1973 with follow-ups on sickness among survivors. He noted, "Among those who had been exposed within one kilometre of the hypocenter, the incidence was reported to be between ten and fifty times above the norm" (1989, p. 104). In addition to leukemia, survivors reported carcinoma of the breast, liver, lungs, reproduction organs, salivary glands, stomach, thyroid, and urinary tract as well as anemia, cataracts, endocrine disorders, mutations, and chromosome aberrations.

1951 After a drop from 15,000 U.S. malaria cases in 1947, the number declined to 2,000 in

1950. The following year, the National Malaria Eradication Program declared the disease eradicated in the U.S.

March 1951 Thousands of Solomon Islanders contracted polio, which paralyzed 1,280 and killed 156. The disease advanced from Guadalcanal to Malaita and the western islands, where medical and rehabilitative facilities were too limited to aid victims. In Tahiti, as islanders recovered from measles, paralytic polio began to assail older children and teens in mid–March. Eight of the 128 victims died.

April 1951 WHO reported 2,070 cases of hemorrhagic fever, a new disease produced by a Hantaan virus that Japanese doctors had first seen in Manchuria. The fatality rate was nearly 6 percent. Simultaneously, U.S. troops reported 1,694 cases with a death rate above five percent. In Korea, Allied soldiers from many nations weakened rapidly from the disease and died of renal or cardiac arrest or shock. Within three years, the fever afflicted around 3,000 military personnel. To contend with the new threat to health, the Army Medical Service prioritized the disease and helicoptered new cases to a specialty hospital, the Hemorrhagic Fever Center established in a former surgical facility outside Uijongbu, South Korea.

winter to summer 1951 During the Korean War, the U.S. military attempted to control outbreaks of scrub typhus or tsutsugamushi disease and rickettsial pox by delousing Communist prisoners of war with DDT powder.

early summer 1951 Following a hurricane, measles carried by visitors infected 99.9 percent of the 4,000 residents of the Julianehaab District of southern Greenland, killing 72. Four of 83 pregnant women died of infection.

mid–August 1951 At Pont Saint-Esprit, France, a baker made baguettes from flour stained gray. The tainted loaves afflicted 230 villagers with Saint Anthony's fire or ergotism, marked by fainting, nausea, vomiting, vertigo, diarrhea, and chills. The most seriously ill experienced hallucinations, depression, delirium, gangrene, and suicidal tendencies from bread that villagers called "le pain maudit" (cursed bread) (Aronson).

November 1951 In the mountains of northeastern Tanganyika, 1,600 suffered plague, which

produced a death rate of nearly 10 percent. The government prevailed on the skeptical Maasai to accept inoculation.

1952 At age 38, Maria Rita Lopes Pontes became Sister Dulce, a volunteer health-care provider in Salvador de Bahía, Brazil, where she rescued starving fever and tuberculosis victims from the streets of Itapagipe Town. By petitioning donors and collecting food from local stores, she set up a reception center in a chicken coop while touring alleys in her Volkswagen van to locate more indigent homeless and incurables. By 1960, her outreach evolved into a 150-bed sanitarium, the Albergue Santo Antonio (Shelter of Saint Anthony), underwritten by American altruists for the care of 37,000 needy patients. According to Nathan Havestock's biography, *Give Us This Day: The Story of Sister Dulce, Angel of Bahía* (1965), she earned two folk names, the Poor's Mother and Bahía's Saint.

1952 Under Communist Chairman Mao Tsetung, China reformed national health care by supplanting Western methods with military-style health initiatives. The regimented system reduced parasitic infestation and tuberculosis and nearly ended opium addiction, venereal disease, and outbreaks of cholera, leprosy, plague, smallpox, typhoid fever, and typhus.

1952 During an epidemic in Mozambique and Tanzania, chikungunya fever, a togovirus spread by the bite of the *Aedes aegypti* mosquito, received the Makonde name for the contortions of patients suffering severe joint pain.

1952 After emerging during World War II, malaria took hold of boggy areas of the Gold Coast (modern Ghana). Between 1952 and 1953, some 52,000 residents contracted the protozoan infection. The epidemic concentrated at Accra, Cape Coast, and Sekondi, striking infants and toddlers. Nearly half died. Reginald Leslie Cheverton, a director of medical services for the colony who helped wipe out the disease in Cyprus in 1940, coordinated efforts to suppress *Anopheles* mosquitoes by draining swamps and reducing standing pools.

1952 Epidemiologist E. Cuyler Hammond, a long-time statistician for the American Cancer Society, published a report linking cigarette smoking with sharp increases in lung cancer and heart disease.

June 1952 An upsurge in polio infections in Auckland, New Zealand, preceded ten months of epidemic. Of 1,298 cases, most were children and teens. Urban areas reported low incidence of paralysis and death.

fall 1952 In Copenhagen, Denmark, where 2.4 percent of the population contracted bulbar polio, doctors treated over 3,000 patients, reaching 50 new cases a day at the epidemic's peak. A severe chest-centered form of paralysis exceeded the capabilities of the iron lung and elevated the number of fatalities. To assist paralyzed patients at Blegsdam Hospital in breathing and swallowing, anesthetist Bjørn Ibsen suggested inserting tubes into the trachea to force oxygen directly into lungs just as doctors do to assist patient breathing during surgery. The hand-powered apparatus required volunteers from the local university's student body. Carl-Gunnar Engstroem, a Swedish physician, designed a mechanical respirator that provided the support that Ibsen proposed. The machine also saved lives in his own country the next year, when polio assailed Stockholm.

September 1952 In Bangkok, Thailand, a three-month epidemic of largely non-paralytic polio afflicted 388 people, including both natives and tourists.

1953 Brock Chisholm, the first director general of the World Health Organization, promoted a final onslaught against smallpox to eradicate it entirely. The support of Victor Zhadnov, Soviet Vice Minister of Health, helped build the program, but limited funds and complaints from industrialized nations doomed the effort. *See also* May 8, 1980.

January 1953 An upsurge in influenza and pneumonia in Peking and Tientsin afflicted 3,148, mostly Chinese toddlers and infants. Treated at Second Children's Hospital over a four-month epidemic, patients suffered a death rate of nearly 53 percent, worsened by accompanying lung hemorrhage, pleurisy, coma, and seizures.

late April 1953 During the rise in new fevers after World War II, an emerging epidemic of unknown Argentine hemorrhagic fever in Mechita and Bragado, the humid pampas areas outside Buenos Aires, ran its course among some 600 farm workers by August. The upsurge cor-

responded to the end of the sugarcane harvest and the beginning of maize planting, a source of septic dust. Patients suffered chills and fever, ague, anorexia, nausea and vomiting, cough, dizziness, flushing, and malaise. Twelve fatalities of the zoonosis presented shock, pulmonary edema, hemorrhaging kidneys, myocarditis and slow heart rate, liver necrosis, and convulsions. Treatment was limited to hydration, nutrition, rest, and warmth. Research into vectors linked the Junin arenavirus to rat urine and feces in homes and listed Argentine hemorrhagic fever among potential bioterrorism agents.

1954 At Pine Blue Arsenal in Arkansas, the U.S. Department of Defense weaponized its first biological pathogen, brucellosis. Applied by air, the virulent bacteria remained stockpiled in the national arsenal until President Richard Nixon ordered destruction of biological weaponry in 1971.

1954 Among the natives of Xingú National Park in Brazil, measles claimed above 17 percent of 654 victims. According to Noel Nutels, who reported on the epidemic for the World Health Organization, "Among those who received medical care, the death rate was 9.6 percent; among those who could not be treated in time, it reached 26.8 percent" (Neel 1970, p. 426).

January 1954 High rates of typhoid fever in Kenya exceeded the usual number as natives sickened from drinking and washing in polluted water. Spread to Nairobi by Kikuyu raiders called Mau Mau, the infection struck mostly indigenous people, whose standard of sanitation and food storage and preparation differed from that of Asians and Europeans.

February 23, 1954 After publishing in the *Journal of the American Medical Association* promising results of sera eradicating the polio virus, virologist Jonas Salk, author of *Poliomyelitis Vaccine in the Fall of 1955* (1956), put his vaccine to the test. He and Thomas Francis launched the largest medical experiment in U.S. history—a field trial of a killed-virus polio vaccine on one million children sponsored by the National Foundation for Infantile Paralysis. Called polio pioneers, the subjects received three inoculations and a booster shot.

Health workers in Pittsburgh, Pennsylvania, initiated the first mass inoculation of children with the Salk vaccine, a killed virus that prevented the attack of poliomyelitis to the central nervous system and its resulting paralysis and death by suffocation.

March 1954 British soldiers posted at Kinrara, Malaya, sickened with an unidentified fever. Aiding the medical staff were virologists from Walter Reed Army Institute of Research, who observed 852 patients and determined that the main cause of discomfort was leptospirosis, which, according to French epidemiologists, had emerged among soldiers in southern Vietnam the previous year.

March 1, 1954 The detonation of a U.S. thermonuclear bomb on Bikini atoll disseminated radiation ash, injuring 239 Marshall islanders, 28 American observers, and 23 Japanese fishermen aboard the *Lucky Dragon V*. Maladies included hemorrhaging, anorexia, conjunctivitis, and anemia.

April 1954 With the aid of the University of Michigan's School of Public Health, Jonas Salk tested a killed polio virus vaccine on around two million children from Canada, Finland, and the United States. On April 12, 1955, scientists announced the safety and effectiveness of a polio vaccine created by Salk. At the public gathering, TV crews and 150 journalists covered the reactions of the world's 500 outstanding public health doctors and researchers. Volunteers for the March of Dimes celebrated the arrival of the Salk vaccine as the death knell of the polio virus, one of humankind's most devastating scourges, which had struck 21,000 Americans the previous year. Citizens rang church bells in celebration of the end of perennial surges in polio infection. A banner headline in the *Detroit News* proclaimed, "Polio Conquered." By 1992, only five cases were reported.

On April 26, 1955, the inoculations decreased paralytic polio, but paralyzed 200 children in the limbs receiving the injections and killed 11 inoculees. Public health officials traced the faulty dosages to unfiltered particulate in vaccine manufactured by Cutter Laboratories in California. Ten days later, the U.S. Surgeon General George E. Armstrong halted use of the vaccine until medical technologists could separate viruses resistant to formaldehyde. By December 1955, resumption of the vaccination program had protected seven million children. In this same period, Albert Bruce Sabin, author of *Acute Ascending Myelitis*

following a Monkey Bite (1934), joined Russian virologists in testing oral vaccine made in the U.S.S.R. from attenuated live virus. By 1960, community clinic nurses were administering live oral polio vaccine drops in children's mouths. *See also* **July 16, 2001**.

August 1954 At the onset of a polio epidemic in Boston, Massachusetts, patients at the Haynes Memorial Infectious Disease Hospital in Brighton profited from the experience of infectious disease specialist Louis Weinstein, a faculty member at Tufts University. After the disease peaked in 1955, new outbreaks in Cleveland shifted staff to City Hospital, which admitted Vic Wirtz, a mildly infected player for the Cleveland Indians. Teammates visiting him spread cheer and encouragement to children and teens hospitalized in wards for more serious infections. Easing the sufferings of patients were updated iron lungs, cuirass respirators, and rocking beds, which assisted weakened chest muscles while allowing patients to look outward rather than straight ahead at an iron lung.

1955 At the Royal Free Hospital of the University of London, England, Bornholm's disease or pleurodynia afflicted 292 members of the medical staff and 12 patients. In addition to headache, sort throat, and prostration, the sick manifested vertigo, nightmares, and panic. Cranial nerve lesions appeared in nearly half the patients, 90.1 percent of whom were female. Melvin Ramsay of the Department of Infectious Diseases observed patients to determine the cause of the acute outbreak, which extended from July to November.

1955 In South Africa, endemic tuberculosis escalated among black mine laborers. To secure the work force, an upgrade of workers' lodgings and health care was the solution of the South African Chamber of Mines, a consortium of 36 gold mines, 22 coal mines, and 16 diamond, platinum, antimony, asbestos, manganese, lead, and copper mines, including Goldfields Ltd. and AngloGold.

1955 In London, 300 nurses and nursing students at the Royal Free Hospital suffered hysteria during epidemic polio. The physical manifestation of anxiety reflected the mass hysteria that produced Saint Vitus's dance in Holland and Germany in the late Middle Ages in response to the Black Death.

early January 1955 Zambians incurred heavy infection with smallpox, which sickened 3,538, killed 501, and left survivors blind or disfigured.

June 1955 In Japan, a serious outbreak of food poisoning in western Japan sickened 12,000 newborns, killing over 130 from fever, diarrhea, convulsions, and debility. In August, a staff member at the Okayama University School of Medicine identified the toxin as arsenic in powdered milk made and distributed by the Morinaga Milk Industry. Company officials faced criminal charges and lawsuits.

August 1955 New Zealand's last major polio epidemic struck Hamilton and New Plymouth before invading the rest of the island nation. The paralysis rate for 1,485 cases was over 62 percent. In all, 73 died of the virus.

November 10, 1955 Owing to fecal contamination in city water at the Wazirabad pump station during flooding of the Yamuna River, acute viral hepatitis engulfed Delhi, India. The event generated the world's most thorough study of the disease. Into late January 1956, as the short-term sickness invaded civilian homes and military installations, the city's 29,300 patients, many of them young, sought aid at hospitals. Pregnant women in the last trimester died at the rate of one in ten from hepatic failure. In 1988, study of the blood drawn during the epidemic revealed a new strain of the disease, named hepatitis E.

1956 An upsurge in polio in Russia eased the flow of information between the United States and the Soviet Union. Russian medical research lagged because of the destruction of labs during World War II, the paranoia of the Stalinist era, and the government's insistence on funding the military rather than boosting civilian welfare. To end the spread of polio, Soviet officials assigned top medical researchers to the job of prevention and rehabilitation.

1956 English epidemiologist C.E. Gordon Smith of the London School of Hygiene and Tropical Medicine isolated the Langat Flavivirus in Kuala Lumpur, Malaya. His experimentation established transmission of encephalitis to humans by Ixodes ticks on rhesus monkeys and white mice. The advancement of viral study in golden hamsters resulted in a vaccine to prevent tick-borne encephalitis.

May 1, 1956 An unexplained neurological affliction on Minamata Bay, Japan, produced convulsions and speech and mobility impairment. Evidence of environmental pollution derived from bay sludge, dead birds and fish, and seaweed. Medical sleuths linked the crippling anomaly to fish and shellfish contaminated with mercury, lead, manganese, arsenic, thallium, copper, and selenium by the Chisso chemical factory. A parallel outbreak of damage to the central nervous system in people residing and working on the Agano River in Niigata, Japan, in 1965 substantiated the risks of living near mercury-tainted outflow. The final tally of 1,784 deaths among 2,265 victims substantiated the lawsuit that 10,000 Japanese pressed against Chisso for compensation.

July 1956 Monsoon rains coincided with rampant dengue fever in Manila, mostly among Filipino preschoolers. Over the next decade, thousands fell victim to the endemic virus.

early February 1957 A mild but highly contagious Asian or H2N2 flu pandemic, quickly identified in Guizhou, China, spread globally to over a third of the world's population. After striking Hong Kong and Singapore, it advanced to Japan by mid-summer, afflicting 80,000 people and causing 1,000 deaths, some from complications such as enteritis, hepatitis, and pulmonary infection. After the scourge migrated to the Philippines, Indonesia, Australia, and India, the cities of Calcutta and Madras suffered heavy sickness among the poor during multiple waves of infection from both type A and type B viruses. Contagion moved rapidly over the Middle East and North Africa.

In late June 1957, rampant Asian influenza advanced on Iran, Pakistan, Syria, and Turkey. Some 375,000 sickened in Turkey, where the disease reached its height in the fall. Some patients suffered nerve and eye damage as well as emotional disorientation. To the south in Africa, the disease overwhelmed nearly 94,000 Tanzanians, killing 158 mostly aged victims. Beginning at Dar es Salaam, carriers aboard a coastal steamer imported the infection to cities along the shore. By November, the disease had traveled the entire world.

In the United States, contagion required preparation of vaccine and surveillance for outbreaks. In summer, when the first cases were reported, health workers began administering flu shots. The rapid rise in reported infections coincided with heavy migrant labor on farms in the Mississippi River Delta. By fall 1957, New York City and most of the Atlantic coast registered serious infection rates. The opening of public schools spread the disease, which flourished in children and young adults, but claimed the elderly and infirm at the highest rate. The lull that began in December prefaced a new surge in infections in the first months of 1958. Overall, the disease felled half the U.S. population, killing around 70,000.

March 1957 The outbreak of a zoonotic, Kyasanur Forest disease, in southern India derived from the bite of a tick that also spread the viral hemorrhagic disease among monkeys. The first tick-bite fever indigenous to India, it felled young adult men engaged in forest projects and produced pulmonary complications. The fever flourished in the weeks preceding the late summer monsoon and began infecting villagers again in September.

July 1957 Transported from Japan, rubella assailed Taipei, infecting 10 percent of the populace. The outbreak forced Taiwanese officials to examine school-age children, who suffered the brunt of the epidemic. Among pregnant women, one-quarter experienced stillbirths, premature labor, and the birth of babies with congenital defects.

1958 When dengue fever erupted in Southeast Asia, it afflicted North Vietnam during the monsoon. In Hanoi, hundreds collapsed. During the initial outbreak in Thailand, a young pediatrician, Suchitra Nimmannitya, chief of the infectious diseases division at Bangkok's Children's Hospital, battled the sickness, which infected 2,418 in the capital city alone. While treating children dying of the emerging virus, she pioneered clinical diagnosis and management of the disease. Her expertise earned the Prince Mahidol Award. After the World Health Organization adopted Suchitra's protocols, she traveled to Indonesia, Singapore, Colombia, and Brazil at each new outbreak of the fever.

March 1958 In Taiwan, encephalitis, a mosquito-borne zoonotic flavivirus found in birds and pigs, blanketed the Pescadore Islands. The first major government initiative against the chronic malady began in 1974, when the Taiwan Provincial Malaria Research Institute began studying Japanese B encephalitis, dengue fever, and elephantiasis with the intent of eradicating them from the islands.

late June 1958 In Ethiopia and Somalia, a serious outbreak of malaria infected 3.5 million and claimed 175,000 over a six-month period. Few received medical care or registered with public health officials. Half the children infected by the *Anopheles gambiae* died.

August 1958 A serious incidence of polio in Singapore afflicted older children, paralyzing 179 or 43.1 percent. Of the 415 victims, around 76 percent were Chinese. The outbreak spurred health officials to vaccinate 200,000 children. In Vietnam, a three-year outbreak required immediate attention. Officials petitioned the Soviet Union for vaccine, which local medical technologists began manufacturing at the Hanoi Institute of Hygiene and Epidemiology.

October 1958 A resurgence of bronchopneumonia in Chinese infants and toddlers menaced Peking as well as Changchiakou, Changchun, Harbin, Huhehot, Shenyang, and Tientsin. Throughout the winter, some 3,398 cases of adenovirus pneumonia with a fatality rate of 15.5 percent assailed patients at Peking's Second Children's Hospital.

1959 A new pathogen causing o'nyong-nyong fever in Uganda increased the list of the world's viral epidemics. A major eruption in central and eastern Africa invaded five million people with a flu-like malady that sickened but rarely killed victims. Scientists deduced that the disease, like malaria, was borne by the *Anopheles* mosquito, but they were unable to identify the animal reservoir.

July 1959 An upsurge in hemorrhagic fever assailed eastern Bolivia over a five-year span. After its emergence at San Joaquín, the disease, called Machupo virus or Bolivian hemorrhagic fever, spread to El Mojon, where peasants torched the village and fled. Of the 736 afflicted with the arenavirus, over 70 percent survived. When residents understood the transmission of the virus through food soiled with rodent urine, they instituted an effective extermination program on farms and dairies.

1960 To combat tuberculosis among reservation Indians, Navaho nurse Annie Dodge Wauneka of Sawmill, Arizona, combined modern medical techniques with the traditional Blessing Way while treating patients at their hogans. In addition to aiding 2,000 patients, twice a week over Gallup's KGAK-radio, she broadcast tutorials about the spread of tuberculosis. In 1963, President John F. Kennedy presented her the Presidential Medal of Freedom; *Ladies' Home Journal* named her Woman of the Year 1976.

1960 In Pennsylvania, public attention to work-related deaths focused on anthracosilicosis, pneumoconiosis, and silicosis. A survey of causes of death in coal miners determined that 2,772 died of respiratory illnesses in 1960.

1960 Spread by the *Aedes aegypti* mosquito, yellow fever in Ethiopia and Sudan produced 5,000 official fatalities and estimates of six times that number in the unofficial total received by the World Health Organization. The death toll is the largest ever recorded from one outbreak.

October–December 1960 In South Kasai, Congo, epidemic kwashiorkor among 150,000 Baluba refugees weakened children and also adults, a unique result of starvation that usually targeted weanlings. Females were twice as likely to suffer malnutrition as males. Patients suffered edema, loss of appetite, reddish hair color, depigmented skin, and atrophy of body fat and muscle. As natural immunity lessened from a diet of cassava, patients experienced anemia, malaria, dysentery, marasmus, and smallpox. At emergency hospitals staffed by WHO physicians, United Nations deliveries of skim milk powder, rice, maize flour, canned and dried fish, and palm oil as well as plasma, feeding tubes, and IV saline dramatically alleviated depletion and secondary infections.

1960s In South Africa, malaria reached epidemic proportions among the residents of Natal and eastern and northern Transvaal. The numbers revealed a disparity in living conditions and health standards: non-white people fell sick with the disease 2.5 times more often than whites.

1960s Haitians died at alarming rates from tuberculosis. The Pan American Health Organization studied the case analysis and determined that 100,000 islanders required treatment. New cases arrived at clinics at the rate of 130 per day.

To meet the need, Henry Koop, a Canadian pharmacist, launched the Croisade Anti Tuberculose, a crusade that began with the inoculation of children and the identification of all contagious consumptives. Volunteers who mobilized on foot and by mule, boat, or Land Rover moved village by village with injector guns and inoculated 85 percent of Haitian children.

1960s In the United States, sexually transmitted disease constituted a hidden epidemic, especially among the teen population. Some 12 million new cases of gonorrhea and syphilis were reported annually, one-quarter of which occurred in patients age 13 to 19. Health officials warned of collateral damage from infertility, cervical cancer, pelvic inflammatory disease, and deaths of newborns.

1961 An outbreak of flu symptoms in Belém and Brasília, Brazil, resulted from the Oropouche virus. In 1980, epidemiologists connected the virus with the sandfly that bred in stagnant water in discarded cacao hulls.

1961 Serious plague struck Vietnam during the war era. In deteriorating social conditions, rats proliferated in Long Khanh province and on the coast of the South China Sea. By mid-decade, worsening health conditions forced the U.S. military to inoculate, spray DDT, and take precautions against bandicoots, the prime carriers of the rat flea. By the end of the 1960s, some 30,000 people had contracted the disease. Worsening the outbreak was the movement of refugees fleeing napalm and deforestation with Agent Orange to shelter in refugee camps.

1961 After decades of research into *itai-itai* (ouch-ouch) disease linked with the Kamioka mines in Japan, Dr. Noboru Hagino released findings about the saturation of rice fields and the Jintsu-Takahara River bottom with cadmium, the slag from zinc refining. The Mitsui Mining and Smelting Company refuted the data and discredited Hagino. The Japanese government corroborated Hagino's findings and proved that toxic cadmium leaching into groundwater caused the joint and bone pain, osteomalacia (soft bones), fractures, and kidney failure that had assailed rice farmers since 1912. In 1972, lawsuits against Mitsui Mining forced company officers to install pollution prevention systems and to monitor heavy metals in effluent and sedimentation.

January 1961 An Asiatic cholera pandemic reminiscent of the rampages of the 19th century began in Indonesia at Makassar and spread to Java by May, Borneo in August, and on toward Timor. Of the 4,107 reported cases, fewer than 78 percent recovered. Chinese islanders produced the lowest number of infections, owing in part to their diet of stir-fried and steamed vegetables and green tea.

By September, the pathogen engulfed the Philippines, where cholera was endemic. Spread from Manila's harbor, the disease prevailed among natives who drank impure water and ate raw fish. Officials revived the use of San Lazaro as a plague hospital and distributed cholera vaccine donated by China's Red Cross. The U.S. Navy dispatched a five-man team from Taipei to assist Filipino health workers. After a year's rampage, the disease had killed 106 of 2,756 victims.

The virulent *Vireo el tor* pathogen savaged the southwestern Pacific, Southeast Asia, and China. At Taiwan in July 1962, a two-month outbreak produced around 400 reported illnesses. In the mid–1960s, sickness engulfed the Indian subcontinent, spreading to the Middle East and Russia. By the end of a 14-year rampage, the disease had blanketed Asia and Africa, dying out in Indonesia in 1975, where it had begun.

April 1961 The beginning of a lengthy bout of malaria struck itinerant Somali of the Haud area on the African horn. Health officials blamed the *Anopheles gambiae* mosquito, which bred in huge holding tanks along the Ethiopia-Somalia border.

July 26, 1961 The *New York Times* divulged that the Salk vaccine was contaminated with a simian virus. Merck and other manufacturers ceased production until they could filter the contaminant from vaccines. The U.S. Public Health Service contained panic by issuing a statement that the virus posed no danger. Four decades later, the virus was connected with tumors, leading researchers to suspect that the monkey virus was carcinogenic.

1962 In Cuba, as Fidel Castro attempted to copy the Soviet style of public health care, the island's health bureau lacked trained medical personnel to stanch rampant gastroenteritis among infants in Marianao or to control an epidemic of gonorrhea and syphilis. Without penicillin and aureomycin, chief of public health Rafael de la Palma relied on poor quality imported antibiotics from the U.S.S.R. Castro's promise to improve the lives of island peasants began with a nationalized medical service and the same quality of health care found among Communist systems in China, North Korea, the Soviet bloc, and Vietnam. By upgrading water supplies, hygiene, and preventive medicine, by 1967, his staff virtually wiped out diphtheria and malaria. Six years later, health workers had almost eradicated polio. By 1974, Castro's bureaucrats had reduced deaths from tuberculosis by 73 percent.

1962 When dengue fever attacked Thailand, data indicated that 200,000 suffered viral infection, which required hospital care for around 4,200. The next spring, the disease struck South Vietnam along the Mekong River on the border of Cambodia, where river traffic carried trade and passengers. Throughout the summer, small children sickened as the death rate rose above 35 percent.

1962 A Pakistani tourist spread smallpox in Rhondda, Wales. Twenty-five Welsh contracted the virus; the six who perished included a nine-month-old child. Because of an outcry for inoculation, the Welsh government vaccinated 900,000 citizens of South Wales.

July 1962 When hemorrhagic fever earned status in Bolivia as a new epidemic disease, the virus had already been active since 1959 in Beni province. In 1963, epidemiologists isolated the Machupo virus, an Arenaviridae, in San Joaquin, Bolivia, and identified the vesper mouse *Calomys callosus* as its reservoir.

1963 At Zermatt, Switzerland, the contamination of water supplies with sewage generated multiple cases of typhoid. Of the 437 patients, many were European skiers and tourists.

1963 The six nurses and two aides at Christian Medical Centre at Rennie's Mill on Junk Bay outside Hong Kong began treating some 90 Taiwanese refugees daily, many with contagious diseases. Superintended by Anita Ho, the center supplied care through a physician at Haven of Hope Hospital, formerly a mission sanitarium for treatment of tuberculosis.

1963 Australian physician Barry Christophers, head of the Victoria Council for Aboriginal Rights, exposed the racist intent of the nation's

Tuberculosis Act, which allotted no recovery allowances to Aborigines suffering from endemic tuberculosis.

late 1963 Epidemic hemorrhagic dengue fever infected Calcutta, India, simultaneously with upsurges of the virus in Burma, Malaysia, the Philippines, Singapore, and Thailand. The outbreak of dengue fever in Puerto Rico and other Caribbean islands, a first in the Americas, resulted in serious sickness. In Puerto Rico, health reports for the year attested to 25,737 victims, over 16 times the number in Jamaica. In 1964, the disease ranged south, striking Venezuela and felling 18,306.

1964 An Ivory Coast native and Vietnam War martyr, Betty Ann Olsen, a nurse with the Christian Missionary Alliance, set up a leprosarium at Ban Me Thuot. On February 1, 1968, she was captured by the Viet Cong during the Tet offensive and incarcerated in chains with two other inmates at a prison camp. Poorly nourished and suffering exhaustion, exposure, and dysentery, she treated agricultural adviser Michael Benge for endemic cerebral malaria until her death on September 29 from constant shuffling among jungle prisons and from beatings and deadly meals of partially cooked bamboo shoots.

1964 Before the availability of vaccine, a huge epidemic of rubella afflicted 12.5 million U.S. residents. In Maryland alone, health workers reported 3,583 cases. Among pregnant women, the disease caused 30,000 stillbirths or miscarriages and crippled 20,000 infants with cardiac anomalies, deafness, eye disease, malformed organs, nerve damage, and retardation. A sub-population of the deaf community filled spaces in schools for the deaf and provided researchers with subjects for study.

1964 The 507 residents of Aberdeen, Scotland, stricken with typhoid owed their illness to impure water and imported corn beef bearing the *Salmonella typhi* bacterium.

March 1964 Madras and Calcutta, India, suffered an upsurge in the *El Tor* strain of cholera imported from Cambodia. The outbreak hit an urban population that suffered simultaneous sickness from hemorrhagic dengue fever. Additional cholera at Katmandu, Nepal, and in Pakistan contributed to the Asian pandemic.

summer 1964 At Houston, Texas, 60 died from epidemic encephalitis bred in wild birds from fecal contamination of bogs and ditches. Identified as the Saint Louis variety isolated in 1933 by Ralph W. Muckenfuss of the National Institutes of Health, the virus yielded to medical containment by October.

1965 A recrudescence of smallpox hit both eastern Java and parts of Borneo. Because of underreporting, the total of cases and deaths is unknown.

1965 Plague infected seven victims in New Mexico, killing a young Navaho girl. As a means of containing infection, authorities sprayed insecticides to kill fleas, ordered the poisoning of 100,000 prairie dogs, and posted radio messages in Dené and English alerting the public to the threat of an epidemic.

1965 Rampant yellow fever infected 20,000 Senegalese, whose health system failed to keep pace with the epidemic in the years following separation from French colonists. Health workers employed an air gun for rapid inoculation of 130,000 children. At Dakar, health agents vaccinated with syringes. Some 240 recipients acquired encephalitis, which killed over 10 percent.

1966 Epidemiologist Donald Ainsley Henderson of Ohio directed the World Health Organization in a global campaign against smallpox. Gradually, the campaign moved into central and western Africa, Brazil, Indonesia, Ethiopia, and Somalia. Over the next twelve years, nations spent a billion dollars on an intense inoculation program. To distribute 2.4 billion powdered shots with a special two-pronged injector, the effort required a team of 200,000 inoculators, who infiltrated warring tribes, traveled the African horn on foot and by air, and examined refugees and nomads along the Ethiopian-Somali border. Sweetening the hunt was a bounty of $1,000 for anyone isolating a case of smallpox.

The return of smallpox to Brazil spurred a coalition of global leaders to support the World Health Organization's decision to eradicate the scourge. Pox struck urban dwellers at Vitória, where vaccination programs had grown lax. Epidemiologists determined that sickness was underreported and response teams were slow to act. The outbreak was ten times worse at Bahía.

Simultaneously, in Mali, the United States funded a campaign to annihilate smallpox and control measles over a five-year period. Although health workers achieved the former goal, measles continued to emerge in frequent outbreaks. *See also* **January 1967.**

1967 Kawasaki Tomisaku, a Tokyo pediatrician, reported an endemic syndrome infecting 6,000 Japanese victims annually. Known as Kawasaki disease or kawasakibyo, the malady, of a still undetermined source, flourished in small children. It ballooned medium-sized blood vessels throughout the body, especially the coronary arteries. Left untreated, it caused fatal aneurysms and heart attack.

1967 Some 10,000 Moroccans contracted cerebrospinal meningitis, which produced a death rate of over 12 percent. In urban settings at Casablanca, Marrakesh, and Rabat, the disease favored tenements where people encountered inadequate sanitation. A recrudescence during the winter of 1967–1968 added 2,821 to the headcount of victims.

That same year, in South Africa, an upsurge in endemic cerebrospinal meningitis afflicted around 2,000 in the first months and an additional 3,424 in the next two years. The most susceptible were preschoolers, sickly school-age children, and some adults. The death rate ranged upward to 12 percent; some survivors tended to suffer retardation or deafness.

January 1967 On the Guinea coast of West Africa, smallpox raised its final siege before inoculation programs eradicated it. The disease felled the youngest villagers on the Sierra Leone border and remained virulent into spring 1968, killing a total of 126.

March 27, 1967 After six decades of non-exposure to the mumps virus, 119 or 56 percent of the 212 native islanders of St. George, Alaska, presented symptoms. The health service fought the insurgence with hyperimmune y-globulin serum.

August 1967 In the town of Marburg, three West German factory workers at Behringwerke AG suffered a mysterious filovirus that caused vomiting and bloody diarrhea. Later identified as the Creutzfeldt-Jakob virus, the disease caused uncontrollable hemorrhaging, intolerable pain, and vomiting of blood. The disease killed seven out of 25 victims, most of whom had contact with African vervet monkeys from Uganda. Later study linked infection with fruit bats. Within a month, five more patients fell ill in Frankfurt and in Belgrade, Yugoslavia. Labeled the Marburg virus, the disease recurred in Rhodesia in 1976, the Soviet Union in 1988, and Kenya in 1990. In 2004, the death rate from the virus in Angola reached 90 percent.

August 1967 A reemergence of smallpox in Djakarta, Indonesia, infected 400 victims per month. The outbreak spurred a World Health Organization campaign to annihilate the disease. As a result, Indonesia was the first Asian nation to report success.

1968 After decades of complacency, epidemiologists determined that vector-borne diseases were once more a threat to world health. In the late 1960s and early 1970s, a resurgence of malaria rose in Ceylon from 17 cases in 1963 to an epidemic of 440,644 five years later. In 1969, the figure rose to 537,705.

1968 In Brazil, during the Indian Protection Service's removal of the Nambikwara beyond the Cuiabá–Porto Velho highway and in the Guaporé River valley, many sickened from disease. Measles killed all the young of the Yanomami tribe. Multiple diseases infected forty Kreen Indians, killing 15 percent. By 1974, their population fell to under 135.

The devastation continued in northern Mato Grosso and southern Pará. When invaders advanced on foot, natives vacated their villages and fled, but could not elude influenza, which invaded the Kreen-Akarore habitat. Anthropologist Steve Schwartzman of the University of Chicago observed:

> In the first wave of epidemics, so many people died, and the survivors were so sick and weakened, that the living were too few and too debilitated to bury the dead. The vultures and turtles ate the corpses. Nursing infants and children died of starvation when their mothers died [Schwartzmah, "The Panará" 1998).

Reacting to what the Panará thought was witchcraft, their tribal leaders executed suspected sorcerers for unleashing mass death.

As road builders advanced the Cuiabá Santarem highway through Panará territory, they discovered the ragged survivors lining the roadbed and begging for food. In February 1975, the government transferred the 79 remaining Panará to the Xingú National Park. In 1995, Chief Akà led the people to the shreds of an Amazonian forest that was once their ancestral homeland.

January 1968 A five-month rubella epidemic in Taipei infected Taiwanese children too young to have developed immunity from the 1957 outbreak. By 1969, a large number of children were born with congenital measles defects, notably stillbirths, blindness, and deafness.

early 1968 Similar to Asian flu, the Hong Kong or H2N2 flu emerged at Guizhou or Yunnan, China, striking the very young. Advance warning prepared health officials for a global outbreak that spread to Singapore, the Philippines, Japan, Australia, the eastern Mediterranean, Europe, and the British Isles. When troop ships carrying solders from Vietnam docked at San Diego, the disease entered the United States, infiltrating California, New Jersey, and Pennsylvania.

In Russia, urban areas took the brunt of the epidemic, which moved into the Baltic region, Byelorussia, Moldavia, and the Ukraine. The pandemic of influenza A boosted fatalities among the elderly in December and January 1969. At epidemic's end in spring 1970, the disease had killed 33,800, the lowest death rate of any 20th-century pandemic.

February 15, 1968 The symptoms of measles in one boy, Roberto Balthasar, a Yanonami (also Yanomama or Yanomami), presaged a terrible epidemic in a virgin-soil population. One of the Amazon's largest surviving indigenous nations, the Yanonami lived in the forested Alto Orinoco region of Southern Venezuela on the border of Brazil. Arriving three months after an outbreak of bronchopnemonia, according to missionary James Barker, measles killed hundreds. The disease may have derived from a visitor from Manaus, Brazil, the previous fall during inter-tribal feasting and trading among villages. Another account lays blame on the child of a missionary on the Brazilian side at Totootobi. When men of one village abducted a woman from another setting, the raid spread the infection.

The epidemic produced a storm of controversy which academics debated in print and in private. In *Darkness in El Dorado* (2000), investigative journalist Patrick Tierney quotes cultural anthropologist Terence Turner of Cornell University, who charged with genocide two notable scientists, Napoleon Chagnon, an anthropologist and geneticist from the University of California at Santa Barbara, and Chagnon's partner, geneticist James Neel of the University of Michigan. Turner believed that the duo vaccinated thousands with the Edmonston B vaccine, which produced the symptoms of measles, and continued their research on eugenics among primitive people while ignoring the natives' medical needs.

The evidence was damning. According to Willard Centerwall, who accompanied the expedition, Neel admitted that the lack of vaccine for all Indians gave him an opportunity to observe an epidemic in a virgin-soil population. In Turner's estimation, "When the chips were down, the Yanonami as patients took second place to the Yanonami as objects of scientific investigation" (van Arsdale 2001). He charged Chagnon and Neel with academic malfeasance and criminal negligence.

In an article for the *American Journal of Epidemiology*, Neel acknowledged that the vaccine had a history of severe reactions among white children. He admitted,

> When the epidemic of measles struck the Indian populations, however, there was no doubt that it was a different entity of far greater severity in terms of prostration, toxicity and complications. At the Ocamo Mission, six of 22 Indians in the fifth day of measles seen at sick call had unequivocal signs of bronchopneumonia [Neel 1970, p. 425].

Neel concluded that reaction to measles vaccination is more pronounced among the Yanonami than among Africans, Micronesians, or U.S. caucasians. He reported a collapse of village life because, in his words, "The concern of the well Indian for the ill seldom extends outside the immediate family" (ibid., p. 427). He ended his article with an impersonal commentary on "herd diseases as agents of natural selection."

late 1968 The world's first and most widespread episode of polychlorinated biphenyls (PCBs) poisoning affected 14,000 Japanese who consumed rice cooking oil made and distributed in Kita by the Kanemi Soko Company. Individuals and whole families of western Japan in northern Kyushu—1,858 victims—became sick from consuming rice bran oil; the tainted oil killed over 100 people. During lawsuits filed by 44 victims, courts found negligence by the oil producer and Kanegafuchi Chemical, manufacturer of PCBs.

1969 In Slovenia, a long-lived dual epidemic of scabies and syphilis emerged and reached its height from 1972 to 1974.

January 12, 1969 With the infection and death of nurse-missionary Laura Wine, Lassa fever, an acute arenaviral infection, emerged in Nigeria and spread to Guinea, Ivory Coast, Liberia, Mali, Sierra Leone, and the Congo. It took its name from the town of Lassa in the Yedseram River valley on the border of Cameroon. At Jos, Nigeria, an autopsy of another victim, Charlotte Shaw, revealed the circulatory system, heart, kidneys, liver, lungs, and spleen clogged with blood cells and platelets. When nurse Lily Pinneo developed symptoms, hospital staff airlifted her to Lagos for a long flight to Columbia-Presbyterian Hospital in New York City. On May 3, she was well enough to leave the hospital. Epidemiologists used her virus-laden blood for study.

June 1969 An outbreak of acute hemorrhagic conjunctivitis (or Apollo conjunctivitis) started in Accra, Ghana, and soon blanketed Africa. Perhaps spread by Muslim pilgrims from a hadj to Mecca, the disease struck Saudi Arabia and Java, Indonesia, before advancing to Singapore, India, Southeast Asia, Hong Kong, Taiwan, the Philippines, and Korea.

July 1969 Much of the South Pacific experienced an attack of influenza. A three-month epidemic of H3N2 or Hong Kong flu struck Papua New Guinea. The death toll rose above 3,000. In the hill country, the disease slew almost all of its victims. The area suffered a recrudescence in 1970.

fall 1969 Of 130,000 Guatemalan patients, some 12,000 died of shigellosis, a tropical form of bacillary dysentery named in 1898 for epidemiologist Kiyoshi Shiga, a Japanese researcher. After a lull in spring 1970, monsoon rains spread contagion in water supplies and fresh vegetables and fruits, killing the oldest and youngest residents. Surprisingly, death rates were higher for hospitalized patients than those seeking no medical attention. The contagion migrated into El Salvador and Mexico and as far north as the Los Angeles barrios of California.

October 1969 In Japan, an epidemic of subacute myelo-opticoneuropathy or SMON disease struck the optic nerve of victims. Caused by the anti-dysentery drug chinoform (Iodochlorhydroxyquin), the malady at first appeared to be infectious. It began invading the sensory and motor nerves in 1955, and afflicted a total of 11,700 patients, killing 500. The ban of the drug in 1970 halted new outbreaks.

1970 A World Health Organization analysis warned that a release of aerosolized anthrax spores upwind of 5,000,000 people could sicken 250,000 or five percent, and kill 100,000 or 2 percent. The study indicated the seriousness of biological warfare and the threat that stockpiles of anthrax posed to any nation victimized by bioterrorists. In the estimation of the U.S. Office of Technology Assessment, such a dispersal of anthrax would be as lethal as the detonation of a hydrogen bomb.

1970 In Burma, a pair of epidemics spread dengue fever and hemorrhagic fever from Rangoon to Bassein and Moulmein, killing 121 people.

1970 The first monkeypox case in humans led World Health Organization workers to study data gathered from central and west Africans to determine whether the disease was spread like smallpox, from person to person. Coinciding with civil unrest in the Democratic Republic of the Congo, field studies from 1980 to 1986 indicated a far lower fatality rate than that of smallpox. However, the possibility that monkeypox might serve terrorists as a biologic weapon raised questions about the need to stockpile vaccine to ward off a medical catastrophe.

May 21, 1970 The reputed last case of smallpox in western and central Africa infected a 27-year-old Nigerian woman. The halt in the advance of smallpox climaxed a five-year effort by the World Health Organization to eradicate the disease in a region that had suffered previous epidemics. *See also* **May 8, 1980.**

September 1970 Victims of viral conjunctivitis in Singapore developed peripheral breathing problems and eye hemorrhages. Within three months, 60,118 patients required treatment. *See also* **June 1971.**

November 5, 1970 Imported by a trader from Abidjan, Ivory Coast, cholera felled thousands of Malians living near Mopti and along the Niger River. In less than two months, the disease invaded Burkino Faso, Mauritania, Niger, Nigeria, and Senegal. When Guineans contracted the malady, medical workers traced the infection to youths summering along the Black Sea. Because Fanti peddlers spread the disease in infected fish, the Guinean government ousted all Ghanians. Cholera advanced through the fish trade into Sierra Leone and spread to Cameroon, Dahomey, Ghana, and Togo. A total of 28,000 West Africans contracted the disease, which attained a mortality rate of five percent.

December 1970 In Japan, chronic arsenic poisoning afflicted 177 patients in Sasagatani and Toroku, Japan. A neuron affliction, it caused a perforated nasal septum, discoloration of the skin, and endangerment of the unborn. The malady affected workers of and people living near arsenic mines, which later closed.

1970s A humanitarian campaign in Uganda instructed mothers on the danger of weaning young children too soon and feeding them entirely on *matoke*, a carbohydrate-rich mashed banana puree. The diet produced rampant kwashiorkor, which killed children in central Africa. Through home cooking demonstrations, health agents showed how to add beans, fish, peanuts, and fried termites to banana pulp to assure enough nutrients for growth and development.

1970s The emergence of AIDS as a catastrophic threat to central Africa beset Zaire with multiple hardships on women and families. Increasing mortality derived from poverty and poor health and from multiple sex partners among women forced to earn a living by prostitution. Lacking sufficient education in how the disease is transmitted and unwilling to use condoms, people faced a sweeping epidemic that changed forever the outlook for better lives.

1971 After the introduction of mineral prospecting and colonization in the Aripuana Park, Brazil, the indigenous Cinta Largas and Surui Indians suffered violence, starvation, and epidemic influenza and tuberculosis. The American Human Friends of Brazil stated:

> Today a new form of genocide is taking place in Brazil, one less dramatic than the bombing of villages or the wholesale slaughter of Indian women and children, but nevertheless, one similar in intent…. In present day Brazil, more than twenty-five years following the U. N. Genocide Convention, the responsible agents for such crimes are two: the military government and the large national and multinational corporations which it represents [Garfield 1995].

March 1971 Bangladeshi of the newly created country suffered an outbreak of hemorrhagic conjunctivitis, which spread to Bengal through Bombay in a wave of Islamic pilgrims returning from a hadj to Mecca. Highly contagious, within six months, the viral eye infection, dubbed "Joyful Bangla," infected a half million recorded victims in Bombay. By mid-summer, many Indian cities reported outbreaks as victims, typically young adults, shaded painful eyes with sunglasses.

March 1971 Half of the people of Tahiti contracted dengue fever in an outbreak that invaded Bora Bora, Fiji, Huahine, Maupiti, Moorea, New Caledonia, Raiatea, and Tahaa. A year later, the virus spread from Fiji or Samoa and afflicted all but 10 percent of Niue Island.

April 22, 1971 On the Indian subcontinent, a half million East Pakistani residents escaping civil upheaval overran Bengal, spreading cholera. Within six weeks, over 13 percent of 9,500 victims died in hospitals that were already filled to capacity from a simultaneous smallpox outbreak. By early summer, the grand total of deaths reached 9,250 or 18.1 percent of 51,000 stricken with the two scourges. In this same period, Indian authorities closed the borders to refugees, who spread the disease over East Pakistan.

May 1971 Residents of Chad, particularly the rural poor, suffered high incidence of cholera, which killed the uninoculated over the next five months. Of 8,225 cases, 2,337 or over 28 percent died. Spreading pathogens were swimmers and bathers in the contaminated waters of Lake Chad. Médecins sans Frontières (Doctors without Borders) reported that both drought and civil strife worsened an already desperate situation.

June 1971 A recrudescence of viral conjunctivitis afflicted Singapore, Malaysia, and spread to Java. Into early winter, around 30,000 victims required medical care. Unlike the outbreak of September 1970, which was caused by the Coxsackie virus, the second outbreak targeted male children and young adults. The epidemic revealed the first human enterovirus known to infect the eyes. Striking poor neighborhoods at the rate of 50 percent, the highly contagious disease derived from crowded living conditions and poor hygiene. *See also* **June 1980.**

summer 1971 The Soviet Union military tested a biological weapon carrying the smallpox virus, perhaps in airborne form. Experiments accidentally spawned an outbreak of ten infections and three fatalities, two of them infants, among the public at Kazakhstan on the Aral Sea. In response, the government health service organized a huge vaccination program. To halt panic, Yuri Andropov, chief of the KGB, suppressed news of the infections. The incident first came to light in the world press in mid–June 2002, when Alan Zelicoff of Sandia National Laboratory in New Mexico presented official details of the outbreak and interviews with two survivors of the infection to epidemiological experts gathered at the National Academy of Sciences in Washington, D.C.

September 1971 Epidemic polio in Malaysia swept the west with a paralytic virus that infected 16,000. Hardest hit were young children, urban males, and Indians.

November 1971 Smallpox emerged among thousands of Bangladeshi refugees at Salt Lake outside Calcutta, India. An attempt at inoculation failed to reach all who crowded into an unsanitary camp. At the creation of a separate nation of Bangladesh on December 16, the inmates spread the disease in their new state, from which contagion migrated across India and Pakistan.

Among eyewitnesses to the world's last great smallpox epidemic were Alfred Sommer of the Johns Hopkins Bloomberg School of Public Health and James Bartleman, volunteer at a Canadian mission for the World Health Organization and winner of the 1999 Aboriginal Achievement Award. By spring, health authorities reported vaccination of 75 million people, 80 percent of the populace, but still noted 7,000 fatalities from smallpox. The disease raged until February 1973. Simultaneously with the pox, an increased incidence of malaria in Bangladesh affected residents for five years. The epidemic began in forested land, near fresh water, and in rice-growing regions, where disease-bearing mosquitoes thrived.

1972 A three-year scourge of diphtheria struck Seattle, Washington. The disease devastated unimmunized Indians, who suffered complications to hearing, genitals, heart, kidneys, lungs, nervous system, and skin. The U.S. government saw the need to fund the Seattle Indian Health Board's medical clinic.

early 1972 Lassa fever at Sierra Leone gained a hold among 65 patients of Panguma Catholic Hospital and created a parallel threat in Liberia. Medical detectives blamed the *Mastomys natalensis*, a rodent common to residences over most of sub-saharan Africa. Among the 23 fatalities in Sierra Leone were two hospital employees. An American Peace Corps worker who contracted the infection was airlifted home. Leading the investigation of the pathogen was Thomas Monath, an experienced virologist at the Centers for Disease Control and Prevention.

spring 1972 One of the last major smallpox outbreaks in Europe struck Yugoslavia after a Muslim pilgrim on a hadj to Mecca, Arabia, contracted the disease at his native village of Kosovo. When Ljatif Muzza of Djakovica fell ill, doctors passed him among wards until he died in Belgrade of hemorrhagic smallpox, the deadliest form of the virus. He passed the disease to 38 victims, of whom eight died. Communist health officials under Dictator Josip Broz Tito clamped down on contagion with quarantines and held the death rate to 35 out of 175 cases. Specialists who had treated the disease in India arrived in Kosovo to contain the virus with strict regimens, including a ban on weddings, meetings, and civic gatherings. Agents vaccinated 18 million people, almost the entire nation.

1973 Endemic dengue fever returned to Thailand and Southeast Asia. At Vietnam, the pathogen infected 18,581 and killed 1,424 over a two-year period.

1973 The world's final epidemic of smallpox began in Bihar, India, where migrant workers, displaced by flooding and drought, carried contagion. The total dead in 8,664 hamlets reached 188,000.

1974 Malaria spread from Pakistan and Sri Lanka into India, where authorities reported nearly 2.5 million cases, less than half the infection rate of 1975. During the rise in petroleum prices, health officials lacked the budget to buy oil-based insecticide and failed to supply the populace with adequate chloroquine, the standard antidote for fever. Over a three-year period, malaria sickened around 43.5 million people, boosting the Indian subcontinent to the status of world leader in infections.

late December 1974 Around Lake Victoria, epidemic cholera in Kenya felled 2,773, killing around 28 percent of its victims. Imported by traders, Islamic pilgrims, and herders, the disease subsided in its fifth month.

1976 The hantavirus (or Hantaan virus), which causes hemorrhagic fever, emerged in Korean cities in the 1970s after affecting rural areas and killing thousands each autumn, including 12,000 soldiers during the Korean War. In 1976, a virologist, Ho Wang Lee of Korea University in Seoul, located the virus in the lungs of field mice. Additional speculation linked the disease with cases of "trench nephritis," which killed soldiers in rat-infested dugouts and foxholes during the American Civil War and World War I.

early January 1976 Private David Lewis, the first victim of swine flu, fell ill at Fort Dix, New Jersey. He joined his platoon on a night hike and collapsed and died. On shaky evidence, President Gerald Ford identified the virus as the cause of the 1918 flu epidemic and had himself vaccinated to set the example for citizens. With a congressional appropriation of $135 million, he initiated a national inoculation program that raised serious questions about justification for widespread panic. However, two deaths reported on October 11 quelled the public's zeal for vaccination. Only five million Americans received the shot. The demand for inoculation dried up almost completely after the December 14 announcement of 24 Guillain-Barré syndrome cases among people receiving the swine flu vaccine. The disease acquired its name in 1932 after researcher Richard Shope, author of *Swine Influenza* (1931) and *Virus Diseases* (1934), investigated the disease by swabbing the noses and mouths of healthy pigs with secretions from sick pigs.

July 1976 In the first two decades after its emergence, Ebola killed 800 people. The filovirus first rocked Yambuku, Zaire, where 88 percent of 318 victims died. The initial outbreaks of the bizarre bleeding malady caused fever, pain, organ failure, and a swelling of cells that produced hemorrhaging from all body orifices. The disease, studied in Belgium by the Antwerp Institute of Tropical Medicine in October 1976, took its name from a river in the Congo.

Virologists identified Ebola in southern Sudan, where 151 or 58.1 percent of 284 cases were fatal. The first official victim, Ugawa, was a cloth room clerk in the N'zara Cotton Manufacturing Factory in N'zara, Maridi, where medical care was unsterile and lacking in barrier nursing techniques. The ingathering of hemorrhaging patients spread the virus person-to-person to other occupants of the building, infecting 76 members of a staff of 230.

The epidemic peaked in September. After some of the remaining hospital workers fled, a handful of dedicated nurses tended the sick in a mud hut. Strict regulation of all traffic sealed off the region, prohibiting the spread of the Ebola virus to urban areas until the end of the crisis in November. Medical sleuths never pinpointed the source of the contagion. *See also* **early August 1979.**

August 2, 1976 Several attendees of an American Legion convention in Philadelphia from July 21 to 24 died of a sudden respiratory ailment similar to pneumonia. Researchers dubbed the unknown pathogen Legionnaire's disease. Of the 182 cases reported among people staying at the Bellevue-Stratford Hotel, 82 percent bore the same symptoms. Public panic fueled an increased fear of swine flu.

Legionnaire's disease cropped up in 1977 in California, Ohio, Tennessee, and Vermont and at a hospital in Nottingham, England. Charles C. Shepard and Joseph E. McDade of the Centers for Disease Control and Prevention identified the bacterium and named it *Legionella pneumophilis*. Isolation of the bacterium in cooling towers, air conditioning systems, ponds, and water systems established that the pathogen had been around for years, but had produced cases of pneumonia that no epidemiological evidence had connected to the germ.

1977 Three diagnosticians wrote on their observations of illness in 12 adults and 39 children in three Connecticut communities near the town of Lyme. The patients exhibited a periodic inflammation of the knee and other joints. Because there was no malformation of the joints, the trio attributed the outbreak to a previously unidentified disease they called Lyme arthritis.

1977 An epidemic of the Japanese B strain of encephalitis virus spiked the death rate in densely populated Uttar Pradesh, India. Over a two-year period, the disease generated 55,500 reported cases, perhaps reaching near two million in all.

May 1977 The spread of H1N1 flu in northeastern China among children and young adults bypassed older people who had survived Asian flu in 1968. By November, when the disease invaded Russia, it earned the names red flu or Russian flu. Contagion moved from Hong Kong, the Philippines, and Singapore, Malaysia, into Vladivostok and west across Siberia to Moscow. In January 1978, cases in the United States affected mostly children. By February, the pandemic had made its way into central Europe, Greece, Israel, and Scandinavia.

summer 1977 An unfortunate removal of retarded children from a state institution for enrollment in New York City public schools failed to take into account their infection with hepatitis B. The event created a clash between the medical community's recognition of carriers' rights versus regard for risks to public wellbeing.

August 1977 In Egypt, the first major emergence of Rift Valley fever, a human hemorrhagic fever, killed 598 of the 18,000 clinical cases, mostly in the Aswan region and throughout the Nile delta. Epidemiologists surmise that there were 200,000 victims in all. Usually found south of the Sahara, the fever had remained isolated in Kenya among sheep. The bunyavirus advanced through Sudan by camel or on sheep or possibly on migrating flocks of birds, infecting humans and livestock and causing mass bleeding, encephalitis, and liver failure. The virus returned the next year in the Sinai among United Nations forces.

October 22, 1977 Discovery of a case of smallpox by the World Health Organization's eradication team disproved earlier claims to global annihilation of the disease. The patient, Ali Maow Maalin, a 23-year-old hospital cook at Merka south of Mogadishu, Somalia, assumed he would die. He had contracted the pox from a family of nomads whom he transported to the hospital in his Land Rover. The nomads' child did not survive, but Maalin did. WHO declared the disease extinct.

1978 Emergence of icteric viral hepatitis or hepatitis E in the Kashmir Valley produced 52,000 cases and 1,560 deaths. When it first appeared, the disease favored more pregnant women than non-pregnant women or men and tended to infect those women in the second trimester of gestation. The infection migrated to Indonesia, Myanmar, Nepal, and Thailand.

1978 Borne by the *Aedes albopictus* mosquito, dengue fever emerged in Kwangtung, China, killing 14 of over 22,000 victims.

March 29, 1978 A polluted well caused a two-month outbreak of cholera in the Maldive Islands following celebration of a national holiday at Male, the nation's capital. Of the 11,303 reported cases, 252 died. The chlorination of 35,000 wells generated a new health hazard by killing fish that ate mosquito larvae, the source of dengue fever.

May 1978 When epidemic cholera invaded the Kivu province in Zaire, it infected women at a higher rate than men. An inoculation program offered some control of infection rates.

September 11, 1978 Refuting the claim of the end of smallpox outbreaks were the unexplained infection and death of Janet Parker, a 40-year-old medical photographer at a pathology laboratory of the University of Birmingham, England. No one knows how the virus found its way to the building's ventilation ducts over her desk a floor above the lab. Because of Parker's infection and subsequent death and the infection of her mother, who survived, virologist Henry S. Bedsen, director of the lab, killed himself by slashing his throat. An international debate erupted concerning the future of the virus, which some claimed was not worth preserving and others countered should be spared as a source of inoculations if the pox ever reappeared during biological warfare. *See also* **May 8, 1980.**

1979 When African swine fever emerged in Haiti in 1979, Caribbean and U.S. health authorities overreacted in an attempt to halt the spread of infection throughout the Western Hemisphere. Island health authorities forced pig farmers to slaughter all swine. The loss devastated the Haitian peasant economy and culture.

1979 The Centers for Disease Control and Prevention (CDC) in Atlanta, Georgia, began investigating AIDS, a complex disease originally called the gay cancer because of its rapid attack of young male homosexuals and bisexuals. The disease is a syndrome complicated by pneumonia, toxoplasmosis, cryptosporidiosis, and retinitis. It also triggered Kaposi's sarcoma, a disease named for German clinician Moriz Kaposi, author of *Pathol-*

ogy and Treatment of Diseases of the Skin (1895), a classic text on dermatology. The syndrome caused large purple blotches on the skin that earned the slang name "gay cancer."

The CDC traced infection through sexual contacts that led to French-Canadian flight attendant Gaetan Dugas of Montreal, dubbed "Patient Zero." A promiscuous employee of Air Canada, he spread the virus through anal intercourse with 250 sex partners. Although he died of kidney failure on March 30, 1984, the contagion quickly reached Australia, Brazil, Haiti, Holland, Japan, and Uganda. In all, 40 cases caused by sexual contact led to him. In some instances, he deliberately exposed his sex partners to the virus without warning them of the danger. In 2016, scientists at Emory University exonerated Dugas, the victim of a cruel urban legend.

April 1979 An upsurge in Kawasaki disease, a threat to the vascular and cardiac systems in children, produced 6,700 cases in Japan. By May 1982, the puzzling malady more than doubled in occurrence to 16,100 cases. In March 1986, numbers still remained high at 14,700 cases, with arterial damage in 20 percent of cases and fatalities in fifteen. *See also* November 20, 2011.

April 1979 Some 30,000 Fijians developed Ross River fever, the first incidence of the virus outside Australia. By late summer, Cook Islanders, New Caledonians, Samoans, and Tongans were experiencing the symptoms.

April 2, 1979 In the Ural Mountains outside Sverdlovsk (now Yekaterinburg) 850 miles east of Moscow, Russia, from April 2 through April 19, 1979, the spread of anthrax resulted from an accident at a nearby biological weapons laboratory, where workers were aerosolizing *Bacillus anthracis* bacteria for the purpose of creating weapons of bioterrorism. Officials immediately denied the nature of the security breach, but could not refute the 68 fatal cases over the next two months, the worst anthrax epidemic in any modern industrialized nation. Victims included 96 people, most employed by the military, and livestock downwind of a military installation where an explosion released spores into the atmosphere. Eleven people suffered severe illness. Had the wind drifted to Moscow, the death rate could have risen to hundreds of thousands.

The truth of the incident did not surface until after the collapse of the Soviet Union at the end of 1991. In February 1992, Russian President Boris Yeltsin admitted to a cover-up of violations to the 1972 Biological Weapons Convention. To protect U.S. citizens, the government established a National Pharmaceutical Stockpile of over 400 tons of antibiotic Ciprofloxacin and other medical supplies, positioned at eight locations around the country in case of a terrorist attack involving biological or chemical agents.

early August 1979 When the Ebola virus recurred in N'zara, Sudan, only 35 percent of the 56 victims survived. The recrudescence began in the same fashion as the first incident, with an infection in a cotton factory worker. To protect families, an expert dispatched by the World Health Organization attempted to wash corpses for burial, but was too short-staffed to conduct the ritual. Again, medical forensics found no clue to the source of the virus.

December 24, 1979 The Soviet-Afghan War placed serious stress on the Soviet society and its draft army of 642,000. In addition to losing 15,000 in combat, the military reported another 469,685 casualties, which comprised 73 percent of the amassed force. Notably, 115,308 contracted infectious hepatitis and 31,080, typhoid fever, a unique toll for a modern army treated by modern medicine. The loss caused citizens to ostracize and scorn veterans, who found it difficult to readjust to civilian life.

1980 Global incidence of measles reached 4,211,432, with Africa reporting 1,240,993 or 29.5 percent of the total. Ghana listed the virus as the second major killer of children after malaria. In the Americas, Mexican health officials reported a 20 percent increase in infections. In 2010, the World Health Assembly selected measles as a scourge to eradicate by 2015.

May 8, 1980 In December 1979, an independent consortium of scientists attested to the global eradication of smallpox, the result of a vaccination program begun by the World Health Organization (WHO) in 1967. After numerous premature proclamations, on May 8, 1980, WHO's Geneva headquarters declared smallpox no longer a health danger.

June 1980 A re-emergence of viral conjunctivitis in Singapore, Malaysia, afflicted 49,000 victims. Medical sleuths determined that an enterovirus was the cause of eye inflammation and hemorrhaging.

November 1980 California physician Joel D. Weisman of Los Angeles reported the first official AIDS case in the U.S. By 1989, the patient, an ailing advertising manager, preceded 100,000 others suffering the same viral complex. Weisman commented, "All I can think of, day in and day out, is all the people who have died—young people, all dead, in such a short period of time since I first saw that patient. It's like some story from science fiction" (Shilts 1989).

Randy Shilts, a reporter for the *San Francisco Chronicle* who became the most knowledgeable media source on AIDS, died of the disease in February 1994 at age 42. In the intervening period, he produced a best-selling book, *And the Band Played On: People, Politics and the AIDS Epidemic* (1987), which exposed ineptitude of the Reagan administration and callousness toward gay people in failing to take immediate action to contain the virus. By the year 2000, 40 million people worldwide were victims of AIDS. *See also* **1983**.

1981 The hospitalization of 116,000 Cuban patients with dengue hemorrhagic fever resulted when 344,000 came down with the viral disease after its introduction to the Western Hemisphere by the *Aedes aegypti* mosquito. The victims suffered a measles-like rash, bleeding gums, aching limbs, and depression; in infants, bleeding gums, leaking capillaries, and bloody vomit, feces, and urine produced a fatality rate of 15 percent. In Cuba, 158 died. The virus also plagued Ecuador, Venezuela, and Peru sequentially in 1988, 1989, and 1990.

1981 At a European Summit Conference in Maastricht, Holland, a cold buffet prepared in a kitchen contaminated with *Salmonella indiana* produced 700 illnesses.

1982 In Uganda, tsetse flies that proliferated on coffee and tea plantations spread sleeping sickness at an alarming rate. The situation had been developing since the rise of Idi Amin in 1971. On his orders, government agencies ousted Asian planters and shifted from control of health conditions to the bolstering of national military defense. Local people stopped draining swamps and spraying hazardous areas. Plantations left untended ran wild with weeds, the breeding ground for flies, carriers of the deadly *Trypanosoma brucei* protozoa.

Until Médecins sans Frontières (Doctors without Borders) began working in the region, Ezra Gashihiri, medical officer at the district hospital in Iganga, fought sleeping sickness without supplies or assistance. He lacked whole blood, intravenous fluids, spinal-tap kits, and laboratory equipment to combat and monitor the progress of the disease. Comatose patients, mostly very young, lay on crusted mattresses as they gradually slipped away. After international relief agencies donated emergency supplies, local thieves hijacked the goods for sale on the black market.

1982 In the Volta River region of Ghana, incidence of onchocerciasis or river blindness affected demographics. People realized that the rivers bore contagion and moved into the uplands, which lacked the natural resources to support the influx. Brides refused to leave villages untouched by the disease, which caused unbearable itching, disfigurement, debility, and a life expectancy lowered by 15 years. Wifeless young men moved from the lowlands to jobs in Bobo-Dioulasso and Ougadougou. Decimated rural farms lay neglected as the pathogen infected workers, prematurely aging them. In the words of David Lamb's *The Africans* (1982), "Though the land may support him and feed him, the African, in the end, remains the slave of his environment. The land owns him and dictates his destiny" (p. 266).

To improve the lives of Voltans in Benin, Burkina Faso, Ghana, Guinea, Ivory Coast, Mali, Niger, Senegal, Sierra Leone, and Togo, the World Health Organization spent $10 million annually from 1974 to quell onchocerciasis. Health officials battled the scourge by monitoring the biting midge *Simulium damnosum*, studying health profiles, clearing land, draining swamps, and spraying fly-breeding grounds from helicopters and turbo-porters with the biodegradable insecticide temephos and with *Bacillus thuringiensis*, a natural biocide. A team of tropical specialists from the *French Institut de Recherche pour le Développement* in Paris monitored ecological strategies. The river-by-river campaign, which had already reduced disease in Kenya, proved tedious, but effective. As a result, farmland began to flourish and rice fields to produce enough grain to stabilize the economy.

November 16, 1982 At a retirement home in Ottawa, Ontario, 31 of the 353 residents contracted gastroenteritis, possibly from club sandwiches contaminated with *Escherichia coli*. The National Enteric Reference Centre of the Canadian Laboratory Centre for Disease Control isolated the pathogen causing a deadly hemorrhagic colitis that killed 19 people.

1983 French medical theorists Françoise Barre-Sinoussi and Luc Montagnier of the Pasteur Institute in Paris officially named the "gay cancer" AIDS and began piecing together disparate clues linking hemophilia and blood transfusions, sex partners of people infected with AIDS, gay sex, intravenous drug use, Haitians, and medical workers in Zaire. Globally, an understanding of the virus altered the medical practice of dental hygienists, medics, nurses, and physicians and surgeons, all of whom adopted disposable rubber gloves and plastic masks as barriers to intimate contact with body effluvia. Nurse-author Jeanne Parker Martin compiled directives to the hospice caregiver in *The Person with AIDS: Nursing Perspectives* (1983), a handbook on food, hygiene, and housekeeping advice.

1983 The hemorrhaging that invaded brain, lungs, stomachs, and eyes of chickens in Pennsylvania caused the U.S. Department of Agriculture to kill 20 million domestic birds to prevent the avian virus from spreading to humans. Additional outbreaks occurred in turkeys in Great Britain in 1991 and shortly afterward among chickens in Mexico.

1983 In Kos, Greece, tourists incurred 58 cases of typhoid from a tomato salad prepared with foul water.

1983 In Zimbabwe, the Italian Association of Friends of Raoul Follereau joined government health officials in an attempt to exterminate leprosy. The campaign, which the Friends had launched five years earlier in the Cape Verde Islands, honored the altruism of Follereau, a crusader for lepers worldwide. Staff began with the training of local health workers, setting up clinics, and destigmatizing the disease. After five years of progress, the program flagged as a result of an erroneous assumption that the effort had eradicated leprosy.

1983 Of the 684 U.S. Army members taking part in the intervention on Grenada, 56 soldiers

of 20 percent of the deployment reported the pain and diarrhea of hookworm infestation. Military health workers traced the infection to first-wave operations near homes exhibiting poor sanitation.

1984 On transatlantic flights issuing from Heathrow Airport outside London, aspic glaze on hors d'oeuvres served to passengers bore *Salmonella enteritidis*. The contamination generated 750 illnesses. That same year, in Canada's Maritime Provinces, distribution of cheddar cheese contaminated with *Salmonella typhimurium* sickened 1,500 people. In August in The Dalles, Oregon, members of OSHO, a religious cult led by the Guru Bhagwan Shree Rajneesh, polluted salad bars in ten restaurants with *Salmonella typhimurium* to prevent the town's 12,000 citizens from voting in a county election. After the pathogens infected 753, over 1,000 people sought treatment at Mid-Columbia Medical Center. Two female cultists, Sheela and Ma Anand Puja, later confessed to the act of biological terrorism.

The next year, a major U.S. outbreak of *Salmonella typhimurium* affected over 16,000 verified cases. It was most likely the result of unpasteurized milk from a source in Illinois. Investigators identified a bacterial strain resistant to ampicillin, carbenicillin, streptomycin, and tetracycline.

July 1984 Protozoa from the Edwards Aquifer caused 2,006 residents of Braun Station outside San Antonio, Texas, to suffer the dehydration, gut pain, and diarrhea of cryptosporidiosis, the first outbreak linked to drinking water. Data confirmed that sewer pipes leaked oocysts into wells. The infestation rate reached 34 percent.

mid–1980s Cuban dictator Fidel Castro's militaristic public health initiative battled an outbreak of HIV as veterans returned from service in Angola. To stem the rise in viral infection, in 1986, when Cuban health workers identified the first positive HIV test, the island's nationalized health care system began screening teens for HIV and quarantined victims involuntarily in *sanitorias* (isolation camps) without recourse to legal hearings or appeal. World reaction to these gulags resulted in pressures on governments from humanitarian sources demanding that the Cuban medical bureaucracy accord victims their civil rights.

When the Soviet Union collapsed, Cuba no longer had the resources to jail patients for life.

Thus, the Cuban *sanitoria* system became voluntary rather than mandatory. The relaxing of strictures came gradually. In 1989, Jorge Perez, director of the Institute of Tropical Medicine, restored civil rights by granting patients unsupervised furloughs from *sanitorias* for extended periods. In 1993, the island medical bureaucracy established an ambulatory care treatment program to allow patients to chose treatment at a *sanitaria*, or living at home. That same year, Merck pharmaceutical company chose Cuba as a place to test Crixivan, an HIV inhibitor. The U.S. government, in pursuit of its embargo against Cuba, charged and fined Merck $125,000 for violating state policy.

mid–1980s A critical care nurse and president of the California Nurses Association, Helen Miramontes, mother of a son afflicted with AIDS, became an activist while working with AIDS patients at Kaiser Permanente in Santa Clara. She served as Nurse Coordinator of Pacific AIDS Education, an advocacy for health education and public sanitation that won her the 1992 Pearl McIver Public Health Nurse Award for humanitarian involvement. As a member of the Presidential Council on HIV/AIDS, she joined a consortium of 50 volunteers to test an HIV vaccine during its Phase III clinical trials, which evolved from the research of Ronald Desrosiers, a microbiologist at Harvard Medical School's New England Primate Center.

1985 After a half-century of progress in the treatment of tuberculosis, incidence of the disease in the United States rose by 18 percent from 1985 to 1991. Health workers surmised that factors contributing to a new pandemic included loss of immunity through HIV and other forms of debilitation, drug abuse, poverty and homelessness, overcrowding in prisons and nursing homes, and an influx of TB-positive people from Africa, Asia, and Latin America.

January 1, 1985 Staff at Primary Children's Medical Center in Salt Lake City, Utah, reported an upsurge in acute rheumatic fever. When the board of health began logging cases, agents noted 136 infections in the calendar year that spread over 20 of 29 counties. Most clustered near urban areas during early spring.

May 14, 1985 The Pan American Health Organization (PAHO), directed by Carlyle Guerra de Macedo and Ciro A. de Quadros, determined to end polio in the Western Hemisphere through mass campaigns. Guerra de Macedo announced confidently, "The time has come for us to say that it is unacceptable for any child in the Americas to suffer from polio" ("Polio, the Beginning of the End"). Mobile health initiatives, paid for in part by Rotary International, succeeded by publicizing national immunization days. Full distribution of vaccine required PAHO to engineer a ceasefire during combat in El Salvador. In Peru, warriors from the Shining Path halted hostilities and assisted in the inoculation program.

With the combined efforts of PAHO, UNICEF, the U.S. Agency for International Development, the Inter-American Development Bank, and the Canadian Public Health Association, the last recorded case of polio occurred in Junín, Peru, in August 1991, when the virus infected two-year-old Luis Fermín Tenorio Cortez. He survived and returned to normal life. In September 1993, an international certification committee declared the disease eradicated in the Americas. The successful program encouraged the World Health Organization to attempt world eradication in 1988.

1986 Symptoms of mad cow disease in southern England spread from a dairy herd to the dairyman. Epidemiologists deduced that the feeding of sheep meat to cattle transferred ovine scrapie (*spongiform encephalopathy*) to cows. In 1988, a ban on the processing of ruminant protein and bone meal to living cattle ended a cycle that spread the brain-based neurodegenerative anomaly. By 1993, the Creutzfeldt-Jakob virus had peaked and began a steady decline. *See also* **March 1996.**

June 1986 In Nigeria, yellow fever began a five-year siege, the nation's most virulent epidemic of the virus. The death rate for 16,230 reported cases approached 23 percent, but unofficial estimates ranged much higher. Most of the diseased were children and youth. The scourge moved into Cameroon and Niger, but did not invade the rest of Africa.

September 1986 At Xinjiang, China, spread of hepatitis E through polluted drinking water infected around 119,300 over a 20-month period. Most at risk were the Uigur, a minority people living at the base of Mt. Tianshan, who drank from contaminated canals and ponds. One quarter risked chronic liver disease.

October 1986 Epidemic hemorrhagic conjunctivitis at Jamaica, St. Croix, and Trinidad afflicted up to 100 people a month. By December, the number of victims reached 15,396, who suffered the effects of acute inflammation with a coxsackie virus.

1987 After years of dormancy, black sickness, a local name for kala azar or visceral leishmaniasis, struck Bangladesh, India, and Nepal, an area inhabited by 110 million people. The World Health Organization brought in teams of entomologists and biochemists to study the outbreak and propose control measures.

1987 Local mosquito spraying around Bangkok, Thailand, did not stop the crisis outbreak of dengue fever, which sickened 170,000 and killed 126. Health department officials blamed urbanization for the burst of infections among young adults in the capital. Symptoms began with spiking fever, rash, and aching joints before attacking the liver. Epidemiologists anticipated the formulation of a vaccine.

1987 Within two years after South African health authorities required tuberculosis inoculation, the disease surged again among blacks. Most at risk were those living in overcrowded city housing projects amid poor sanitation and limited health care. By 1990, the caseload had risen to 124,000 and continued to spike.

1987 Rift Valley fever broke out in Mauritania and Senegal during completion of the Diama Dam at the Senegal River delta. Construction debris encouraged the breeding of virus-bearing mosquitoes, carriers of the bunyavirus. The death toll rose to 200.

1987 Malaria in Madagascar killed 100,000, in part because the government failed to spray for disease-bearing mosquitoes after islanders gained independence from France. In a joint humanitarian effort, France and Italy airlifted medicines, sprayers, and insecticide. The Swiss bankrolled enough chloroquine to protect citizens through 1988.

February 7, 1987 In Carroll County, Georgia, 13,000 or 20 percent of residents reported nausea, fever, vomiting, dehydration, stomach cramps, and diarrhea from cryptosporidiosis. Public health officials suspected that cattle introduced the protozoa to a river that supplied area water.

1988 Screenwriter and elegist Paul Monette composed from personal experience *Borrowed Time: An AIDS Memoir*, a first-person testimony to 19 months of observation and care of an AIDS patient, his mate Roger Horwitz. The dizzying recounting of clinics, home nursing, pioneering treatment with AZT, compromised human systems, physical degeneration, and collapsing veins explained the catastrophic toll of the disease in physical and emotional pain. Monette listed as "minefields" the distancing of neighbors and relatives from contagion and the tedious protocols of swabbing with alcohol and Betadine and sterilizing with bleach as anti-contagion procedures. The disease killed Roger as well as Monette's subsequent mate and Paul himself, who died February 10, 1995, of AIDS complications.

1988 In Ethiopia, the incidence of HIV increased in three years from one case to an infection rate of 17 percent. At urban Addis Ababa, the rate had reached 24 percent. By the mid–1990s, there were 355 AIDS patients, a contagion rate much smaller than the rest of African nations.

1988 To halt the AIDS epidemic, Japanese lawmakers enacted the Law Concerning the Prevention of AIDS, effective in 1989. It required doctors to explain the manner of transmission and to report to the prefectural governor each patient's name, age, and address, along with how the patient contracted the virus. The law exempted from formal registration patients contaminated through imported blood products. Because the diagnosis preceded a second medical examination and more instructions, critics charged that the procedure violated patient rights.

early 1988 An outbreak of hepatitis A in Shanghai resulted from consumption of four million pounds of fecal-contaminated clams harvested from a part of the shoreline where a river carried sewage to the sea. The epidemic offered medical historians a unique view of incubation and diagnoses in a major urban population. According to observers, the Chinese, who preferred steaming or soaking the shellfish in hot water, failed to raise temperatures high enough to kill the virus. Of some 320,750 cases that resulted, the majority were young adults who produced symptoms between January 20 and February 10. The death toll rose to 50, mostly comprised of people over age 50.

summer 1988 The British government issued warnings of *Listeriosis monocytogenes* or listeria after an epidemic of food poisoning resulted from consumption of imported pâté in England and Wales. Health officials tried to protect vulnerable population groups from contaminants in soft cheese, pâté, and foods that required chilling after cooking. These high-risk foods also demanded more stringent production standards to reduce incidence of disease.

summer 1988 An epidemic kala azar or visceral leishmaniasis, called "dumdum fever," emerged in southern Sudan (Cohen 2001, p. 327). Contributing to the people's misery, civil war hindered medical care, survey of the epidemic and mortality, and control measures. The 100,000 deaths appear to have approached a 70 percent fatality rate. Malnutrition and poor hygiene worsened the chances of survival.

In Duar, an isolated part of southern Sudan on the banks of the White Nile River, Jill Seaman of Moscow, Idaho, a nine-year veteran with Médecins sans Frontières (Doctors without Borders), battled the fever. Because of lack of precedents, she designed protocols and treatments, which she administered to a total of 10,000 patients. Canadian entomologist Judith Schorscher spent six months pinning down the vector—*Phlebotomus orientalis* sandflies, which spread the killer protozoan disease.

Aiding the two-woman team were Dutch nurse Sjoukje De Wit and Francis Galiek, a nurse from Nuer who ran the operation when the outsiders departed. Of the 70 Nuer and Dinka nurses Seaman and her staff trained, over 75 percent suffered kala azar. Five saw their own children die of the disease.

Each case required constant patient monitoring during bone marrow or spleen diagnosis and a 20-day course of injections of a potentially lethal antimony-based drug. Some 1,400 patients arrived on foot to the tent clinic and camped along with families and cattle; 600 more sought help to the south at Ler. Complications of the seven-year epidemic arose from problem pregnancies, spear wounds, fungus, outbreaks of measles and meningitis, and secondary malaria,

pneumonia, and tuberculosis. As thanks for Seaman's dedication, Chief Tongwar named many of his daughters and sons after Jill. Villagers sacrificed a cow to her honor and named her Chotnyang (brown cow without horns), a reference to her pacifism.

summer 1988 In England, Germany, Holland, Ireland, and Scotland along the North Sea, government officials feared for human health after 18,000 harbor seals died of a picornavirus, which resembled hoof-and-mouth disease.

November 1988 In Peru's first occurrence of botulism, twelve residents of Huancayo east of Lima fell ill from a restaurant meal of salchipapas, a combination of hot dogs, eggs, fried potatoes, and mayonnaise. The death rate from the type B botulin toxin reached 16.7 percent.

1989 Canada's public health service lowered the incidence of HIV/AIDS among intravenous drug abusers through a needle-exchange program. In Toronto, the control method produced the world's lowest rate of new infections.

1989 Meningitis struck southern and western Ethiopia as far south as Wollega in Oromiya. The region is part of the African meningitis belt, which includes Burkina Faso, Mali, Niger, and Nigeria. The area had been visited by outbreaks in regular eight-to-twelve-year cycles since 1902. Epidemiologists explained the phenomenon as the result of environmental factors, particularly a unique strain of the *Neisseria meningitidis* virus.

To protect the healthy, agents of the Ethiopian government vaccinated 1.6 million residents. Least protected, the rural populace, 70 percent of the headcount, lived over three days' walk from health centers having the required electricity and refrigeration for storing vaccine. Attacking 133 of every 100,000 people, the disease surged among young males and produced over 2,000 fatalities among 46,000 cases.

February 1989 Rampant cholera in Africa killed 12 percent of its 210,000 victims. Hard hit, Kitwe and Lusaka, Zambia, lacked controls of privies, garbage disposal, and human burial. The water-borne disease spread over sub–Saharan Africa through contaminated grain cooked at street stalls. Victims ingested pathogens from raw or partially cooked seafood caught in contaminated waters. In this same period, the Centers for Disease Control and Prevention in Atlanta, Georgia, noted heavy infection in Xinjiang, China, with milder outbreaks in Japan, Kuwait, Macao, Myanmar, and Nepal.

February 3, 1989 The death of a Nigerian mechanical engineer in Chicago, Illinois, after a visit to his homeland introduced Lassa fever, a viral hemorrhagic disease, in the United States. Identified in Nigeria in 1969, the fever, which is transmitted by rat saliva and urine, killed 5,000 in 1989. Quick action by the Centers for Disease Control and Prevention in Atlanta, Georgia, halted potential spread in Illinois.

summer 1989 Botulism, which is generally rare in Great Britain, produced the nation's largest community outbreak. In northern Wales and northwest England, one person died among 27 infected by hazelnut yogurt contaminated with *Clostridium botulinum* type B toxin.

1990 Globally, according to the World Health Organization, malaria remained pandemic, causing more sickness than Chagas's disease, filariasis, leishmaniasis, schistosomiasis, and sleeping sickness combined:

Disease	Cases
malaria	800,000,000
filariasis	300,000,000
leishmaniasis	200,000,000
schistosomiasis	200,000,000
Chagas's disease	12,000,000
sleeping sickness	1,000,000

1990 As civil unrest and lawlessness overran Malawi and Mozambique, peasants struggled to grow vegetable gardens. The increase in refugees and displaced persons produced high incidence of pellagra from a deficiency of B-complex vitamins in the diet. After investigation, disaster relief teams reported over 18,000 cases of the disease.

1990 Iraqi children incurred high mortality rates during the U.S. embargo of their country. According to the World Health Organization, "The vast majority of the country's population has been on a semi-starvation diet for years" (Adebajo 1998). Within five years, the death rate for infants doubled; the mortality rate for pre-schoolers rose six-fold.

At the outbreak of the Persian Gulf War, enteritis and respiratory ills, exacerbated by malnutrition and lack of vaccination and basic health care, boosted deaths in toddlers and preschoolers. The health care bureaucracy stinted on control of cholera, kwashiorkor, measles, tetanus, and typhoid. In comparison with children in Jordan, the Iraqi had more mothers unscreened during pregnancy, more children born at home without medical intervention, limited measles immunization, a lower rate of breastfeeding, and a higher rate of foods and formula prepared with impure water. Iraqi females born in rural areas to families from low educational and income levels faced greater survival risks.

1990 Because of Peru's vulnerability to epidemic, cholera raged through the populace. Fatal infections swept the outback; urban dwellers of Arequipa, Lima, and other Pacific coast cities fared better because of the location of major medical facilities. Whereas the highlands had a doctor for every 12,000 Peruvians, the numbers were one per 400 in the capital city. The advantage of sewer systems and potable water also boosted survival in metropolitan areas.

Those Peruvians living in rural squalor bore the brunt of chicken pox, gastroenteritis, influenza, malaria, measles, pertussis, respiratory ailments, and tuberculosis, all of which killed off young children at a rate of 40 percent of all recorded deaths. Additional advantage in cities came from wellness programs that limited barrels, drums, open cisterns, tin cans, and tires where stagnant water bred the *Aedes aegypti* mosquito. Similar lopsided situations spread the virus among the unvaccinated poor of Bolivia, Brazil, Colombia, Ecuador, Panama, and Venezuela.

1990 WHO reported 630,000 deaths from 1,374,083 cases of measles globally, chiefly in developing countries. Africa led the total with 481,024 infections compared to the Western Hemisphere with 218,579.

1990 A virgin population in the Andes Mountains of Peru suffered its first outbreak of *Bartonellosis bacilliformis*, a potentially lethal fever that killed 88 percent of victims, most of them children. Carried in sandflies, the endemic flourished in stone and adobe residences, where the insects nested in walls. Patients suffered stiffness, chills and fever, and pustules that exuded fluid and matter. Advanced cases resulted in migraines, seizures, and endocarditis or inflamed heart valves.

early August 1990 Some 2,000 fair-goers at a North Dakota threshing exhibition on July 28–

29 appeared to have contracted enteritis from roast beef served at a buffet.

December 1990 After identification of the first HIV victim in Vietnam, the disease advanced at an alarming rate. Worsening the toll on human health, an economic detente spurred hordes of tourists and the flow of capitalist investments. Spreading infection derived from shared needles in hospitals, illicit drug use, and the daily work of some 600,000 prostitutes in a sex trade that favored visiting Westerners.

1990s Over half the reported cases of leprosy globally occurred in India, where the disease had ravaged the populace for centuries. The Leper Act of 1992 banned some four million sick and disfigured patients from public transportation. By the end of the 20th century, new cases fell to a half million.

1990s Across the United States, some four million people contracted hepatitis C, an infection rate four times that of AIDS. Striking mostly non-white adults, the disease flourished among drug addicts sharing needles and patrons of tattoo parlors and body-piercing salons as well as patients of dialysis clinics. Paralleling the epidemic, a mounting problem with sexually transmitted disease afflicted 12 million per year. The spread of chlamydia, genital warts, gonorrhea, hepatitis B, herpes, and syphilis endangered women and babies and precipitated new HIV infections.

1990s In South Africa, although health workers virtually eradicated diphtheria, leprosy, and smallpox, they failed to control typhoid fever, which rose significantly among blacks as compared to whites and Asians. The number of cases was 4.4 times higher among urban blacks living in crowded, foul-smelling slums with limited drainage and sewage treatment. Likewise, the headcount for measles infection found blacks over twice as likely to contract the disease, which sickened up to 220 per 100,000.

early 1990s When malaria and typhoid fever invaded Tajikistan during civil war, Medical Emergency Relief International treated the sick in Kulyab. Volunteers also distributed medicines, launched rehabilitation programs, and trained technicians in epidemiology at eight laboratories and 17 infectious disease hospitals.

early 1990s Around a third of the world's population, some 1.7 billion people, carried *Myco-bacterium tuberculosis,* a bacteria causing TB. Of these, 95 percent lived in developing countries.

1991 The collapse of the Soviet Union generated social and economic ills that turned despairing and unemployed Ukrainians to unsafe sex, alcoholism, and intravenous drug use as escapes from national upheaval. The increase promoted the spread of HIV from contaminated drugs and shared needles. By 2000, the World Health Organization reported the extent of the epidemic at 700,000.

A simultaneous resurgence of diphtheria in the Ukraine forced the health service to request aid from the United Nations and U.S. relief agencies to treat the 1,103 victims. With health volunteers armed with adequate vaccines, the Russian government began inoculating citizens. Despite a coordinated effort, the disease spread among children and young teens in Ukraine and sickened 1,869 in Russia. The infection advanced on Belarus, Latvia, Lithuania, and Norway. The outbreak continued until 1995, when contagion rates began to decline.

1991 The World Health Assembly committed its membership to eradicating dracunculiasis or guinea worm infection, one of humanity's oldest parasitic infestations. The scourge once inhabited the sub–Saharan nations of Nigeria, Ghana, Cameroon, Benin, Burkina Faso, Central African Republic, Chad, Ethiopia, Ivory Coast, Kenya, Mali, Mauritania, Niger, Senegal, Sudan, Togo, and Uganda in a geographic band lying between the equator and the Sahara Desert. At one time, infestations overwhelmed the workers of Mali, creating famine. Still active in Yemen, Korea, Pakistan, Thailand, and some states of India, the parasite victimized 1.3 million people. Work on the problem reduced the number five years later to nine cases recorded in three villages in Jodhpur, Rajasthan.

Dracunculiasis followed smallpox as the second disease wiped out globally. The Carter Center, an altruistic foundation established by former U.S. president Jimmy Carter and his wife Rosalyn, supported the defeat of the parasite by filtering drinking water, disinfecting standing pools, and protecting wells and pumps from infestation. On December 4, 1995, the Carters joined the World Health Organization in celebrating an international collaboration to eradicate dracunculiasis as a Third World scourge. In 1996, data revealed no new cases.

1991 Already threatened by dengue fever, hepatitis, pneumonia, tuberculosis, and yellow fever imported by some 50,000 predatory *garimpeiros* (gold miners), the Yanonami (or Yanomami) of the Amazon Basin of Venezuela suffered 110 deaths from malaria. For their considerable losses in the previous quarter century, they stood on the brink of extinction.

1991 The World Health Organization estimated that the world's population included around nine million tuberculosis victims, most living in developing countries. Intravenous drug addiction and epidemic AIDS increased infections in Europe and the United States.

January 31, 1991 After a century free of cholera, a serious spread of gastroenteritis and severe diarrhea in Chancay, Peru, proved to be the *El Tor* variety of *Vibrio cholerae*, a virulent form that may have derived from bilge pumped into the harbor from a Chinese freighter. Contaminants began the outbreak at Chimbote, a fishing village north of Lima, on January 31, apparently from peasants consuming seafood contaminated with sewage. The pathogen infected the entire coastline before spreading to the mountains and forests, sickening 55,000 and killing 258. In Ecuador, 20 cases indicated the disease had spread beyond Peruvian borders. The Pan American Health Organization in Washington, D.C., mobilized an international relief effort.

Health care workers surmised that contaminated water in shantytowns endangered the poor. In Lima, the failure of infrastructure to keep up with population growth endangered people. At overtaxed water purification plants, human feces dirtied drinking water that had too little chlorine added to kill microbes. In the Andes, stagnant water for irrigation and fertilization of crops growing around communities also fostered bacteria-rich water, which people used to prepare raw vegetables and fish. Farther south in Argentina, widespread cholera induced the press and churchmen to acknowledge the growing underclass. To save face, President Carlos Menem turned attention from social priorities by increasing publicity of his accomplishments.

summer 1991 Western epidemiologists learned of epidemic Guillain-Barré syndrome, a form of polyneuritis dubbed "Chinese paralytic syndrome," in children of rural families in northern China. The disease, precipitated by the *Campylobacter jejuni* pathogen, disabled the axons of nerve cells, causing creeping paralysis.

August 2, 1991 Iraq admitted to a United Nations Special Commission Team the military's research into germ warfare. Planners of terrorism favored anthrax and botulism.

1992 In New York City, Clara McBride Hale, a caregiver to babies infected with AIDS, planned a hospice for dying children. After her death, her daughter, Lorraine Hale, joined Mayor David Dinkins to make Clara's dream come true. By 1998, there were 2,000 facilities for terminally ill children infected with the virus.

1992 Epidemics of visceral leishmaniasis, called kala azar (black fever) in Hindi, killed over 100,000 in India and the Sudan. The malady struck bone marrow, liver, lymphatic system, mucous linings of the upper respiratory system, skin, and spleen. A protozoan disease, it ulcerated and killed if left untreated. In Patna alone, 10,000 or 37 percent of 27,000 victims died of the disease.

January 1992 During a U.S. Army training exercise with the Botswana Defense Force outside Shoshong, Botswana, 169 soldiers contracted spotted fever rickettsiosis or typhus while undergoing mock combat in flat, semi-arid land. Around 30 percent of the soldiers developed headache, fever, and sore lymph glands as well as infected insect bites. Investigated by epidemiologists from the Walter Reed Army Institute of Research, 126 of the 169 participants showed symptoms of typhus.

January 1992 An eight-month epidemic of plague beset the Ituri forest of northeastern Zaire, producing a death rate of nearly 41 percent among 191 victims. Civil strife reduced chances of survival in an area lacking communication service and highways.

June 1992 According to John R. Lumpkin, director of the Illinois state public health agency, Chicago and suburban Cook County reported epidemic syphilis among drug abusers, especially poor blacks selling sex for crack cocaine. No longer the endemic disease of primarily white male homosexuals, the disease responded to greater use of condoms among non-white gay males seeking to avoid HIV infections.

mid-summer 1992 For nearly a year, patients in Great Britain displayed the effects of *Salmonella virchow*, an invasive serotype derived from poultry. Over 350 people fell victim to *Salmonella wangata*, a rare version common in immigrants from the Indian subcontinent or in diners at ethnic restaurants serving chicken-based menu items.

mid–December 1992 Six-year-old Lauren Beth Rudolph contracted *Escherichia coli* from an undercooked cheeseburger purchased at a California Jack in the Box. After severe food poisoning, three heart attacks, and coma, she died on December 28. Two weeks later in Washington State, 230 persons and another suspected 80 victims suffered infections with *Escherichia coli* contracted from hamburgers cooked by the same fast-food chain. The outbreak recurred in Washington, Idaho, California, and Nevada.

On January 17, 1993, Brianne Kiner of Redmond, Washington, was one of over 600 people sickened by eating undercooked hamburgers from a Jack in the Box in Seattle. Three died. Hospitalized with hemolytic uremic syndrome from infection with the *Escherichia coli* bacterium, she survived after a six-week coma and two months of intensive care. Because the ordeal caused reproductive sterility, brain damage, diabetes, and threats to her colon and pancreas, she sued the restaurant chain for $15.6 million.

In August 1993, Foodmaker Incorporated of San Diego settled a class-action suit for $44.5 million; in 1995, a shareholder class-action suit cost the restaurant chain an addition $8 million plus the cost of 500 victim claims. Brianne entered a private boarding school in New Mexico to learn to walk and talk. The Seattle law firm of Marler & Clark set aside $25,000 to establish the Brianne Kiner Foundation and donated $25,000 to a hospital. Over five states, 732 cases were reported, four of them fatal. Nine beef suppliers reached a $58.5 million settlement with the fast-food chain, which made restitution to victims' families and initiated a strict regimen of pure food handling. Nationwide, public outrage fueled changes in the way the U.S. inspected meat.

February 19, 1993 Some 23,900 of the 390,000 residents of Kitchen-Waterloo, Ontario, incurred cryptosporidiosis over a three-month endemic. Epidemiologists traced the protozoa to the Grand River sand and gravel aquifer and to an Ontario River well.

April 5, 1993 Endemic cryptosporidium in Milwaukee, Wisconsin, emptied offices and schools as 403,000 sufferers fought dehydration, fever, abdominal cramps, and persistent diarrhea. City officials advised citizens to boil their drinking water and closed a water filtration plant to rid the source of the protozoa. The epidemic hospitalized 1,000 and killed 104, most of them AIDS or cancer patients. City water inspectors replaced chlorine decontaminant with ozone, which purified and deodorized the city supply.

May 14, 1993 Near Gallup, New Mexico, in Four Corners where Utah, Arizona, Colorado, and New Mexico meet, Merrill Bahe, a long-distance runner, died a few days after the death of his girlfriend. Public health and doctors identified a mysterious pathogen that struck 13 people, several of whom were Navaho. Because of quick isolation and analysis of symptoms, by June, researchers identified the culprit as a rodent-borne hantavirus and averted a potential epidemic. Officials credited the teamwork between research facilities and government agencies, notably, Richard Malone of the Office of the Medical Investigator, the Centers for Disease Control and Prevention, New Mexico State Department of Health, Indian Health Service, University of New Mexico, and U.S. Army.

Contributing to academic knowledge was the wisdom of Navaho elders who recognized that, after a five-year drought, increase in snow and rain that winter and spring produced an oversized crop of piñon nuts, which boosted tenfold the population of deer mice, the vectors whose urine and feces spread a virus that infected humans who inhaled it. The elders recalled that similar emergence of disease occurred in 1918 and 1936 during environmental changes favoring deer mice. In November 1993, the CDC and the Army developed a diagnostic test. By 1997, around 172 cases of hantavirus pulmonary syndrome (HPS) were positively identified in 20 states and another 25 from an outbreak in Chile. Quick response in the U.S. held the death rate at 45 percent. Shifts in the food chain cycle slowed HPS, which could have killed more people if it had continued spreading.

September 1993 Epidemic measles in Queensland, Australia, raged into 1994, particularly among the young. The disease infected mainly those lacking inoculation, who were ten times more susceptible.

1994 An outbreak of intestinal disease struck hundreds of Australians and killed one young child after victims ate a fermented salami contaminated with *Escherichia coli*. Epidemiologists warned that *E. coli* can survive the dry curing of meat.

1994 In South Africa, the resurgence of tuberculosis among poor urban blacks produced 47,800 new cases and an annual death toll of 6,000 in people infected with TB and secondary ailments. Malaria also threatened life at the rate of 4,194 cases annually. Of those infected, non-white people were 6.75 times more likely to suffer the disease than blacks.

Malaria also dogged the Indian subcontinent. In western India, 4,000 deaths from malaria attested to a rise in infection since the early 1990s. Inadequate health facilities stymied efforts to train agents, distribute chloroquine, and spray for insects following heavy seasonal rains. Adding to mosquito breeding grounds, an irrigation canal doubled contagion among migrant laborers.

1994 Nicaragua experienced a two-year scourge of dengue fever, which befell 29,469 the first year and 19,260 the second. The scourge spread to Mexico.

February–March 1994 Four people in Helena, Montana, were infected by a rare form of *Escherichia coli* that caused hemorrhagic colitis. Investigated by the Lewis and Clark County Department of Health and Environmental Sciences, the Montana Department of Health and Environmental Sciences, and the Centers for Disease Control and Prevention, the outbreak appeared to have derived from milk and dairy products, which sickened 18 more cases.

April 1994 The adenovirus, introduced into wells by a river flood at rural Noormarkku in southwestern Finland caused gastroenteritis in some 3,000 people, nearly half the population.

July 1994 An epidemic of *El Tor* cholera, which first emerged at the El Tor quarantine camp in Egypt on the Red Sea in 1905, struck Rwandan refugees fleeing to Goma, Zaire. The International Red Cross and other relief workers required the aid of the French military to cope with filth, sickness, and deaths among one million homeless people. Worsening the picture were famine and enteritis as mounds of garbage began to ring the grounds. In one month, of 70,000 illnesses, 12,000 resulted in death at the rate of 17.1 percent. Overall, the annual total for cholera worldwide was 384,403 cases with 10,692 deaths for a fatality rate of 2.8 percent.

July 17–30, 1994 Some 13 percent of pre-teen campers and counselors in Virginia developed *Escherichia coli* from contaminated campfire meals. At a summer enclave, participants frequently ate meals cooked over a fire in rural areas. According to the Virginia Department of Health and the Centers for Disease Control and Prevention, most consumed rare hamburgers. The infection caused the hospitalization of one patient with hemolytic uremic syndrome and a risk of renal failure.

August 5, 1994 After decades of absence, bubonic plague epidemics recurred in India, Malawi, and Mozambique. The outbreak began in Mamla village in Maharashtra with 596 cases and spread to Surat in Gujarat, where 54 of 146 victims died by October. By August 26, the emergence of bubonic and pneumonic or pulmonary plague on the west coast of India produced 693 cases. Precipitating the outbreak, rodents flourished in grain sent to relieve victims of a 1993 earthquake. Because of Hindu homage for the sacred rat, the devout refrained from killing mice in their homes.

When plague struck the slums of Surat, Gujarat, on September 22, the city closed as 55 of 6,344 victims died. Param Pujya Pramukh Swami Maharaj issued a call for tetracycline to distribute to citizens. At the city hospital, volunteers served tea and meals to some 3,000 patients.

The World Health Organization convened an international session to recommend methods of prevention and control. In October, Hiroshi Nakajima, WHO's director general, sent a team of experts from India, Russia, Switzerland, and the United States. Under WHO International Health Regulations, workers screened travelers at airports for symptoms of plague, fumigated aircraft cabins and cargo holds, and searched for rodents. In the harbors, inspectors checked outgoing vessels for rats and insects.

September 1994 A large outbreak of *Salmonella typhimurium* struck Hmong immigrants in Tulare County, California, after 200 attendees at a family celebration ate raw beef in *lahb*, a traditional meat and rice salad flavored with lime juice, mint, and fish sauce. Of the 130 victims, one two-year-old died. Food inspectors determined that the family bought a steer slaughtered on-site at a nearby ranch and left the meat unchilled for seven hours.

October 4, 1994 The Schwann's Ice Cream plant at Marshall, Minnesota, withdrew frozen ice cream, ice cream sandwiches, sherbet, sundae cups, and yogurt carrying raw egg bearing *Salmonella enteritidis* bacteria, which sickened at least 224,000 Americans between September 19 and October 10, 1994. The spread of illness, one of the largest on public record, derived from a premix of unpasteurized eggs hauled in unsanitary tanker trucks. Three days later, the company voluntarily stopped production at the Marshall plant and ceased distribution of questionable stock until health officials could determine the source of contamination.

1995 Three years after 108 people suffered cryptosporidiosis in Torbay, England, a new outbreak infested another 557 with the protozoa. Public health officials traced the oocysts to water intake from the Littlehempston River.

1995 The five-year battle with diphtheria in Asia resulted in disastrous rates, with 35,652 cases in Russia alone. The total of 50,412 attested to greater contagion in Tajikistan, central Asia, and Ukraine. By year's end, totals fell to 20,215, a downward trend of 60 percent fewer new infections.

1995 After two decades of *Aedes aegypti* mosquito control in Latin America and Asia, surveillance relaxed in the 1970s. Data pinpointed cases in Baluchistan, Pakistan, Tortola, Florida, Washington, D.C., and the Tex-Mex border. By 1995, dengue fever was hyperepidemic in Puerto Rico. Of a half million cases in 25 countries, 13,000 took the deadlier hemorrhagic form.

1995 The isolation of Greenland enabled researchers to chart the upsurge of HIV, which struck 180 Nuuk per 100,000 islanders and 78 Sisimiut. The initial infection came from homosexual contact of a Greenland man in Denmark. While the rate of infection dropped slightly among the Sisimiut by 2000, the Nuuk reported 140 cases.

1995 A five-month streptococcus epidemic among the mentally handicapped in Copenhagen, Denmark, infected 90 patients, killing four people who were severely limited in ability to perform normal hygienic procedures. The disease spread over 13 of the city's 111 institutions. Nurses who recognized the outbreak in its early stage represented victims who could not speak for themselves.

1995 Epidemic malaria in northern India affected three million residents. A new strain of parasitic vector worsened the outbreak, as did a corrupt government that misappropriated funds for disease control. Into 1996, heavy rains and the failure of the Rajasthan dam system expanded mosquito breeding grounds. The spread of infection to 2.85 million people swamped government agencies, which reported only 2,300 deaths.

April 1995 In northern Colombia and Venezuela, a reemergence of the Venezuelan equine encephalomyelitis virus infected some 100,000 people, causing convulsions in children and ending pregnancies with spontaneous abortions and fetal deaths.

April 1995 Ebola, a mystery virus, re-emerged in Kikwit, Zaire, where a man smoked infected monkey meat for a meal. The victim spouted blood from every body orifice and died in four days from the degeneration of internal organs. The disease spread from his nurse and a nun to three more religious workers. To stem a global epidemic, the Geneva-based World Health Organization mobilized microbiologists and researchers to the abandoned hospital. Fearful natives fled the scene, carrying the filovirus to others. In all, 316 died.

March 20, 1995 Twelve Japanese commuters died after the release of the lethal nerve gas sarin in the Tokyo subway system during the 8:00 a.m. rush. The subway sabotage was an act of bioterrorism unprecedented in Japanese history. The work of Asahara Shoko, leader of the 10,000-member Aum doomsday cult, the plot anticipated the simultaneous arrival of all affected trains at the Central Tokyo station. The puncturing of plastic bags of gas by 41 fanatics on the Hibiya, Chiyoda, and Marunouchi train lines afflicted 6,000 passengers at 15 underground stops with eye irritation, respiratory difficulties, weakness, and unconsciousness.

Paramedics, doctors, and nurses set up first aid stations in nearby streets and temples. Beginning at 8:35 a.m., ambulance crews and fire companies transported the sickest 550 victims to 41 hospitals for treatment with atropine. Others walked or traveled by car to clinics and emergency rooms. Of 3,227 seeking aid, 493 required hospitalization and 17 required intensive care. Some survivors were permanently brain damaged. By 11:00 a.m., police identified the poisonous agent as a potent chemical weapon capable of degrading neuromuscular connections and causing asphyxiation.

June 23–26, 1995 In summer 1995, the Centers for Disease Control recorded new outbreaks of illness from infectious meals. Three north Georgians and two eastern Tennesseans contracted illness from fast-food hamburgers contaminated with *Escherichia coli* bacteria. One victim was hospitalized with hemolytic uremic syndrome. Two weeks later, in Winnebago County, Illinois, health workers studied five cases of *Escherichia coli* among Rockford children hospitalized with intestinal distress. All had swum in a lake in an Illinois state park on June 24–25.

The next year in Osaka, Japan, around 10,000 people sickened and 12 died after consuming unclean radish greens in school lunches. The meals, held at room temperature for three hours, were contaminated with *Escherichia coli*. Around 90 percent of victims were school-age children. In that same period in Scotland, unwary diners suffered 400 illnesses and 21 fatalities from the same bacteria, which contaminated burgers at a butcher shop. A total of 1,039 cases required treatment in Scotland, England, and Wales. As a result of late 20th-century threats to life and wellness, the World Health Organization launched the division of Emerging Viral and Bacterial Disease Surveillance and Control.

November 1995 Over a two-month period, butchers in Jeddah in southwestern Saudi Arabia accounted for 40 cases of Alkhurma hemorrhagic fever (AHF). One-quarter of victims died, beginning with patient zero. Caused by the tick-borne flavivirus, the zoonotic disease links to the herding and rendering of camels, goats, and sheep, contact with infected animal blood on skin wounds, and consumption of unpasteurized milk and urine. Patient zero developed the disease after slaughtering a sheep in Alkhumra, Makkah. Symptoms of fever, headache, pain, anorexia, liver enzyme elevation, and vomiting preceded death from encephalitis in 20 percent of victims. WHO declared AHF an emerging infectious disease along with the hantavirus and Crimean-Congo hemorrhagic fever.

winter 1995 American epidemiologists aided a half million Muscovites and people from other areas of Russia suffering an outbreak of influenza. Before the disease peaked at the beginning of 1997, around one million people or 40 percent of the city of Moscow had incurred the virus, with 50,000 new cases reported daily. Health workers in Volgograd estimated 18,000 cases; similar illness struck Czech cities. The closure of Russian schools and public institutions and the loss of factory labor illustrated the seriousness of the blow to a people already suffering poor health, economic stress, irregular immunization, and the effects of urban pollution.

1996 In Henan, China, Gao Yaojie, a retired doctor, began examining patients suffering an inexplicable disease that afflicted over half of some villages. She disclosed epidemic HIV-AIDS in 100,000 people, who contracted the virus from dirty needles and recycled blood. With her own funds, she provided palliative medicines.

For casting suspicion on the province, the government urged Gao to keep silent, tapped her phones, seized her mail, and placed her under surveillance. She defied the entrenched bureaucracy by distributing close to a half-million brochures and pamphlets to warn villagers of the virus. In 2002, the Global Health Council awarded her the Jonathan Mann Award, which carried a $20,000 purse. China's repressive government refused her travel papers to journey to the U.S. to receive the honor. In her place at the ceremony was U. N. Secretary-General Kofi Annan.

1996 Two years after a mast year or bumper crop of acorns in Duchess County, New York, the Centers for Disease Control and Prevention recorded a 40 percent rise in incidence of Lyme disease over the previous decade. The 1994 figure for the United States reached 11,700; by 1997, the number had risen to 16,455. Of the total, 90 percent occurred in California, Connecticut, Maryland, Massachusetts, Minnesota, New Jersey, New York, Pennsylvania, Rhode Island, and Wisconsin.

1996 Among indigenous people in Canada, an upsurge in tuberculosis proved difficult to combat. As reported by the British Columbia Centre for Disease Control Society, the rate reached 140 cases per 100,000 in northern and remote areas, notably the Dene in the Northwest Territories. The rate was half that number among the Inuit.

1996 West Africans suffered unprecedented epidemic cerebrospinal meningitis. At Kaduna, Nigeria, lack of leadership and poverty contributed to the decline in health care. To prevent a spread to the Middle East, Saudi Arabian officials banned Nigerians from the annual Islamic hadj to Mecca, a perennial source of contagion. In Burkina Faso, the nation's worst upsurge of endemic meningitis infected 40,000, killing 4,500.

1996 Over a 17-year period, health workers in Taiwan, Japan, and the U.S. identified serious ailments endemic to foundries, mines, glass manufacturing, and blasting. Around 4.5 percent of coal miners suffered pneumoconiosis or black lung, which killed 14,156 from severe scarring leading to suffocation. Silicosis, derived from exposure to crystalline silica, threatened 1.6 million workers, weakened 60,000, and killed 2,694, some from secondary lung cancer.

1996 In South Africa, the first AIDS death in 1982 was followed by a steady growth in new AIDS diagnoses, climbing from 613 in 1991 to 1,123 in 1992. In four years, the number surged to 10,351, with a total of one million residents testing positive to HIV. Health officials surmised that as many as 700 South Africans acquired the infection each day and projected eight million primarily urban residents would be infected by 2000. Increasing the mortality rate was widespread tuberculosis, which killed in a shorter time. To stem the damage to the labor force, the South African Chamber of Mines established treatment, random blood tests, and counseling on the dangers of sexually transmitted disease. Backed by the World Health Organization and international experts, the National AIDS Convention of South Africa stepped up a tutorial campaign.

January 1, 1996 Cholera at Ibadan in Oyo, Nigeria, devastated 302 children, who constituted 21.8 percent of diagnosed cases. The victims lived in dilapidated housing lacking sanitary water supplies and suffered a death rate of 5.3 percent. The victimization of poor children encouraged humanitarian efforts by Médecins sans Frontières (Doctors without Borders) and UNICEF to boost hygiene for urban Nigeria.

February 1996 At Mayibout, Gabon, Ebola struck 31 villagers and killed 65 percent, mostly young people. Residents along the Ivindo River east of Libreville contracted the virus from a dead chimpanzee that 19 men retrieved from the jungle for its meat despite warnings from the government that citizens should not handle sick or dead forest animals. Health workers noted that the resulting hemorrhagic fever was deadly to both humans and primates. Ebola spread to seven more suspected victims in nearby villages and to Makokou, capital of Ogooue-Ivindo province.

March 1996 A scientific disclosure that eight out of ten infected people had died of Creutzfeldt-Jakob brain disease proved that bovine spongiform encephalopathy in cattle endangered the human brain and nervous system. The disease, which first appeared in rural settings of Great Britain in 1986, caused cattle to drool, stumble, lose weight, and menace other animals before dying. The odd behavior resulted in the common name "mad cow disease."

In 1996, research at the Institute for Animal Health in Edinburgh, Scotland, linked the mysterious anomaly to the feeding of healthy animals with parts of cattle stricken with mad cow disease. The European Union halted exports of cattle and bovine products from the United Kingdom, but farmers continued allowing contamination with human food. Stalling on the part of the Tory majority produced a surge in Labour sympathies and the election of Tony Blair as prime minister. By 1999, conclusive evidence from the universities of Edinburgh and Oxford indicated that a ban on the cows-fed-on-cows cycle stopped the spread of the disease.

May 1996 Along the border of Venezuela, onchocerciasis, called river fever, infected around 87 percent of the indigenous Yanonami (or Yanomami) of Roraima, Brazil. Infection endangered their eyesight and threatened their pathetically underserved culture with extinction. A team of 120 volunteers worked at removing tumors raised by parasites inflicted by the bite of the female *Simulium* blackfly. In a study conducted in 2000, health agents discovered that nearly all the

Yanomami still tested positive for onchocercia-sis.

May 1996 The World Health Organization assessed the number of HIV infections in India at 1.75 million. Staff, blaming unsafe sexual practices and unscreened blood products, predicted that, by the 21st century, the nation would suffer more cases than any other part of the world.

May 3, 1996 In southern Thailand on the Myanmar border, an outbreak of elephantiasis swelled limbs and genitals in 8,000 of some 600,000 Burmese laborers employed in Ranong Province. To protect Thais, the government established immigration policies that denied temporary permits to migrant workers afflicted with tuberculosis, leprosy, syphilis, or elephantiasis or displaying symptoms of alcoholism, drug addiction, retardation, or mental illness.

May 13, 1996 In four villages at Irian Jaya in the highlands of Indonesia, dengue fever, spread by virus-carrying *Aedes aegypti* mosquitoes, afflicted many and killed two during an annual pandemic that felled 6,000. Compounding problems with insects were unsanitary conditions and heavy rains. At Jakarta, the annual rate of infection doubled to 6,000.

May 15, 1996 A nosocomial outbreak of Lassa fever at Kenema, Sierra Leone, generated 46 deaths among 76 hospital patients, mostly among female teens. The disease, spread by a thriving rat population, caused spontaneous abortion and death to mothers. The government set up an isolation ward at a state hospital and collected blood samples to send to the U.S. Centers for Disease Control and Prevention for assessment.

June 1996 River sediment in Ogose, Japan, hibernated oocysts that infected 8,705 with cryptosporidiosis, an intestinal infestation with protozoa. A total of 2,880 patients sought medical care. Health officials surmised that heavy rainfall in May heightened water turbidity and increased fecal matter in potable water.

July 13, 1996 Spreading from Libreville to Booué, the deadly Ebola virus emerged in Gabon for the second time in five months, killing 19 of 27 victims by late fall. Numerous world relief agencies converged on the area to offer the nation's health officials additional supplies, medical expertise, and advice about containment.

Despite their efforts, the disease traveled southeast to Johannesburg, South Africa, where it felled a medical worker who had aided patients in Gabon. He survived, but the nurse who treated him died of the virus.

July 15, 1996 In the largest outbreak of arboviral illness in Europe of the decade, West Nile fever altered when it generated the first urban epidemic, which occurred in Bucharest and southeastern Romania. The zoonotic virus, which researchers located in mosquitoes, birds, and horses, inflicted over 500 people with meningial encephalitis over a three-month period and produced a mortality rate of 10 percent. Neurological damage occurred in 352 survivors.

October 1996 A seasonal surge of cholera in Kathmandu, Nepal, appeared to derive from river-borne *El Tor* variety of *Vibrio cholerae*. The epidemic began in contaminated waterways. Infection prostrated young children, but produced few fatalities.

October 31, 1996 The U.S. Food and Drug Administration identified Odwalla apple juice as the source of a flare-up of *Escherichia coli* in British Columbia, Colorado, Nevada, New Mexico, and Texas. The following October, health workers linked unpasteurized apple cider or apple juice to three more outbreaks of bacteria-bearing food. Lawsuits pressed in May 1998 resulted in a settlement of $15.6 million for one victim of an outbreak infecting five families. As a result, Odwalla had to begin pasteurizing its juices.

November 1996 At Wishaw, Lanarkshire, England's public health authorities reported an outbreak of *Escherichia coli*. Of the 400 victims, 60 required hospitalization and 21 died. Incubated in cooked turkey, ham, and beef sold by John M. Barr & Son and served at a pensioners' lunch at a church and at an 18-year-old's birthday party at the Cascade Bar, the disease sent many to the hospital for treatment. Among the fatalities were 18 elderly diners. The government launched a £500,000 public-awareness campaign on food contaminants. Barr's food business survived a lawsuit because of lack of evidence.

November 1996 In Malawi, Amnesty International delegates inspecting Zomba Central Prison found 180 children crammed into four cells while awaiting trial. Lack of water, overcrowding, and accumulated human excrement worsened infestations of fleas and lice. A scabies epidemic increased discomfort and infection from untended sores. After local officials equipped the prison with showers and toilets and supplied soap and clean clothing, the United Nations Children's Fund shipped medicines to cure sick inmates.

November 2, 1996 An eight-month surge in leptospirosis infected 1,390 urban patients living in slums near Salvador, Brazil, where seasonal rains produced flooding in sewer drainage basins. The environment boosted populations of *Rattus norvegicus,* a vector of the zoonotic. The mostly adult, mostly male victims suffered jaundice, anemia, and reduced kidney output. 50 patients died despite undergoing dialysis for renal failure.

1997 The Asmat of the Ayam region of West Papua New Guinea faced near extinction from epidemic cholera, enteritis, gonorrhea, influenza, measles, pertussis, and syphilis. The people followed animistic tenets, which blamed illness on spirits angered by law-breaking and taboo infractions. To rid themselves of *arau pok* (sickness), the Asmat applied magic charms or propitiated spirits. Because of widespread forest fires, tribe members faced death from inhalation of smoke and toxic fumes, poor health, and starvation.

1997 Staff of the World Health Organization and a United Nations program estimated that, worldwide, over 30 million people suffered from HIV or AIDS. After the U.S. made significant inroads against the epidemic, infections began increasing in 1995 among teen and adult females from 7 percent of the total cases to 19 percent. The disease gained ground at a phenomenal rate in sub–Saharan Africa, primarily Malawi, Rwanda, Tanzania, and Uganda, in part because men working away from home had intimate contact with a larger number of sexual partners and prostitutes. The open lesions caused by venereal disease offered entry points for the AIDS virus. Because Africans had a lower rate of circumcision, which hardens the glans of the penis and makes it less permeable to the virus, males faced a greater risk of infection.

January 1997 After a respite of 15 years, dengue fever returned to Santiago, Cuba, the result of a lapse in public health controls and an increase in citizen migration. Of 2,946 cases, 205

suffered the hemorrhagic variety, which killed 12 before health workers could control the epidemic.

January 13, 1997 Aided by Médecins sans Frontières (Doctors Without Borders), the Pan American Health Organization (PAHO) launched a program to end endemic Chagas's disease, a protozoan infection from *Trypanosoma cruzi* that killed over 43,000 people a year in Bolivia, Brazil, Guatemala, Honduras, and Mexico. The disease infected up to 18 million of 205 million people. Gabriel Schmunis, a communicable disease specialist with PAHO, remarked that infestations of the triatomine "kissing bug" or "assassin bug" decreased from insecticide spraying of rural dirt-walled dwellings at a cost of $207 million. The screening of blood donors also helped to stem infections.

March 1997 Chickens died of whole-body hemorrhaging outside Yuen Long, Hong Kong. At two of the three farms involved, nearly all chickens perished in a bloody heap. Kennedy Shortridge, chairman of the microbiology department at Hong Kong University, identified the virus as lethal and highly contagious. Unlike swine flu, the pathogen appeared to move from chickens to humans, but not from human to human.

The avian virus remained an agricultural problem until May 9, when a small boy in Hong Kong came in contact with virus-bearing chicken feces, suffered symptoms, and died after his blood stopped clotting. In November and December, five more residents contracted the avian flu. Hong Kong officials inoculated residents with amantadine, an antiviral drug. To counter a disease with potential for a massive lethal epidemic, officials of the Department of Agriculture and Fisheries slaughtered 1.3 million chickens.

March 1997 Russian health officials recorded 3,327 HIV infections, but totals fell short of an actual infection rate closer to 10,000. The Russian Ministry of Health predicted that, by 2000, around one million Russians would be infected with the virus.

May–July 1997 A collaboration of the Centers for Disease Control and Prevention, Medical Emergency Relief International, the International Medical Corps, and CARE International fought epidemic sleeping sickness among the poor in Tambura province in southern Sudan. Volunteers provided treatment to the sick, established a primary health care system, and reduced the tsetse fly population by setting 1,000 traps.

June 1997 Simultaneously with a major AIDS infection, cholera in Kenya overran villagers of western Nyana near Lake Victoria at the nexus of Kenya, Uganda, and Tanzania. The outbreak claimed 228 of 4,571 patients. Some of the sick traveled for miles and arrived overland by wheelbarrow or on relatives' backs. In addition to rehydrating them with intravenous fluid, Joyce Akinyi, a health care worker at the Migosi center, observed the fever protocol set up by Médecins sans Frontières (Doctors without Borders) by spraying chlorine on victims' feet and disinfecting hands.

June and July 1997 In Michigan and Virginia, *Escherichia coli* sickened people who ate alfalfa sprouts permeated with the bacteria. Grown by a single sprouter, the shoots derived from contaminated seeds from Idaho, which caused diarrhea, cramps, bloody stool, vomiting, and fever.

July 1997 As reported by the Dirección Nacional de Estadísticas del Ministerio de Salud Pública de Cuba, an outbreak of acute hemorrhagic conjunctivitis afflicted 137,136 islanders. The result of an enterovirus, the disease derived from fecal contamination and proved highly contagious.

July 13, 1997 An isolated incidence of Ebola at a logging camp near Booué, Gabon, afflicted 27, killing 17 or 63 percent. After the death of a 37-year-old hunter, the virus killed his close friend, a traditional healer, and the healer's assistant.

August 20, 1997 Retired Swedish-American virologist Johan Hultin of the University of Iowa exhumed the remains of 1918 Spanish influenza victims. He examined bodies buried in a mass grave in the permafrost of the Alaskan tundra at a mission on the Seward Peninsula, where 72 of 80 residents died. His intent was to isolate the lethal virus, a feat he had attempted unsuccessfully in 1941, when injection of tissue into a ferret's nostrils failed to produce Spanish influenza. Over a half century later, rekindled with enthusiasm by the writings in *Science* magazine of Jeffrey Taubenberger, military pathologist for the Armed Forces Institute of Pathology in Washing-

ton, D.C., Hultin made the five-day journey to Alaska and petitioned tribal elders for permission to open the grave once more.

For a shoestring investment of $3,000, Hultin and four native assistants located preserved tissue. With frozen slices from the lungs of Lucy, an obese 30-year-old Inuit female well preserved by layers of fat, Hultin studied DNA sequencing to determine reasons for the virulence and speed of infection of the Spanish flu. He hoped that Taubenberger could use the tissue to gain more information and to formulate a vaccine against a killer flu that might sweep the globe again. World expert Robert G. Webster of Saint Jude Children's Research Hospital in Memphis, Tennessee, surmised that humans may have contracted the virus directly from birds.

October 1997 In a millennium when around 300 million people died of smallpox, the World Health Organization proclaimed that a cook, 23-year-old Ali Maow Maalin of Somalia, was last known victim of smallpox on the globe. A rapid vaccination effort prevented an epidemic. Months before the announcement, the United States government resolved to keep stores of the variola virus. The decision indicated concerns that terrorists might be maintaining their own stockpile to develop as a weapon of biological warfare. *See also* **August 1998.**

late October 1997 Among Japanese hemophiliacs treated in 1989 with HIV-tainted blood products, an outbreak of AIDS afflicted 2,000, over 400 of whom died. Families sued the government, the Green Cross Corporation, and four other drug companies. Kan Naoto, a member of the House of Representatives and subsequent minister of health and welfare, settled the case. A court indicted Abe Takeshi, former chief of the AIDS research team at the Ministry of Health and Welfare, with careless handling of infectious materials.

November 1997 Pandemic tuberculosis resistant to major drugs threatened the world's population. Around three million adults died of the disease in the previous year. As reported by actuaries of the World Health Organization and the U.S. Centers for Disease Control and Prevention, the disease flared in one-third of the world's countries, with serious outbreaks in Argentina, the Dominican Republic, Estonia, India, the Ivory Coast, Latvia, and Russia. In the United

States, 42 states reported cases of tuberculosis. Canada incurred a serious upsurge in infections among indigenous people, who died at higher rates because of inadequate housing and health care. Worldwide, the high caseload threatened to swamp local health agencies, particularly where tuberculosis paralleled high incidence of HIV-AIDS and contributed to a majority of deaths from AIDS.

November 24, 1997 According to researchers from the Albert Einstein College of Medicine in Bronx, New York, the 9,000 Yanonami (or Yanomami) living in the Amazon Basin of Brazil on the border of Venezuela faced annihilation from a high incidence of tuberculosis resistant to medication. In a virgin-soil population, the disease infected 6.4 percent of natives.

November 25, 1997 At Nairobi, the Kenyan Ministry of Health and the World Health Organization studied 478 deaths in Kenya and Somalia from an unidentified fever. The mysterious malady also caused spontaneous abortion and deadly hemorrhage in livestock. The U.S. Centers for Disease Control and Prevention identified the acute infection as Rift Valley fever, a mosquito-borne virus afflicting people involved in herding, milking, slaughtering, and tending domestic animals. Thousands died.

Several elements prohibited immediate aid to isolated areas. Government officials feared a drop in tourism, particularly in light of unfavorable media reports concerning violence and political turmoil at Mombasa. Because flooded highways and fallen bridges hindered Médecins sans Frontières (Doctors without Borders) and the International Red Cross, volunteers called for helicopter backup. Experts on global warming blamed El Niño, a climate variation that produced warm, moist weather triggering an explosion in the mosquito population.

November 27, 1997 Peter Piot, director of the joint United Nations Programme on HIV/AIDS (UNAIDS), reported that the global AIDS epidemic was far worse than projected. While the rate held steady in the United States and declined in some industrialized countries, infections rose 9 percent, 90 percent of them in developing countries. Projections for the year predicted 5.8 million new cases globally and 2.3 million deaths, raising the total deaths to 11.7 million. Peripheral damage included lost skills from sick and dying

workers and the orphaning of 8.2 million uninfected children. *See also* **July 3, 2002.**

late 1997 Uganda's AIDS commission reported 400,000 deaths from HIV/AIDS since the first tabulation of cases in 1984, a welcome decline compared to the rates in Burundi, Malawi, Rwanda, and Zambia. Health officials concluded that 90 percent of Ugandans had heard radio information on condom use and on the practice of safe sex to prevent infection.

December 1997 In Japan, the height of infections from Minamata disease, a malady similar to acute anterior poliomyelitis, produced 2,262 recorded cases. Centered in the city of Minamata, Kyushu, the disease derived from the consumption of fish tainted with methyl mercury that the Chisso Corporation's acetaldehyde-synthesizing plant ejected into local waters. The anomaly resulted in degeneration of neurons and caused mental retardation in the children of farmers and fishers. The disease had appeared in modern fiction by Ishimure Michiko, a Minamata native, who published *Kugai Jodo: Waga Minamatabyo* (*Paradise in the Sea of Our Sorrow*) (1972), based on interviews with local victims. When lawsuits were settled in 1995, parties responsible for the toxic effluent compensated victims.

In the same period, *itai itai* (ouch-ouch) disease afflicted 156 in the town of Fuchu in central Japan. The victims, mostly middle-aged women, suffered back and joint pain and frequent bone fractures. Caused by accumulation of cadmium in the kidneys, the outbreak was the result of the toxic wastes from the Kamioka mine of the Mitsui Mining & Smelting Co., Ltd. Contributing to the disease was rice harvested from land watered with cadmium-contaminated water.

late 1990s Drug cocktails began extending the lives of AIDS patients. As the death rate decreased in the United States, the disease spread among urban and minority adolescents at the rate of 40,000 new cases per year. Meanwhile, the virus advanced worldwide, setting off a slow-motion version of the Black Death across sub–Saharan Africa. Medical workers and health officials distributed condoms, set up clinics monitoring pregnancies, and arranged needle exchange for drug addicts. In Bangkok, the Thai Ministry of Health discouraged breast-feeding by HIV-positive mothers and offered free tins of infant formula mix to protect children from contagion.

1998 Epidemic tuberculosis paralleled the rise in HIV infections in Namibia's urban and refugee populations. Exacerbating the situation were alcoholism, poverty, promiscuity, and unemployment. At Katatura Hospital in Windhoek, the six contagion wards were overrun. The previous year, staff treated 1,500 patients, of whom 15 percent suffered infection in bones, urinary tract, or meninges. Globally, around 50 million people carried the same multidrug-resistant strain of tuberculosis.

1998 The World Health Organization noted more than double the previous reported figures for co-infections of leishmaniasis and HIV in southwestern Europe. Spain experienced 835, Italy 229, France 259, and Portugal 117, some 71 percent of whom were intravenous drug users. The disease also thrived in Brazil and was anticipated to worsen in Africa, where population migrations, displacement, civil uprisings, and war altered demographic patterns. Treatment was successful in over 80 percent of cases, but relapses occurred in 52 percent of cases.

1998 At Negeri Sembilan and Selangor in Malaysia's largest swine-producing area, the Nipah virus caused a two-year outbreak of epizootic encephalitis on pig farms. The pathogen emerged from fruit-eating bats from the rain forest that fed on mango groves near pigpens, where bat saliva spread the virus to the animals. Among farm folk, butchers, and meat handlers, 265 people who touched infected swine tissue or body fluids contracted the brain infection; 104 or 39.2 percent died from it. Health officials blamed the high incidence of infection on poor hygiene and old-fashioned farm production systems.

As a result of global concern that the Nipah virus would spread to other regions, the United States halted imports of pork from Malaysia. The region's military supervised the culling of one million pigs in the infected area. Loss of livestock proved disastrous to the Malaysian agricultural economy.

In April 1999, after encephalitis reached epidemic numbers in Malaysia, farmers killed over 300,000 pigs. The disease, which hospitalized 99 and killed 67, afflicted only people linked to pig farming. The Centers for Disease Control and Prevention in Atlanta identified the pathogen as similar to the Hendra virus, which derives from fruit bats and passes to swine.

In December 1999, another death and 200 more infections proved that the situation in Malaysia was still out of control. The export of infected animals also endangered Singapore, the major buyer of Malaysian pork. A dozen abattoir workers contracted the virus; one died from brain fever. Medical research surmised that fruit bats incubated the virus, for which there was no antidote.

January 1998 Reports from Islamabad indicated that Afghani children suffered rampant scabies. Worsening their hygiene, the absence of community pumps in refugee settlements limited sources of potable water.

February 1998 When malaria devastated Kenya, Medical Emergency Relief International and Oxfam set up three mobile clinics to treat 20,000 nomads and settled people, imported mosquito netting, supplied hospitals with medicines, and held tutorials to explain how people contracted infection. The total program extended to distribution of vitamin A and vaccination of 66,000 for measles.

February 1998 WHO declared war on the buruli ulcer, a bacterial infection of the limbs that tends to strike children and teens. Caused by the mycobacterium, the necrotic tissue rapidly cripples victims at the rate of 6,000 cases annually in tropical and subtropical regions of Australia, Benin, Brazil, Cameroon, China, Congo, Gabon, Ghana, Guyana, India, Indonesia, Ivory Coast, Japan, Mexico, New Zealand, Nigeria, Papua New Guinea, Sudan, Togo, Tanganyika, and Uganda. Worsening lesions destroyed eyes and breasts and forced surgical excision or limb amputation.

February 6, 1998 At Maputo, Mozambique, a cholera epidemic in force since mid–August 1997 slew over 400 people. Worst hit was the city of Beira. Poor sanitation may have cost 109 lives among 2,000 cases. At Maputo, 234 of 9,000 died of the water-borne disease.

June 18, 1998 Denmark's first outbreak of multidrug-resistant *Salmonella typhimurium* DT104 derived from indigenous pork, which infected 25 patients and one meat worker. Their ailment confirmed at Statens Serum Institut, victims appear to have been infected by foods from a Zealand Island abattoir obtained from one herd. Health investigators linked the illnesses to a drug-resistant strain of bacteria found in a pork slaughterhouse.

The spread of the disease was indicative of the new bacterium's virulence, which moved from consumers of tainted pork to a hospital nurse, who infected an 82-year-old cancer patient who died of the contaminant. The health department diagrammed transmission of the disease:

> herds 1 & 2—>slaughterhouse—>pork—>18 consumers and one slaughterer—>seven hospitalized—>one infected nurse, who survived, and one infected female patient, who died.

In the aftermath, Alicia Anderson of the U.S. National Antimicrobial Resistance Monitoring System in Atlanta, Georgia, noted that indiscriminate use of antibiotics in the rations of farm animals threatened human health from superbugs.

June 15, 1998 In west-central Wisconsin, an outbreak of eight confirmed and four suspected cases of *Escherichia coli* appeared to have begun in contaminated fresh cheese curds made from unpasteurized milk at a dairy plant.

July 14, 1998 A power outage in Austin, Texas, allowed 179,000 gallons of untreated sewage to contaminate Brushy Creek, causing gastroenteritis in an estimated 1,500 residents. Diagnosed as cryptosporidium, the endemic debilitated patients from July 21 to August 2. Water from the city of Round Rock replaced local supplies until monitors declared the Austin water safe for consumption.

July 15, 1998 During Vietnam's annual three-month rainy season, dengue fever produced five times its average infections in 19 provinces. After the scourge peaked in the fall, around 119,429 caught the fever and 342 died. Hardest hit were children, 17 of whom died in Dong Nai province. Analysts attributed the outbreak to El Niño and predicted that increased incidence of fever would impact social and economic wellbeing.

August 1998 In the United States, 20 people died from *Listeria monocytogenes* after the contamination of a Michigan meat packing plant by work on the air-conditioning system. The worst outbreak in a string of listeria reports, the 100 cases spread over Connecticut, Georgia, Massachusetts, Michigan, New York, Ohio, Oregon, Tennessee, Vermont, and West Virginia suggested that more resistant bacterial strains could threaten ready-to-eat food products. At the North Zeeland, Michigan, plant of the Sara Lee Corporation, workers ceased producing Bil Mar brand hot dogs and deli meats after health investigators traced 70 illnesses to the Sara Lee food line. The upsurge cost the company some $70 million in income and shut down its Bil Mar line until late February 1999.

August 1998 In a major flood of the Yangtze, a half million Chinese fled central Hubei province as rainfall reached seven times its standard level. Amid stinking muck, landslides, and homelessness from damaged buildings, Médecins sans Frontières (Doctors without Borders), the Red Crescent, and the International Red Cross fought outbreaks of cholera, dysentery, enteritis, hemorrhagic fever, influenza, leptospirosis, pneumonia, schistosomiasis, and typhoid. Increasing rampant infections derived from unsafe drinking water, sewage spills, vermin, and putrefying livestock corpses.

September 1998 Worldwide, health data reported over 300 million new cases of gonorrhea, syphilis, and chlamydia. The United States reported 14 million of these new infections. Evidence of unprotected sexual intercourse increased concerns that patients also risked infection with HIV.

September 1998 An upsurge in gastroenteritis among 2,200 residents and 200 transients at La Neuville, Switzerland, preceded a diagnosis of *Escherichia coli*. An inadequate purification system allowed sewage to contaminate potable water, forcing officials to cancel a grape festival. The installation of chlorination and ultraviolet disinfection rid supplies of fecal pollutants.

September 11, 1998 An influx of 125 Danish and 13 Swedish tourists to a hotel in Marmaris, Turkey, suffered bacterial gastroenteritis from nine different microbes, including *Escherichia coli* and *Cryptosporidium parvum*. Of those stricken, 75 sought medical attention and 20 entered a hospital for treatment of dehydration. An investigation of samples from the hotel's water tank proved that water contamination caused the outbreak.

October 1998 An endemic malaria that had stalked India throughout the decade besieged Gujarat. Doctors estimated that some 400 new infections occurred each day.

November 1998 Chen Zhengming, a Chinese researcher at Oxford University, reported on an epidemic of smoking-related respiratory and cardiac ills in China. In male smokers, one in three suffered tobacco-related deaths from such ailments as chronic emphysema, lung cancer, and bronchopneumonia.

November 1998 During an era of hunger and protracted warfare in Somalia, cholera and dysentery afflicted residents of Mogadishu and its environs. Into March 1999, the disease spread, overwhelming starving children in refugee camps. UNICEF upgraded conditions by purifying drinking water. Around 7,000 cases of cholera continued to generate misery and death among some of the world's most desperate people.

December 1998 The U.S. Food and Drug Administration's approval of LYMErix, a Lyme disease vaccine, permitted use in certain adults between 15 and 70 to kill a spirochete that invaded the human bloodstream. A second breakthrough in Lyme disease vaccines, ImuLyme, had yet to be approved.

early December 1998 At Tegucigalpa, Honduras, damage from Hurricane Mitch included upsurges in cholera, enteritis, leptospirosis, and malaria. As described by Public Health Minister Marco Antonio Rosa, an epidemic siege of illness struck 260,000. Structures lay under mud and water, leaving residents homeless and rats and mosquitoes well served with new breeding grounds. Of the total sick, 208,000 suffered enteritis, which was deadly to infants. Malaria infected 31,000 and cholera, 20,000. Of the 62 confirmed cases of leptospirosis, four died.

1999 West Nile fever, which epidemiologists thought was limited to Africa, Europe, Oceania, and Asia, reached the United States, killing seven of the 62 New Yorkers it infected. In Volgograd, Russia, the same virus produced 318 infections and 40 deaths. Additional outbreaks occurred in the Czech Republic, Israel, Tunisia and elsewhere. In all, hospitals treated around 8,200 cases with 140 developing meningoencephalitis.

The outbreak followed by two years a similar surge in infections in the Czech Republic. By 2001, the disease spread to Connecticut, the District of Columbia, Florida, Georgia, Maryland, Massachusetts, New Hampshire, New Jersey, New York, Pennsylvania, and Rhode Island, causing a total of eleven deaths out of 81 cases. The U.S. Centers for Disease Control and Prevention (CDC) analyzed the strains and found them comparable to a type common to the Middle East, where the disease tends to turn into meningoencephalitis.

Experts theorized that an infected bird transported to a New York City zoo carried the virus. To typify sources of infection, authorities examined a large-scale bird die-off and found over 70 species carrying the virus. They also identified bats, cats, chipmunks, horses, rabbits, raccoons, skunks, and squirrels as carriers.

1999 Continued battle with malaria in India proved futile as 90,000 new infections and 5,000 deaths in Hazaribagh resulted from government ineptitude. Failure of mosquito control and inadequate drug distribution lessened residents' chances of surviving the endemic fever.

1999 A study of African children in rural Guinea determined the endemic nature of intestinal parasitism, which afflicted 53 percent in ages one to 18 years. Researchers determined that patients contracted hookworm through geophagia (consuming dirt) in half the cases as opposed to acquiring nematodes from skin penetration.

1999 An influx of immigrants to Canada renewed fears of multidrug-resistant tuberculosis, which threatened airline staff, border agents, and health care providers. In April, Ottawa tightened immigration surveillance after uncovering rampant x-ray sharing among people who knew they carried the pathogen and feared they would be denied admission. To assure control of disease, border patrols began flagging new arrivals from suspect countries, including Russia, Somalia, Tibet, and Vietnam. In response to ethnic profiling, humanitarians and religious workers warned that stereotyping the dispossessed as diseased would further endanger at-risk refugees and orphans.

February 1999 An epidemic of syphilis among white slaves recruited from Latin America, Russia, Eastern Europe, Nepal, and Burma aroused international outrage. The Argentine Societe de Protection et de Secours aux Femmes queried:

> And what is the end of their career? When their health is broken down, their bodies utterly ruined, their minds poisoned and dulled, they are thrust out into the streets to perish there, unless some hospital ward opens its door to

them. What else could happen to them? [Doe-zema 1999].

February 28, 1999 During the ten-day West-friese Flora trade show in Holland, Legionnaire's disease from a whirlpool tub at Bovankarspel out-side Amsterdam infected hundreds of attendees. By March 13, the Dutch health minister was aware of the outbreak, which produced 242 ill-nesses and 28 deaths. New regulations concern-ing plumbing at public facilities began eliminat-ing sources of *Legionella* bacteria.

March 1999 A meningitis epidemic overran the Sudan in Northern Darfur, killing 262 and threatening tens of thousands. Returning in ten-year cycles, the disease targeted children, teens, and young adults who were already at-risk from drought, hunger, and 15 years of civil war. The Ministry of Health appealed to the World Health Organization, International Red Cross, and UNICEF for medical aid and vaccines, which arrived at Khartoum by emergency airlifts. The Sudanese Red Crescent petitioned for donations to fund immunization of over 1.3 million resi-dents and to supply medical centers. Municipal authorities burned 60 infected villages as 10,000 Sudanese fled to Chad.

March 1999 Endemic cholera assailed Mada-gascar with huge numbers of victims. Worsened in February 2000 by cyclones, mudslides, and floods, the pathogen bloomed once more from poor sanitation and contaminated drinking water. By March some 6 percent of 23,000 cases proved fatal, although official figures underreported in-fections. Rain isolated the sick from relief work-ers, who focused on relieving the misery of 10,000 homeless islanders.

World relief agencies responded to the need. Because famine resulted from the destruction of coffee plantations and of stands of bananas, oranges, avocadoes and cocoa trees, the World Health Organization and the World Food Pro-gram joined forces to prevent deaths from hunger. The French military made food drops. To halt a mounting outbreak of malaria, UNICEF inoculated children and sprayed for insects. Complaining of government interference, the Swiss sector of Médecins sans Frontières (Doc-tors without Borders) abandoned its mission to cholera victims.

March 25, 1999 Epidemic viral Japanese B encephalitis inundated Malaysia with 58 fatalities among 157 cases. Caused by the *Culex tritae-niorhynchus* mosquito, the disease attacks some 50,000 Asians annually. Contagion causes paral-ysis, coma, and death, especially among male pig keepers. While U.S. and Australian experts teams investigated the upsurge in Kuala Lumpur, med-ical workers sprayed 150,000 farms, inoculated 64,767 people, and destroyed 300,000 pigs.

April 14, 1999 As reported in a press release from Smith DeWaal, Director of Food Safety for the Center for Science in the Public Interest, food-borne diseases sickened 33,000,000 citi-zens at unpredictable intervals and places in North America and killed 9,000.

April 19, 1999 The United Nations Children's Fund reported an emergency polio vaccination campaign among refugee camps in Luanda, Angola, and Caxito, South Africa. Staff inocu-lated 634,368 young children after Luanda reported 487 cases and 30 fatalities. Benguela incurred only seven victims; Icolo-e-Bengo re-corded two cases, which bore evidence of menin-gitis.

June 9–11, 1999 In Texas, the Tarrant County Health Department investigated an outbreak of *Escherichia coli* among teenagers at a cheerleading camp. Illness from food contamination sent two to the hospital with hemolytic uremic syndrome and required appendectomies in two other teens.

June 14, 1999 Illegal Chinese immigrants, all inmates of a Taiwan jail, displayed symptoms of beriberi, a nutritional deficiency caused by a diet based on polished rice. The sixteen hospitalized patients reported weak limbs, heart failure, leg swelling, anorexia, and labored breathing. Thi-amine injections relieved symptoms the first day of treatment.

June 17, 1999 An article in the *Cleveland Free Times* pondered the scientific battle over the hypothesis that bacteria in milk may cause or contribute to the misery of Crohn's disease, a chronic intestinal complaint that became en-demic globally. As explained by Rodrick J. Chio-dini, a microbiologist at the University of Con-necticut, in the 19th century, the *Mycobacterium paratuberculosis* microbe that debilitated dairy cows with chronic intestinal inflammation, diar-rhea, weight loss, reduced milk output, and early death was identified as the cause of Johne's dis-ease. Chiodini located the same microbe in the intestines of victims of Crohn's disease and postulated that contaminated milk may be the cause. *See also* **May 21, 2001.**

July 1999 Scientific evidence published in the *American Journal of Psychiatry* linked the incidence of schizophrenia with enteroviruses of intestines and stomach that parturient women encountered during pregnancy. The connection led researchers Janna Suvisaari and colleagues of the National Public Health Institute, Helsinki, Finland, to conclude that polio may be linked to schizophrenia, which decreased after the widespread vaccination against polio. Another clue to Suvisaari's theory was the rise in incidence of schizophrenia in people born to women exposed during the fourth month of gestation in late summer and early fall, peak times for the polio epidemics from 1951 to 1969.

July 1999 As described by the World Health Organization, epidemic pellagra at Kuito, Angola, surged during two decades of war. In Bie province of the central highlands, 155,000 refugees suffered nutritional deficiency because they were unable to grow food and depended on rations from the World Food Program. Classic skin lesions on the neck, enteritis, and dementia attested to a need for more B-complex vitamins in the diet. By January 2000, 898 cases required hospital care. Workers at a supplementary feeding center distributed nicotinamide tablets, vitamins, and dried fish.

September 3, 1999 Ten children hospitalized near Albany, New York, with *Escherichia coli* appear to have eaten tainted food at the Washington County Fair during the last week of August. Other fair-goers suffered *Campylobacter jejuni*.

November 23, 1999 An outbreak of Japanese B encephalitis at Andhra Pradesh in southern India weakened 660 children, killing 138. To halt the spread of the virus, health agents slew hundreds of pigs and vaccinated thousands in the area surrounding Hyderabad, the state capital.

December 31, 1999 As the millennium waned, Benjamin Ip of Nevada, a volunteer physician with Médecins sans Frontières (Doctors without Borders) (MSF), canoed along the Niger River Delta into the Nigerian outback during a malaria epidemic to treat children and pregnant women. He recognized common faults of hygiene, namely, the simultaneous use of the river as toilet and bathing facilities and drinking water. Of his numerous assignments with MSF, he told interviewers from *Biography* magazine,

> I don't want an ordinary, boring life. I have been dirty, dusty, and freezing cold. I have also been angry, frustrated, depressed, close to the verge of tears, and sometimes absolutely terrified—but those were the times I felt most alive [Spencer 2000].

His frustration with epidemic situations was understandable. According to *Smithsonian* magazine, in 2000, pandemic malaria infected 500,000,000 people annually and killed around 2.7 million, endangering 40 percent of the global population.

2000 While the world incidence of measles dropped by two-thirds over the past decade—from 4,211,431 to 1,374,083 cases—Africa experienced an upsurge to 520,102 victims. The western Pacific also incurred a rise from 155,490 to 176,493. The Americas exulted in the largest drop, from 218,579 to 1,755, a reduction of 99.2 percent. Decreased vaccination with MMR in 20 percent of Irish children resulted in 1,000 infections and three deaths.

2000 After a nine-year campaign to eradicate leprosy, the World Health Assembly reached the goal of fewer than one case for 10,000 people, a drop from 21.1 per 10,000 in 1983. By 2014, the number declined by 75 percent to 0.24 per 10,000.

early 2000 After over a decade of progress in preventing HIV infections in Uganda, health officials reported a decrease in the spread among rural citizens, the people least likely to receive wellness counsel.

January 2000 When 79 Americans suffered an upsurge of salmonella in 13 states, the Centers for Disease Control and Prevention investigated and found that they had all consumed mangoes in the previous two months. Medical detectives traced the bacterium to a contaminated cooling tank in Brazil in which farm workers dipped fruit.

January 14, 2000 After a devastating mudslide in Venezuela, health officials at Caracas reported an outbreak of scabies in shelters where limited living conditions crowded people into close contact.

April 25, 2000 In Australia's largest incidence of Legionnaire's disease, 101 people encountered pathogens at the new Melbourne Aquarium in New South Wales. The outbreak, which killed four people, resulted from bacteria growing in the aquarium's cooling towers. Officials replaced the water-cooled air conditioner with an air-cooled system.

April–May 2000 Among 125,000 Somali refugees in camps at Dagahaley, Ifo, and Hgadera, Kenya, volunteers reported an alarming increase in visceral leishmaniasis, a parasitic disease causing ulceration, disfigurement, anemia, and death if not treated. Cattle and sheep herders faced risk from exposure to the *Phlebotomus* sandfly that carried the disease. Widespread drought and famine complicated the problem, which authorities battled by spraying for the flies.

May 2000 Beginning in spring 2000 and continuing for a year, three cryptosporidiosis upsurges in Belfast, Ireland, produced a total of 476 cases of acute gastroenteritis from protozoan oocysts. Sources ranged from pasture runoff, leaking septic systems, and blocked drains, all evidence of inadequate monitoring of urban water supplies in Northern Ireland.

June 24, 2000 At a public bath in Ibaraki, Japan, Legionnaire's disease killed one person and sickened 14 others aged 58 to 85 after the victims patronized a facility outside Tokyo between May and June 24.

summer 2000 Martha's Vineyard, Massachusetts, reported pneumonic tularemia, a repeat of an epidemic in 1978. Health officials warned that lawn mowing and brush cutting contributed to infection by the *Chrysops* tick or deer fly. A more serious outbreak in Kosovo, Serbia, followed a decade of war. Advanced cases caused mouth and throat infection, swollen glands, and disfigurement from ulcerating lymphadenitis. Diagnoses of 900 cases required investigation of households for food and water pollutants and infestations of rats and voles in areas overwhelmed by combat.

summer 2000 On the Sharon plain, epidemic West Nile fever afflicted thousands of Israelis, killing 13. Health investigators connected the virus with flight patterns of migrating birds, which passed the pathogen to mosquitoes. Deaths were rare, except among the elderly and patients with diminished immune systems.

mid–July 2000 Brianne Kiner, an adult survivor of catastrophic illness from *Escherichia coli* contracted from an under-cooked hamburger purchased at Jack in the Box, comforted young Brianna Kriefall. The three-year-old was a victim of the same food-borne illness and suffered 10,000 seizures, three strokes, brain damage, and failure of heart, kidneys, respiratory system, and liver. When Kriefall died at Wauwatosa, Washington, on July 22, 2000, an expert at gastroenterology and hepatology at the Medical College of Wisconsin acknowledged that the bacterium affects children more seriously than it does adults.

mid–August 2000 South Africans in the eastern and northeastern parts of the country suffered a fierce onset in cholera. Over a six-month span, the disease afflicted the predominantly black KwaZulu-Natal area, killing 229 of 105,874 victims. The health ministry called on the World Health Organization for assistance in containing infection.

Officials determined that piped and municipal water supplies were uncontaminated with *Vibrio cholerae*. The people at high risk drew water from springs, streams, dams, and rivers, and those whose immune systems were compromised by HIV. Similarly affected by resurgent cholera were Malawi, Mozambique, Swaziland, Zambia, and Zimbabwe, where drinking water was typically substandard.

By June 4, 2001, in southeastern South Africa, a recrudescence of cholera struck over 100,028 with 208 fatalities. The death toll was one of the lowest on record. In KwaZulu-Natal, the disease advanced among those living in unhygienic conditions, infecting up to 250 per day.

September 2000 When Ebola infected 428 at Gulu, Masindi, and Mbarara in northern Uganda, it was the first time local people had witnessed the disease. A more infectious mutation of the original virus killed 53 percent of victims with a quick, painful death from vomiting, bleeding gums and nose, and bloody diarrhea. Health workers set up an isolation ward at Locar Hospital in Gulu, 225 miles north of Kampala. One of the supervisors of contagion, Matthew Lukwiya, medical superintendent of Saint Mary's Hospital, died of the disease. In late October, the health ministry reported 182 sick with the hemorrhagic fever, which killed 60. The government designated two graveyards for Ebola victims close to Gulu's two hospitals.

To contain the virus, the government closed schools, instituted curfews, and halted the ritual cleansing of corpses before funerals. When the disease made its way southwest to Mbarara, Ugandan health officials requested aid from the World Health Organization and humanitarian agencies. One observer, Dean R. Hirsch, president of World Vision International, who visited the area in December, discovered that Ugandans had replaced hugs, pats, and handshakes with a no-touch greeting consisting of swinging the fists from knees to shoulders. By February 2001, WHO declared Uganda Ebola free.

September 19, 2000 A. Martin Lerner of William Beaumont Hospital in Royal Oak, Michigan, disclosed that patients suffering chronic-fatigue syndrome, an endemic malady of unknown origin, rallied from treatment with regimens of the potent antiviral drug valacyclovir or ganciclovir. He concluded to a gathering of the American Society of Microbiology that the Epstein-Barr virus and cytomegalovirus weakened the heart. Because the body's immune system suppresses rather than exterminates the viruses, enough survived to damage cardiac fiber.

September 25, 2000 Information in the *American Journal of Surgical Pathology* from pathologists at Raigmore Hospital in Inverness, Scotland, connected the bite of a bacteria-bearing sheep tick to B-cell lymphoma, a deadly form of human skin cancer, which might be treatable with antibiotics. The bacterium *Borrelia burgdorferi*, which causes Lyme disease and is also related to syphilis and cancer, elevated Scotland's rate of Lyme disease to the highest in northern Europe.

October 2000 In Galicia, Spain, 28 cases of Legionnaire's disease resulted in three fatalities. The victims came in contact with a local hospital where the *Legionella* bacteria may have derived from the facility's cooling towers. A second outbreak in Valencia produced 70 cases and two deaths from the bacteria, but officials did not locate the source. Within weeks, a third onset in Barcelona struck 40 people, leaving four critically ill. All patients resided in Barceloneta, a seacoast community.

October 22, 2000 At Riyadh, Saudi Arabia, an upsurge in Rift Valley fever killed 84 out of 423 victims. The number plus 70 more in Yemen brought the area death total to 154 within a four-week period. The fatality rate was lower for some 500 Yemeni victims. Government agents countered the virus by spraying mosquito-infested areas from airborne tanks.

November 2000 In the worst case of food poisoning in Canada's history, food or water contaminated with *Escherichia coli* sickened 2,500 in Walkerton, Ontario, a town of 5,000. Of the total stricken, 11 died. Out of 784 treated for infection, 78 adults and twelve children entered hospitals, where five children required dialysis. A series of investigations and lawsuits forced the province to provide monetary relief to the community, which struggled with the costs of the incident.

An eight-week cleanup assured the town of sanitary water, but left them dependent on other sources, notably, ten million liters of bottled water, donated in part from an Australian company. Meanwhile, municipal authorities went door to door and business by business to flush the city's water system. Because tainted runoff seeped into local wells, officials proposed piping in water from Lake Huron.

November 9, 2000 When Joseph J. Burrascano, Jr., of East Hampton, New York, faced nine state charges of medical misconduct from the New York State Office of Professional Medical Conduct, some 400 victims of Lyme disease and their primary physicians joined thousands rallying for the defendant on Fifth Avenue opposite Central Park South. Charged with unprofessional treatment of advanced or persistent cases of Lyme disease, Burrascano found solace in "Voices of Lyme," a collection of patients who improved under his care. Because he deviated from the standard three-week course of antibiotics with large intravenous doses over a longer period of time, he and 17 other pioneers came under scrutiny. One physician, Perry Orens, a cardiologist from Quogue, New York, lost his medical license for administering a curative measure that is common among Swiss and German doctors.

Burrascano based his variant treatment on differences in stages of the disease from first appearance to advanced stages, which caused sufferers to be homebound, some bedfast. Supporters demanded that he and others be allowed a clinical license to rescue resistant cases from misery, even if his methods violated the standards of HMOs. Burrascano also earned the backing of some 85 New York doctors as well as the International Lyme and Associated Disease Society.

December 11, 2000 An upsurge of listeriosis or *Listeria monocytogenes* assailed Hispanics of Forsyth County, North Carolina. Extension agents traced ten cases involving two stillbirths to homemade soft cheese and unpasteurized milk. Workers disseminated warnings through 111 Latino churches and community resources that the disease could kill people with diminished immune systems, the elderly, infants, pregnant women, and the unborn.

2001 In its second decade in the United States, the AIDS epidemic produced one million illnesses and 450,000 fatalities from the disease or from AIDS-related complications. Adding to the increase in infections was a rise in black, Latino, and female victims infected by heterosexual contact.

2001 An unforeseen spike in asbestosis and mesothelioma began a fifteen-year increase in mortality, mostly to U.S. laboring men. Statisticians identified four asbestos-related occupational diseases:

lung cancer	4,800
mesothelioma	2,509
asbestosis	1,398
gastro-intestinal cancer	1,200
total	9,907

Asbestosis caused lung scarring and plaque from inhaled fibers and gasping for breath. Cancer linked to asbestos remains incurable. Leading the U.S. surge were the states of California, Florida, New York, Pennsylvania, and Texas and isolated counties in Arizona, Maryland, Massachusetts, and Michigan. The total number of asbestos-related deaths ranked fourth for adults behind AIDS, alcoholism, and gunshot wounds.

January 1, 2001 According to *Popular Science*, in the calendar year 2000, 3.8 million new cases of AIDS afflicted Africans, making the continent the hardest hit of any part of the world's population. The doubling of diagnoses of HIV every 12 to 18 months in the Caribbean placed the region second only to sub–Saharan Africa in infections. Worst hit was Haiti, where 8 percent of urban islanders carried the virus.

February 6, 2001 When scientists began studying Creutzfeldt-Jakob disease in the early 2000s, they returned to frozen blood samples containing kuru, a fatal neurodegenerative affliction of the brain that spread among the Fore celebrants of cannibal funerary rituals in Papua New Guinea in the 1940s and 1950s. Study of DNA from the tissue at the National Institutes of Health in Bethesda, Maryland, determined that kuru could evolve after over two decades of incubation.

March 2001 U.S. CDC agents responded to evidence of lead poisoning in pre-school children in Torreón, Coahuila, Mexico. The 11,181 children aged six and under living and playing near a lead smelter of Met-Mex Peñoles, the largest metal processing factory in Latin America, ingested heavy metal contamination from water, soil, dust, and toys. Environmental contaminants threatened hearing, mental and physical development, and behavior in growing children.

March 30, 2001 From a two-year Georgia study begun in 1997, Norman Carvalho of Children's Healthcare of Atlanta determined that rickets was making a comeback. Long in abeyance through emphasis on outdoor play and vitamin D–fortified milk, the bone softening anomaly appeared to derive from multiple stresses on childhood—lack of vitamin D in breast milk and soy formula, long hours in daycare, long bus rides to school, vegetarian and macrobiotic diets, smog, and extreme protection from the sun. Most at risk are breast-fed black infants with significant amounts of melanin in their skin.

The first ten cases prepared Carvalho for a lengthy study of malnutrition, which he reported in the April 2001 issue of *Pediatrics*. His findings substantiated a claim that one in 200,000 young Georgians was hospitalized with rickets. Because the disease became rare in the previous 70 years, doctors no longer reported it to boards of health and ceased instructing mothers on the essential vitamins for growth and development. Corroborating Carvalho's findings, Kelley S. Scanlon and her team's article in the July issue of the *American Journal of Clinical Nutrition* in which they discovered that 42 percent of the 1,546 black women in the study were deficient in vitamin D as compared to only 4.2 percent of the 1,426 white women in the study.

April 2001 After cholera recurred in Dhaka, Bangladesh, bacteriologist Andrew Camilli of Tufts University School of Medicine in Boston, Massachusetts, studied the bacterium's infectivity. He and his colleagues determined that the microbe is impotent in sterile laboratory surroundings, but that human digestive juices activate ten different genes that boost its potency seven hundred times. With newly acquired strength, *Vibrio cholerae* became more lethal. The knowledge encouraged a new formulation of vaccine from activated bacteria rather than lab-reared stock.

April 10, 2001 A Johns Hopkins cardiologist studied ultrasound readings of the hearts of Puerto Ricans living in Ponce and on the island of Vieques, a U.S. Naval testing site for ordnance training. The data established that people on Vieques had a high incidence of vibroacoustic disease, which resulted in a thickening of the carotid artery to the neck and cranium. However, the creator of the test drew no conclusions that would charge the U.S. military with jeopardizing islanders' health from repeated shellings and explosions.

April 18, 2001 A shipment of five varieties of almond snacks packed by Hughson Nuts in California and sold by Bulk Barn stores contained *Salmonella*, which caused sickness in some 160 Canadians in Ontario, New Brunswick, and Nova Scotia. A recall reduced the spread of infection.

May 21, 2001 Gilles Thomas of the *French Institut National de la Santé et de la Recherche Medicale* in Paris and Gabriel Nuñez of the University of Michigan defeated bacterial theories of Crohn's disease and identified the cause as Nod2, a defective gene causing an immune malfunction. By attacking itself, the body afflicted with Crohn's disease generated inflammation and ulceration of the gut in one of every 1,000 citizens of western nations.

June 2001 "Dark Winter," a 48-hour simulation exercise staged at Andrews Air Force Base in Washington, D.C., replicated an Iraqi attack with smallpox toxins. Staged at Oklahoma, the imaginary 14-day epidemic exhausted stocks of 12 million doses of smallpox vaccine. The "disease" left 1,000 dead as it swept over 16,000 in 25 states and on to ten other countries. The war-game scenario predicted that, within 21 days,

300,000 would die. The exercise alerted the U.S. House Subcommittee on National Security of the potential for catastrophic bioterrorism.

June 2001 E. E. Stobberingh of the University Hospital of Maastricht, Holland, reported in the *Journal of Antimicrobial Chemotherapy* that humans commonly acquire antibiotic-resistant *Escherichia Coli* from animals. After studying the development of resistance in turkeys, broilers, and laying hens infrequently dosed with antibiotics, he compared fecal samples of their handlers and slaughterers. The results accounted for an elevated human resistance to almost every antibiotic, particularly among those in contact with broilers and turkeys.

June 15, 2001 In Iowa, *Shigella sonnei* infected 26 people who had entered a wading pool in a large city park. Investigators blamed the absence of regular monitoring, disinfection, and recirculation of the water, where a child's leaky diaper may have spread bacteria.

July 2001 In Mumbai, India, an epidemic of leptospirosis killed 19 of 272 victims. Unofficial blame for the outbreak moved from rats and dogs to buffaloes and cows, which Hindus consider sacred.

July 16, 2001 According to testimony to the U.S. House Government Reform Committee, fear mounted among medical researchers that an insidious simian virus that contaminated polio vaccine in the 1950s and 1960s may cause rare human cancers in brain, bone, and lungs, including mesothelioma. Because the virus appears in tumors of people who weren't part of the widespread inoculation, experts surmise that a virus derived from the Asian rhesus monkey may spread from human to human.

July 16, 2001 In Murcia southeast of Madrid, Spain, the largest recorded incidence of Legionnaire's disease produced 800 cases of pneumonia. Four died; 12 required intensive care. Most of the sickness occurred in the same part of town, but medical sleuths were unable to pinpoint the source. A month later in Pamplona, Spain, 18 people sickened and three died from Legionnaire's disease incurred from a hospital hot water system. Staff disinfected the waterway by superchlorination.

July 24, 2001 In *Biology of Plagues* (2001), Susan Scott and Christopher Duncan of the University of Liverpool theorized that the diagnosis of the Black Death as bubonic plague may have been erroneous. The new hypothesis proposed that the hemorrhagic plague might be related to current outbreaks of the Ebola virus. Scott warned that a similar virus with a long but symptomless incubation could spread worldwide before it began to appear in human victims.

July 28, 2001 In Utah County, Utah, the identification of two cases of tularemia in hunters raised concerns for an outbreak. The disease, which is rare in the United States, is dangerous enough to be a choice of bioterrorists, along with plague, botulism, brucella, and anthrax. Spread by the bite of deer flies in a camp at Pinedale, Wyoming, *Franciscella tularensis*, also called rabbit fever, was potentially fatal but responded to antibiotics.

August 5, 2001 At the Ozaukee County Fair in Cedarburg, Wisconsin, 141 people became ill from *Escherichia coli* contamination. Eleven confirmed cases of the pathogen included two children treated at Children's Hospital of Wisconsin. State public health director Glenda Madlom noted that the cause of the outbreak was undetermined during tests of 39 specimens. However, cancellation of events during a quarantine at the fairgrounds held firm until health workers completed environmental sampling.

September 4, 2001 At Stavanger in western Norway, 28 people suffered Legionnaire's disease, which killed seven. Officials tentatively traced the infection to a decorative outdoor fountain and nearby hotel cooling towers at the city center.

October 2001 After the first American strikes against the Taliban in Afghanistan, the health of refugees declined alarmingly. Local people suffered the second highest maternal mortality rate in the world and high infant mortality exacerbated by malnutrition, poor hygiene, overcrowding in camps, fetid drinking water, and lack of medical care. Among children, epidemic enteritis caused bloody diarrhea, dehydration, and death. Additional threats to their health derived from rampant scurvy resulting from limited variety in the diet and malnutrition.

October 6, 2001 Some 90,000 U.S. military service members were examined and treated for a mysterious body of ills dubbed Gulf War syndrome. As reported in *Annals of Epidemiology*, a U.S. government questionnaire to 21,000 veterans of the Gulf War from all four branches of military service determined that children born to veterans were up to three times as likely as those of other military veterans to suffer birth defects. According to Veterans Affairs epidemiologist Han Kang, of the 14,700 who replied to the queries, a large number reported miscarriages and fears of birth defects.

November 1, 2001 Hawaiians experienced a slowing of dengue fever, which infected 89 people—64 on Maui, 20 on Oahu, and five on Kauai. Teams traveled door to door pointing out to residents such dangers as pools of stagnant water, breeding grounds for the virus-bearing *Aedes aegypti* mosquito.

December 11, 2001 In the Ogooue-Ivindo province of Gabon, ten people died of the onset of Ebola in West Africa, causing concern in the nation's ministry of health and among an international team from the World Health Organization. The death rate held steady at 70 percent, rather than the usual 90 percent. *See also* **June 13, 2002.**

December 16, 2001 United States officials offered scarce supplies of anthrax vaccine to some 3,000 people whom an unknown terrorist exposed to the bacterium in October, including mail workers in Trenton, New Jersey, who sorted letters containing powdered *Bacillus anthracis*. Because spores have been known to encapsulate and lie dormant in the lung before rejuvenating, the Centers for Disease Control and Prevention studied the problem of prophylaxis and treatment. Some health professionals worried that the spores of dry anthrax, which killed five and compromised the health of 20 mail handlers and two U.S. senators, might hibernate beyond the effectiveness of antibiotics administered over a 60-day period.

In this same period, the FBI used genetic fingerprinting to trace the deadly dry anthrax bacteria to one U.S. military source, the U.S. Army Medical Research Institute of Infectious Disease (USAMRIID) at Fort Detrick, Maryland. Scientists surmised that the terrorist had to access microbes housed at Fort Detrick, Utah's Dugway Proving Ground military research facility, Porton Down in Wiltshire, England, or germ banks maintained by Louisiana State University and Northern Arizona University. Further investigation at Canada's Defense Research Establishment Suffield, the University of New Mexico in Albuquerque, and the Battelle Memorial Institute in Columbus, Ohio, tracked additional samples obtained from USAMRIID. *See also* **May 10, 2002**.

December 27, 2001 While malaria subsided, dengue fever re-emerged in Vietnam during a devastating flood that left 1,780,000 homeless and 1,650,000 in danger of starving. In the midst of environmental havoc, the disease felled nearly 40,000 and claimed 81. In the high humidity of the Mekong Delta, mosquitoes perpetuated conditions preceding fever. *See also* **April 2002**.

2002 As reported by the Centers for Disease Control and Prevention, the United States curbed an upsurge in gonorrhea, particularly in Birmingham, Alabama; Buffalo, New York; Detroit, Michigan; Jacksonville, Florida; and Kansas City, Missouri. While rates dropped in Atlanta, Chicago and Washington, D.C., high rates of infection covered the Southern states, especially Mississippi.

2002 Widespread malaria in Africa robbed the nation of health care funds. The shortfall caused caregivers to switch to the cheaper and less effective drug chloroquine, which helped only 20 percent of patients. Volunteers with Médecins sans Frontières (Doctors without Borders) warned that the disease remained lethal in sub–Saharan nations, generating 1.8 million deaths annually. Of these, 90 percent were children. The disease struck 3.25 million or half the population of Burundi; in Tanzania, 300,000 died in 2001 from malaria.

January 1, 2002 A resurgence of African trypanosomiasis, or sleeping sickness, afflicted travellers to East African and patients examined by Médecins sans Frontières (Doctors without Borders). Thought to have been exterminated in the 1960s, the disease gained a new foothold because of the erosion of disease control by war and political instability in Africa. The disease threatened some 60 million residents of Angola, Congo, Sudan, Tanzania, and Uganda.

In February 2002, the Organisation of African Unity pledged a unified fight against sleeping sickness, which killed up to 100 people a day in 36 sub–Saharan countries for a total of over 35,000 Africans as well as three million livestock each year. The worst hit countries were Angola, Cameroon, Sudan, Uganda, and the Central African Republic. A pilot program to eradicate the tsetse fly in Zanzibar involved releasing sterile male flies among females.

January 17, 2002 The World Health Organization (WHO) reported an onset of 1,500 cases of leishmaniasis among Afghan refugees at the Kurram Agency on Pakistan's northwest frontier with Afghanistan. These new occurrences plus 1,027 among locals totaled 2,527 people suffering severe skin ulceration. A common disorder in the region, the disease thrived in a female sandfly and spread among animals and people. Because of a scarcity of needed drugs and the lack of medical know-how among health workers, untreated patients risked permanent scars and facial deformity.

In May, WHO determined that an epidemic of leishmaniasis in Kabul, Afghanistan, threatened 100,000 victims with the stigma of permanent scars, the result of large unhealed sores. Officials feared that the return of some 3,420,000 Afghan refugees from western Pakistan would spread the disease in camps. Health workers issued an appeal to international donations of insect spray and netting for sleeping quarters and drugs for the 28-day treatment, which requires daily injections of Glucantime or Pentostam. Girls and women whom men shunned faced psychological damage. As summer approached, the increase in sandflies spread the disease at an exponential rate.

January 18, 2002 Dengue hemorrhagic fever returned to Brazil and Cuba, killing a slum dweller of Rio de Janeiro and three others. Worsening mosquito infestation of stagnant waters spread the virus. By February, 18 had died and 1,600 new cases per day boosted the total, which was projected around 430,000. *See also* **April 2002**.

January 22, 2002 As reported in *USA Today*, the death of a Nairobian immigrant in Mobile, Alabama, alerted public health workers to a global epidemic of multi-drug resistant tuberculosis. After the identification of 48 infected people in the Mobile area, experts began formulating details for travelers and refugees. Leaflets explained that the debilitating lung disease could take up to two years of complex, costly drug treatment. Some victims never recovered completely.

February 2002 An escalation of hepatitis A in Polk County east of Tampa, Florida, affected 138 of the area's half million residents, killing one and destroying the liver of another. Health officials linked one branch of the contagion to an infected restaurant cook who may have passed along the virus by failing to wash his hands after going to the toilet. The man claimed to have contracted hepatitis A in prison.

Hepatitis A appeared to have originated in a large population of methamphetamine users, who transmitted contagion to friends and others through sexual contact and shared drug paraphernalia. The ailment became common among day laborers, food-service employees, and migrant workers. At the rate of twelve new cases per week, the disease infected ten times more people in 2001. The year's total rose six times higher than the national average.

February 2002 In Uvira, East Congo, a half million refugees fled violence and conflict and huddled at an Action against Hunger therapeutic feeding center. Children displayed rampant marasmus and kwashiorkor in swollen limbs and peeling skin. Secondary infections of upper respiratory infection, scabies, and malaria deepened their misery.

April 2002 The reemergence of dengue fever in India, South America, Southeast Asia, and the Western Pacific produced a high number of cases. The World Health Organization estimated that data reported only a half million of the 50 million annual cases. Called "malaria's poor cousin," the disease debilitated, reducing productivity and quality of life. Worst hit, South America and Southeast Asia ranked the fever the number one deterrent to health. During the year, Brazil and Cuba suffered the most devastating flareups. As the fever burgeoned in May in the South Pacific, Polynesia, Easter Island, and Hawaii, tens of thousands of Tahitians contracted the disease in 2001. On Fiji, some victims died. Hawaii experienced over 100 cases, its most serious onset in a half century.

WHO officials urged a thorough study of the disease and its causes to the World Health Assembly in hopes of formulating a vaccine. Standard public health regimens call for recording incidence of dengue fever and controlling *Aedes aegypti* mosquitoes that carry the disease, particularly in heavily populated areas. Former vigilance over stagnant water, open containers, and litter lessened in the 1970s during phenomenal urban growth. The faltering of public sanitation safeguards left vulnerable populations of poor nations, especially the young, whose immune systems lacked development. The formulation of new vaccines at Mahidol University in Bangkok, Thailand, and at Walter Reed Army Institute of Research in Silver Spring, Maryland, showed promise.

April 2002 A new strain of meningitis, W135, overran Burkina Faso, killing 672 out of 4,578 victims. First reported in January, the disease originated in Saudi Arabia in pilgrims returning from a hadj to Mecca. Brain fever invaded Burkina Faso and afflicted 15 of the nation's health care districts. Over 90 percent of malaria victims displayed symptoms of W135.

April 1, 2002 According to Action against Hunger, some 20,000 displaced persons in South Kivu, eastern Democratic Republic of Congo, fled heavy combat. In December 2001, refugees did not escape cholera, which killed half of its victims. Through contaminated food and water, infection spread the following April to Kazimia south of the Ubwari peninsula.

mid–April 2002 The illness of hundreds of Americans contracting foodborne *Salmonella enteriditis* resulted in a class-action suit filed on behalf of all victims. The three-week scourge at a convention held at the Wyndham Anatole in Dallas, Texas, the Southwest's largest convention hotel, exposed more than 3,000 individuals to bacteria from tainted food or water.

April 18, 2002 As reported in the *New England Journal of Medicine*, doctors at Children's Hospital in Pittsburgh, Pennsylvania, reported an outbreak in 46 school-age children of strep throat that was resistant to antibiotics. Reported among other community members as well, the group A streptococcus bacteria failed to yield to erythromycin, a standard treatment from the macrolides drug family. The findings indicated that physicians should limit prescriptions of the drugs.

April 22, 2002 Up to four million people in the U.S. endured the hepatitis C virus (HCV), which a medical consortium discovered in 1988. Spreading silently through tainted blood transfusions, intravenous drug use, and tattoo and body-piercing parlors, the disease was not always fatal. Around 15 percent of carriers had strong enough immune systems to quell HCV without medical intervention. Hardest hit were drug users who shared needles and hemophiliacs, who exhibited close to a 100 percent infection rate. Lesser numbers of infections occurred among dialysis patients, people who had required blood transfusions in the 1960s, infants born to infected mothers, and health-care workers pricked by needles or nicked through rubber gloves. Open to question, high rates of hepatitis among Vietnam War veterans may have dated to vaccination with inoculation guns. The high incidence of liver failure from HCV boosted demand for transplants.

May 2002 Resistant strains of gonorrhea migrated from Asia to Hawaii and California. No longer treatable with antibiotics, cephalosporins, or fluoroquinolones, the disease threatened public health. The Centers for Disease Control and Prevention warned that the disease caused pelvic inflammatory disease, ectopic pregnancies, and infertility in both males and females.

May 10, 2002 Microbial forensics began producing answers to the investigation of bioterrorism through the mail. Anthrax researcher Claire Fraser of the Institute of Genomic Research in Rockville, Maryland, explained that her team analyzed the bacteria that killed the first victim after the October 2001 attack with dry spores sent in a letter. Identified by DNA perusal of 5.2 million DNA letters as the Florida strain of the Ames isolate, the *Bacillus anthracis* originated in tissue taken from a cow carcass in Texas in 1981.

May 16, 2002 An unidentified enteric fever struck 18 British soldiers from the medical corps serving with allied forces at Bagram 30 miles north of Kabul, Afghanistan. Three of the victims became seriously ill and required hospitalization in England and Germany. An American soldier received advanced care in Uzbekistan. To stem the spread of contagion, authorities quarantined 30 people.

May 20, 2002 A United Nations initiative distributed polio vaccine to the world's poorest nations in an effort to eradicate the crippling virus. The World Health Organization and the United Nations Children's Fund scheduled systematic immunization to continue for 20 months. Health workers targeted the underclass in south-ern Asia and Africa, where 480 cases of polio were diagnosed in 2001.

Among the most susceptible people, refugees in Afghanistan and slum dwellers in Nigeria lacked basic hygiene and immunization. WHO polio expert Ciro A. de Quadros of Brazil, director of the Pan American Health Organization's Division of Vaccines and Immunization and recipient of the Albert B. Sabin Gold Medal, remained hopeful of a victory over a major crippler. He predicted that the massive campaign would eliminate polio by 2003.

June 2002 Chinese health leaders established a plan to halt endemic hepatitis B that assailed 60 percent of the populace by vaccinating all newborns between 2002 and 2007. The project, administered by the Global Alliance for Vaccines and Immunization and funded by Bill and Melinda Gates, fought the steady rise of the virus by distributing unreusable syringes.

June 12, 2002 As reported by B. K. Baishya, Assam's chief malaria control officer, 73 of some half-million malaria victims in Guwahati in the oil-rich and tea-growing state of Assam, India, perished within a six-week period. Health inspectors attributed the outbreak to stagnant pools that filled during extensive seasonal rains. Authorities formed 150 teams to spray insecticide and distribute chloroquine tablets to villagers.

June 13, 2002 At Oloba in northwestern Congo, five victims of Ebola died. The deceased appeared to have eaten contaminated monkey meat and removed internal organs from human corpses as a funeral rite. The deaths raised to 73 the death toll for the Congo and Gabon in an epidemic that erupted in December 2001.

The escalation of hemorrhagic fever set off panic in central Africa. To halt the spread of the virus, government officials banned the consumption of chimpanzees, gorillas, and other primates, which are the primary source of protein for natives. Residents objected to changing their diet because they had few substitute sources of protein.

June 26, 2002 The outlook for relief from global AIDS looked grim. The continent of Africa suffered 2.3 million deaths and 3.4 million new HIV infections in 2001. Michel Sidibe, director of UNAIDS, reported that the viral epidemic in sub–Saharan Africa reduced economic growth of some areas from 5 percent to 1 percent and cut worker productivity by half. In Botswana, where 24,000 died from AIDS in 1999, the loss of taxes from sick workers limited public and social services. The loss of seven million farmers to disease in Burkina Faso increased famine in southern Africa. Likewise compromised, the military reported that half the soldiers in some countries were too ill to provide national security. Kenyan police forces sustained a loss of 25 percent of their officers because of AIDS. Stephen Lewis, the U.N.'s special HIV/AIDS envoy, challenged wealthy nations to raise their contributions to the Global Fund to Fight AIDS, Tuberculosis, and Malaria.

In this same period, China's medical authorities announced a potential explosion of AIDS infections and predicted ten million victims by 2010. Since the reporting of the first case in 1985, the number of HIV-positive victims reached an official 30,736, but estimates of 1.5 million seemed closer to reality. From January to July 2001, the number of patients rose by two-thirds. In China, intravenous drug use and the sale of human blood boosted infection.

mid-summer 2002 According to Minister for Medical Services Hussein Maalim Mohamed, a malaria epidemic in the highlands of western Kenya killed 294 out of over 158,000 reported cases. In a population lacking resistance, the disease struck victims who followed no specific regimen against insects propagated by unusually heavy rains.

July 1, 2002 A growing number of mysterious lung infections resembling tuberculosis caused U.S. doctors to suspect mycobacteria. A noncontagious pathogen, the disease, called NTM for nontuberculous mycobacteria, began appearing in the 1980s. A rapid grower, it thrived in the warm, moist atmosphere of ponds and in homes in the slime coating faucets and showerheads. The disease produced a mounting epidemic subject to frequent relapses that were tricky to cure, taking up to 18 months of treatment. NTM thrived in thin, white middle-aged women and devastated AIDS patients. The energy saver's reduction of temperature in hot-water heaters increased microbes in households.

July 3, 2002 Of the 37 million people with AIDS worldwide, 70 percent lived in sub–Saharan Africa, where 10 percent of the populace was HIV positive. One key to slowing the pandemic, according to experts attending the XIV International AIDS Conference in Barcelona, demanded the freeing of young women from gender constraints. In patriarchal African societies where men controlled intimacy and determined when and with whom to have intercourse and whether to use condoms, women remained powerless to avoid infection. As summarized in the *Wall Street Journal*, "Women are infected more easily and more often than men, but men drive the spread of the virus" (Schoofs and Zimmerman, 2002, p. D4). Rapes by military and police personnel of women in refugee camps illustrated the gender bias in Angola.

Peter Piot, director of the joint United Nations Programme on HIV/AIDS (UNAIDS), warned that the virus would unsettle Africa, China, Eastern Europe, India, and Russia economically and politically, spilling chaos into world affairs. By 2010, the average life expectancy in Botswana would fall to 26.7 as population growth dropped to -2.1 percent; Mozambique's life expectancy would decline to 27.1. The survival picture was equally bleak for Lesotho, Malawi, South Africa, and Zimbabwe.

By 2020, Piot predicted the death toll would rise to 68 million, producing the worst pandemic in world history. The loss would cost sub–Saharan Africa its most productive people, leaving the elderly and orphans to cope with a crumbling society. Contributing to gradual social destabilization, the high rate of infection and death among women threatened family survival. The gender imbalance forced males to look outside their region for mates. The distant wife-search spread the virus farther afield. Stephen Lewis, U.N. envoy on AIDS in Africa predicted a demographic rupture: "We're going to have all kinds of men without partners, wandering the landscape on a continent where there is already substantial instability" (Brown 2002, p. 14A).

Simultaneous with Africa's catastrophic health, an AIDS crisis advanced in Burma, Cambodia, China, Russia, and Thailand. Thai health data reported a quarter million people infected with HIV, which spread at the rate of 25,000 per year. Projections of costs for prevention and care of children orphaned by AIDS reached $10 billion annually. Piot blamed heads of countries for waiting too long to check the explosive spread of the virus. He expanded: "One of the reasons we have such a big epidemic is a failure of leadership. We've lost precious time, time that will translate into millions and millions of deaths" (Sternberg July 3, 2002, p. 10D).

July 12, 2002 An outbreak of salmonella poisoning among at least 70 patrons of a Red Lobster restaurant at Northgate Mall in the Hixson district of Chattanooga, Tennessee, resulted in 160 illnesses, 35 confirmed cases, and one death. Diners who patronized the establishment on June 21 or 22 suffered fever, diarrhea, and vomiting. Authorities were unable to identify the food that contained the contaminant.

July 28, 2002 In South Africa, one-tenth of infants born in the Transkei region suffered severe, chronic hunger, the source of kwashiorkor and marasmus. A survey of 3,120 local children found 50 percent living on less than half the recommended allotment of nutrients and incurring stunted growth and deficiency diseases. Lowering their chances of survival were the high number of unemployed parents or families living with HIV/AIDs.

One source of hope for the poor was the African Children's Feeding Scheme, which provided families with porridge, corn meal, and fruit. The U. N. World Food Program headquartered in Johannesburg also fought famine. Strategists intended to raise life expectancy among Southern Africa's 12 million residents, in whom famine elevated learning disability and susceptibility to bacterial infection.

August 6, 2002 A study from the Tabriz University of Medical Sciences in northwestern Iran proposed supplying dog collars to poor countries to rid pets of sandflies, the vector for visceral leishmaniasis. Tests of the collars reduced endemic disease in children by 42 percent. In Brazil, where the disease was rampant, the collars replaced the unsavory chore of exterminating 20,000 dogs per year and spraying 200,000 residences for flies. While these approaches appeared to have helped in China, the article warned, they had little success elsewhere.

August 21, 2002 During a rise in West Nile fever infections in Europe and the Middle East, the fever resurged in the United States, where it was first detected in 1999. By 2001, the virus had infected 100 types of birds, 29 varieties of mosquitoes, animals, and humans. In the opinion of

Paul R. Epstein, associate director of the Center for Health and the Global Environment at the Harvard Medical School, global warming and weather extremes encouraged mosquitoes that carried the virus while suppressing populations of amphibians and dragonflies, the natural predators of mosquitoes.

Summer produced an epidemic of brain infection, either encephalitis or meningitis, in 251 cases reported in 36 states and Washington, D.C. In Louisiana, the eye of the viral storm, 85 cases resulted in seven deaths. Identical to the Israeli strain, the virus killed birds, raising the question of whether the virus is gaining in virulence.

fall 2002 In the Blue Mountains of New South Wales, Australia, 59 cases of psittacosis comprised the island nation's largest scourge of parrot fever. The lung disease derived during lawn mowing from inhaling dust and dander of wild king parrots, crimson rosellas and sulphur-crested cockatoos. Most victims reported fever, cough, headache, and body pain accompanied by nausea and diarrhea. Hospitalized patients suffered atypical pneumonia caused by *Chlamydia psittaci* pathogens.

September 17, 2002 Over 200 people—adults and children—sickened after ingesting rat poison at the Heshenyuan Soybean Milk Shop, a snack café and caterer in Tangshang near Nanking, China. Police launched a criminal investigation after over 40 unsuspecting breakfast diners died of the mass poisoning of fried dough sticks, sesame cakes, and servings of rice. The poison caused people to collapse after only a few mouthfuls of food and to spit blood and mucus before lapsing into shock. *The China Daily* and *Hong Kong's Ta Kung Pao* reported that soldiers, construction laborers, migrant workers, elderly citizens, and students at a nearby school were among the victims. Agents for Xinhua, the repressive government news agency, concealed an exact count. The control of media information provoked public protests of limitations on public knowledge of health issues.

September 22, 2002 In "A Pox on Our House," an essay for the *New York Times Magazine*, Gregg Bourland, the youngest tribal chairman of the Cheyenne River Sioux in Eagle Butte, South Dakota, summarized growing concern among Native Americans about epidemic disease. They feared that threats of bioterrorism with smallpox could revive the devastating scourges that wiped out whole tribes during the European settlement of the Americas. He explained that fear of pestilence is vivid in the minds of Indians, who lack the centuries of immunity that Europeans have evolved. In describing the speed with which pathogens circulate, he noted: "One carrier could travel to 10 major cities in the United States in just a couple of days' time and spread it God knows where" (2002, p. 100). He proposed that, in an international situation where vaccine is limited, health agents should place native elders and native children at the top of the priority list to insure the perpetuation of culture. He concluded, "I don't want to seem selfish just for Native Americans. But I have to defend my people first."

November 3, 2002 Some 300 reports of pertussis overwhelmed the Darwaz district of Badakhsan province in northeastern Afghanistan on the Tajik border, killing 61 children under the age of twelve within four weeks from coughing and suffocation. Traveling by plane and on horseback, teams from the Ministry of Health and the World Health Organization departed on October 24 to carry supplies to the stricken area. Fadela Chaib, a WHO spokeswoman, reported another 68 children ill with shortness of breath, coughing, fever, and bleeding from the nose. The disease, which is highly contagious, tends to generate a 15 percent mortality rate in unvaccinated populations. Chaib noted that only 40 percent of Afghanis were inoculated.

November 11, 2002 At Rostov-on-Don, contaminated milk, sour cream, and yogurt from a dairy in Kropotkin in the Krasnodar region of southern Russia sent 1,200 people to the hospital with dysentery. Of the total stricken with dire intestinal discomfort, diarrhea, fever, nausea, and vomiting, 900 under the age of 15 and two dairy employees produced symptoms. As a result of the widespread sickness, after a criminal investigation, the dairy's owner fired the plant director.

November 16, 2002 The initial four cases of Severe Acute Respiratory Syndrome (SARS) developed at Fosham City, Quangdong, in southern China on the border of Hong Kong. The explosive pestilence, the first pandemic of the twenty-first century, resulted from a zoonosis with a ten-day incubation period following human contact with a bridge species that sup-

ported the host. The toxin emerged from the cages of Himalayan palm civets, ferret badgers, red foxes, Syrian hamsters, rhesus macaques, domestic cats, mice, horseshoe bats, and raccoon dogs in meat markets and the blood and feces of animal handlers. Symptoms of the "super spreader" ranged from fever above 100.4 degrees F, headache, confusion, low white cell count, and body pain to respiratory distress, diarrhea, rash, and dry cough. Among 305 cases, the outbreak spread by sputum droplets and produced atypical pneumonia and respiratory, heart, and liver failure, which killed five victims or .016 percent. Faulty and deliberately misleading data from China at first identified the lung disease as *Chlamydia pneumoniae*, but failed to conceal the SARS epidemic from the outside world.

November 27, 2002 Over a two-month period, 945 passengers and crew aboard U.S.-based cruise ships contracted a sickness that may have been the Norwalk virus, a brief, but highly contagious gastrointestinal ailment. It spread quickly to dense populations confined on ships, including over 500 on Holland America's *Amsterdam*, 398 on the Disney Cruise Line 2,400-passenger ship *Magic*, 47 aboard Holland America's *Statendam*. The virus moved quickly from person-to-person contact, from drinking water and swimming pools containing the microbe, from eating tainted salads and shellfish, and from touching contaminated handgrips, counter tops, and serving utensils.

The Centers for Disease Control and Prevention in Atlanta, Georgia, required that all liners experiencing illness in at least 2 percent of people on board report the outbreak 24 hours before reaching a U.S. port. To stem bad publicity from lawsuits and rumors of unclean staterooms and dining areas, cruise lines took ships out of service and scheduled aggressive cleaning and sanitizing. Holland America also scrubbed its terminal with a chlorine solution.

December 8, 2002 State health officials pondered the case of a chemotherapy nurse at Dr. Tahir Javed's hematology and oncology clinic in Fremont, Nebraska. The woman reused the same hypodermic syringe to flush a group of chest catheters with saline. Her unprofessional behavior caused 81 of the clinic's 612 patients to contract hepatitis C in March 2000 from one virus-bearing patient. The practice continued into

December 2001, but produced no obvious symptoms in patients. The damage did not come to light until the state epidemiologist, Dr. Tom Safranek, began tracking the rare genotype 3A in the clinic's clientele, many of whom already suffered from compromised immune systems. In July 2002, Dr. Javed apparently fled to Pakistan.

January 22, 2003 In Djolu, Bosobolo, Karawa, Gbadolite, and Genema, Congo, over 2,000 people died from an epidemic of influenza that began in November 2002. Coinciding with political tensions, an upsurge in meningitis, and widespread starvation, the outbreak of flu sickened 500,000 or 47 percent of the population and proved fatal primarily to the elderly and children under age five. Aid to the beleaguered country came from the World Health organization, Médecins sans Frontières, and the Pasteur Institute in Paris, where laboratory tests helped to isolate the H5N1 avian strain.

February 22, 2003 WHO dispatched some of the 800 experts from the Centers for Disease Control and Prevention to China to investigate the mysterious Southeast Asian epidemic of atypical pneumonia, which killed Dr. Liu Jianlun, patient zero. Chinese authorities barred the researchers from entering Quangdong. Health officials studied methods of transmission, which seemed to vary from the typical aerosol spread of droplets from sneezing and coughing. Unlike H5N1 or bird flu, the unidentified infection produced marked swelling and inflammation in the lungs from a new strain of the coronavirus related to the common cold. Epidemiologists determined that the best antigen is a serum made from the blood of survivors of the disease.

February 23, 2003 In China, a re-emergence of SARS toxins in 1,190 members of an agrarian population in Quangdong killed 47 victims. The sickness spread to the Metropole Hotel in Hong Kong and infected twelve guests and 99 hospital workers. Contagion also migrated to Singapore; Taipei, Taiwan; and Hanoi, Vietnam, where 38 medical professionals developed the infection. At the French Hospital of Hanoi, Dr. Carlo Urbani, an Italian specialist in parasitic disease for WHO, classified SARS as an emerging disease.

March 12, 2003 WHO named SARS as a new form of pneumonia that infected victims in Singapore, Malaysia, and Canada. Alerts informed global travelers and medical workers of the deadly contagion. The CDC warned travelers to avoid China, Hong Kong, Singapore, and Vietnam.

March 17, 2003 A world network of 80 clinicians at eleven top laboratories began isolating causes and treatments for SARS, which had spread to the United States. WHO declared SARS a threat to global health security, especially to clinics, hospitals, and nursing homes.

March 25, 2003 Nine tourists from Amoy Gardens in Kowloon, Hong Kong, caught SARS at their homes from septic U-traps in bathroom floors. The travelers spread the disease during an Air China flight to Beijing. Health officials enforced quarantines of the sick at the ten-block Amoy Gardens housing complex. Authorities in Hong Kong and Singapore closed kindergartens, public schools, and junior colleges.

March 29, 2003 Italian epidemiologist Carlo Urbani traveled to a convention in Singapore, Malaysia, where he died of complications from SARS. His death alerted medical researchers and hospital staffs of the severity of the virus.

March 31, 2003 WHO tabulated 1,622 SARS cases in thirteen countries in the Americas, Asia, and Europe, including Germany, Spain, Slovenia, and Great Britain. The death rate reached 58 or .036 percent.

April 3, 2003 In an era edgy with fears of world terrorism by bomb or biological attack, the emerging global epidemic of SARS spurred seventeen nations to collaborate on anti-viral measures at borders, airports and hotels, hospitals, and urban streets. Of the total victims in Asia, Australia, Europe, and North America, the aggressive virus sickened 2,300 and killed 80, with China, Hong Kong, Singapore, eastern Taiwan, Thailand, and Vietnam reporting 72 of the dead. Border entry screening suppressed contagion in Australia. In San Jose, California, authorities quarantined a plane originating in Hong Kong and passing through Tokyo, Japan, and observed five air passengers showing symptoms of SARS.

April 9, 2003 Simultaneous with the SARS epidemic, an outbreak of monkeypox in the Midwest, the first incidence outside Africa, penetrated Illinois, Indiana, Kansas, Missouri, Ohio, and Wisconsin. Medical detectives linked the rise in swollen lymph glands, weakness, fever, and pustules in 47 patients to the importation of exotic pets. Vendors marketed from Ghana poached rats, African squirrels, porcupines, dormice, and striped mice, which spread the monkeypox virus to prairie dogs. Smallpox vaccine proved 85 percent effective against transmission of the orthopox virus. *See also* June 13, 2003.

April 11, 2003 Despite global precautions, Africa reported its first SARS diagnosis in South Africa. While the CDC attempted to halt the outbreak of Asian SARS in North America, WHO officials identified world air travel as a source of spreading contagion. The alarm suppressed the tourism industry, disrupted health care, and elevated health and sanitation costs in affected areas.

April 14, 2003 As deaths from SARS rose to thirteen in Canada, Dr. Marco Marra at the Michael Smith Genome Centre in Vancouver, British Columbia, announced a breakthrough in identifying the genetic code of the coronavirus that caused the infection.

April 15, 2003 In Taiwan, the board of Taipei Municipal Hospital quarantined 1,170 people exposed to the SARS virus and investigated sanitation on its premises. The infection flourished among intensive care staff involved in intubation, endotracheal suction, nebulizing, cardiopulmonary resuscitation, and nasogastric feeding.

April 16, 2003 From East Asia, SARS advanced west to Dinnur, a garbage-strewn village east of Bangalore, India, infecting two victims. Chinese premier Wen Jiabao fired two dilatory health professionals and revealed a rigorous policy requiring medical authorities to report infections promptly.

April 20, 2003 China disclosed a rise in cases of SARS in Beijing from 37 to 339 and canceled a national holiday. Within 24 hours, the number of victims rose by 35.7 percent to 407. A day later, Beijing schools and part of Peking University closed. Despite caution, the contagion emerged farther west at Anhui.

April 26, 2003 As the SARS epidemic worsened in Beijing to some 740 cases, Chinese officials closed theaters, karaoke bars, cinemas, and dance clubs. Additional closure of state banks and offices, restaurants, markets, and universities resulted from the rise in SARS fatalities.

April 29, 2003 In response to SARS, the Association of Southeast Asian Nations held a forum on the microbial epidemic in Bangkok, Thailand, to assess the danger of more infections and issued health-screening protocols for Asian airports. Canada initiated travel measures to evaluate passengers at airports and on cargo vessels and ocean liners.

May 1, 2003 The spread of a virulent strain of SARS in Toronto, Canada, coincided with containment of the disease in Taiwan and the waning of illness through Southeast Asia. Infections in Ontario forced a postponement of all elective surgery as 7,000 people entered quarantine at home or in hospitals. Some medical centers closed; others provided only emergency care. In Beijing, China, 7,000 workers erected the Xiaotangshan Hospital in eight days as a treatment center for urban patients.

May 15, 2003 In addition to quarantining the populace of Nanjing and outlawing spitting in public in Quangzhou, Chinese authorities issued a ban on violating quarantine of SARS patients on pain of life in prison or execution.

May 17, 2003 Seasonal influenza A (H1) and B (H3N2) spread over the continental U.S., reaching a height in the plain states. Mortality peaked by March 1. Regional infections varied from east to west:

Region	% Influenza A	% Influenza B
Pacific Coast	76.8	23.2
Western states	77.9	22.1
Northern Plains	48.9	51.1
Lower Plains	42.3	57.7
Great Lakes	64.3	35.7
Central South	21.9	78.1
South Atlantic	41.1	58.9
Lower New England	85.9	14.1
Upper New England	57.8	42.2

May 24, 2003 When 20 people presented SARS symptoms in Toronto, Canada, WHO made the city the first Western location on its global travel advisory. The virus, which incubated in two to seven days, killed eighteen Ontarians; one-fourth of patients were health professionals. Fatalities increased among the elderly and sufferers of diabetes, cardiac disease, and hepatitis. Suppliers raced to increase production of face masks and protective gowns and gloves.

May 28, 2003 In its sixth month, the lung-ravaging SARS epidemic spread to the Russo-Chinese border and to 25 other countries, infecting 4,800 and killing over 293. Doctors fought the disease with the antibiotics cefotaxime, clarithromycin, levofloxacin, and steroids and the antiviral Ribavirin. As panic spread, the pathogen played havoc with international business, schools, churches, sport events, health care, and tourism.

May 29, 2003 Canadian health officials quarantined 5,000 people as the result of 34 more cases of SARS at Toronto hospitals.

June 6, 2003 While reported cases of SARS decreased in Inner Mongolia and Shanxi and Tianjin, China, Ontario infections grew to 82 cases.

June 13, 2003 Monkey pox, an orthopox virus that flourished in the Congo since 1997, infected 40 victims in the U.S. Midwestern states of Wisconsin, Indiana, and Illinois. Health workers traced the outbreak to exotic pets—Gambian rats and prairie dogs—that Illinois and Wisconsin pet distributors sold to unwary citizens. Experts feared that the disease, an incurable virus that mimics smallpox, could spread to other species, such as rabbits, gerbils, and hamsters.

July 10, 2003 Four months after the World Health Organization warned the world about SARS, Taiwan was the only nation that suffered a widespread infection among the general population because of tardy quarantine measures. Of the 8,439 cases reported worldwide, Singapore, Malaysia, suffered 206, Canada 251, Taiwan 674, Hong Kong 1,755, and China 5,327. One explanation of the emergence of the virus from China is the area's agrarian culture, which tends to exist in close contact with domestic fowl and swine. Daily interaction with animals gives viruses an opportunity to recombine genetically into new strains.

July 31, 2003 WHO tabulated the cumulative cases of SARS country by country between November 1, 2002, and July 31, 2003:

COUNTRY	total	deaths	percentage of fatalities
Australia	6	0	0
Canada	251	43	17
China	5,327	349	7
France	7	1	14
Germany	9	0	0
Hong Kong	1,755	299	17
India	3	0	0
Indonesia	2	0	0
Ireland	1	0	0
Italy	4	0	0
Korea	3	0	0
Kuwait	1	0	0
Macao	1	0	0
Malaysia	5	2	40
Mongolia	9	0	0
New Zealand	1	0	0
Philippines	14	2	14
Romania	1	0	0
Russia	1	0	0
Singapore	238	33	14
South Africa	1	1	100
Spain	1	0	0
Sweden	5	0	0
Switzerland	1	0	0
Taiwan	346	37	11
Thailand	9	2	22
U.K.	4	0	0
U.S.	27	0	0
Vietnam	63	5	8
TOTAL	8,096	774	9.6

August 15, 2003 A resurgence of measles at Majuro in the Marshall Islands halted air and sea travel. When the 35 cases reported in July increased to 87, hospital staffs conducted house-to-house immunization. With vaccine from the U.S., public health teams chartered inter-island flights to remote areas to immunize 15,000—a quarter of the population—between the ages of infancy and 30 years.

December 19, 2003 The spread of seasonal influenza A over the U.S. began earlier than usual and reached 36 states with symptoms of fever, lethargy, respiratory problems, and pale, bluish skin. According to CDC tabulations, infection from H3N2 and secondary pneumonia killed 42 children, 40 percent of whom already had conditions that compromised their immunity.

December 31, 2003 The SARS pandemic infected 8,096 in 14 countries, killing 774 or .096 percent. The U.S. reported eight cases out of 156 suspected SARS infections, all of them linked to air travel.

December 31, 2003 The U.N. explored the explosion of HIV/AIDS in sub–Saharan Africa. By comparison to 46,000 cases in Oceania; 760,000 in the Caribbean; and 1,400,000 in

North Africa and Middle East, the staggering 27,900,000 cases in black Africa—two-thirds of the world's victims—foretold mass deaths of men and women and the orphaning of millions of children. Increasing the toll on females, sexual violence damaged vaginal tissues, raising the risk of HIV transference through infected saliva and seminal fluids.

January 1, 2004 UNAIDS numbered the world's HIV patients at 37.8 million with a skewed infection rate by global region:

Sub-Saharan Africa	25,000,000
Southern Asia	6,500,000
Latin America	1,600,000
East & Central Asia	1,300,000
North America	1,000,000
East Asia	900,000
Western Europe	580,000
North Africa & Middle East	480,000
Caribbean	430,000
Oceania	32,000

Additional data recorded 4.8 million patients infected during the 2003 calendar year, with child victims reaching 630,000. In the same period, 2.9 million died of the virus, 490,000 of them under age fifteen. Experts warned that women were more susceptible to infection with HIV than men.

February 2004 An outbreak of parotitis or the mumps virus in England and Wales targeted teens and young adults, a group bypassed by MMR vaccination programs. By year's end, 16,367 cases placed survivors at risk of complications from encephalitis, spontaneous abortion, deafness, inflamed testicles or ovaries, or pancreatitis.

April 2004 German doctors saw four times the usual number of cases of conjunctivitis over the first four months of 2004. An outbreak of 6,378 cases numbered mostly kindergartners and young men from the German military infected with the adenovirus. Patients complained of redness, swelling, pain, bloody conjunctivae, and discharge.

April 26, 2004 The mumps virus gained hold in southwestern Sweden, with 42 cases and six hospitalizations reported over an eight-week period. *See also* **May 10, 2005.**

May 18, 2004 The SARS epidemic ended in China.

June 2004 Seasonal rains at Bangui, Central African Republic, overran wells and other clean water sources with polluted drainage. A 16-month outbreak of hepatitis E, an acute liver disease, affected 411 residents with jaundice, vomiting, distended abdomen, anorexia, diarrhea, and debility. Contagion peaked in September and dropped rapidly in mid-spring 2005.

July 2004 An upsurge of 131 influenza C virus in Miyagi and Yamagata, Japan, produced a fourteen-year height in infections. Simultaneous with seasonal influenza A, the influenza C strain of *Orthomyxovirus* echoed the usual symptoms—fever, runny nose, head pain, body ache, and dry cough. Complications included bronchitis and pneumonia.

July 14, 2004 H5N1, called Fujian flu, emerged in Aksu, China, near the border of Kyrgyzstan. A highly virulent strain, it incubated in poultry and spread over Japan, South Korea, China, Thailand, Malaysia, and Indonesia.

October 2004 The world's deadliest onset of Marburg hemorrhagic fever engulfed northwestern Angolans at Uige, killing 329 or 88 percent of patients. Similar to Ebola, the virus caused severe vomiting, diarrhea, and gut pain as well as cough, sore throat, and chest discomfort. The age span skipped from preschoolers to a scattering of adult medical staff. War and socioeconomic obstacles kept the sickness undiagnosed until March 2005. WHO declared the contagion ended on July 27, 2005.

2005 Among the Fore tribe, the last victim of kuru, a human variant of Creutzfeldt-Jakob disease, died in Papua New Guinea. Contributing to the extermination of a tissue-eating horror, a ban on mortuary feasting ended consumption of deceased humans. Epidemiologists cautiously considered the disease eradicated, but remained on alert for new cases until 2010.

2005 A simultaneous scourge of scabies in rural Nigeria and Mexico and urban Brazil and Turkey affected school children and slum residents in hot tropical settings. A total of 3,386 patients required clinical scrapings and treatment. Additional tabulations in Thailand, Sierra Leone, and Malaysia tied the burrowing *Sarcoptes scabiei* parasite to close associations in refugee camps, nurseries, and orphanages. Symptoms of itchy papules and crusting on skin and scalp resulted from infection with staph or strep pathogens.

2005 The year opened on an initial spread of H5N1 avian flu over Vietnam, Thailand, and Cambodia before reaching Jakarta, Indonesia, in July. Public health agencies in Sichuan, China, reported 38 fatalities out of 200 infections by August. The end of year report listed more deaths in all five countries, with heightened lethality in early 2006 in Turkey, Egypt, and Nigeria.

2005 The start of a five-year spike in invasive pneumococcal disease in western Canada produced 1,048 cases, mostly in middle-aged males. Targeting Native Americans and the homeless, the disease slipped under health surveillance for months. Epidemiologists named *Streptococcus pneumonia* as the villain bacterium endangering patients in British Columbia, Alberta, Saskatchewan, and Manitoba, particularly smokers, substance abusers, and alcoholics. Health workers reduced risks among drifters by vaccinating the jobless and homeless and by monitoring complications from hepatitis and heart disease.

January 2005 Survivors of the tsunami that inundated Aceh, Indonesia, on December 26, 2004, reported post-disaster illness within two weeks. Some 106 cases of tetanus felled adults with a death rate of 19 percent, heightened by lack of globulin serum and antitoxins. The anaerobic *Clostridium tetani* pathogen gained ground after entering superficial wounds. Advanced symptoms of painful muscular spasms, difficulty swallowing, palsy, drooling, sweating, incontinence, and nerve damage required tracheotomies for the sickest patients to prevent asphyxiation and aspiration pneumonia.

May 2005 A five-month scourge in chikungunya fever in the Comoros Islands affected Madagascar, the Maldives, Mauritius, Mayotte, the Seychelles, and La Réunion Island, where it infected one-third of the populace. Out of 244,000 arboviral infections by *Aedes aegypti* bites, 123 victims suffered severe illness and 203 died following the onset of fever, rash, and joint pain and complications from vascular, cardiac, and neurological impairment. Births by infected women produced 41 instances of mother-to-child transmission.

May 10, 2005 Continued mumps infections in Great Britain reached 28,470 since the first of 2005. Most affected were young adults ages 19 to 23, many of them university students. Campus health services stepped up immunization and ordered more vaccine. In London, family doctors reported the noncompliance of some 20 percent of parents of infants and toddlers in required MMR vaccination. *See also* March 11, 2006.

July 18–19, 2005 Virologists surmised that Typhoon Haitang hit Taiwan with heavy rainfall that spread endemic melioidosis, an infection caused by *Burkholderia pseudomallei*. The 54 cases and eleven deaths boosted counts above the usual one to three victims per year. Patients reported fever, cough, septic shock, difficulty breathing, chest pain, and pneumonia. The endemic pathogen typically strikes survivors of crisis weather fronts, floods, typhoons, and tsunamis.

September 2005 A crisis outbreak of dengue fever in Singapore, Malaysia, in fall 2005 yielded 14,006 cases and 27 deaths. By year's end, virologists identified viral mutation and global warming as causes of unusually toxic contagion. Public health officials fogged vacant land and cleared drains to suppress the *Aedes aegypti* mosquito, the carrier of infection to humans. Families stocked their homes with citronella and Cameron Highlands plants to repel insects. Infections spiraled around the globe:

India	90,000
Indonesia	80,037
Trinidad/Tobago	35,000
Malaysia	32,950
Thailand	31,000
Philippines	21,537
Cambodia	20,000
Vietnam	20,000
Costa Rica	19,000
Martinique	6,000
Pakistan	4,800
Sri Lanka	3,000

Global upsurges contributed to a total patient count of 50 million victims and 3,256 deaths.

December 2005 A joint report from WHO-UNAIDS on increased facets of HIV/AIDS as a threat to life worldwide disclosed a total of 40,300,000 victims and 3,100,000 deaths for the calendar year:

Sub-Saharan Africa	25,000,000
Southern Asia	7,400,000
Latin America	1,800,000
East & Central Asia	1,600,000

North America	1,000,000
East Asia	870,000
Western Europe	720,000
North Africa & Middle East	510,000
Caribbean	300,000
Oceania	74,000

The virus gained momentum, overwhelming 4,900,000 for the year. The news for healthy women was unsettling: sub–Saharan Africa maintained a 57 percent infection rate for females with Oceania approaching the total with 55 percent.

2006 Troubling increases in measles epidemics rose in the Ukraine from 25,000 in 2002 to 44,534 four years later. The soar in contagion exposed lax vaccination policies requiring MMR inoculation among school-age children.

2006 In an unusual outbreak, 200,000 Brazilians incurred acute hemorrhagic conjunctivitis, a disease previously limited to Asia, Egypt, and the U.S. A typically mild infection, in rare cases, the eye disease caused the sickest victims to suffer aseptic meningitis and paralysis.

2006 A four-year epidemic of yellow fever in the Central African Republic began from swarms of *Aedes aegypti* carrying the flavivirus. Generating a death rate of 30,000 out of 200,000 cases or 6.7 percent, the disease thrived in the rural regions of Basse-Kotto, Haute-Kotto, Ombella-Poko, and Ouham-Pendé and among urbanites in Bangui, where international trade in used tires provided breeding grounds for the vectors.

February 2006 Epidemic chikungunya struck one million residents of Indian Ocean islands, particularly northern Mauritius, Comoros, Mayotte, and La Réunion with fever, rash, headache, digestive upset, skin discoloration, and debilitating joint pain. Derived from contagion in Lamu and Mombasa, Kenya, the togovirus afflicted sufferers in India and the Maldives as well as Gabon and Ravenna, Italy. Epidemiologists found the *Aedes albopictus* mosquito the primary vector during the rainy season, when infections peaked at 265 cases per day with a 32 percent hospitalization rate.

March 11, 2006 A growth of cases of mumps among university students in Iowa to 233 yielded the largest outbreak in the U.S. since 1988. Epidemiologists failed to isolate a source, but linked the toxin to a variety of virus thriving in the U.K. and began vaccinating with MMR.

March 15, 2006 Across Africa's meningitis belt, infections rose to 5,719 with 580 deaths. To the west, outbreaks threatened Burkina Faso, Ivory Coast, Mali, and Niger. To the east, infection rose in Kenya, Sudan, and Uganda, with serious impact on the West Darfur refugee camps. Health officials administered trivalent vaccine with the aid of Médecins sans Frontières and WHO.

August 2006 A nine-month outbreak of gastroenteritis came to an end in Australia and New Zealand. The result of rampant infection with the norovirus, the disease followed a spring outbreak at retirement homes in Hong Kong that peaked in June and July 2006 with 2,263 and 3,444 cases. The infections of the intestinal tract spread across Europe.

October 6, 2006 A month-long rise in foodborne *Escherichia coli* beset residents of 26 U.S. states. Linked to raw spinach, the infection struck in Wisconsin, Illinois, Nebraska, Ohio, and New Mexico. The gastrointestinal complaint afflicted 199 with dehydration and bloody diarrhea, sending 141 to hospitals. Three victims died and 31 suffered hemolytic uremic syndrome, a form of kidney failure. Canadian food inspectors banned contaminated bag salad and loose leaf spinach from the U.S., where Angus cattle manure tainted groundwater in the spinach fields of California.

November 2006 A startling outbreak of measles among school-age children emerging over all of Switzerland produced 4,415 cases within a 2.5 year epidemic. In Europe's highest rate of infection, epidemiologists identified 656 cases of complications and hospitalization. At least 91 cases outside the country derived from the serious contagion. Incidence remained high into 2007 and 2008, reaching 20 times the average number of cases. *See also* 2007.

December 31, 2006 An annual UNAIDS update disclosed 39.5 million victims of HIV worldwide, 2.3 million of them children. Data indicated a steady growth of infection and death:

Sub-Saharan Africa	24,700,000
Southern Asia	7,800,000
Latin America	1,700,000
East & Central Asia	1,700,000
North America	1,400,000
East Asia	750,000
Western Europe	740,000

North Africa & Middle East	460,000
Caribbean	250,000
Oceania	81,000

The year's totals denoted higher incidence among teens and young adults and lower counts of people living with HIV as a result of 2.9 million deaths. The preponderance of suffering fell on black Africa, where 63 percent of the world's HIV victims clustered in the south of the continent. The breakdown by gender disclosed that 59 percent of patients were female. No other region came close to the disproportionate threat to women.

2007 During the previous twelve months, Europeans suffered seven deaths out 12,132 cases of measles. Outbreaks occurred primarily in Germany, Italy, Romania, Switzerland, and Great Britain among unvaccinated populations.

2007 The beginning of a three-year epidemic of Q fever in southern Holland advanced to 4,000 cases. Rural residents associated with dairy goat farming risk infection by the zoonosis, caused by *Coxiella burnetii* bacteria. Some 800 patients required hospital care; deaths of 15.8 percent of victims usually resulted from endocarditis, gastrointestinal bleeding, or heart failure. Veterinarians interceded by vaccinating goats and sheep to prevent spontaneous abortions, the source of transmission to herders.

March 31, 2007 A spike in the Puumala hantavirus in northern Sweden caused 972 cases of nephropathia epidemica, which can cause renal collapse. Health researchers linked the contagion to a mild winter from climate shift and a rising population of bank voles, the carriers of the hemorrhagic fever. Deaths from kidney failure followed milder symptoms of fever, head and body pain, nausea, vomiting, and blurred vision.

April 2007 Over a period of five and a half years, residents and visitors to Point Lonsdale and Queenscliff, Australia, incurred 79 cases of Buruli ulcer, the largest outbreak of soft tissue infestation in the nation's history. Tests of *Aedes camptorrhynchus* mosquitoes and the saliva of water bugs found the pathogen causing lesions, which flourished in patients over age 55.

April 2007 The upsurge of the Zika virus in the Yap islands of Micronesia represented the first incidence of the emerging pathogen outside Africa since its discovery in 1947. A total of 108 islanders manifested headache, rash above the waist, fever, back pain, and malaise. Zika remained dormant until a major onslaught in Brazil in 2013–2014.

September 12, 2007 News from Baghdad reported a cholera surge at Kirkuk and Sulaimaniya in northern Iraq that infected 7,000. As the contagion spread to Erbil and Nineveh, WHO investigators blamed an inadequate decontamination of potable water with chlorine allowed *Vibrio cholera* to sicken residents. Another source of pollution derived from sewage systems that permeated water wells, causing vomiting, muscle cramps, and diarrhea in victims. The most vulnerable risked shock and death.

October 2007 One of the largest hepatitis E pestilences in global history struck a virgin population in Kitgum, Uganda, causing 160 fatalities among 10,196 patients. Because of the uncertainty of vaccine risks, the greater number of deaths occurred among pregnant women. Incidence of acute jaundice peaked in April 2008 and ebbed in January 2009.

October 30, 2007 Spread of Chagas disease in Caracas, Venezuela generated high absenteeism from public schools for nine weeks. Biological sampling of cafeteria meals suggested that the overnight cooling of guava juice might have allowed contamination with insect feces. The 103 patients suffered gut and muscle pain, fever, headache, facial swelling, malaise, and irregular heartbeat.

December 2007 Year-end data that recorded 2.1 million AIDS deaths also reported 33.2 million patients living with HIV. The infection rate continued to climb, especially in the South Pacific:

Sub-Saharan Africa	22,500,000
Southern Asia	4,000,000
Latin America	1,600,000
East & Central Asia	1,700,000
North America	1,300,000
East Asia	800,000
Western Europe	740,000
North Africa & Middle East	380,000
Caribbean	230,000
Oceania	75,000

Out of a horrendous count of 2.5 million people newly afflicted with HIV, 2.1 million died of resultant AIDS. Infections of women rose in the Caribbean by 6 percent to 43 percent. For maximum surveillance, health authorities in Angola, Benin, Cambodia, Central African Republic, Haiti, Liberia, Malawi, Mali, Sudan, and Swaziland added new accounting measures.

2008 During an upsurge in hepatitis C in Alexandria, Egypt, the ministry of health made hospital staffs the first line of defense, control, and public information and education.

2008 The first ever cases of rabies at Bukit, Bali, engulfed the island, killing 130 dog-bite victims. Health officials surmised that a fisherman imported the strain from Sulawesi, Indonesia. After the extermination of 108,000 dogs, a rapid inoculation program rid more than 200,000 dogs and puppies of the virus.

2008 Limited to the salt marshes of northern Australia, the Barmah Forest virus produced a seven-year high of 8,050 infections. Spread from 1995 by the *Aedes vigilax* and the *Culex annulirostris* mosquitoes, the arbovirus caused muscle and joint pain, rash on palms, malaise, and low fever extending over six months. A subsequent uprise three years later produced 1,855 diagnoses in Australia. Port control attempted to halt the spread of the fever to the *Aedes polynesiensis* of Papua New Guinea.

2008 Tests of tube wells in Bangladesh by UNICEF found 61 percent of groundwater contaminated with arsenic and only 12,000,000 or 39 percent safe for drinking. Pervasive arsenicosis from tainted drinking water caused skin lesions, cognitive disabilities, stillbirths, and cancer of the bladder, liver, lungs, and skin. The problem threatened 30 million Bangladesh residents and six million Indians in West Bengal with fatal cancer.

2008 Shifts in world HIV/AIDS tabulations indicated dramatic change over a five-year period. While North America remained essentially unchanged in number of infections, Oceania stopped a rampant surge that began in 2006. In this same 60-month span, Eastern Europe and Central Asia produced the largest spiral in infections, keeping close to the growth of the epidemic in Latin America, the Caribbean, the Middle East, North Africa, and sub–Saharan Africa. Increased public health oversight in Australia, China, parts of South America and Africa, and India enhanced monitoring, with 31 countries conducting surveys of HIV prevalence. Urgent need of diagnosis and drug intervention rose, particularly among women in Haiti, Swaziland, South Africa, Zimbabwe, Central African Republic, and Uganda.

January 24, 2008 The report of 2,000 cases of mumps in Moldova indicated need for more childhood vaccination with MMR.

March 2008–2012 Rampant hand, foot, and mouth disease in Fuyang, China, infected 25,000 victims, killing 42 by the combined onslaught of the coxsackie virus and enterovirus 71. Simultaneous outbreaks in Singapore, Malaysia, sickened 2,600, with additional reports of 2,300 cases in Vietnam, 1,600 in Mongolia, and 1,053 in Brunei. The disease thrived into 2009, killing 50 of 115,000 patients. Still virulent in 2010, the pathogen sickened 70, 756 and killed 537. The Asian epidemic continued its upsurge in China, producing 1,340,259 infections in fall 2011 and 1,520,274 by mid-summer 2012.

March 27, 2008 At Southampton, England, 3,000 passengers returning from a three-month world cruise to the Western Hemisphere, Samoa, Tonga, New Zealand, Australia, Hong Kong, Thailand, Singapore, Malaysia, India, Egypt, Greece, Madeira, and Spain incurred exposure to hepatitis E. Analysis of seafood consumption—prawns, squid, cod, mussels, hake, salmon, lobster, scallops, and shrimp—and swimming in pools and seawater confirmed exposure in Honolulu, Hawaii, and Pago Pago, Samoa. Passengers seeking medical care exhibited malaise, dark urine, loss of appetite, nausea, vomiting, and jaundice. Acute infections occurred in 33 males reporting a high consumption of alcohol.

April 2008 The sudden upsurge in mumps infections in Ireland infiltrated young adults from secondary schools and colleges. Irish health officials blamed delays in child immunization on government budget cuts and media misinformation about vaccination causing autism.

May 2008 The noncompliance of Israel's Orthodox Jews with MMR vaccination caused a sharp jump in cases of mumps from twelve to 1,000.

July 2008 The prevalence of enteritis in Chungnam, South Korea, derived from the echovirus and coxsackievirus in infants and babies during the hottest summer months. The outbreak subsided by November. Into 2009, the pathogens resulted in diagnoses of aseptic meningitis, septicemia, tonsillitis and pharyngitis, bronchial pneumonia, herpangina, and hand, foot, and mouth disease.

December 2008 The end-of-year report on HIV/AIDS from UNAIDS listed sobering data—the orphaning of 12 million children in black Africa and the stability of infection rates in East Asia and North America:

Sub-Saharan Africa	22,000,000
Southern Asia	4,200,000
Latin America	1,700,000
East Europe & Central Asia	1,500,000
North America	1,200,000
East Asia	740,000
Western Europe	730,000
North Africa & Middle East	380,000
Caribbean	230,000
Oceania	74,000

Progress reports affirmed the efforts that 147 countries made on stemming infections and preventing HIV

Endemic cholera spiked around East Pokot, Kenya, with a high rate of fatality. *Vibrio cholera* bacteria flourished in flooding caused by the El Niño weather pattern. In 11,425 cases, 264 died, often from inadequate supplies of rehydration fluids. *See also* **November 23, 2015.**

March 16, 2009 A scourge of rabies in Luanda, Angola, killed 93 children. The problem of stray dogs in urban slums and the lack of veterinary experts worsened the situation. Because of a global dearth of vaccine for the five sequential treatments, victims died from paralysis, coma, hallucinations, and encephalitis.

April 2009 Bulgarians incurred 23,791 measles infections and 24 fatalities from the pathogen. The contagion set off pestilence in Germany, Greece, Macedonia, and Turkey.

April 6, 2009 A boat explosion at Ashmore Reef off Perth in northwestern Australia imperiled 47 Afghan asylum seekers and crew members from Indonesia and sent 29 to hospitals. Because of serious burns and wounds, the 44 survivors incurred exposure to hepatitis B from an Afghan carrier. Medical staff at Royal Perth Hospital questioned 23 patients through interpreters about any exposure via intimacy, injected drugs, tattooing, razors, and biohazards.

April 19, 2009 A nine-month rise in cholera infections in Zimbabwe sickened 98,592 and killed 4,288. The national emergency began in Chitungwiza the previous August 20 and dispersed contagion to Botswana, Malawi, Mozambique, South Africa, and Zambia, in part following the movement of migrant workers. Growing refugee crises and insufficient sanitation and health care worsened the risk of contamination with *Vibrio cholerae*, which spread from sewage during the rainy season into water supplies. The poorest residents lacked enough fuel to boil drinking water. Worsening the situation were poverty, hunger, and HIV/AIDS.

August 2009 Amid heavy metal contamination in Wenping, Hunan, China, 1,300 children suffered lead poisoning at an elementary school adjacent to a manganese factory. Since the plant opening in May 2008, children suffered anemia, muscle atrophy, and irreversible cerebral damage. A similar threat at Shaanxi in northern China compromised the health and cognitive faculties of work crews and 615 children from lead poisoning exuding from the Dongling Lead and Zinc Smelting Company at Fengxiang. More than 140 children entered hospitals for treatment.

September 2009 An uptick in infectious blastomycosis from soil-borne spores in Marathon County in central Wisconsin continued for nine months. Of the 55 victims, 39 entered hospitals; three died. Most of the sick had backpacked, camped, and fished in the area. The inhalants disproportionately felled Hmong. *See also* **June 2013.**

November 11, 2009 After a 15-year pause, Nicaragua experienced a new surge in dengue fever with 17,140 cases. According to Médecins sans Frontières, Africa's largest dengue fever epidemic assailed Cape Verde with six deaths in 13,187 cases, a first for the islands. Diagnoses proceeded at the rate of 1,000 per day. In this same period, Bolivians fought the same pestilence.

December 2009 UNAIDS made its annual accounting of HIV/AIDS and reported 33.4 million people living with HIV with 2.7 million of the total newly infected. AIDS deaths for the year reached 2 million, 280,000 of them children. Stasis in Eastern Europe and Central Asia and a downward trend in contagion in Oceania, Southern Asia, North Africa, and the Middle East gave health workers hope of containing the global threat:

Sub-Saharan Africa	22,400,000
Southern Asia	3,800,000
Latin America	2,000,000
East Europe & Central Asia	1,500,000
North America	1,400,000
East Asia	850,000
Western Europe	850,000
North Africa & Middle East	310,000
Caribbean	240,000
Oceania	74,000

December 31, 2009 After the total of trypanosomiasis or sleeping sickness fell 73 percent below 10,000 cases in the 36 nations of sub–Saharan Africa, WHO announced its intent to eradicate the disease by 2020. Of resistant pockets, the Democratic Republic of the Congo reported 70 percent of diagnosed illnesses. Health officials in Benin, Botswana, Burundi, Equatorial Guinea, Ethiopia, Gambia, Ghana, Guinea Bissau, Kenya, Liberia, Mali, Mozambique, Namibia, Niger, Nigeria, Rwanda, Senegal, Sierra Leone, Swaziland, and Togo reported none for more than ten years.

January 1, 2010 In the highest incidence of pertussis or whooping cough in six and a half decades, California public health personnel recorded 9,154 cases with 809 hospitalizations and ten deaths, most among infants. The *Bordetella pertussis* pathogen assaulted Hispanic babies, of whom 91 percent were inoculated. The dissemination of DPT vaccine to infants and postpartum women and reinnoculation of older children and adults eased the epidemic in a vulnerable population. Sickness in families refusing the standard four-dose immunization for their children resulted in rates 2.5 times higher than in vaccinated communities.

June 2010 A three-month siege of lead poisonings in northern Zamfara, Nigeria, felled 355 victims and killed 163 illegal miners, 68 percent of them children. Digging for gold ore in the Sahel contributed to blood contamination with toxic lead, which polluted soil, water, hands, shoes, and tools and compromised hearing, brain, and kidney and bone marrow function. Médecins sans Frontières, the Blacksmith Institute, and WHO staffed two treatment camps, where the sickest patients suffered convulsions and coma.

July 2010 The onset of endemic pappataci fever in Kahramanmaras, Turkey, initiated diagnosing and treatment over a month-long phlebotomine sandfly scourge, which carried both the Cyprus and Sicilian strains. The 40 urban victims felt achy and feverish and displayed gut ailments and vomiting. All recovered without incidence of meningitis or encephalitis.

October 2010 The rise of paratyphoid fever from *Salmonella paratyphi* assailed 601 rural patients in Yuanjiang, China, and 368 in Songkla, Thailand. The cause in China, eating contaminated cold, raw green onions, peppermint, and coriander, derived from an outflow of hospital wastewater to agrarian irrigation systems. In Songkla, medical detectives found pathogens in coconut milk and raw chicken and pork sold in local markets. Patients responded to doses of amoxicillin, ampicillin, cefotaxime, ceftriaxone, co-trimoxazole, and ciprofloxacin.

October 14, 2010 A cholera upsurge in Haiti assailed 800,000 and killed 10,000 or 1.25 percent. Detection of sources of *Vibrio cholerae* identified contamination of the Artibonite River downstream from U.N. peacekeepers from Nepal, where residents of Kathmandu were battling their own epidemic. Islanders used the sewage-polluted water for cooking and washing.

December 2010 After flourishing for twelve months in Korea, pandemic H1N1 or Influenza A struck India and advanced around the globe. By the seventh month of infection, 1,763 of 20,604 victims had died. After the vaccination initiative waned, the virulent scourge recurred in winter 2012–2013 and again in December 2014, when 2,000 or 6.7 percent of 30,000 victims died. Virologists studied the two-year intervals as well as the rate of mutation and surmised that a drift virus or escaped mutant gained dominance over the original H1N1 virus.

December 2010 UNAIDS celebrated the drop in new infections with HIV by 19 percent. Regional reports on people living with HIV exhibited reason for hope that the virus was losing its clout worldwide:

Sub-Saharan Africa	22,500,000
Southern Asia	4,100,000
Latin America	1,400,000
East Europe & Central Asia	1,400,000
North America	1,500,000
East Asia	770,000
Western Europe	820,000
North Africa & Middle East	460,000

Caribbean	230,000
Oceania	57,000

The prevalence of HIV in sub–Saharan Africa held steady at 68 percent, afflicting more women than men.

January 13, 2011 The highly contagious H5N1 virus produced two deaths out of 130 cases in Egypt. Caused by pestilence in poultry, the disease also infected 13 patients in Phnom Penh, Cambodia, and 174 in West Java. By June, numbers increased to 178 in Indonesia, 150 in Egypt, and sixteen in Cambodia.

May 2, 2011 The world's largest upsurge in hemolytic uremic syndrome from *Escherichia coli* bacteria swept Hamburg and Lübeck in northern Germany for a month, causing bloody diarrhea, anemia, low platelet count, and kidney failure in 857 patients. Health inspectors blamed fresh vegetables, particularly bean sprouts and cucumbers from Spain contaminated with *E. coli,* for sickening 3,167 people and killing 51. The outbreak affected travelers from Canada, Denmark, Holland, Poland, Switzerland, Great Britain, and the U.S. A subsequent outbreak in Bordeaux, France, derived from fenugreek seed from Egypt. By June, human loss cost Europeans $2.84 billion and an agricultural loss of $417 million from banned vegetables left unharvested in the fields.

July 2011 A rising rate of pertussis or whooping cough among school-age children in Western Australia reached 766 cases by November and maintained unprecedented contagion into December and January 2012. Half the victims required hospitalization and treatment with clarithromycin, azithromycin, erythromycin, or trimethoprim/sulfamethoxazole.

July 31, 2011 A seven-month upsurge in highly contagious scarlet fever swamped Hong Kong doctors with 996 cases. Additional endemic scarlet fever struck Macao and Mainland China. Complicating treatment, the *Streptococcus pyogenes* strain developed resistance to erythromycin.

November 3, 2011 An outbreak of shigellosis, a food-borne gastrointestinal complaint among school children in Georgia and South Carolina, and in an orthodox Jewish neighborhood in Williamsburg, New York, infected 385 people and hospitalized four children. Patients complained of fever, nausea, and bloody diarrhea. The upsurge caused schools to close at Honea Path, South Carolina, and Worth County, Georgia.

November 10, 2011 At the University of California at San Diego, Dr. Jane C. Burns linked the mysterious patterns of Kawasaki disease to global wind currents carrying dust from the Gobi Desert. Patients exhibited swollen, dry lips, red tongue, peeling hands and feet, rash, conjunctivitis, aneurysms, and enlargement of coronary arteries, possibly from *Candida* fungi or a respiratory virus. The predominance of outbreaks in Japan coincided with northwesterly tropospheric winds from Central Asia, which spread the infection to Hawaii and San Diego. South Korea also recorded 134 cases in preschoolers.

December 2011 Analysts at UNAIDS reported a 21 percent decline in new HIV infections and a decline in AIDS-related deaths. A cause for celebration in Ethiopia, Nigeria, South Africa, Zambia, and Zimbabwe, the decrease in HIV/AIDS in black Africa resulted from free antiretroviral therapy. Cautious optimism followed declines in Botswana, Lesotho, Mozambique, and Namibia. In the Caribbean, a drop of 25 percent in new infections marked health reports from the Dominican Republic and Jamaica. Rates remained stable in Latin America, the Middle East, North Africa, North America, and Western and Central Europe.

More good news from southern Asia announced a 40 percent drop in infection since the peak of contagion in 1996. Even better reports from India cited a 56 percent decline in the spread of HIV. Countering too rosy a view of disease, IV drug users in Eastern Europe and Central Asia elevated infections rates by 250 percent.

December 12, 2011 An undiagnosed chronic kidney ailment among men in El Salvador and Nicaragua threatened thousands of day laborers. Males in hot, sultry sugarcane fields who handled toxic herbicides and pesticides and indulged in alcohol aged at a rapid rate and died of kidney failure. Sugar companies denied responsibility for the unexplained debility and deaths.

December 31, 2011 A 26-week school outbreak of measles in Quebec affected 776 patients, 47 of whom imported the contagion from the Caribbean, France, and Pakistan. Incidence targeted teenagers. Hospitals admitted 57 pediatric cases, most incurred in the child's neighborhood or daycare center.

The WHO year-end report also calculated cholera infections in Cameroon at 33,192, with 1,440 deaths. Around 10 percent of the cases struck the southwest in the monsoon zone bordering the Gulf of Guinea. The epidemic of *Vibrio cholera* reached its height from April 16 to May 20 and produced a second surge over June 26–27.

2011–2012 After a two-year total of 32,091 cases of giardiasis reported in the U.S., the CDC noted a rapid decline of protozoan parasites, the most common cause of intestinal infection. The lull appeared to result from informed surveillance and treatment of groundwater and recreational water to suppress fecal contaminants

2012 Nigeria claimed more than half of global polio cases. The infection of 122 patients out of 223 worldwide preceded a lull in contagion, with only one child diagnosed as of September 6, 2016.

May 2012 Of the 22,000 measles cases the French tabulated over the previous 4.5 years, 5,000 patients required hospitalization. Worsening infections, secondary pneumonia, encephalitis, and myelitis caused ten deaths.

June 13, 2012 In Saudi Arabia, patient zero, suffering acute pneumonia from Middle Eastern Respiratory Syndrome (MERS), survived only eleven days. Egyptian microbiologist Ali Mohamed Zaki of Ain Shams University in Cairo, identified the coronavirus, which derived from close contact with camels, goats, cows, sheep, water buffalo, pigs, and wild birds. The virus rapidly spread to Europe, Great Britain, North Africa, Malaysia, China, Korea, and the Philippines. *See also* December 31, 2014.

June 25, 2012 Contaminated river water in Maharashtra, India, carried hepatitis E, a viral liver infection that killed 18 of 4,000 patients. Restaurants and schools closed until health officials could quell industrial pollutants and seal leaking sewer pipes. Victims reported fever, nausea, dark urine, joint pain, and jaundice. Travelers received instructions on avoiding unboiled water and ice and eating raw or partially cooked food.

July 2012 At Phnom Penh, 78 Cambodian toddlers diagnosed with a hand, foot, and mouth form of enterovirus 71 received hospital treatment. Death rate reached 69.2 percent.

August 2012 At Orientale, Democratic Republic of Congo, WHO investigated ten Ebola cases and six deaths. The infection resulted from the Bundibugyo strain, a variant from the Ebola-Sudan outbreak at Luweero in central Uganda, which resulted in 17 deaths or 70.8 percent of 24 cases.

August 2012 Of the 1,118 cases of West Nile virus in the U.S., 80 percent of the emerging scourge struck Mississippi, Oklahoma, Louisiana, South Dakota, and Texas, the disease epicenter. By August 23, Arkansas and Michigan health authorities reported deaths from the fever bringing the total to 44. The increase in vector-borne contagion resulted from hot weather and environments favorable to mosquitoes. Patients over age 50 risked the most severe illness and possible hospitalization.

August 15, 2012 In Kenya, the hepatitis E virus inundated 460,000 Somali refugees at the Dadaab camp, the world's largest refugee enclave, and advanced on the villages of Biyamadow and Darkanley. Until the building of new latrines and hygiene enforcement in November, the infection raged for five months, during which aid workers experienced terrorism and abductions. The outbreak slowed resettlement to the U.S. and other countries and increased vaccinations and health screenings.

fall 2012 An unusually high mortality rate accompanied epidemic yellow fever in Darfur, Sudan, which felled the sick with malaise and fever and chills. The deaths of 32 out of 84 yielded a 38.1 percent mortality rate. WHO distributed protective garb and bed nets to shield the Sudanese from mosquitoes carrying the contagion.

September 2012 Among 1.5 million refugees fleeing war in Syria for camps in Lebanon, 948 exhibited symptoms of cutaneous leishmaniasis. Flourishing in youth age 18 or younger, the protozoa borne by sandflies caused ulcerous bites, nodes, and skin lesions. In spring, WHO distributed megulamine antimonite for intramuscular injection into 1,275 patients.

September 5, 2012 Virulent cholera in Sierra Leone produced a national emergency. WHO and Médecins sans Frontières aided villagers with heightened sanitation and water purification tablets for the 45 percent who relied on unreliable water sources.

late September 2012 Terra Haute in Vigo County, Indiana, documented an outbreak of chicken pox. The 92 victims complained of head and stomach pain, fever, and itchy blisters. Some of the patients who had received the standard two inoculations suffered a milder case than those who had only one shot or none. Contact with the pathogen threatened adults with pneumonia, blood and bone sepsis, toxic shock, and encephalitis.

October 19, 2012 Marburg hemorrhagic fever infected 20 Ugandans with the filovirus, which killed nine patients or 45 percent of hospitalized victims.

December 2012 Dramatic results in the battle against HIV/AIDS resulted from a 60 percent increase in HIV treatment, which reached eight million victims. The UNAIDS report for the year cited stubborn rates in black Africa, which accounted for 69 percent of the world's total of 34 million people living with HIV.

December 31, 2012 Over the previous twelve months, the U.S. reported twice the usual cases of pertussis or whooping cough resulting from the highly contagious *Bordatella pertussis* pathogen. A 1,000 percent increase in Wisconsin and Washington State required government aid to inoculate children and administer boosters for adults.

2013 South Sudan led the world in endemic dracunculiasis, followed by Chad, Ethiopia, and Mali. The four countries incurred 148 infections by the guinea worm, a reduction of more than 99 percent worldwide. Over the next three years, the incidences again fell to 22 infections in Chad, Ethiopia, Mali, and South Sudan. In the same report, Syria experienced the most cases of leishmaniasis, with an upsurge of 22 percent to 71,996 infections.

2013 The 91 cases of blastomycosis in northeastern Wisconsin resulted in 54 hospitalizations of victims for cough, chest and muscle pain, fever, fatigue, and weight loss. The state veterinarian linked the pneumonia-like endemic to inhaling spores of *Blastomyces dermatitidis*, a common fungus in soil.

2013 Near a lead mine and smelter in Kabwe in Zambia's copperbelt, more than 1,000 preschoolers tested positive for lead poisoning from the leaching of heavy metals into water and soil. Patients with high heavy metal count in the blood underwent lead chelation therapy along with behavioral and learning intervention to reduce cognitive stunting.

January 27, 2013 At Jamam, an Upper Nile refugee camp in South Sudan sheltered 5,080 people exhibiting the yellow eyes caused by acute jaundice syndrome. Analysis of patient blood confirmed that 54.3 percent of the inmates were susceptible to hepatitis E. More than 10 percent of victims were pregnant women. In cramped conditions worsened by rains and flooding, workers distributed supplemental food and multivitamins, soap, and instructions on hygiene. The sickest patients suffered electrolytic imbalance, high fever, vomiting, diarrhea, bleeding, agitation, and coma.

February 2013 War in Syria increased the dangers of measles, which infected 7,000. Many fled to refugee camps, spreading the highly contagious *Morbillivirus*. Another uptick in Nigeria produced 4,000 measles infections and 36 fatalities, most of them unvaccinated children. Europe reported 8,499 cases, 432 of them in southwestern Wales, where 51 victims sought care in hospitals. Increasing mortality, ear infection, pneumonia, and encephalitis threatened survival.

February 22, 2013 According to WHO and Reuters, more anguish for troubled Syrians resulted from 2,500 cases of typhoid in Deir al-Zor, causing high fever, enlarged liver or spleen, headache, and diarrhea. Loss of water supplies in a region controlled by rebels forced residents to rely on the Euphrates River, which bore *Salmonella typhi* from sewage.

March 22, 2013 The first ever infection of humans in Hong Kong with microsporidiasis began on the muddy playing field at a rugby tournament and spread to 160 athletes from Australia, Hong Kong, Malaysia, and the United Arab Emirates. The soil-borne parasite attacked corneas, causing eye pain, discharge, itching, and swelling. The spore of the one-celled vector also threatened the intestines, lungs, brain, urinary track, muscles, and nerves.

April 2013 A dearth of pesticides plunged Paraguay into a dengue fever epidemic, which killed 45 patients in the first four months.

April 2013 The initial onset of H7N9 avian flu in China sickened victims in Shanghai, Hangzhou, Nanjing, and Anhui Province, killing 47 of the first 147 cases. As diagnoses increased, WHO virologists linked the human contagion to chickens, ducks, and pigeons and warned of complications from respiratory crisis and severe pneumonia.

May 1, 2013 Over the first third of the year, Japan tabulated 5,442 rubella cases exhibiting fever, rash, and swollen lymph glands. The endemic virus followed a five-year cycle, heightening in spring and summer, particularly in Osaka and Tokyo. By March 2014, more than 15,000 infections and 43 cases of congenital rubella syndrome weakened males aged 20–40 because vaccination campaigns centered on school girls.

June 2013 A religious minority in Veluwe, Holland, suffered 161 measles cases because of parents' refusal to have their families vaccinated with MMR. Five children entered hospitals for treatment of pneumonia and meningitis.

June 15, 2013 A U.S. outbreak of 631 incidents of cyclosporiasis, caused by the *Cyclospora cayetanensis* virus, ranged over 25 states in the Midwest. The parasite, which thrived in raw vegetables, caused diarrhea and weight loss, gastric cramping, and fatigue. The culprit, bagged salad from Mexico, reached Olive Garden and Red Lobster restaurant patrons in Iowa, Illinois, Nebraska, Texas, and Wisconsin.

August 15, 2013 For 1,492 patients at Chappa, Badakhshan, in mountainous northeastern Afghanistan, cholera from one polluted spring killed one resident and threatened survival of an entire village. Aid from WHO and UNICEF brought antibiotics and potable water.

November 2013 A three-month surge of cholera, the product of the *Vibrio cholera* bacterium along the polluted Hidalgo River running through south central Mexico, bore similarities to the disease in Cuba, Dominican Republic, and Haiti. Heavy flooding displaced families fleeing tropical storms, which reduced the quality of drinking water. The sick responded to rehydration, chloramphenicol, and doxycycline, which reduced cramping, vomiting, and diarrhea. Depletion of electrolytes and shock killed one patient and threatened 176 others suffering from compromised immune systems.

December 2013 The UNAIDS evaluation of HIV/AIDS effects on the world noted a one-third decline in new infections with HIV and a downturn in AIDS deaths to 1.6 million. Thwarting success in Africa, Asia, and Latin America was the availability of antiretroviral treatment to only 10 percent of victims in Angola, Brazil, China, Cameroon, Central African Republic, Chad, Colombia, Democratic Republic of the Congo, Ethiopia, Ghana, India, Indonesia, Ivory Coast, Kenya, Lesotho, Malawi, Mozambique, Myanmar, Nigeria, Russia, South Africa, Sudan, Tanzania, Thailand, Togo, Uganda, Ukraine, Vietnam, Zambia, and Zimbabwe.

2014 The H58 strain of *Salmonella enterica typhi* advanced south from Kenya over Tanzania and Mozambique to Malawi, spreading a gastrointestinal form of typhoid fever. Patients complained of malaise, diarrhea, nausea, rash, headache, and enlargement of the liver. The most at risk experienced intestinal perforation and hemorrhage. Of the 782 infections, 97 percent resisted drug treatment. The outbreak continued in Malawi into April 2016 among villagers lacking clean water and sanitation.

2014 The CDC tabulation of 350,062 cases of gonorrhea indicated a rise of 5.1 percent, especially among youth and women. Females who ignored infections risked pelvic inflammatory disease, urethral blockage, and infertility caused by scarred fallopian tubes. Males developed cardiac damage, inflamed spine or brain, and arthritis. The data on sexually transmitted disease continued to rise, especially in California.

2014 After significant drops in measles cases over 34 years, the Western Hemisphere reported a startling rise—from 66 to 19,898 cases. Some of the 610 infections in 20 U.S. states may have been imported from the Philippines. Other communicable diseases on the increase by country and number included these:

buruli ulcer	Australia	110
	Cameroon	133
	Congo	234
	Guinea	72
	Nigeria	113

onchocerciasis	Benin	2,952,152
	Burkina Faso	202,009
	Cameroon	7,203, 643
	Congo	26,049,139
	Ivory Coast	1,865,273
	Tanzania	3,338,320
	Uganda	3,338,320
rabies	Philippines	236

On a positive note, Colombia became the first nation to eradicate onchocerciasis or river blindness, a tropical skin and eye disease caused by parasites in the bites of infected blackflies.

2014 In an era when tuberculosis infections dropped annually by as much as 47 percent over a quarter century, contamination of lungs with *Mycobacterium tuberculosis* remained the world's top infectious killer. Contagion flooded the poorest countries and targeted women in their prime. India, Indonesia, Nigeria, Pakistan, China, and South Africa suffered 58 percent of the world's infections. Of the one million tubercular children, 140,000 succumbed to the bacillus. For people battling HIV, one-third died of TB, a brutal stalker in Africa. WHO set a containment date of 2030 for curbing outbreaks. Despite the resolve of epidemiologists, within the year, the CDC tabulated a rise of TB in the U.S. to 9,563 cases, one third of them resistant to antibiotics.

January 3, 2014 Madagascar, the source of 80 percent of the world's incidents of bubonic plague, suffered 60 cases of pneumonic plague, 24 of bubonic plague, and 42 deaths over a month's time. Rat infestation threatened homeowners who stored rice and the survival of prisoners, who occupy overcrowded spaces.

January 5, 2014 WHO reported a measles epidemic in the Philippines that killed 110 patients out of 58,010. Epidemiologists blamed unvaccinated Amish tourists visiting the island nation. A simultaneous upsurge in Vietnam sickened 1,048 in late December and 993 in January. The outbreak killed seven. Medical statisticians cited the failure of parents to insist on a second dose of the measles serum, which WHO and UNICEF distributed.

January 14, 2014 Portland, Oregon, felt the brunt of H1N1 swine flu, the first pandemic since April 2009. Of the 179 hospitalizations, there were seven deaths. Most victims suffered fatigue, chills, body pain, headache, and runny nose. The CDC recommended vaccination with a three-part serum targeting H1N1, H3N2, and influenza B.

February 4, 2014 Endemic scarlet fever, a streptococcal infection, raised contagion levels on Prince Edward Island, Nova Scotia. Victims reported a patchy rash, peeling of toes and fingers, sore throat, headache, and nausea. The sickest risked kidney inflammation, pneumonia, and swollen joints.

March 14, 2014 The emergence of an unidentified virus in Macenta, Guinea, signaled the beginning of an unprecedented epidemic, which took its name from the Ebola River in Zaire. The disease also savaged the unsuspecting in Liberia and Sierra Leone and reached a total of 964 illnesses and 603 fatalities by July 15. Within a year, 24,957 caught the hemorrhagic fever, which killed 10,350 or 41.5 percent. Health statisticians calculated that the *Zaire ebolavirus* produced an 80 percent death rate. The numbers of nosocomial infections caused the sick to avoid hospitals.

A massive health crusade sapped health care budgets and forced West Africans into self-quarantine. Epidemiologists named two-year-old Emile Ouamouno patient zero, who may have had contact with virus-bearing fruit bats. He infected his mother and sister and eleven residents of Meliandou village. Suspicions fell on bats as the vector. By summer, cases in Mali, Nigeria, Senegal, Spain, the Western Pacific, and the U.S. resulted from travel to infected regions. The horrific death rate forced Liberia to close its borders on July 28 and to oblige restaurants, hotels, and theaters to air tutorial film clips on prevention of infection. In August, experts in the Democratic Republic of Congo identified a separate strain requiring more volunteers to augment the few native physicians. By year's end, Ebola had killed 7,842 of its 20,081 victims and raised concerns for employees of hospitals, clinics, aid stations, and mortuaries.

March 31, 2014 A rainfall ten times the average at Guadalcanal in the Solomon Islands spread the rotavirus, with 4,087 cases in Hoiara alone. One-third of the patients were preschoolers. The death total reached 27, mainly from dehydration caused by severe vomiting and diarrhea.

April 2014 In Tahiti, the world's largest epidemic of the Zika virus threatened 32,000 patients with rash, fever, headache, body pain, and conjunctivitis. The disease increased incidence of the Guillain-Barré syndrome in 42 Tahitian adults, triggering facial palsy and difficult breathing and swallowing. Sixteen victims required intensive care. The virus moved through the Pacific to New Caledonia, Cook Islands, Easter Island, and east to Brazil.

summer 2014 A seasonal swine flu epidemic of the H1N1 virus invaded Russia, killing 232 victims, 57 of them babies. The cycle began in 2009. By the 2015 peak, the disease flourished among urbanites in Moscow and St. Petersburg. The 2016 repeat of winter infection sent more people to drug stores to buy facemasks and more patients to hospitals to avoid pneumonia.

August 2014 The lack of electricity, potable water, sewage plants, garbage collection, medical facilities, and burial of the dead left Palestinians in Gaza in danger of epidemic. Displaced residents suffered scabies and diarrhea.

August 16, 2014 The CDC warned of a 30 percent surge in pertussis or whooping cough cases in the U.S. The total number for the first 7.5 months reached 17,325, with 9,935 or 57.3 percent of instances centered in California. To protect offspring, concerned parents and pregnant women sought immunization.

September 29, 2014 In 34 countries of the Western Hemisphere, doctors crusaded against chikungunya, a viral infection carried by mosquitoes. The tropical fever was unknown in the region until December 2013, when it overran St. Martin in the Caribbean and sickened 50 islanders. The epidemic moved on to the British Virgin Islands, Dominica, Guadeloupe, and St. Bart's before inching south, west, and north to the Americas. The spread reached dangerous proportions in the Dominican Republic, where a half million people incurred fever and joint pain. Data ranged downward in number from El Salvador to Venezuela:

El Salvador	30,000
French Guiana	2,656
Puerto Rico	2,022
Colombia	1,600
St. Martin	1,515
Venezuela	398

From coastal South America, the virus struck its first U.S. victim in Florida, where an additional 57 infections sickened travelers from the outside world.

October 18, 2014 The village of Tungri Tola in Smdega, India, lost seven residents over a two-month period to endemic anthrax caused by consumption of two infected bullocks. Symptoms ranged from chest and stomach pain to vomiting blood. Health officials quarantined 30 households and advised villagers to restrict sick cows and to burn cattle carcasses with kerosene.

November 10, 2014 In a fall period of humid, moist weather, health officials in North Vietnam reported a wildfire epidemic of pink eye or conjunctivitis, which reddened and inflamed the conjunctiva inside the eyes of 6,000 in Hanoi. Symptoms consistedof tearing and discharge, irritation, and crusted eyelashes. Officials advised against sharing towels and washcloths or handkerchiefs.

November 11, 2014 Al-Sumaria Television reported that epidemic mumps was spreading over the Middle East, producing 18,535 cases in Palestine and 4,600 victims in Iraq, beginning in October 2015 and continuing for four months.

December 2014 The annual UNAIDS tabulation cited belief that global action could end the HIV/AIDS epidemic. New infections declined by 13 percent. AIDS-related deaths and fatal tuberculosis infections fell significantly, especially in South Africa, the Dominican Republic, Ukraine, Kenya, Ethiopia, and Cambodia. Arousing concern, the nine million HIV patients also infected with hepatitis A, B, or C required a concerted effort at diagnosis and treatment.

December 15, 2014 A twelve-state spread of *Salmonella enteritis* produced 111 cases of food poisoning from California to Colorado, Iowa, and Minnesota. CDC detection named mung bean sprouts from Brooklyn, New York, as a source of two-thirds of the illnesses. The other one-third derived from tainted eggs, nut butter, chia powder, cucumbers, and raw cashew cheese. A parallel pandemic in northern France derived from frozen hamburger shipped from Poland. Another scourge of food poisoning in Sweden resulted from eating a spice mix containing contaminated dried vegetables.

December 16, 2014 A viral anomaly in northwestern Cambodia spread HIV to 226 villagers. Ages varied from elders to infants. Authorities jailed a local physician for treating multiple patients with the same septic syringe.

December 18, 2014 The sweep of the rare enterovirus D68 over 49 states afflicted 1,125 U.S. children and threatened patients with complications from acute flaccid paralysis. Severe respiratory distress killed a dozen patients, many of them asthmatics.

December 31, 2014 Emerging viruses threatened global populations. Epidemic H7N9 afflicted Chinese poultry and 454 bird handlers and sellers, causing severe pneumonia and respiratory crises. Fowl pens and markets in Taiwan and Malaysia caused similar outbreaks of avian Influenza A.

December 31, 2014 WHO tabulated MERS infections, which inflicted 938 patients and caused 343 deaths on five continents, largely at Jeddah and Riyadh, Saudi Arabia. Two infections were evident in the U.S. by May 2015, when the coronavirus emerged in Florida and Indiana. By June, Algeria reported its first two cases and six deaths in Jordan. To protect devout Muslims, the government monitored the Hajj to Mecca in October and enforced sanitation and precautions against an epidemic.

2015 Incidence of sleeping sickness (trypanosomiasis) fell dramatically in sub–Saharan Africa. The disease continued to plague the Congo, with 2,351 infections. In contrast, Benin, Equatorial Guinea, Ghana, Mali, Niger, Nigeria, and Togo had no infections. In this same period, India reported a three-year record high of 127,326 cases of leprosy.

2015 The transference of dengue fever worldwide reached alarming rates: Brazil, 1,500,000, Philippines, 169,000, Malaysia, 111,000, Taiwan, 40,000, 20,000 in Vietnam, and 12,000 in Thailand with a 100 percent uptick in mortality. Even the U.S. totaled 190 cases in Hawaii. Increasing health concerns were higher numbers of child hospitalizations and a death rate of 2.5 percent. The flu-like symptoms preceded ominous complications from bleed-out, organ collapse, and respiratory failure.

January 9, 2015 The rapid sweep of Influenza B (H3N2) over 46 states in the continental U.S. killed 26 children. Because of viral drift, the year's vaccine mix offered limited protection from the strain, which remained virulent until the peak of contagion in February.

January 28, 2015 Report of a rapidly worsening form of HIV in Cuba reduced the progression rate by 40 percent. The CRF19 recombinant virus rushed the onset of AIDS from the typical five years to three years. The quick-killing variety threatened 15,000 adults and 1,000 islanders already diagnosed with HIV.

February 2015 Zika infections in French Polynesia spread to the Solomon Islands, Vanuatu, Samoa, Tonga, New Zealand, and the Marshall Islands.

February 2015 Reports of increased infection rates with seasonal Kyasanur Forest disease in Pali southeast of Jodhpur, India, listed a death rate of 10 percent in 50 victims. Patients suffered chills and fever, headache, seizures, sensitivity to light, and bleeding from stomach, gums, and nose. Spread by langurs, red-faced bonnet monkeys, cattle, Blanford's rats, flying squirrels, fruit-eating bats, jungle fowl, and shrews infected with tick bite, the contagion responded to vaccination, but immunity was shortlived, requiring booster shots.

February 9, 2015 The mumps virus in Idaho and Washington State affected 21 students. Upsurges in childhood diseases on the Pacific Coast reflected the results of an anti-vaccination campaign centered in California.

February 11, 2015 As WHO, UNICEF, and the Carter Center collaborated on the eradication of the guinea worm or dracunculiasis, the 3.5 million annual infestations in Africa and Asia in 1986 dropped to 126 cases spread over South Sudan, Mali, Chad, and Ethiopia. WHO declared Asia and Ghana free of the disfiguring parasite.

February 11, 2015 Measles advanced over the U.S. from the New Year at the rate of more than 100 victims in six weeks. Each patient typically infected from twelve to eighteen others with the scourge. Investigation of 403 cases on the Pacific Coast over a two-month span confirmed that the outbreak linked to five company workers and 42 children visiting Disney theme parks in California. Positive diagnoses extended from the park to Arizona, Colorado, Nebraska, Oregon, Utah, and Washington with eleven more in Canada and Mexico. Half of the children lacked MMR vaccinations. Anti-vaccination forces defied public health officials for charging them with child neglect. Other indicators connected the California endemic with the Filipino epidemic in January 2014.

mid–February 2015 A descendant of swine flu, pandemic H1N1 influenza in India overwhelmed Gujarat, Rajasthan, Delhi, and Maharashtra with the preponderance of 33,761 cases and 2,035 fatalities, some of whom already depleted by cancer or diabetes. At Jodhpur, Aligarh Muslim University and National Law University closed their campuses after one student died and eight others sickened with the contagion. In Jaipur, a 1,953-bed hospital stretched supplies to accommodate 3,000 patients. The flu strain was the same as the cause of the 2009 pandemic.

March 2015 English schoolchildren incurred 17,586 cases of aggressive scarlet fever, more than ten times the number from previous years. In nurseries and schools, a super strain of *Streptococcus pyogenes* resistant to treatment followed the typical spring outbreak, marked by a textured rash on the torso, headache, and fever. Complications ranged from ear infection to throat abscess and pneumonia. The upsurge in infections continued at the beginning of the 2016 school year with 6,000 cases at the rate of 600 cases per week.

March 2015 Ongoing crises with public epidemiology in Lebanon advanced on Palestinian and Syrian refugees living in unsanitary camps. According to former mayor Dr. Abdul-Rahman Bizri, concurrent epidemics—measles, leishmaniasis, and hepatitis A—threatened the overburdened health care system.

March 8, 2015 Pakistan led the world in poliomyelitis, a highly infectious virus that spread by oral contact with tainted feces and potable water. As the rest of the global population enjoyed a 99 percent drop in contagion, Pakistanis in Peshawar in the Khyber region suffered 319 cases since 2014 that produced irreversible paralysis. Police intervened by arresting parents who failed to protect their children with immunization. The Taliban tried to halt vaccinations by killing 60 door-to-door field agents and by circulating rumors that the serum reduced fertility.

March 14, 2015 An explosion of Ross River virus infections hit Brisbane, Australia, felling 2,835 patients in a period of fourteen weeks. Spread by mosquitoes, the virus appeared to infest possums and wallabies. Human victims suffered swollen glands, fever, achy joints, headache, rash, and fatigue. High tides following Cyclone Marcia in February may have exacerbated the contagion in Queensland. Health officials recommended cleaning water from gutters and screening windows and doors

April 5, 2015 A crisis in Indiana arose from identification of 81 HIV cases around Scott County, particularly among poor people addicted to injectable Oxymorphone, Oxycontin, and heroin. The town of Austin met the challenge with a needle exchange program and an emergency clinic.

April 13, 2015 A concentration of elephantiasis among residents of Kamwenge, Uganda, swelled victims' legs, breasts, and genitals to grotesque proportions and inflicted sweating, itching, gripping pain in wounds, and disability. The parasitic disease spread during the rainy season from mosquito bites contaminated with the wuchereria bancrofti worm. The infestation with lymphatic filariasis grew more insidious across the Indian subcontinent, where 60 percent of the population was a risk. District health efforts to eradicate the parasite increased in mid–June 2015, but advanced cases were irreversible.

April 21, 2015 The largest botulism outbreak in the U.S. in forty years sickened 29 guests at a potluck supper sponsored by the Cross Pointe Free Will Baptist Church in Fairfield, Ohio. The sick shared symptoms—blurred and double vision, drooping eyelids, and impaired swallowing. Vaccination with anti-toxin, intubation, and ventilation saved eleven from death; the neuroparalytic disease killed one woman from respiratory failure. Laboratory findings isolated potato salad made from improperly home-canned potatoes as the source of the nerve toxin.

April 29, 2015 According to UNICEF, the CDC, and the United Nations Foundation, the eradication of rubella in the Americas with MMR vaccinations ended resulting cataracts, retardation, blindness, deafness, miscarriages, heart defects, and stillbirths, all elements of chronic rubella syndrome.

April 29, 2015 Reports from Kilimanjaro, Tanzania, resurrected fears of schistosomiasis or snail fever, a parasitic worm penetrating skin and infesting organs. Adventurers traveling, tubing, wading, or swimming on African waterways risked infection with the blood fluke, which physicians treated with praziquantel.

May 13, 2015 Over the first 14 months of the year, Brazilians suffered 746,000 cases of fever and 229 deaths. The epidemic constituted a rise of 234 percent in infections. The contagion centered on Sao Paulo, where public health workers, backed by soldiers, urged homeowners to cover swimming pools and fill plant pots with sand.

June 2015 The combination of 151 cases of Guillain-Barré with a spread of the Zika virus and incidents of microcephaly in newborns in Salvador, Brazil, ranged 19 times higher than epidemiologists predicted. As symptoms of eye pain, conjunctivitis, fever, headache, body pain, and rash reached their height, vectors of the contagion remained undetermined.

July 14, 2015 Pertussis or whooping cough infected 240 residents of Clark County, Washington, and 350 cases in Oregon, largely because of lax inoculation of school-age children. Health authorities feared a repeat of 2012, when 5,000 victims in Washington State weathered infection by *Bordetella pertussis* bacteria.

August 2015 WHO identified 273 instances of seasonal Lassa hemorrhagic fever in Nigeria with a mortality rate of 54.6 percent. Additional cases in Benin, Togo, Liberia, and Sierra Leone and exportation to Germany and Sweden attested to the virulence of the Lassa virus. The vector, the mastomys rat, spread the virus through human contact with feces. Among medical staff, 20 percent died of the fever. Heightened surveillance hinged on increased facemask protection, clean gowns, and sterile gloves for nurses, doctors, and laboratory technicians. Ample rainfall from the El Niño weather phenomenon into November 2015 increased crop yield and supplied plentiful food for rats. A burgeoning rodent population heightened demand for house cats.

September 18, 2015 The CDC cited fast growing Lyme disease (borelliosis) for infecting 334,610 people in the U.S. The brunt fell on New England, Detroit, and the Great Lakes area, with considerable contagion in California, Florida, Georgia, Arizona, and Texas spread by the black-legged tick. Pennsylvania documented 10,000 cases. To the north, Canada reported 917 cases, primarily in woods populated by white-tail deer. Since CDC data collection began in 1982, the incidence has exploded 25-fold as the range of infected ticks spreads geographically with a 320 percent increase. Scientists charged urban sprawl, forest depletion, and climate change with the endemic.

September 18, 2015 In northern England at Leeds, Macclesfield, Oldham, and Scunthorpe, a drug-resistant H041 strain of gonorrhea developed in heterosexual patients, most under age 25. As the number skyrocketed 19 percent from 29,419 to 34,958 over a year's time, symptoms included painful urination, breakthrough menstrual bleeding, ectopic pregnancies, and a green or yellow discharge from genitals as well as infection of newborns. The health ministry feared the mutation of gonorrhea into a scourge untreatable by injectable ceftriaxone, the most common antibiotic. Cases of incurable gonorrhea also emerged in Japan, France, and Spain.

October 12, 2015 Global accounting of MERS cases reached 1,599 with 574 deaths. A total of 186 cases of MERS in South Korea over a five-month outbreak escalated to 36 deaths, one caused by sharing a hospital room with a MERS-infected patient.

October 28, 2015 A sudden onset of shigella in Santa Clara, California, affected 190 people and required intensive care for eleven victims. Health department agents traced the infections to the Mariscos San Juan restaurant in San Jose.

November 17, 2015 Record levels of chlamydia infections in the U.S. reached 1.4 million or 456 cases per 100,000. A state-by-state tally listed Alaska with 788 patients, the American Virgin Islands with 755, and Mississippi with 655. West Virginia and Maine posted the lowest counts at 255 and 266 respectively. Evidence of risky sexual behavior derived from higher numbers of syphilis and gonorrhea infections, at 20,000 and 350,062 respectively, particularly among homosexual and bisexual males. Left untreated, chlamydia could cause blindness, paralysis, stroke, and death.

November 23, 2015 On the southwestern border of Somalia, the Dadaab camp in Kenya incurred 1,566 infections with endemic cholera and ten deaths among 330,000 Somali refugees. Because of a shortfall in donations, inadequate latrines and soap during the rainy season increased sanitation concerns as hospital wards struggled to admit double their usual patient load. Medical staff combatted vomiting and diarrhea with oral and intravenous rehydration and tetracycline or doxycycline. The Ministry of Health treated 8,360 patients. By May 9, 2016, morbidity from cholera in Kenya reached 15,103 cases and 238 deaths.

December 7, 2015 Epidemiologists in Brazil made the connection between Zika infections of parturient women and microcephaly in 739 infants. Of the babies deformed by the flavivirus, 487 lived in Pernambuco.

December 10, 2015 One-fourth of students at Brunswick North West Primary School in Melbourne, Australia, contracted chicken pox. Staff linked the puzzling uprise of 80 cases to the 26.8 percent of students who lack immunization. Overall, Victoria led the area in contagion, with 432 cases presenting symptoms of malaise, fever, itch, and skin blisters. The infections aroused new antipathies between conservative parents and anti-vaxers.

December 15, 2015 A U.S. leap in *Listeria monocytogenes* contagion struck Arizona, California, Minnesota, Missouri, New Mexico, North Carolina, Nevada, Texas, Utah, Washington, and Wisconsin and two additional infections in Canada. The 32 hospital admissions included eleven pregnant women, whose ailments threatened their unborn babies. Most of the sick reported eating caramel-coated Gala and Granny Smith apples, the source of body aches, stiffness, chills and fever, headache, and diarrhea. The food-borne outbreak resulted on a wrongful death suit against Safeway grocery stores in California.

December 31, 2015 With syphilis on the increase over the U.S. by 15 percent, Las Vegas, Nevada, led the upturn with 700 cases, mostly among homosexual males involved in anonymous sex scheduled through social media.

2016 Thai health monitors braced for a dengue fever flareup to top the previous year's viral count of 140,000 patients and 106 deaths. By April, the count had reached 14,825. The critical phase advanced from diarrhea, vomiting, dehydration, headache, and oral bleeding to liver damage, stomach bleeds, abdominal and lung fluids, leaking vessels, and seizures. The failure of spraying to control the mosquito vector suggested that Dengvaxia, approved in Brazil, Mexico, and the Philippines, held the only answer to epidemics.

January 4, 2016 WHO tabulated the four-year spread of MERS to 1,644 cases and 590 deaths. The last four cases occurred in Saudi Arabia. Health department count continued to report case clusters daily. To deter the disease, Dr. Gabriel Defang at the U.S. Naval Medical Research Center in Maryland bioengineered antibodies in cows.

early January 2016 At Manzanares in east central Spain, 230 people contracted Legionnaires' disease. Of the 23 patients hospitalized under intensive care, three died of acute pneumonia and respiratory failure. A decorative fountain near the city bus station may have harbored the legionella bacteria. Closure of the public landmark ended the outbreak.

January 10, 2016 A WHO tabulation of epidemic malaria noted an 18 percent decline in cases from a height of 316 million at the end of the twentieth century to 214 million in 2015. Africa still bore the burden of 88 percent of illnesses. As of January 10, South Sudan faced the most severe contagion, which infected 1.6 million, the majority residents of Bahr el Ghazar. Worsening the health crisis, supplies of paracetamol and other anti-malarial drugs fell short of demand.

February 2016 At Piracicaba, Brazil, the release of genetically modified mosquitoes reduced the probability of epidemic chikungunya, dengue, yellow fever, and Zika by preventing the female *Aedes aegypti* mosquito from reproducing viable offspring. Test results gave hope of preventing the birth of microcephalic babies, which had already reached a population of 1,709 children requiring life-long care. A concurrent epidemic of Zika in Venezuela produced more than 5,000 cases. Angry residents accused President Nicolas Maduro of concealing the contagion in Venezuela from Zika, dengue, malaria, and chikungunya, all carried by insect vectors. Worsening the situation, a water shortage forced homeowners to store water indoors, giving mosquitos a place to breed. By February 2, 2016, WHO director Dr. Margaret Chan declared the explosive Zika epidemic and resultant microcephaly a world public health crisis.

February 3, 2016 A boost to virulent tuberculosis diagnoses and three deaths in Marion, Alabama, required 2,200 tests of a population of

3,600. The result—127 positive tests—proved difficult for health department staff examining people suspicious of medical questions about their health and intimate contacts. The offer of $20 per test, another $20 for a return visit, and $100 for taking antibiotics spawned a rush to the clinic for people outside the area. News of the outbreak destroyed businesses with the stigma of contagion.

late February 2016 Out of some 8,800 flu cases in the U.S., the CDC identified a predominance of H1N1 or Influenza A.

February 29, 2016 New Ebola infections in Korokpara, Guinea, ended a two-month hiatus of the disease, which caused internal hemorrhaging and multiple organ failures. The government declared a three-week quarantine of 816 people who had contact with the sick.

early March 2016 The diagnosis of the mumps virus among 40 students at the Harvard University campus in Cambridge, Massachusetts, threatened end-of-the-semester events and spring graduation. Additional infections at Tufts University, Boston University, the University of Massachusetts, and Sacred Heart in Fairfield, Connecticut, spread rapidly in warm dormitory, dining hall, and fraternity house environments by kissing, coughing, sneezing, and conversing and by sharing food, water bottles, and utensils.

March 15, 2016 At Montevideo, Uruguay, 587 cases of dengue fever prostrated an urban population. Inspectors of victims' homes found larvae of the *Aedes aegypti* mosquito in flower pots and pet water bowls. Fumigation reduced the insect count.

April 20, 2016 A three-decade upsurge in yellow fever sickened 2,420 in Luanda, Angola, killing 298 from shock, hemorrhage, and organ failure. Doctors first misdiagnosed the pestilence as food poisoning. Carried by the *Aedes aegypti* mosquito, the fever flourished during a vaccine shortage. Epidemiologists feared the commercial flights to Beijing, China, and Dubai would spread contagion. A travel ban stopped un-immunized outsiders from crossing the borders.

May 2, 2016 In the Cusco area of the Amazon rainforest of Peru, 57 residents fell ill from Oropouche fever, the result of an arbovirus. Symptoms of fever, headache, body pain, and vomiting advanced in some to aseptic meningitis, a nerve inflammation. The vector, swarming *Culicoides paraensis* midges, thrived around water, forcing residents to screen doors and windows, spray repellents, and wear long clothing.

June 3, 2016 A mumps scourge of 41 cases in Whistler, British Columbia spread to Vancouver and Squamash. Infection flourished in adults born between 1970 and 1996 before the MMR inoculation entered the vaccination schedule.

June 23, 2016 Under siege by terrorists of Boko Haram, 1,200 people at Bama in northeastern Nigeria died of disease and starvation. Kwashiorkor, caused by inadequate protein, vitamins, and trace elements, reached epidemic proportions in 800 children, of whom 62.5 percent died of nutritional depletion and secondary infections. Fed on porridge and gruel, newly weaned babies subsisted on carbohydrates. Atrophied muscles and the accumulation of fluid in the abdomen produced potbellies in anemic babies. Parasites, dehydration, and diarrhea left the sickly children susceptible to shock, gastroenteritis, pneumonia, measles, septicemia, tuberculosis, and HIV. Recovery depended on the gradual introduction of milk or soya feedings, cereals, antibiotics, and vitamin and mineral supplements followed by iron tonics.

June 11, 2016 With the world's highest incidence of dengue hemorrhagic fever, the Philippines continued to lead the world in patient numbers with a 41 percent increase. Of 52,177 cases, 207 died. Public health officials expected a spike in contagion from the virus between July and November after the onset of seasonal rains. Fourth-graders had the best outlook on wellness after the world's first public dengue immunization effort with Dengvaxia, begun in January 2016.

June 23, 2016 The infection with Zika of 7,500 islanders of Cape Verde off Africa's west coast yielded eleven births of babies with microcephaly. The incipient pestilence introduced the continent to the Asian strain of the virus.

July 1, 2016 Within weeks of the 2016 Olympic Games, an outbreak of H1N1 influenza in Brazil caused 1,233 deaths, mostly around Sao Paolo. Complications produced 6,569 cases of acute respiratory syndrome. Mass public health initiatives immunized 95.5 percent. A concurrent epidemic in Panama hospitalized 671, of whom 22 died.

July 25, 2016 At the end of the rainy season, Colombia became the first Latin American country to declare Zika eradicated. Over ten months, the nation's count had reached 99,721 victims and 21 microcephalic births, making Colombia patient count the world's second-highest.

August 2, 2016 During a heatwave that thawed permafrost with 95 degree Fahrenheit temperatures, the infection of 40 nomads and the death of a twelve-year-old boy in Yamalo-Nenets in northern Siberia coincided with anthrax from a reindeer carcass interred in 1941. The outbreak above the Arctic Circle in the Yamal Peninsula coincided with the deaths of 2,300 reindeer from spores of *Bacillus anthracis*, which permeated groundwater. Animals feeding on contaminated grass yielded spoiled venison and blood, which the boy drank. Health officials examined 54 children and 31 adult herders at hospitals for gastrointestinal anthrax and quarantined 63 herders until September.

To decontaminate the area, the Russian government sent 200 army specialists to vaccinate uninfected animals, incinerate toxic reindeer remains, and ban export of venison, antlers, and pelts from the region. Poachers increased the danger by selling toxic meat on the black market. Scientists conjectured that Khanty and Nenet corpses buried in shallow graves across the tundra could reactivate bubonic plague, smallpox, and Spanish flu, one of the insidious effects of global warming.

mid–August 2016 In Japan, fast-spreading measles attacked staff at Kansai International Airport. In addition to 32 of 15,000 workers, three medical workers contracted the virus. Epidemiologists traced the outbreak to a man passing through the facility in late July. Further sleuthing in September linked 41 victims to a Justin Bieber concert.

September 5, 2016 WHO declared Sri Lanka malaria free. Worldwide, infections with the zoonotic disease decreased by 37 percent, primarily from preventing the bite of the *Aedes aegypti* mosquito. Deaths from malarial fever fell by 60 percent to 292,000 in Africa, killing mostly preschoolers.

September 8, 2016 Food-borne hepatitis A sickened 131 people in Arkansas, Virginia, Maryland, New York, North Carolina, Oregon, West Virginia, and Wisconsin. Around 46 percent entered hospitals for treatment of the viral liver infection, gut pain, and jaundice, which the CDC linked to frozen strawberries imported from Egypt and served in drinks at Tropical Smoothie Cafés. The company replaced North African fruit with berries from California and Mexico, but still incurred lawsuits for negligence in dispensing adulterated food.

In a separate endemic, Hawaii's department of health reported 284 cases of hepatitis A, its worst tabulation in 20 years. The source of the contagion, scallops imported from Koha Oriental Foods in the Philippines, forced the closure of Genki Sushi restaurants on Hawaii, Kauai, Maui, and Oahu after patrons collapsed with pain and vomiting after eating the seafood raw. The rapid effects of contagion sent 71 to hospitals and sapped the supply of vaccine.

September 16, 2016 In the Karkardooma section of south Delhi, India, doctors treated mosquito-borne chikungunya virus, which caused sudden fevers, headache, nausea, rash, fatigue, and joint and muscle pain in 1,000 victims. The seasonal epidemic was the subcontinent's worst, infecting some 12,250 residents and killing eleven. Sanitation officials blamed open sewer drains, where mosquitoes thrived.

October 5, 2016 Signaled by deaths and spontaneous abortions in cattle from phlebovirus, over an eight-week period, Rift Valley Fever afflicted 90 nomadic herders at Tchintabaraden, Tahoua, in western Niger on the border of Mali, killing 28 or 31 percent of human victims. Stockbreeders accounted for 84 percent of patients, who risked retinitis, blindness, encephalitis, and bleeding to death. WHO aid workers in a mobile health laboratory at Dakar supported diagnosis and intervention and advised local citizens to avoid handling infected carcasses, bury dead animals carefully, and boil raw milk before drinking. Observers kept watch over herding from Niger into Benin, Mali, and Nigeria.

November 13, 2016 With a total of 5.9 million preventable deaths per year, India maintained a hold on child mortality from pneumonia and rotavirus. Next in order of fatalities, Indonesia, Chad, China, and Somalia continued to lose children under age five because of failure to vaccinate.

December 21, 2016 NBC news divulged the death count from global AIDS infections reached 35 million. Some 36 million more carried the HIV infection, which threatened the health of 1.2 million U.S. citizens. New Orleans headed the list of HIV contagion along with high counts of gonorrhea and syphilis. Dozens of pharmaceuticals offered relief from HIV/AIDS symptoms, but no vaccine or cure existed. Contributing to deaths in the Philippines among homosexual males, punitive laws restricted purchase of condoms and prohibited safe sex education and HIV testing in teenagers without parental consent.

December 25, 2016 The *New York Times* reported on a resurgence of black lung in the Appalachian coal mines of Virginia, West Virginia, Pennsylvania, and Ohio. Since 2006, clinics recorded 962 cases of progressive massive fibrosis, a more complicated form of the wasting disease from inhalation of quartz and coal dust. In Kentucky, pulmonologists treated 60 patients suffering from labored breathing and suspected more undiagnosed cases.

Glossary

The following technical terms and agencies appear in discussions of wellness, epidemics and world diseases, preventive medicine, immunology, and medical research.

adenovirus one of a family of pathogens named for the adenoids. An adenovirus causes fever or infection of the eye and of the mucus membrane of the respiratory tract, such as the microbe that causes the common cold or sore throat.

AIDS an acronym for *a*cquired *i*mmuno*de*ficiency *s*yndrome, a viral infection of the immune system characterized by diminished production of T cells and by a reduction in natural protection against threatening pathogens, such as pneumonia, tuberculosis, and Kaposi's sarcoma

anomaly an irregularity, peculiarity, or deviation from the body's normal function, for example, the paralysis caused by poliomyelitis or the blindness produced by trachoma

antibiotic a drug that is selectively toxic to prevent or inhibit the growth of pathogens or to destroy susceptible microbes, such as penicillin, which kills the pathogens that cause gonorrhea, syphilis, and streptococcal infection

antibody a protective protein in the blood that recognizes, surrounds, and neutralizes an antigen, for instance, the immunity against tetanus created by injecting the patient with tetanus vaccine

antigen a foreign substance in the body that induces a spontaneous immune response, for example, salmonella bacteria contained in raw or undercooked foods or cholera bacilli found in impure water

arbovirus (also **arborvirus**, from *a*rthropod-*bo*rne virus) a virus carried by an arthropod, such as the *Aedes aegypti* mosquito that carries the yellow fever virus

archeopathology *see* paleopathology

bacillus, pl. **bacilli** a disease-producing microbe or bacterium, commonly called a germ, for example, leprosy or anthrax, which can infect humans or animals

bactericide a drug or substance such as hexachlorophene that kills harmful bacteria, such as bacteria-killing agents found in pump sprayers and kitchen hand soaps and lotions to suppress or eliminate viruses, fungi, and the microbes causing skin infections, intestinal diseases like *E. coli* and salmonella, and respiratory illness spread by human touch.

bacterium, pl. **bacteria** a single-celled microbe, for instance, the pathogen that causes gonorrhea, syphilis, tuberculosis, or streptococcal infection

balneology the study of therapeutic bathing in warm water, heated vapors, or mineral springs as a treatment of ailments, an element of treatment at Lourdes, a religious hospice in France that receives patients suffering chronic and life-threatening disease

biological weapon a toxic weapon capable of spreading lethal pathogens to cause widespread death or to disable a number of people with disease, as found in the stockpile of anthrax, bubonic plague, and other antigens in Japan's experimental laboratories in Harbin, China, during the first half of the 1940s

body effluvia any liquid issuing from the human body, including perspiration, tears, mucus, saliva, urine, feces, blood, vaginal fluid, semen, or menses

bridge species an animal that spreads a zoonotic toxin to humans, as with fruit-eating bats and the Nipah virus in 1999 and Himalayan civets and SARS in 2002–2003.

CDC (Centers for Disease Control and Prevention) a major federal agency formed in 1946 and headquartered in Atlanta, Georgia, to protect the health and well-being of citizens at home and abroad, to advise on health matters, and to promote wellness. The CDC develops and applies disease prevention and control, boosts environmental quality, and educates the public on health issues, particularly current threats to the populace from HIV/AIDS, Ebola, West Nile fever, hantavirus, and MERS.

congenital acquired during development in the womb and existing from the time of birth, such as congenital syphilis or deafness, blindness, cardiac malfunction, or mental retardation caused by an attack on the fetus by rubella during the first trimester of development

consumption an early synonym for pulmonary tuberculosis, a progressive wasting disease that literally consumed the body's strength and natural defense against illness

contagium, pl. **contagia** a pathogen, such as a virus or bacterium, that produces a communicable disease, such as influenza, pertussis, or measles

cordon sanitaire a protective shield or barrier, for example, a quarantine line or harbor sequestration of ships, passengers, or cargo to prohibit the importation of plague, cholera, trachoma, or other communicable disease

drift virus an escaped mutant that gains dominance over the original virus, a troubling pattern in outbreaks of influenza

emergent disease a disease previously unrecognized among humans, as was the case with the Marburg virus and Ebola in the late 20th century

empiricism conclusions of medical science based on observation or experimentation, for example, the development of methods of variolation of the skin with cowpox matter to prevent infection by smallpox

endemic a disease such as malaria, cholera, or elephantiasis that is particular to and recurring in a specific locale, for example, the tropics. In another form, pellagra or scurvy are common diseases derived from nutritional deficiency among specific populations, notably people whose diet relies heavily on polished rice or ships' crews who lack access to fresh vegetables and fruit

enterovirus a picornavirus, for example, poliomyelitis, that lives in and infects the intestines and may attack the lungs, nerves, or spinal cord

enzootic present in an animal community at all times but occurring in only small numbers of cases, as with the bubonic plague pathogen found in rats or the hantavirus in deer mice **cf. epizootic**

epidemic a disease that spreads rapidly over an area to large numbers of victims, who become ill simultaneously, as found in outbreaks of yellow fever in large cities during summer months when mosquitoes flourish in stagnant water

epidemiology the study of the causes and spread of disease and how to contain and cure them, which is the work of such agencies as the Centers for Disease Control and Prevention and the World Health Organization

epizootic present in an animal community at all times and occurring in huge numbers of cases, causing an epidemic among animals or humans, such as the Black Death, a deadly instance of bubonic plague in Asia and Europe in the late 1340s

eugenics methods of producing strong, healthy offspring through selective breeding, genetic screening for diseases like Tay-Sachs or epilepsy, sterilization of the insane or retarded, and reduction or execution of weak strains, for example, the intent of Nazi Germany to annihilate all humans except for the pure-bred Anglo-Saxon blue-eyed blonds

filaria slender nematodes that can invade the blood stream, particularly those causing elephantiasis or onchocerciasis or the pathogens discovered by tropical medicine expert Sir Patrick Manson in his search for ways to prevent malaria

flavivirus a family of deadly viruses that infect vertebrate animals causing dengue fever, encephalitis, hemorrhagic fevers, and yellow fever

fomes, pl. **fomites** an inanimate particle that carries and transmits infection by absorbing microbes and dispersing them, notably from unclean towels or bedding, used drinking cups, or blood-soaked cutting boards and kitchen implements

forensic medicine argumentative or conjectural study and investigation of medical puzzles or quandaries, specifically, exhumation of victims of the Spanish influenza pandemic to determine the exact type of pathogen that killed the patients or the study of living tissue extracted from polio victims to determine how the human body contracts poliomyelitis

geophagia chronic ingestion of dirt or clay soil, a peculiar behavior that introduces the hookworm nematode into the body

heliotherapy treatment of disease through sunbathing or exposure to the sun's rays, a method preferred by designers of late 19th-century sanatoria for the curing of pulmonary tuberculosis

hematology the chemical analysis of blood samples, evolved in 1833 by French professor of internal pathology Gabriel Andral as a means of determining the source of influenza

herbalism the practice of healing or curing disease through applications of herbs as food supplements, medicines, teas, salves, vapors, or unguents, historically, those nostrums prescribed by Aesculapius, the first Greek doctor, or mugwort burned for moxibustion, a healing method advanced by Chinese medicine

HIV an acronym for the *h*uman *i*mmunodeficiency *v*irus, which causes AIDS by destroying T cells within the immune system, thus lowering the body's natural shield against infection **cf. AIDS**

homeopathy a system of healing that applies minute doses of natural substances to stimulate the body to heal itself, a curative method founded in Germany at the beginning of the 19th century by Samuel Christian Hahnemann of Saxony, who prescribed belladonna, a member of the poisonous nightshade family, as an antidote to scarlet fever

host an organism that provides food or shelter for another organism, internally or externally, for example, a deer on which a tick fastens before transferring Lyme disease to humans

hyperendemic a common disease present at a constantly high incidence within a population, for example, yaws in Ecuador and impetigo in Oceania

hyperepidemic an epidemic disease present at a constantly high incidence within a wide geographic area, as with dengue fever and malaria throughout Latin America and enteritis among most infants in poor nations where drinking water is polluted with sewage

immunity natural resistance or counteraction against infection or disease caused by bacteria or viruses, for example, the life-long protection acquired after an attack of diphtheria or measles or the natural immunity a fetus gains in utero to pertussis or hepatitis B

immunization the process of shielding the body against disease by application of serum or vaccine, such as the Salk polio vaccine or a tetanus or smallpox vaccination

immunology the study of natural and induced resistance to infection, a scientific discipline that has increased the life expectancy of victims of HIV/AIDS

incidence the rate of infection of a disease over a known period of time, a form of demographic study that followed the world epidemic of Spanish influenza

infection the acquisition of a microbe or an attack of pathogens that are capable of producing an infection such as typhus, a common scourge among soldiers living in close quarters or sailors on ships

infectivity the ability to infect, weaken, or sicken through attacks of pathogens, for example, the high infectivity of smallpox, plague, and influenza and resulting pandemics

lazaret (also **lazarette** or **lazaretto**) a plague hospital or pesthouse restricted to victims of contagion, such as those built in France during the early 12th century to house lepers and prevent their contact with healthy people

Médecins sans Frontières (Doctors without Borders) an apolitical international humanitarian aid task force that provides crisis relief and emergency medical assistance to endangered people in over 80 countries where basic sanitation and health structures are lacking or insufficient and where people are in eminent danger of death from war, famine, or natural upheaval

microbe a microscopic organism, typically, a bacterium, virus, protozoa, and some algae

moral environmentalism an assumption that sin and degeneracy cause disease, a belief maintained by pious settlers of the Massachusetts Bay Colony during the mid–17th century who blamed impiety and sin for the spread of smallpox

morbidity incidence of disease or susceptibility to infection, an element that causes refugees and prisoners of war to sicken and die at a higher rate than that of the rest of a population

mortality fatality rate or proportion of a population that succumbs to a disease. For example, conjunctivitis and dracunculiasis have a much lower mortality rate than rabies or tetanus.

National Institutes of Health a national agency that supports the medical and behavioral research and health legislation of the United States. Founded in 1887 in Staten Island, New York, the NIH developed into a scientific effort to analyze illness and disability and to nurture the wellbeing of citizens through research, preventive medicine, public accountability, and social responsibility.

necrosis death of human tissue in a wound or infected area, for example, gangrene or the loss of sensitivity and blood supply to extremities of lepers

nosocomial spread within the confines of a hospital, for instance, the outbreaks of puerperal fever in maternity wards among women examined by health workers using septic hands and instruments

oocyst an encased parasite, such as the protozoa that causes cryptosporidiosis

Oxfam International a confederation of 12 organizations that aids 100 countries in solving problems of poverty, suffering, unhealthful living conditions, disease, and injustice

paleopathology the study and diagnosis of the diseases or causes of trauma on human bones, skeletons, teeth, mummies, and other tissues that survive from ancient times, a science that enables researchers to determine the diseases that plagued early humans

palliative a drug or curative method that reduces the virulence of disease, for example, willow bark or aspirin to reduce the ache and fever of influenza or chamomile tea to soothe the misery of enteritis

Pan American Health Organization (PAHO) an international public health agency based in Washington, D.C., that improves health and living standards in the 27 countries of the Western Hemisphere by promoting primary health care and extending health services to vulnerable areas lacking basic hygiene, sanitation, immunization, and medical assistance

pandemic a disease such as malaria, smallpox, or influenza that spreads over a wide geographic area and afflicts large numbers of people in more than one continent

Pasteur Institute a nationally subsidized research agency set up in France in 1888 as an outgrowth of the research of epidemiologist Louis Pasteur and his students into the infectious nature of diphtheria, plague, rabies, tuberculosis, and typhus. At locations in Europe and Asia, the institute's "Pastorians" extended their scrutiny into the prevention and cure of yellow fever, diphtheria, and tetanus.

pathogen a virus or bacterium causing disease, for example, tuberculosis bacilli and the adenoviruses that cause colds and sore throats

pathogenic any infective agent causing disease, as with the worm larva causing trichinosis or the coal dust producing pneumoconiosis and silicosis

pathological concerning a structural or functional change in the body or deviation from normalcy as the result of a disease, as with the study of septicemia and how it depletes the blood system

pathology the study of disease and the identification of its method of infection, for instance, the examination of victims of hemorrhagic fevers to determine what hosts carry them and how they spread

patient zero the initial or primary case of the investigation of an epidemic, for example, Dr. Liu Jianlun, the first Chinese victim of SARS in Quangdong in February 2003.

pharmacopeia or **pharmacopoeia** drugs, chemicals, and medical preparations that fight disease, notably, quinine and its synthetic imitators, atabrine and chloroquine, which were essential to Allied troop health during World War II

predisposition a natural inclination or susceptibility to disease, for example, auto-immune deficiency or skin cancer

prevalence the number of infections of a disease in a population at a given time, often expressed in the number of cases per 100,000 people, a valuable mathematical tool in determining why some groups are more or less susceptible to disease than others

prion disease a fatal neurodegenerative disorder contracted from consumption of a rogue prion protein, as with kuru, a brain-wasting disease similar to Creutzfeldt-Jakob disease and found in the people of Papua New Guinea who practice funerary cannibalism

prodrome a telltale symptom of the outbreak of infectious disease, as with the fever and ache that precede full involvement in an influenza epidemic

prophylactic discouraging or preventing the spread of disease, notably, the use of condoms for sexual intercourse to halt the spread of chlamydia, genital herpes, gonorrhea, syphilis, or HIV

prophylaxis prevention of disease, as found in port cities that spray incoming planes and ships to rid them of disease-bearing insects, such as the *Anopheles* or *Aedes aegypti* mosquito

prosector a pathologist who gives medical demonstrations by dissecting human cadavers, a visual method relied on by medical science before the evolution of microbiology, blood analysis, x-ray, and other forms of medical technology

psychogenic originating in the emotions or mind, a phenomenon of groups of people suffering false symptoms of disease from heightened fear of epidemic polio, AIDS, or rabies

recrudescence the re-emergence of a disease after a period of inactivity or abatement, particularly the phenomenon occurring in London nearly two years after the Great Plague of 1665

regimen a systematic plan of diet, medication, therapy, and exercise to improve and maintain well being, especially those treatments intended to improve the health of victims of kwashiorkor, marasmus, or other forms of starvation

reservoir an organism that nourishes an antigen without suffering contagion and which serves as the source of infection for other host organisms, as with rats that nurture bubonic plague and dog ticks that carry Rocky Mountain spotted fever bacteria

rhinovirus a picornavirus that causes the common cold and other ailments of the upper respiratory tract

sanatorium a live-in health facility or hospital that treats tuberculosis or other chronic disease through holistic health regimens, including wholesome diet, rest, exercise, group activities, therapeutic sun baths or mineral waters, and medical care

sanitarium a spa or health-building facility that promotes wellness and rehabilitation from disease, such as convalescents recovering from tuberculosis; a sanatorium

septicemia invasion of the bloodstream by microbes located in an isolated part of the body, for example, the spread of puerperal infection from examination of parturient women with unclean instruments or hands

sequela, pl. **sequelae** an aftereffect of a disease, notably, the debility that follows influenza, cardiac valve damage from scarlet fever, deafness from measles, or the paralysis of polio

serum, pl. **sera** an injectable fluid that bolsters the body's immune system against such infectious disease as diphtheria, pertussis, pneumonia, or hepatitis

shaman a practitioner of animistic or pagan medicine, divination, and the casting of spells, for example, a priest or practitioner of voodoo

simian virus a virus that thrives in primates, who are first cousins to humans, notably Ebola, which afflicts humans who hunt and kill monkeys for food

spirochete a thin spiral pathogen that causes bejel, relapsing fever, syphilis, and yaws

streptococcus a round bacterium that causes inflammation and throat infection, as with the sulfa-resistant strep infections that sickened U.S. soldiers during the Korean War and felled retarded citizens in Denmark in the mid–1990s

super spreader an infection that spreads from one patient to clusters of people in a hospital, school, airplane, or hotel, a characteristic of Ebola, tuberculosis, rubella, and SARS.

susceptible having no immunity to a disease, a characteristic of colonial forces during the American Revolution during a smallpox epidemic

sylvatic occurring among wild animals, for instance, dengue fever, Chagas disease, or rabies, which flourishes among bats, raccoons, and foxes

syphiloid a disease resembling syphilis, for example, yaws and bejel, two non-venereal forms of syphilis

syphilologist a specialist in the diagnosis, treatment, and containment of syphilis, a disease that emerged in Europe after the return of Christopher Columbus from his first expedition to the Caribbean

tarantism epidemic hysteria called St. Vitus's dance that causes people to dance, jump, twitch, jerk, and convulse, sometimes producing permanent mental and physical dysfunction and death, a phenomenon reported in Wales in 1188 and throughout Europe a quarter century after the Black Death

theory a body of facts or principles that explain phenomena in the natural world, notably, French epidemiologist Louis Pasteur's study of microbes to determine what pathogens caused rabies and other contagious diseases

therapeutic emigration travel from an unhealthful location to a more salubrious setting, a popular treatment for tubercular English patients whom doctors dispatched to Australia and the Swiss Alps for rest and rehabilitation

toxin any poisonous or harmful substance that can be produced by an animal, plant, or microbe, including the pathogens that cause sleeping sickness, cholera, and hepatitis

trivalent vaccine a protective serum composed of three strains of a pathogen, specifically, seasonal flu vaccines composed of variant types of the virus

UNICEF an international children's welfare bureau created by the United Nations General Assembly in 1946 to aid children after World War II as the United Nations International Children's Emergency Fund. In 1953, it evolved into the United Nations Children's Fund, a task force reducing the effects of poverty, ignorance, war, and unsanitary living conditions in developing countries through rescue, nutrition, education, health services, and supplies

vaccination immunization by injecting vaccine into the body to produce immunity against a particular pathogen, for example, the diphtheria-pertussis-tetanus shot that most infants receive in the first year of life

variolation an early form of inoculation against variola (smallpox) by injecting or implanting matter from pox pustules in the bodies of healthy people to create immunity

vector an animal or insect that bites a victim, ingests virus-bearing liquid, and hosts the virus in its own body before passing it through a bite to another animal or human, as with the tick that carries Lyme disease bacteria from deer to humans or the sandfly that spreads leishmaniasis

venereologist a specialist in the diagnosis, treatment, and containment of venereal disease, a significant booster to the health of Europeans after syphilis overran populations in the fifteenth century and remained deadly until the discovery of penicillin

viremic bearing a virus, a characteristic of vectors such as the mosquito, which carries malaria, Murray Valley encephalitis, sleeping sickness, and yellow fever

virgin-soil population a community, tribe, or nation that has had no previous contact with a pathogen, for example, Mesoamericans who had no experience with smallpox, measles, influenza, and other European diseases before the arrival of the Spanish in 1519

virulence the capacity of a microbe to infect or to generate disease, notably, the pathogens causing Ebola, smallpox, and Marburg virus

World Food Program an agency of the United Nations set up in 1963 to battle global hunger as a result of harvest failure, famine, internal conflict, or natural disaster

World Health Assembly an international consortium on health and well being convened by the World Health Organization

World Health Organization (WHO) a global consortium which the United Nations organized in Geneva, Switzerland, in 1946 to mount an advanced system of monitoring disease prevention and the outbreak of epidemics

zoonosis a disease such as ringworm, conjunctivitis, or Rift Valley fever, that can pass directly from animals to humans

Appendix A
Epidemic Diseases and Sources

Disease	Significant Research	Source
acrodynia	Germany, 1923, by Alfred Stock	mercury poisoning
adenovirus	New Jersey, 1953, by Wallace Rowe	viruses carried by fecal matter into drinking water
African fever *see* malaria	England, 1897, by Ronald Ross	
ague *see* malaria	England, 1897, by Ronald Ross	
AIDS *see* HIV	Kinshasa, Zaire, 1983 by Peter Piot	contact with virus-bearing body effluvia or unprotected sexual intimacy
alastrim *(Variola minor)* *see also* smallpox	Congo, 1925, by Lucius van Hoof	contact with virus-bearing lesions or droplets breathed into the air
Alkhurma hemorrhagic fever (AHF)	Jeddah, Saudi Arabia, 1995, by Mikhail Chumakov	a flavivirus carried by ticks
amoebiasis	South Africa, 1906, Joan Scraggs	infestation of the large intestine by the amoeba *Entamoeba histolytica*
ankylostomiasis *see* hookworm	Germany, 1852, by Theodor Bilharz	
anthracosilicosis	Pennsylvania, 1960, by Frederick Banting	respiratory exposure to coal dust causing fibrosis in the lungs
anthrax *(Bacillus anthracis)*	Poland, 1875, by Robert Koch	meat and milk from bacillus-bearing cows, goats, horses, pigs, sheep
Argentine hemorrhagic fever *(Junin virus)*	Buenos Aires, 1958, by Izak Pirosky	contact with fluids of virus-bearing corn rats
arsenic poisoning (arsenicosis)	Thailand, 1987, by Nakhon Si Thammarat	consumption of arsenic through contaminated water or cosmetics
asbestosis	Manchester, England, 1924 by Montague Murray	chronic inflammation and scarring of lungs from inhaled asbestos fibers
Barmah Forest virus or epidemic polyarthritis	Australia, 1974, by Stephen L. Doggett	bite of alphavirus-bearing mosquito
Bartonellosis bacilliformis (cat scratch fever)	Peru, 1885, by Daniel Alcides Carrión	bite of a bacteria-bearing female *Phlebotomus* sandfly
bejel or endemic syphilis	England, 1947, by Cecil J. Hackett	a non-venereal syphilis caused by the effect of the *Treponema* spirochete on bones and skin
beriberi	Indonesia, 1897, by Christiaan Eijkman	lack of vitamin B_1 or thiamine in the diet
bilharziasis *see* schistosomiasis	Germany, 1851, by Theodor Bilharz	

Disease	Significant Research	Source
bilious remittent fever *see* malaria	England, 1897, by Ronald Ross	
black bane *see* anthrax	Poland, 1875, by Robert Koch	
black fever *see* leishmaniasis	Calcutta, 1890, by Charles Donovan	
black lung *see* pneumoconiosis	Germany, 1867, by Friedrich Albert Zenker	
black typhus *see* Machupo virus	Bolivia, 1959, by Karl Johnson	
blastomycosis (*Blastomyces* *dermatitidis*)	Chicago, 1894, by Thomas Casper Gilchrist	inhalation from spore-bearing soil to human lungs
bloody flux *see* cholera	Italy, 1854, by Filippo Pacini	
Bolivian hemorrhagic fever *see* Machupo virus	Bolivia, 1959, by Karl Johnson	
Bornholm disease (*Coxsackie B.*)	Norway, 1872, by Anders Daae	contact with virus-bearing human feces
botulism (*Clostridium* *botulinum*)	Germany, 1820, by Justinus Kerner	from bacteria-bearing soil to food
breakbone fever *see* dengue fever	Australia, 1934, by John Burton Cleland	
Brill's disease *see* typhus	New York, 1935, by Hans Zinsser	
brucellosis (*Brucella* *melitensis*)	Malta, 1884, by David Bruce	meat and milk from protozoa-bearing cows, goats, horses, pigs, sheep
bubas *see* yaws	Brazil, 1995, by Lorenzo Breda	chronic infection of the skin in crowded living conditions
Buruli ulcer from (*Mycobacterium ulcerans*)	Tanzania, 1864, by James Augustus Grant	an infectious pathogen that develops a nodule on the limbs into an ulcer eating into the bone
camp fever *see* typhus	Poland, 1939, by Rudolf Weigl	
Campylobacter jejuni	Germany, 1886, by Theodor Escherich	handling and cooking of bacteria- bearing chicken, barbecued meat, raw milk, and shellfish
Canicola fever *see* leptospirosis	Germany, 1886, by Adolf Weil	
cat scratch fever *see* Bartonella henselae	France, 1889, by Henri Parinaud	bite of bacteria-bearing arthropod or cat flea
celiac disease *see* sprue	Barbados, 1759, by William Hillary	
Chagas disease (*Trypanosoma cruzi*)	Brazil, 1909, by Carlos Chagas	protozoa spread by the urine or feces of a sucking insect, the *Triatoma infestans*
chicken pox (*Varicella zoster*)	Chicago, 1984, by Maurice Hilleman	contact with virus-bearing human or from airborne droplet
chikungunya fever (*Alphavirus*)	Tanzania, 1881, by James Christie	bite of a virus-bearing *Aedes aegypti* or *Aedes albopictus* mosquito
childbed fever *see* puerperal fever	Hungary, 1847, by Ignaz Semmelweis	
chlamydia (*Chlamydia* *trachomatis*)	Chicago, 1944, by Maurice Hilleman	intimate contact with bacteria-bearing fluids from rectum, urethra, or vagina
cholera (*Vibrio cholerae*)	England, 1854, by John Snow Massachusetts, 1955, by Lloyd Conover	food/water contaminated by bacteria in human feces
common cold (*rhinovirus*)	Tennessee, 2007, by E. Kathryn Miller	linked to camels

Disease	Significant Research	Source
conjunctivitis *(adenovirus)*	India, 1914, by Kirk Patrick	contamination of eyes by virus-bearing hands or linens
consumption *see* tuberculosis	Italy, 1546, by Girolamo Fracastoro	
Creutzfeldt Jakob virus *(Spongiform encephalopathy)*	Germany, 1920, by HansGerhard Creutzfeldt	degenerative brain disease caused by prion transmitted by eating parts of diseased animals
cryptosporidiosis	Massachusetts, 1807, by Ernest Edward Tyzzer	protozoa-bearing feces in water, undercooked food, household items, and livestock
dengue fever *(Flavivirus)*	Australia, 1934, by John Burton Cleland	bite of virus-bearing *Aedes aegypti* mosquito
diarrhea *see* cryptosporidiosis; rotavirus	Massachusetts, 1807, by Ernest Edward Tyzzer	
diarrhea alba *see* sprue	Barbados, 1759, by William Hillary	
diphtheria *(Corynebacterium diphtheriae)*	Germany, 1901, by Emil von Behring	contact with infected food and bacteria-bearing human or by airborne droplets
dracunculiasis *(Dracunculus medinensis)*	Uzbekistan, 1860, by Alexei Pavlovich Fedchenko	drinking water containing parasitic filaria of the *Cyclops* larva
dumdum fever *see* leishmaniasis	Calcutta, 1890, by Charles Donovan	
dysentery *(Shigella dysenteriae)*	Japan, 1906, by Kiyoshi Shiga	food or water contaminated by bacillus in human feces
Ebola *(filovirus)*	Zaire, 1976, by Peter Piot	contact with virus-infected blood, secretions, or human tissue
Egyptian ophthalmia *see* trachoma	Malta 1948, by Vincent Trabone	
elephantiasis *(Wuchereria bancrofti)*	France, 1863, by Jean-Nicolas Demarquay	bite of anopheles mosquito carrying the microfilaria that infest lymph glands
encephalitis	Russia, 1837, by Lev Zilber	bite of virus-bearing mosquito or tick
encephalitis lethargica	Austria, 1917, by Constantin von Economo	viral infection through mouth or nose
English sweating disease *see* sudor Anglicus	England, 1551, John Caius	
enteric fever *see* typhoid fever	England, 1838, by William Budd	
enteritis *(Campylobacter enteritis)*	France, 1963, by Madeleine Sebald and Michel Verón	contact with humans, animals, raw poultry, fresh produce, unpasteurized milk, or water contaminated with the bacteria
enterovirus D68 *(Picornavirus)*	Germany, 1897, by Friedrich Loeffler and Paul Frosch	acute respiratory distress from a polio-like flaccid myelitis
epidemic jaundice *see* Hepatitis A	Massachusetts, 1979, by George Hitchings and Gertrude Elion	
ergotism	France, 1670, by Jean Thuillier	ingestion of food containing ergot fungus
erysipelas *(hemolytic streptococcus)*	Germany, 1883, by Friedrich Fehleisen	bacteria from infected human
Escherichia coli (E. coli)	Germany, 1885, by Theodor Escherich	consumption of bacteria-bearing water or raw meat

Disease	Significant Research	Source
filariasis *(Brugia malayi)*	Scotland, 1877, by Patrick Manson	nematode from the bite of the *Culex quinquefasciatus* mosquito
flu *see* influenza	Germany, 1894, by Richard Pfeiffer and Robert Koch	
flux or bloody flux *see* dysentery	Japan, 1906, by Kiyoshi Shiga	
German measles *see* rubella	Washington, 1966, by Harry Martin Meyer and Paul D. Parkman	
giardiasis *(Giardia lamblia)*	Holland, 1681, by Anton van Leeuwenhoek	water contaminated by the protozoa-bearing feces of wild animals
glanders *(Burkholderia mallei)*	Germany, 1882, by Friedrich Löffler and Wilhelm Schutz	close contact with animals infected with the bacilli
gonorrhea *(Neisseria gonorrheae)*	Germany, 1879, by Albert Neisser	intimate contact with bacteria-bearing fluids from urethra or vagina
granular conjunctivis *see* trachoma	Malta 1948, by Vincent Trabone	
green monkey disease *see* Marburg virus	Germany, 1967, by Christian Kunz	
grippe *see* influenza	Germany, 1894, by Richard Pfeiffer and Robert Koch	
Guillain-Barré syndrome *(Campylobacter jejuni)*	France, 1916, by Jean Alexandre Barré and Georges Guillain	an auto-immune deficiency triggered by high fever or by swine flu inoculation
guinea worm *see* dracunculiasis	Uzbekistan, 1860, by Alexei Pavlovich Fedchenko	
H1N1 (Russian flu, swine flu)	St. Petersburg, Russia, December 1, 1889	influenza A killed one million victims
H2N2 (Asian flu)	Guizhou, China, February, 1956	a mixed strain of avian and human influenza viruses
H3N2 (Hong Kong flu)	Hong Kong, July 13, 1968	a mild pandemic derived from swine and afflicting 750,000
H5N1 (Fujian flu)	Aksu, China, July 14, 2004	virulent influenza incubated in poultry and spread over the Middle East and Southeast Asia.
hand, foot, and mouth disease	England, 2012	a result of the Picornavirus causing rash, fever, and stiff neck and back
Hansen's disease *see* leprosy	Norway, 1873, by Gerhard Henrik Armauer Hansen	
hantavirus *(Bunyavirus)*	Korea, 1776, by Ho-Wang Lee	breathing virus-bearing rodent feces, urine, or saliva
Heine-Medin disease *see* polio	Sweden, 1890, by Oskar Karl Medin	
hemorrhagic conjunctivitis *(enterovirus)*	Germany, 1897, by Friedrich Loeffler and Paul Frosch	human contact with bearer of the enterovirus 70 from saliva, stool, or blood
hemorrhagic fever *see* Lassa Fever	Korea, 1951, by Langdon Brown	
hemorrhagic jaundice *see* leptospirosis	Germany, 1886, by Adolf Weil	
hepatitis A (HAV) *(Picornavirus)*	Massachusetts, 1979, by George Hitchings and Gertrude Elion	virus-bearing water or food contaminated by fecal matter

Disease	Significant Research	Source
hepatitis B (HBV) *(Hepadnavirus)*	New York, 1980, by Baruch Blumberg	intimate contact with virus-bearing effluvia, injection with contaminated needle, transfusion, or through the placenta to a fetus
hepatitis C (HCV) *(Flavivirus)*	Australia, 1960, by Baruch Blumberg	virus-bearing body fluids; sexual transmission
hepatitis D (HDV)	Italy, 1977, by Mario Rizzetto	intimate contact with humans carrying African deltavirus
hepatitis E (HEV)	Georgia, 2010, by Michael A. Purdy	waterborne or contact with virus-bearing human
HIV (human immunodificiency virus) *(retrovirus) see also* AIDS	Maryland, 1983, by Robert Gallo France, 1983, by Luc Montagnier	intimate contact with virus-bearing blood, semen, or other body fluids
hookworm *(Ancylostoma duodenale* and *Necator americanus)*	Germany, 1852, by Theodor Bilharz	penetration of the skin by a larval parasite, usually through walking on soil contaminated by human feces
hospital fever *see* typhus	Poland, 1939, by Rudolf Weigl	
hot fever *see* relapsing fever	Angola, 1857, by David Livingstone	
infectious jaundice *see* leptospirosis	Germany, 1886, by Adolf Weil	
influenza C *(Orthomyxovirus)*	England, 1997, by A. R. Douglas and A. J. Hay	contact with virus-bearing animal or human or by airborne droplets
intermittent fever *see* malaria	Greece, 410 B.C.E., by Hippocrates	alternating fevers as a symptom of disease
Irish ague or Irish fever *see* typhus	Poland, 1939, by Rudolf Weigl	
itai-itai (cadmium poisoning)	Japan, 1955, by Noboru Hagino	cadmium in the kidneys and bones
itch *see* scabies	Italy, 1687, by Giovanni Cosimo Bonomo	
jail fever *see* typhus	Poland, 1939, by Rudolf Weigl	
jake paralysis or "jake leg"	California, 1931, by Frank G. Crandall	consumption of the toxin tri-ortho-cresyl phosphate
Japanese B encephalitis *(flavivirus)*	Italy, 2000, by Paolo Ravanini	virus from the bite of the *Culex tritaeniorhynchus* mosquito
Japanese river fever *see* tsutsugamushi	Australia, 1942, by S.W. Williams	
kala azar *see* leishmaniasis	Calcutta, 1890, by Charles Donovan	
Kawasaki disease (or kawasakibyo)	California, 2011, by Jane C. Burns	inflammation of vascular system, lymph nodes, and mucous membranes by unknown cause
Korean hemorrhagic fever *(Bunyavirus) see* hantavirus	Korea, 1776, by Ho-Wang Lee	
kuru, a form of bovine spongiform encephalopathy *see also* Creutzfeldt-Jakob virus	Germany, 1920, by Hans Gerhard Creutzfeldt	eating the remains of dead humans during mortuary feasts
kwashiorkor	Jamaica, 1935, by Cicely Williams	edema and enlarged liver from lack of protein
Kyasanur Forest disease *(flavivirus)*	China, 2009, by J. Wang	bite of a virus-bearing tick in humans and small animals
Langat virus	Kuala Lumpur, Malaya, 1956 by C.E. Gordon Smith	a tick-borne encephalitis flavivirus

Disease	Significant Research	Source
Lassa fever *(arenavirus)*	New York, 1970, by John D. Frame	contact with saliva or urine of a virus-bearing rat or by airborne droplets
lead poisoning	Massachusetts, 1943, by Randolph Byers	consumption of lead-bearing paint or from food served in dishes glazed with lead
legionnaire's disease *(Legionella pneumophilis)*	Pennsylvania, 1976, by Stephen B. Thacker	bacteria spread through water and water vapor
leishmaniasis *(Leishmania donovani)*	Calcutta, 1890, by Charles Donovan	bite of the protozoa-bearing female phlebotomine Phlebotomus or *Lutzomyia* sandfly that is hosted by dogs and rodents
leprosy	Norway, 1873, by Gerhard Henrik Armauer Hansen	contact with bacilli from droplets from mouth or nose or to a fetus through the placenta
leptospirosis *(Leptospira interrogans)*	Germany, 1886, by Adolf Weil	contact with fresh water contaminated with the bacteria from the urine or tissues of infected rats
listeria *(Listeriosis monocytogenes)*	Switzerland, 1961, by H.P.R. seeliger	consumption of processed meat, soft cheese, unpasteurized milk, or other bacteria-bearing foods, especially those that are chilled after cooking
lockjaw *see* tetanus	Italy, 1884, by Antonio Carle and Giorgio Rattone	
Lyme disease *(Borrelia burgdorferi)*	Montana, 1998, by Wilhelm Burgdorfer	bite of bacteria-bearing *Ixodes* tick from a deer or rodent
Machupo virus *(arenavirus)*	Bolivia, 1959, by Karl Johnson	contact with the excreta of virus-bearing vesper mouse
mad cow disease *see* Creutzfeldt-Jakob virus	Germany, 1920, by Hans Gerhard Creutzfeldt	
malaria *(Malaria plasmodium)*	England, 1897, by Sir Ronald Ross	bite of protozoa-bearing *Anopheles* mosquito
Malta fever *(Micrococcus Melitensis) see* brucellosis	Malta, 1884, by David Bruce	
marasmus	France, 1993, by Jean-Gerard Pelletier	edema caused by a lack of protein and calories
Marburg virus *(filovirus)*	Germany, 1967, by Christian Kunz	handling of the vervet (or green) monkey or contact with virus-infected effluvia or human tissue
marsh fever *see* malaria	England, 1897, by Ronald Ross	
measles *(Paramyxovirus)*	Chicago, 1964, by Maurice Hilleman	contact with virus-bearing droplets of human saliva from talking, coughing, sneezing
Mediterranean fever *see* brucellosis	Malta, 1884, by David Bruce	
melioidosis *(Burkholderia pseudomallei)*	Burma, 1912 by Alfred Whitmore and Krishnaswami	infection of cuts and abrasions with bacteria in soil and water
meningitis *(Neisseria meningitidis)*	Scotland, 1768, by Robert Whytt	contact with diseased and bacteria-bearing human
mercury poisoning *see* Minamata disease	Japan, 1956, by T. Takeuchi and Hasuo Ito	
MERS (Middle Eastern Respiratory Syndrome)	Saudi Arabia, 2012, by Ali Mohamed Zaki	betacoronavirus spread from camels to humans and complicated by pneumonia

Disease	Significant Research	Source
microsporidiosis (*microsporidia*)	Louisiana, 2006, by Elizabeth S. Didier	protozoa-forming spores that attack corneas of the eye, lungs, brain, and intestines
miliary fever *see* sudor Anglicus	England, 1551, John Caius	
Minamata disease (mercury poisoning)	Japan, 1956, by T. Takeuchi and Hasuo Ito	neurological crippling by consumption of mercury in fish and shellfish or passed through the placenta to the fetus
Molbay disease	Quebec, 1773, by James Bowman	unidentified lung infection spread by human contact
monkey fever *see* Kyasanaur Forest disease	China, 2009, by J. Wang	
monkeypox (*orthopoxvirus*)	Texas, 2003, by Ted Rosen and Jennifer Jablon	the bite or contact with the blood of a virus-bearing animal
mononucleosis	Maryland, 1920, by Thomas P. Sprunt and Frank A. Evans	contact with saliva bearing the Epstein-Barr virus
mumps (infectious parotitis)	Chicago, 1963, by Maurice Hilleman	contact with saliva of virus-bearing human or airborne droplets
murine typhus (*Rickettsia typhi*)	Crimea, 1854, by Joseph Désiré Tholozan	a form of typhus born by rodent fleas
Murray Valley encephalitis (*flavovirus*)	Australia, 2014, by Linda A. Selvey and Lynne Dailey	bite of a virus-bearing *Culex annulirostris* mosquito
myelo-optico-neuropathy chinoform	Japan, 2010, by Akihiro Igata	damage of the optic nerve from consumption of the drug
nipah virus (*paramyxovirus*)	China, 2008, by Y. Li	contact with virus-bearing pigs
nontuberculous mycobacteria (NTM)	Washington, 1959, by Ernest H. Runyon	possibly from pathogens found in soil and water
Omsk hemorrhagic fever	Russia, 2010, by Daniel Ruzek	a virus spread by ticks on muskrats and fruit bats to eyes and skin
onchocerciasis (*Wolbachia*)	Congo, 1932, by Jean Hissette	microbes in filiariae invade the body, causing blindness, or infect by the bite of the nematode-bearing *Simulium* blackfly
o'nyong-nyong fever (*togovirus*)	Georgia, 2005, by Drew L. Posey	bite of virus-bearing *Anopheles* mosquito
ophthalmia *see* trachoma	Malta 1948, by Vincent Trabone	
Oropouche virus (*Bunyavirus*)	England, 2013, by Osman Dar	bite of the virus-bearing sandfly or mosquito
Oroya fever *see* Bartonellosis bacilliformis	England, 1838, by William Graham	
pappataci fever (*Bunyavirus*)	Afghanistan, 1978, by I.L. Yavorsky	bite of the virus-bearing *Phlebotomus papatasii* sandfly
paratyphoid (*Salmonella paratyphi*)	Germany, 1880, by Karl Joseph Eberth	food or water contaminated by feces of bacteria-bearing humans
parrot fever *see* psittacosis	France, 1930, by Edmond Nocard	
pellagra	Spain, 1735, by Gaspar Casal	insufficient niacin or vitamin B in the diet
pertussis (*Bordetella pertussis*)	Belgium, 1906, by Jules Bordet and Octave Gengou	contact with diseased human or by airborne droplets
Peruvian wart *see* Bartonellosis *bacilliformis*	England, 1838, by William Graham	

Disease	Significant Research	Source
petechial fever *see* typhus	Poland, 1939, by Rudolf Weigl	
phagedaena	England, 1795, by John Hunter Swediaur	ulceration resulting from spreading necrosis caused by anaerobic pathogens
phthisis *see* tuberculosis	New Jersey, 1943, by Selman A. Waksman	
pink disease *see* acrodynia	Germany, 1923, by Alfred Stock	
plague *(Yersinia pestis)*	New Jersey, 1943, by Albert Schatz	bite of bacteria-bearing flea from wild rat
pleurodynia *see* Bornholm disease	Norway, 1872, by Anders Daae	
pneumoconiosis	Germany, 1867, by Friedrich Albert Zenker	exposure to coal dust
pneumonia *(Streptococcus pneumoniae)*	France, 1881, by Louis Pasteur	contact with bacteria-bearing human or by airborne droplets
polio *(poliomyelitis)*	London, 1784, by Michael Underwood	contact with virus-bearing water or food or by air-borne droplets
polyneuritis endemic perniciosa *see* beriberi; jake paralysis	Indonesia, 1897, by Christiaan Eijkman	
Pontiac fever *see* Legionnaire's disease	Pennsylvania, 1976, by Stephen B. Thacker	
psilosis *see* sprue	Barbados, 1759, by William Hillary	
psittacosis *(Chlamydia psittaci)*	France, 1930, by Edmond Nocard	contact with bacteria-bearing bird or bird feces
puerperal fever *(Streptococcus pyogenes)*	Hungary, 1847, by Ignaz Semmelweis	contact with streptococcal-contaminated hands during labor and delivery
putrid fever *see* diphtheria	Germany, 1901, by Emil von Behring	
Puumala virus *(nephropathia epidemica) see also* hantavirus	Finland, 1996 by Olli Vapalahti	inhalation of urine or feces of the bank vole
pyaemia (or pyemia)	Persia, 11th century, by Avicenna	a general term for septicemia or toxemia
Q fever *(Coxiella burnetii)*	Australia, 1935, by Herald Rea Cox and Macfarlane Burnet	inhalation of the dried feces of an infected tick in barnyard or slaughter-house
rabbit fever *see* tularemia	Japan, 1925, by Hachiro Ohara	
rabies *(rhabdovirus)*	France, 1882, by Louis Pasteur	bite of virus-bearing animal
relapsing fever *(Borrealia recurrentis)*	Angola, 1857, by David Livingstone	spread by the spirochete-bearing *Ornithodoros* tick
remitting fever *see* malaria	England, 1897, by Ronald Ross	
rheumatic fever *(Streptococcus pharyngitis)*	Massachusetts, 1944, by T. Duckett Jones	contact with humans carrying the bacteria; ingestion of contaminated food
rickets	England, 1921, by Edward Mellanby	insufficient vitamin D in the diet
rickettsial pox *(Rickettisa akari)*	New York, 1946, by Robert J. Huebner and Charles Armstrong	bite of the bacteria-bearing *Liponyssiodes sanguineus* mouse mite
Rift Valley fever *(Bunyaviridae)*	England, 2013, by Osman Dar	bite of a virus-bearing sandfly or gnat
ringworm *(Tinea capitis)*	England, 1901, by Allan MacFadlyen	caused by fungal spores spread by shared towels, walking barefoot,

Disease	Significant Research	Source
		touching infected pets, or close contact sports, such as wrestling
river blindness *see* onchocerciasis	Congo, 1932, by Jean Hissette	
Rocky Mountain spotted fever *(Rickettsia rickettsii)*	Ohio, 1900, by Howard Taylor	bite of the bacteria-bearing *Dermacentor* or other Ixodid tick from a mouse or rabbit
Ross River fever or epidemic polyarthritis	Australia, 1943, by I.D. Marshall and J.A.R. Miles	bite of a virus-bearing mosquito
rotavirus	Australia, 1973, by Ruth Bishop	contact with virus-bearing human feces from contaminated food or water or person-to-person
rubella (German measles)	Washington, 1966, by Harry Martin Meyer and Paul D. Parkman	contact with virus-bearing human or to a fetus through the placenta
rubeola *see* measles	New York 1856, by Henry Koplik	
sacroptic itch *see* scabies	Italy, 1687, by Giovanni Cosimo Bonomo	
St. Anthony's fire *see* erysipelas	Germany, 1883, by Friedrich Fehleisen	
Salmonella enteritidis	Germany, 1879, by Karl Joseph Eberth	from poultry, especially during handling of bacteria-bearing chicken eggs, or ground meat
sandfly fever *see* pappataci fever	Afghanistan, 1978, by I.L. Yavorsky	
SARS (Severe Acute Respitory Syndrome)	Italy, 2003, by Carlo Urbani	a *coronavirus* inflammation of the lungs caused by contact with infected human or body fluids
Savill's disease	Scotland, 1986, by Margaret Anne Crowther	a skin inflammation of an unknown source
scabies	Italy, 1687, by Giovanni Cosimo Bonomo	caused by the *Sarcoptes scabiei* mite, contracted from infested bedding or intimate contact with victims
scarlet fever (or scarlatina) *(Streptococcus aureus)*	Chicago, 1924, by Gladys Henry Dick and FrederickDick	contact with strep-bearing human or body effluvia
schistosomiasis *(Schistosoma mansoni, S. haematobium, and S. japonicum)*	Germany, 1851, by Theodor Bilharz	contact with a trematode-bearing snail
scrofula *see also* tuberculosis	Ireland, 18th century, by Elizabeth Pearson	tubercular infection of glands in the neck and throat
scrub typhus *see* tsutsugamushi	Australia, 1942, by S.W. Williams	
scurvy	Scotland, 1747, by James Lind	lack of vitamin C
septicemia *see* puerperal fever	Persia, 11th century, by Avicenna	decay of blood
shank or shinbone fever *see* trench fever	England, 1838, by William Graham	
Shigella sonnei	Denmark, 1908, by Carl Olaf Sonne	contact with bacteria-bearing manure water, or sewage or with infected human
shigellosis *(shigella flexneri* and *shigella sonnei)*	Japan, 1875, by Kiyoshi Shiga	food or water contaminated by bacillus-bearing human feces

Disease	Significant Research	Source
ship fever *see* yellow fever	Cuba, 1881, by Carlos Juan Finlay	
silicosis	Texas, 1970, by Steven Weisenfeld	exposure to silicon fiber, which creates nodules in the upper lung
sleeping sickness (*Trypanosoma brucei rhodesiense* or East African variety)	Germany, 1907, by Paul Ehrlich	bite of the protozoa-bearing *Glossina* tsetse fly
sleeping sickness (*Trypanosoma brucei gambiense* or West African variety)	Gambia, 1902, by Joseph Everett Dutton	bite of the protozoa-bearing *Glossina* tsetse fly or via transfusion with contaminated blood or from organ transplant or birth from a disease-bearing mother
smallpox (*Variola major*) *see also* alastrim	England, 1796, by Edward Jenner	contact with virus-bearing lesions or droplets breathed into the air or to a fetus through the placenta
SMON disease *see* myelo-opticoneuropathy	Japan, 2010, by Akihiro Igata	
spotted fever *see* typhus	Poland, 1939, by Rudolf Weigl	
sprue (also celiac disease or psilosis)	Barbados, 1759, by William Hillary	may be caused by a virus that interferes with normal digestion
Stittgart disease *see* leptospirosis	Germany, 1886, by Adolf Weil	
strep throat (*Streptococcus A*)	Germany, 1939, by Gerhard Domagk	contact with humans carrying the bacteria; ingestion of contaminated food
sudor Anglicus or English sweating disease	England, 1551, John Caius	unknown
syphilis (*Treponema pallidum*)	England, 1928, Alexander Fleming	intimate contact with bacteria-bearing blood, semen, or fluids from a chancre or through the placenta to a fetus
syphilitic septicemia *see* plague	New Jersey, 1943, by Albert Schatz	
tarantism (choreomania)	Italy, 1959, by Ernesto de Martino	epidemic hysteria causing participants to dance and twitch frenetically
tertian fever *see* malaria	England, 1897, by Ronald Ross	
tetanus (*Clostridium tetani*)	Italy, 1884, by Antonio Carle and Giorgio Rattone	contact with animal feces containing bacterial spores, which enter the skin through cuts and wounds
toxic shock syndrome (*Staphylococcus aureus*)	Denver, 1978, by James K. Todd	blood poisoning from the bacteria colonized in mucous linings
trachoma (*Chlamydia trachomatis*)	Malta 1948, by Vincent Tabone	eye granulation spread by poor sanitation in crowded areas
trench fever (*Bartonella quintana*)	England, 1838, by William Graham	bite of bacillus-bearing lice
trichomoniasis (*Trichomonas vaginalis*)	France, 1836, by Alfred Francois Donné	intimate contact with protozoa-bearing fluids from prostate, urethra, or vagina
tsutsugamushi (*Orientia tsutsugamushi*)	Australia, 1942, by S.W. Williams	bite of bacteria-bearing *Leptotrombidium* larval red mite on field mice
tuberculosis (*Mycobacterium tuberculosis*)	New Jersey, 1943, by Selman A. Waksman	contact with contaminated milk or with saliva from bacteria-bearing human or by airborne droplets

Disease	Significant Research	Source
tularemia (*Franciscella tularensis*)	Japan, 1925, by Hachiro Ohara	contact with bacteria-bearing rabbit or the bite of the bacteria-bearing *Chrysops* tick or deer fly
typhoid fever (*Salmonella typhi*)	England, 1838, by William Budd	food or water contaminated by feces of bacteria-bearing human
typhus (*Rickettsia prowazekii*) *see also* murine typhus	Poland, 1939, by Rudolf Weigl	feces of bacteria-bearing *Pediculus humanus* louse or contaminations by louse feces
uncinariasis *see* hookworm	Germany, 1852, by Theodor Bilharz	
undulant fever *see* brucellosis	Malta, 1884, by David Bruce	
Venezuelan equine encephalomyelitis (*Togavirus*)	Venezuela, 1975, by Jose Esparza and A. Sánchez	contact with virus-bearing *Culex tarsalis* mosquito
Weil's disease *see* leptospirosis	Germany, 1886, by Adolf Weil	
West Nile fever (*Flavovirus*)	France, 1998, by Vincent Deubel	contact with virus-bearing birds
whooping cough *see* pertussis	Paris, 1578, by Guillaume de Baillou	
xerophthalmia	Maryland, 1979, by Charles Andrew Sommer	lack of vitamin A in the diet causing night blindness
yaws (*Treponema pertenue*)	England, 1679, Ceylon, 1905 by Aldo Castellani	contact with skin or clothing bearing the spirochete or by flies contaminating abraded skin
yellow fever (*Flavivirus*)	Cuba, 1881, by Carlos Juan Finlay	bite of a virus-bearing *Aedes aegypti* mosquito or of an infected ape, monkey, opossum, or other jungle animal
Zika virus (*Flavivirus*)	London, 1952, by George Dick	a virus born by the *Aedes stegomyia* mosquito

Appendix B
Historic Writings on Disease

The following list of publications on the history of disease, arranged alphabetically by title, derives from the text and is not intended as an all-inclusive overview of writings on the subject. Appendix C presents the same information in chronological order by date of publication. Appendix D reprises the data alphabetically by author.

An Account of the Contagious Epidemic Yellow Fever, Which Prevailed in Philadelphia in the Summer and Autumn of 1797 (1798), Felix Pascalis-Ouvière

Account of the Plague (1357), Michael Platiensis

"An Account or History of the Procuring of the Small Pox by Incision of Inoculation" (1714), Emanuel Timoni

Acute Ascending Myelitis Following a Monkey Bite (1934), Albert Bruce Sabin

Die Aetiologie der Tuberculose [On the Nature of Tuberculosis] (1882), Robert Koch

The Air and Ventilation of Subways (1908), George A. Soper

Akute Poliomyelitis (1911), Ivar Wickman

The American Red Cross Textbook on Home Hygiene and Care of the Sick (1918), Jane Arminda Delano

And the Band Played On: People, Politics and the AIDS Epidemic (1987), Randy Shilts

The Angel of Bethesda (1722), Cotton Mather

Angina Maligna (1749), Martino Ghisi

Arte Chirurgia Copiosa [On the Complete Surgical Art] (1514), Giovanni de Vigo

The Backwash of War: The Human Wreckage of the Battlefield as Witnessed by an American Nurse (1916), Ellen N. La Motte

Biology of Plagues (2001), Susan Scott and Christopher Duncan

A Boke or Counseill Against the Disease Commonly Called the Sweate or Sweatyng Sicknesse (1552), John Caius

Brief Description of the Pest (1568), Gilbert Skeyne

A Brief History of Epidemic and Pestilential Diseases (1799), Noah Webster

A Brief Rule to Guide the Common-People of New England How to Order Themselves & Theirs in the Small-Pocks or Measels (1678), Thomas Thacher

Capitulum Singulaire [A Singular Chapter] (1350), Guy de Chauliac

The Cause and Prevention of Beri-beri (1907), W. Leonard Braddon

A Chronological History of the Weather and Seasons on the Prevailing Diseases in Dublin (1770), John Rutty

Clinique Medicale [Medical Clinic] (1833), Gabriel Andral

Collections of Oribasius Sardianus (ca. 365 C.E.), Rufus of Ephesus

The Coming of the Spirit of Pestilence: Introduced Infectious Diseases and Population Decline Among Northwest Coast Indians, 1774–1874 (1999), Robert Boyd

Commentarii de Morborum Historia et Curatione [Commentaries on the History and Cure of Diseases] (1802), William Heberden

The Compendium (ca. 925 C.E.), Rhazes of Baghdad

Concilium contra Pestilenciam [Advice on the Plague] (1348), Gentile da Foligno

Cura Mali Francici [Cure of the French Pox] (1496), Theodericus Ulsenius

The Cure and Prevention of Scarlet Fever (1801), Samuel Christian Hahnemann

Darkness in El Dorado (2000), Patrick Tierney

De Alimentorum Facultatibus [Concerning the Workings of Digestion] (ca. 180 C.E.), Galen

De Contagione et Contagiosis Morbis [On Contagion and Contagious Diseases] (1546), Girolamo Fracastorius

De Febre in Inflammatoria [On Fever in Inflammations] (1810), Moritz Wilhelm Müller

De Febris Epidemicae, et Novae Quae Latine Puncticularis [On Epidemic Fevers and New Ones Particularized in Latin] (1574), Alonso de Torres

De l'Auscultation Médiate, ou Traité du Diagnostic des

A Jail Fever (1767), Stephen Theodore Jansen

Khitab ash-Shifa [Healing of the Soul] (early 1000s), Avicenna of Kazakhstan

Korori Chijun [Treatise on Cholera] (1858), Ogata Koan

Kugai Jodo: Waga Minamatabyo [Paradise in the Sea of Our Sorrow] (1972), Ishimure Michiko

Lazaretti della Città, Li [City Pesthouses] (1658), Antero Maria Micone da San Bonaventura

Leaves from a Nurse's Life's History, 1906 (1906), Jean S. Edmunds

Libellus de Morbo Gallico [A Pamphlet on the French Disease] (1497), Niccolo Leonicen

Lord Have Mercy Upon Us (1636), Thomas Brewer

Lösch Wien [Vienna Blotted Out] (1680), Abraham a Sancta Clara

Ma'mar Be-qadahot Divriot U-minei Qadahot [A Treatise on Pestilential Fevers and Other Kinds of Fevers] (1326), Abraham ben David Caslari

Medical Histories and Reflections (1792), John Ferriar

Medical History of the Niger Expedition (1843), James Ormiston McWilliam

Medical Works (1498), Francisco López de Villalobos

Meichuang Mi Lu [Secret Records on Syphilis] (1632), Chen Sicheng

Memoir on the Cholera at Oxford (1856), Henry Wentworth Acland

Memoir on the Yellow Fever (ca. 1805), José María Mociño

Mémoires sur les Hôpitaux de Paris [Memoirs of Paris Hospitals] (1788), Jacques-René Tenon

Memorandum on Cholera (1866), Joseph-Charles Taché

Memoria in Occasione del Cholèra-Morbus nell'Anno 1855 [Recalling the Cholera Epidemic of the Year 1855] (1855), Antonio Fachinetti

Menschliche Auslese und Rassenhygiene [The Principles of Human Heredity and Race Hygiene] (1931), Eugen Fischer

Merk's, Wien! [Pay Attention, Vienna] (1680), Abraham a Sancta Clara

Miscellaneous Ideas in Medicine (1643), Yu T'ien-Chih

Moral and Physical Condition of the Working Classes Employed in the Cotton Manufacture in Manchester (1832), James Phillips Kay-Shuttleworth

Narrative of the Proceedings of the Black People, During the Late Awful Calamity in Philadelphia (1793), Richard Allen and Absalom Jones

Natural and Political Observations Made Upon the Bills of Mortality (1662), John Graunt

Natural Resistance to Infectious Disease and Its Reinforcement (1910), Simon Flexner

Nei Ching (2700 B.C.E.), Huang Ti

Observationes Medicae [Clinical Observations] (1652), Nicholas Tulp

Observationes Medicae [Clinical Observations] (1676), Thomas Sydenham

Observations Made During the Epidemic of Measles on the Faeroe Islands in the Year 1846 (1846), Peter Ludwig Panum

Observations on Diseases of the Army in Camp and Garrison (1752), John Pringle

Observations on the Changes of the Air and the Concomitant Epidemical Diseases in the Island of Barbados (1759), William Hillary

Observations on the Climate and Diseases of New York (1738), Cadwallader D. Colden

Observations on the Diseases Incident to Seamen (1785), Gilbert Blane

Observations on the Scarlet Fever (1808), Samuel Christian Hahnemann

On Cholera (ca. 11 C.E.), Areteus

On the Mode of Communication of Cholera (1855), John Snow

On the Most Effectual Means of Preserving the Health of Seamen (1757), James Lind

On the Typhus Fever, Which Occurred at Philadelphia in the Spring and Summer of 1836 (1837), William Wood Gerhard

On Ulcerations About the Tonsils (ca. 110 C.E.), Areteus

Opera Medica [Medical Works] (1715), Thomas Sydenham

Opera Medicinalia, in Quibus Quam Plurima Extant Scitu Medico Necessaria [A Medical Encyclopedia Containing Everything Known to Science] (1570), Francisco Bravo

The Organization of Permanent Nation-Wide Anti-Aedes Aegypti Measures in Brazil (1943), Fred Lowe Soper

Pathology and Treatment of Diseases of the Skin (1895), Moriz Kaposi

The Person with AIDS: Nursing Perspectives (1983), Jeanne Parker Martin

Plague: A Manual for Medical and Public Health Workers (1936), Wu Lien-teh

Plague Fighter: The Autobiography of a Modern Chinese Physician (1959), Wu Lien-teh

Pneumopathies Infectieuses Communautaires chez la Personne Âgée [Infectious Lung Disease in the Aged] (1858), Nicolás Manzini

Poliomyelitis Vaccine in the Fall of 1955 (1956), Jonas Salk

Pox Americana: The Great Smallpox Epidemic of 1775–82 (2001), Elizabeth A. Fenn

La Pratique des Accouchements [The Practice of Obstetrics] (1694), Philippe Peu

Praxis Medica [The Practice of Medicine] (1707), Thomas Sydenham

Précis d'Anatomie Pathologique [Summary of Pathological Anatomy] (1829), Gabriel Andral

The Prevention of Pellagra (1915), Joseph Goldberger

La Propagation de la Peste [The Spread of Plague] (1898), Paul Louis Simond

A Prospect of Exterminating the Smallpox (1800), Benjamin Waterhouse

Public Health Nursing (1916), Mary Sewall Gardner

Pyretologia: A Rational Account of the Cause and Cures of Agues (1672), Robert Talbor

Recherches sur le Fievre Jaune [Research on Yellow Fever] (1820), Jean Andre Rochoux

Appendix C
Timeline of Writings on Disease

Antiquity

Nei Ching (2700 B.C.E.), Huang Ti
Sushruta Samhita (600 B.C.E.), Sushruta
Des Airs, des Eaux, et des Lieux [On Air, Water, and Occasions] (ca. 310 B.C.E.), Hippocrates
The Epidemics (ca. 310 B.C.E.), Hippocrates
On Cholera (ca. 11 C.E.), Areteus
On Ulcerations About the Tonsils (ca. 110 C.E.), Areteus
De Alimentorum Facultatibus [Concerning the Workings of Digestion] (ca. 180 C.E.), Galen
De Mortalitate [On Mortality] (252 C.E.), Cyprian
Collections of Oribasius Sardianus (ca. 365 C.E.), Rufus of Ephesus

Middle Ages

De Variolis et Morbillis Commentarius [On Smallpox and Measles] (910 C.E.), Rhazes of Baghdad
The Compendium (ca. 925 C.E.), Rhazes of Baghdad
Ishinho (982 C.E.), Yasuhori Tanbo
Khitab ash-Shifa [Healing of the Soul] (early 1000s), Avicenna of Kazakhstan
Ma'mar Be-qadahot Divriot U-minei Qadahot [A Treatise on Pestilential Fevers and Other Kinds of Fevers] (1326), Abraham ben David Caslari
Concilium contra Pestilenciam [Advice on the Plague] (1348), Gentile da Foligno
Capitulum Singulaire [A Singular Chapter] (1350), Guy de Chauliac
Account of the Plague (1357), Michael Platiensis

Renaissance

Vaticinium in Epidemicam Scabiem [Revelation on Epidemic Scabies] (1496), Theodericus Ulsenius
Cura Mali Francici [Cure of the French Pox] (1496), Theodericus Ulsenius
Libellus de Morbo Gallico [A Pamphlet on the French Disease] (1497), Niccolo Leonicen
Medical Works (1498), Francisco López de Villalobos

De Morbo Gallico [On the French Disease] (1502), Juan Almenor
Arte Chirurgia Copiosa [On the Complete Surgical Art] (1514), Giovanni de Vigo
De Morbo Gallico [On the French Disease] (1519), Ulrich von Hutten
Treatise on the Serpentine Malady (ca. 1520), Rodrigo Ruy Díaz de Isla
Syphilis sive Morbus Gallicus [Syphilis or French Disease] (1530), Girolamo Fracastorius
De Contagione et Contagiosis Morbis [On Contagion and Contagious Diseases] (1546), Girolamo Fracastorius
A Boke or Counseill Against the Disease Commonly Called the Sweate or Sweatyng Sicknesse (1552), John Caius
A Dialogue Against the Fever Pestilence (1564), William Bullein
Informacion y Curacion dela Peste [Information and Treatment of Plague] (1565), Juan Tomás Porcell
Brief Description of the Pest (1568), Gilbert Skeyne
Opera Medicinalia, in Quibus Quam Plurima Extant Scitu Medico Necessaria [A Medical Encyclopedia Containing Everything Known to Science] (1570), Francisco Bravo
De Febris Epidemicae, et Novae Quae Latine Puncticularis [On Epidemic Fevers and New Ones Particularized in Latin] (1574), Alonso de Torres
Handbook of the Healing Art (1578), Bengt Olsson
Summa y Recopilación de Chirugia [Treatise and Recompilation of Surgery] (1578), Alfonso Lopez de Hinojosos
De Officio Fidelis [On the Duty of the Faithful] (1582), Johann von Ewich
Discourses on the Plague (1595), Simon Forman

Seventeenth Century

Tongui Pogam (1610), Ho Jun
Disputationes Medicae [Medical Controversies] (1611), Alfonso Ruizibus de Fontecha
Discourse of the Whole Art of Chyrurgie (1612), Peter Lowe

The Surgeon's Mate, or Military & Domestique Surgery (1617), John Woodall

A Short Instruction (1619), Caspar Bartholin

De Medicina Indorum [On Medicine in the Indies] (1629), Jacobus Bontius

Meichuang Mi Lu [Secret Records on Syphilis] (1632), Chen Sicheng

Lord Have Mercy Upon Us (1636), Thomas Brewer

Epidemiorum et Ephemeridium [Epidemics and Journals] (1640), Guillaume de Baillou

Wenyilun [On Pestilence] (1642), Wu Youxing

Miscellaneous Ideas in Medicine (1643), Yu T'ien-Chih

Observationes Medicae [Clinical Observations] (1652), Nicholas Tulp

Ephemerides Morborum [Diaries of Diseases] (ca. 1655), Theodore Turquet de Mayerne

Lazaretti della Città, Li [City Pesthouses] (1658), Antero Maria Micone da San Bonaventura

Diatribae Duae: De Febribus [Two Texts: On Fevers] (1659), Thomas Willis North Hinksey

Natural and Political Observations Made Upon the Bills of Mortality (1662), John Graunt

Historical Account of the Plague in London in 1665 (1665), Nathaniel Hodges

Pyretologia: A Rational Account of the Cause and Cures of Agues (1672), Robert Talbor

Observationes Medicae [Clinical Observations] (1676), Thomas Sydenham

A Brief Rule to Guide the Common-People of New England How to Order Themselves & Theirs in the Small-Pocks or Measles (1678), Thomas Thacher

De Peste Tractatus [Treatise on Pestilence] (1678), Paul Barbette

Lösch Wien [Vienna Blotted Out] (1680), Abraham a Sancta Clara

Merk's, Wien! [Pay Attention, Vienna] (1680), Abraham a Sancta Clara

Die Grosse Totenbruderschaft [The Great Brotherhood of the Dead] (1681), Abraham a Sancta Clara

The English Remedy, or Talbor's Wonderful Secret for Curing of Agues and Feavers (1682), Robert Talbor

Tractatus de Avertenda et Profliganda Peste [Treatise on Averting and Avoiding Plague] (1684), Hieronymi Gastaldi

La Pratique des Accouchements [The Practice of Obstetrics] (1694), Philippe Peu

Eighteenth Century

De Morbis Artificum Diatriba [Diseases of Workers] (1700), Bernardino Ramazzini

Praxis Medica [The Practice of Medicine] (1707), Thomas Sydenham

Dissertatio de Nativis [Dissertation on Natives] (1711), Giovanni Maria Lancisi

"An Account or History of the Procuring of the Small Pox by Incision of Inoculation" (1714), Emanuel Timoni

Opera Medica [Medical Works] (1715), Thomas Sydenham

A Historical Relation of the Plague at Marseilles in the Year 1720 (1720), Jean Baptiste Ber-trand

The Angel of Bethesda (1722), Cotton Mather

An Historical Account of the Small-Pox Inoculated in New England (1725), Zabdiel Boylston

Description of the Cape of Good Hope; with Matters Concerning It (1726), François Valentyn

An Essay on Inoculation (1738), James Kilpatrick

Observations on the Climate and Diseases of New York (1738), Cadwallader D. Colden

Domestic Medicine; or, the Family Physician (1749), William Buchan

Angina Maligna (1749), Martino Ghisi

Observations on Diseases of the Army in Camp and Garrison (1752), John Pringle

A Treatise on Scurvy: In Three Parts. Containing an Inquiry into the Nature, Causes, and Cure, of the Disease (1753), James Lind

On the Most Effectual Means of Preserving the Health of Seamen (1757), James Lind

An Essay on Fevers (1757), John Huxham

Observations on the Changes of the Air and the Concomitant Epidemical Diseases in the Island of Barbados (1759), William Hillary

Historia Natural y Medica de el Principado de Asturias [Biology and Medical History of Asturias Province] (1762), Gaspar Casal y Julian

Reflexions on Variolation (1764), Angelo Gatti

A Jail Fever (1767), Stephen Theodore Jansen

An Essay on Diseases Incidental to Europeans in Hot Climates (1768), James Lind

An Enquiry into the Nature, Cause, and Cure, of the Angina Suffocativa, or, Sore Throat Distemper (1769), Samuel Bard

A Chronological History of the Weather and Seasons on the Prevailing Diseases in Dublin (1770), John Rutty

Utility of Vaccination (1771), Jonathan Potts

Description of the Plague Which Attacked Kherson (1784), Danilo S. Samoilovich

Direction pour la Guérison du Mal de la Baie Saint-Paul [Methods for Treating Baie Saint-Paul Disease] (1785), James Bowman

Observations on the Diseases Incident to Seamen (1785), Gilbert Blane

Remarks on the Distemper Generally Known by the Name of the Molbay Disease (1786), Robert Jones

Mémoires sur les Hôpitaux de Paris [Memoirs of Paris Hospitals] (1788), Jacques-René Tenon

Treatise on Diseases of Children (1789), Michael Underwood

Medical Histories and Reflections (1792), John Ferriar

Narrative of the Proceedings of the Black People, During the Late Awful Calamity in Philadelphia (1793), Richard Allen and Absalom Jones

A Short Account of the Malignant Fever Lately Prevalent in Philadelphia (1793), Matthew Carey

A View of the Diseases of the Army in Great Britain,

America, the West Indies, and on Board of King's Ships and Transports (1793), Thomas Dickson Reide

Treatise on the Epidemic Puerperal Fever of Aberdeen (1795), Alexander Gordon

The Inoculator (1796), Daniel Sutton

An Account of the Contagious Epidemic Yellow Fever, Which Prevailed in Philadelphia in the Summer and Autumn of 1797 (1798), Felix Pascalis-Ouvière

An Inquiry Into the Causes and Effects of the Variolae Vaccinae (1798), Edward Jenner

A Brief History of Epidemic and Pestilential Diseases (1799), Noah Webster

Nineteenth Century

A Prospect of Exterminating the Smallpox (1800), Benjamin Waterhouse

The Cure and Prevention of Scarlet Fever (1801), Samuel Christian Hahnemann

Commentarii de Morborum Historia et Curatione [Commentaries on the History and Cure of Diseases] (1802), William Heberden

Istituzioni di Chirurgiche [The Profession of Surgery] (1802), Giovanni Battista Monteggia

Disertación de la Fiebre Epidémica, Que Padecio Cádiz [Treatise on the Epidemic Fever That Assailed Cadiz] (1804), José María Mociño

Memoir on the Yellow Fever (ca. 1805), José María Mociño

Observations on the Scarlet Fever (1808), Samuel Christian Hahnemann

Trattato di Vaccinazione [Treatise on Vaccination] (1809), Luigi Sacco

De Febre in Inflammatoria [On Fever in Inflammations] (1810), Moritz Wilhelm Müller

De l'Auscultation Médiate, ou Traité du Diagnostic des Maladies du Coeur [Mediate Auscultation or a Treatise on the Diagnosis of Heart Disease] (1818), René Théophile Hyacinthe Laënnec

Recherches sur le Fievre Jaune [Research on Yellow Fever] (1820), Jean Andre Rochoux

An Exposition of the Natural System of the Nerves of the Human Body (1824), Charles Bell

Treatise of Pathology and Therapeutics (1828), John Esten Cooke

Précis d'Anatomie Pathologique [Summary of Pathological Anatomy] (1829), Gabriel Andral

Moral and Physical Condition of the Working Classes Employed in the Cotton Manufacture in Manchester (1832), James Phillips Kay-Shuttleworth

Clinique Medicale [Medical Clinic] (1833), Gabriel Andral

On the Typhus Fever, Which Occurred at Philadelphia in the Spring and Summer of 1836 (1837), William Wood Gerhard

Report on the Sanitary Condition of the Labouring Population of Great Britain (1839), Edwin Chadwick

Medical History of the Niger Expedition (1843), James Ormiston McWilliam

Observations Made During the Epidemic of Measles on the Faeroe Islands in the Year 1846 (1846), Peter Ludwig Panum

Reports on the Typhus Epidemics of Upper Silesia (1848), Rudolf Ludwig Karl Virchow

A History of the Ravages of the Yellow Fever in Norfolk, Virginia, A.D. 1855 (1855), George D. Armstrong

Memoria in Occasione del Cholèra-Morbus nell'Anno 1855 [Recalling the Cholera Epidemic of the Year 1855] (1855), Antonio Fachinetti

On the Mode of Communication of Cholera (1855), John Snow

Memoir on the Cholera at Oxford (1856), Henry Wentworth Acland

Typhoid Fever: On Its Nature, Mode of Spreading, and Infection (1856), William Budd

The Wonderful Adventures of Mrs. Seacole in Many Lands (1857), Mary Jane Seacole

Korori Chijun [Treatise on Cholera] (1858), Ogata Koan

Pneumopathies Infectieuses Communautaires chez la Personne Agée [Infectious Lung Disease in the Aged] (1858), by Nicolás Manzini

Étude sur l'Hygiène et la Topographie Médicale de la Ville de Saint-Etienne [Study on Hygiene and Medical Topography of the City of Saint-Etienne] (1862), Philippe Béroud

Memorandum on Cholera (1866), Joseph-Charles Taché

Études sur la Phthisie [Studies in Tuberculosis] (1868), Jean Antoine Villemin

Report on Epidemic Cholera and Yellow Fever in the Army of the United States, During the Year 1867 (1868), Joseph Janvier Woodward

The Ship Captain's Medical Guide (1868), Harry Leach

The Yellow Fever Epidemic of 1878 in Memphis, Tennessee (1879), John McLeod Keating

Investigations into the Etiology of Traumatic Infective Diseases (1880), Robert Koch

Die Aetiologie der Tuberculose [On the Nature of Tuberculosis] (1882), Robert Koch

Elementary Anatomy, Physiology and Hygiene for the Use of Schools and Families (1882), Edward Playter

Handbuch der Hygiene und der Gewerbekrankheiten [Handbook of Hygiene and Occupational Disease] (1883), Max von Pettenkofer

Drusenfieber [Glandular Fever] (1889), Emil Pfeiffer

Deutscher Bakteriologe [German Bacteriology] (1890), Robert Koch

Pathology and Treatment of Diseases of the Skin (1895), Moriz Kaposi

Über die Pestepidemie in Formosa [Concerning the Plague Epidemic in Formosa] (1897), Ogata Masanori

La Propagation de la Peste [The Spread of Plague] (1898), Paul Louis Simond

Report on the Origin and Spread of Typhoid Fever in U.S. Military Camps During the Spanish War of 1898 (1898), Walter Reed, Edward O. Shakespeare, and Victor C. Vaughan

Tropical Disease (1898), Patrick Manson

Traité de la Syphilis [Treatise on Syphilis] (1899), Jean Alfred Fournier

Twentieth Century (first half)

Diary of a Nurse in South Africa (1901), Alice Bron

Trachoma (1904), Julius Boldt

Leaves from a Nurse's Life's History, 1906 (1906), Jean S. Edmunds

The Cause and Prevention of Beri-beri (1907), W. Leonard Braddon

The Air and Ventilation of Subways (1908), George A. Soper

Natural Resistance to Infectious Disease and Its Reinforcement (1910), Simon Flexner

Akute Poliomyelitis (1911), Ivar Wickman

The Prevention of Pellagra (1915), Joseph Goldberger

Sanitation in Panama (1915), William Crawford Gorgas

The Tuberculosis Nurse (1915), Ellen N. La Motte

The Backwash of War: The Human Wreckage of the Battlefield as Witnessed by an American Nurse (1916), Ellen N. La Motte

Public Health Nursing (1916), Mary Sewall Gardner

The Treatment of Infantile Paralysis (1916), Robert W. Lovett

The American Red Cross Textbook on Home Hygiene and Care of the Sick (1918), Jane Arminda Delano

Die Encephalitis Lethargica (1918), Constantin von Economo

Studies in Paleopathology of Egypt (1921), Marc Armand Ruffer

What Every Emigrant Should Know (1922), Cecilia Razovsky

Experimental Studies of Yellow Fever in Northern Brazil (1924), Hideyo Noguchi

Health Problems of the Empire (1924), Andrew Balfour

Infantile Paralysis in Vermont: 1894–1922 (1924), Charles Solomon Caverly

Report on the Outbreak of Plague in Kumasi, Ashanti (1926), Percy Selwyn Selwyn-Clarke

Human Adaptation to the Parasitic Environment (1929), Sheldon F. Dudley

Menschliche Auslese und Rassenhygiene [The Principles of Human Heredity and Race Hygiene] (1931), Eugen Fischer

Swine Influenza (1931), Richard Shope

Acute Ascending Myelitis Following a Monkey Bite (1934), Albert Bruce Sabin

Virus Diseases (1934), Richard Shope

Ultrafiltration of the Virus of Vesicular Stomatitis (1934), Herald R. Cox

Infantile Paralysis (1935), George Draper

History of Chinese Medicine (1936), Wu Lien-teh

Plague: A Manual for Medical and Public Health Workers (1936), Wu Lien-teh

The Organization of Permanent Nation-Wide Anti-Aedes Aegypti Measures in Brazil (1943), Fred Lowe Soper

Estudios sobre la Fiebre Amarilla [Studies on Yellow Fever] (1945), Carlos Juan Finlay

Virus as Organism (1945), Frank MacFarlane Burnet

Transfusion of Defibrinated Blood in P.O.W. Camps (1946), Jacob Markowitz

50 Years in Starch (1948), Anne Williamson

Twentieth Century (second half)

"The Epidemiology of Deafness Due to Maternal Rubella" (1954), Henry Oliver Lancaster

Poliomyelitis Vaccine in the Fall of 1955 (1956), Jonas Salk

Plague Fighter: The Autobiography of a Modern Chinese Physician (1959), Wu Lien-teh

Russia and the Cholera 1823–1832 (1965), Roderick E. McGrew

Kugai Jodo: Waga Minamatabyo [Paradise in the Sea of Our Sorrow] (1972), Ishimure Michiko

With the Armies of the Tsar: A Nurse at the Russian Front, 1914–1918 (1974), Florence Farmborough

The Person with AIDS: Nursing Perspectives (1983), Jeanne Parker Martin

And the Band Played On: People, Politics and the AIDS Epidemic (1987), Randy Shilts

The Grit Behind the Miracle: A True Story of the Determination and Hard Work Behind an Emergency Infantile Paralysis Hospital, 1944–1945, Hickory, North Carolina (1998), Alice E. Sink

The Coming of the Spirit of Pestilence (1999), Robert Boyd

Twenty-First Century

Darkness in El Dorado (2000), Patrick Tierney

The Invisible Enemy: A Natural History of Viruses (2000), E. Margaret Crawford

Island Epidemics (2000), Andrew Cliff, Peter Haggett, and M.R. Smallman-Raynor

Snowshoe & Lancet: Memoirs of a Frontier Newfoundland Doctor (2000), Robert Skidmore Ecke

Biology of Plagues (2001), Susan Scott and Christopher Duncan

Encyclopedia of Plague and Pestilence (2001), George C. Cohen

In the Wake of the Plague: The Black Death and the World It Made (2001), Norman F. Cantor

Pox Americana: The Great Smallpox Epidemic of 1775–82 (2001), Elizabeth A. Fenn

An American Plague (2003), Jim Murphy

Epidemics and Pandemics: Their Impacts on Human History (2005), Jo N. Hays

Ebola Epidemic (2014), Ghazi Mokammel Hossain

Epidemics in Modern Asia (2016), Robert Peckham

Pandemic: Tracking Contagions (2016), Sonia Shah

Appendix D
Authors of Major Works on Disease

Abraham a Sancta Clara, Die Grosse Totenbruder-schaft [The Great Brotherhood of the Dead] (1681) *Lösch Wien* [Vienna Blotted Out] (1680) *Merk's, Wien!* [Pay Attention, Vienna] (1680)

Acland, Henry Wentworth, *Memoir on the Cholera at Oxford* (1856)

Allen, Richard, and Absalom Jones, *Narrative of the Proceedings of the Black People, During the Late Awful Calamity in Philadelphia* (1793)

Almenor, Juan, *De Morbo Gallico* [On the French Disease] (1502)

Andral, Gabriel, *Clinique Medicale* [Medical Clinic] (1833) *Précis d'Anatomie Pathologique* [Summary of Pathological Anatomy] (1829)

Areteus, *On Cholera* (ca. 11 C.E.) *On Ulcerations About the Tonsils* (ca. 110 C.E.)

Armstrong, George D., *A History of the Ravages of the Yellow Fever in Norfolk, Virginia, A.D. 1855* (1855)

Avicenna of Kazakhstan, *Khitab ash-Shifa* [Healing of the Soul] (early 1000s)

Baillou, Guillaume de, *Epidemiorum et Ephemeridium* [Epidemics and Journals] (1640)

Balfour, Andrew, *Health Problems of the Empire* (1924)

Barbette, Paul, *De Peste Tractatus* [Treatise on Pestilence] (1678)

Bard, Samuel, *An Enquiry into the Nature, Cause, and Cure, of the Angina Suffocativa, or, Sore Throat Distemper* (1769)

Bartholin, Caspar, *A Short Instruction* (1619)

Bell, Charles, *An Exposition of the Natural System of the Nerves of the Human Body* (1824) *A Historical Relation of the Plague at Marseilles in the Year 1720* (1720)

Béroud, Philippe, *Étude sur l'Hygiène et la Topographie Médicale de la Ville de Saint-Etienne* [Study on Hygiene and Medical Topography of the City of Saint-Etienne] (1862)

Blane, Gilbert, *Observations on the Diseases Incident to Seamen* (1785)

Boldt, Julius, *Trachoma* (1904)

Bontius, Jacobus, *De Medicina Indorum* [On Medicine in the Indies] (1629)

Bowman, James, *Direction pour la Guérison du Mal de la Baie Saint-Paul* [Methods for Treating Baie Saint-Paul Disease] (1785)

Boyd, Robert, *The Coming of the Spirit of Pestilance: Introduced Infectious Diseases and Population Decline Among Northwest Coast Indians, 1774–1874* (1999)

Boylston, Zabdiel, *An Historical Account of the Small-Pox Inoculated in New England* (1725)

Braddon, W. Leonard, *The Cause and Prevention of Beriberi* (1907)

Bravo, Francisco, *Opera Medicinalia, in Quibus Quam Plurima Extant Scitu Medico Necessaria* [A Medical Encyclopedia Containing Everything Known to Science] (1570)

Brewer, Thomas, *Lord Have Mercy Upon Us* (1636)

Bron, Alice, *Diary of a Nurse in South Africa* (1901)

Buchan, William, *Domestic Medicine; or, the Family Physician* (1749)

Budd, William, *Typhoid Fever: On Its Nature, Mode of Spreading, and Infection* (1856)

Bullein, William, *A Dialogue Against the Fever Pestilence* (1564)

Burnet, Frank MacFarlane, *Virus as Organism* (1945)

Caius, John, *A Boke or Counseill Against the Disease Commonly Called the Sweate or Sweatyng Sicknesse* (1552)

Carey, Matthew, *A Short Account of the Malignant Fever Lately Prevalent in Philadelphia* (1793)

Casal y Julian, Gaspar, *Historia Natural y Medica de el Principado de Asturias* [Biology and Medical History of Asturias Province] (1762)

Caslari, Abraham ben David, *Ma'mar Be-qadahot Divriot U-minei Qadahot* [A Treatise on Pestilential Fevers and Other Kinds of Fevers] (1326)

Caverly, Charles Solomon, *Infantile Paralysis in Vermont: 1894–1922* (1924)

Chadwick, Edwin, *Report on the Sanitary Condition of the Labouring Population of Great Britain* (1839)

Chauliac, Guy de, *Capitulum Singulaire* [A Singular Chapter] (1350)

Chen Sicheng, *Meichuang Mi Lu* [Secret Records on Syphilis] (1632)

Christian, Samuel, *The Cure and Prevention of Scarlet Fever* (1801)

Colden, Cadwallader, D., *Observations on the Climate and Diseases of New York* (1738)

Cooke, John Esten, *Treatise of Pathology and Therapeutics* (1828)

Cox, Herald R., *Ultrafiltration of the Virus of Vesicular Stomatitis* (1934)

Cyprian, *De Mortalitate* [On Mortality] (252 C.E.)

de Hinojosos, Alfonso Lopez, *Summa y Recopilación de Chirugia* [Treatise and Recompilation of Surgery] (1578)

Delano, Jane Arminda, *The American Red Cross Textbook on Home Hygiene and Care of the Sick* (1918)

de Torres, Alonso, *De Febris Epidemicae, et Novae Quae Latine Puncticularis* [On Epidemic Fevers and New Ones Particularized in Latin] (1574)

de Vigo, Giovanni, *Arte Chirurgia Copiosa* [On the Complete Surgical Art] (1514)

Díaz de Isla, Rodrigo Ruy, *Treatise on the Serpentine Malady* (ca. 1520)

Draper, George, *Infantile Paralysis* (1935)

Dudley, Seldon F., *Human Adaptation to the Parasitic Environment* (1929)

Ecke, Robert Skidmore, *Snowshoe & Lancet: Memoirs of a Frontier Newfoundland Doctor* (2000)

Edmunds, Jean S., *Leaves from a Nurse's Life's History, 1906* (1906)

Fachinetti, Antonio, *Memoria in Occasione del Cholèra-Morbus nell'Anno 1855* [Recalling the Cholera Epidemic of the Year 1855] (1855)

Farmborough, Florence, *With the Armies of the Tsar: A Nurse at the Russian Front, 1914–1918* (1974)

Fenn, Elizabeth A., *Pox Americana: The Great Smallpox Epidemic of 1775–82* (2001)

Ferriar, John, *Medical Histories and Reflections* (1792)

Finlay, Carlos Juan, *Estudios sobre la Fiebre Amarilla* [Studies on Yellow Fever] (1945)

Fischer, Eugen, *Menschliche Auslese und Rassenhygiene* [The Principles of Human Heredity and Race Hygiene] (1931)

Flexner, Simon, *Natural Resistance to Infectious Disease and Its Reinforcement* (1910)

Foligno, Gentile da, *Concilium contra Pestilenciam* [Advice on the Plague] (1348)

Fontecha, Alfonso Ruizibus de, *Disputationes Medicae* [Medical Controversies] (1611)

Forman, Simon, *Discourses on the Plague* (1595)

Fournier, Jean Alfred, *Traité de la Syphilis* [Treatise on Syphilis] (1899)

Fracastorius, Girolamo, *De Contagione et Contagiosis Morbis* [On Contagion and Contagious Diseases] (1546)

Fracastorius, Girolamo, *Syphilis sive Morbus Gallicus* [Syphilis or French Disease] (1530)

Galen, *De Alimentorum Facultatibus* [Concerning the Workings of Digestion] (ca. 180 C.E.)

Gardner, Mary Sewall, *Public Health Nursing* (1916)

Gastaldi, Hieronymi, *Tractatus de Avertenda et Profliganda Peste* [Treatise on Averting and Avoiding Plague] (1684)

Gatti, Angelo, *Reflexions on Variolation* (1764)

Gerhard, William Wood, *On the Typhus Fever, Which Occurred at Philadelphia in the Spring and Summer of 1836* (1837)

Ghisi, Martino, *Angina Maligna* (1749)

Goldberger, Joseph, *The Prevention of Pellagra* (1915)

Gordon, Alexander, *Treatise on the Epidemic Puerperal Fever of Aberdeen* (1795)

Gorgas, William Crawford, *Sanitation in Panama* (1915)

Graunt, John, *Natural and Political Observations Made upon the Bills of Mortality* (1662)

Hahnemann, Samuel Christian, *Observations on the Scarlet Fever* (1808)

Heberden, William, *Commentarii de Morborum Historia et Curatione* [Commentaries on the History and Cure of Diseases] (1802)

Hillary, William, *Observations on the Changes of the Air and the Concomitant Epidemical Diseases in the Island of Barbados* (1759)

Hinksey, Thomas Willis North, *Diatribae Duae: De Febribus* [Two Texts: On Fevers] (1659)

Hippocrates, *Des Airs, des Eaux, et des Lieux* [On Air, Water, and Occasions] (ca. 310 B.C.E.); *The Epidemics* (ca. 310 B.C.E.)

Hodges, Nathaniel, *Historical Account of the Plague in London in 1665* (1665)

Ho Jun, *Tongui Pogam* (1610)

Huang Ti, *Nei Ching* (2700 B.C.E.)

Hutten, Ulrich von, *De Morbo Gallico* [On the French Disease] (1519)

Huxham, John, *An Essay on Fevers* (1757)

Ishimure, Michiko, *Kugai Jodo: Waga Minamatabyo* [Paradise in the Sea of Our Sorrow] (1972)

Jansen, Stephen Theodore, *A Jail Fever* (1767)

Jenner, Edward, *An Inquiry into the Causes and Effects of the Variolae Vaccinae* (1798)

Jones, Robert, *Remarks on the Distemper Generally Known by the Name of the Molbay Disease* (1786)

Kaposi, Moriz, *Pathology and Treatment of Diseases of the Skin* (1895)

Kay-Shuttleworth, James Phillips, *Moral and Physical Condition of the Working Classes Employed in the Cotton Manufacture in Manchester* (1832)

Keating, John McLeod, *The Yellow Fever Epidemic of 1878 in Memphis, Tennessee* (1879)

Kilpatrick, James, *An Essay on Inoculation* (1738)

Koch, Robert, *Die Aetiologie der Tuberculose,* [On the Nature of Tuberculosis] (1882) *Deutscher Bakteriologe* [German Bacteriology] (1890) *Investigations into the Etiology of Traumatic Infective Diseases* (1880)

Laënnec, René Théophile, *De l'Auscultation Médiate, ou Traité du Diagnostic des Maladies du Coeur* [Mediate Auscultation or a Treatise on the Diagnosis of Heart Disease] (1818)

La Motte, Ellen N., *The Tuberculosis Nurse* (1915) *The Backwash of War: The Human Wreckage of the Battlefield as Witnessed by an American Nurse* (1916)

Lancaster, Oliver, "The Epidemiology of Deafness Due to Maternal Rubella" (1954)

Lancisi, Giovanni Maria, *Dissertatio de Nativis* [Dissertation on Natives] (1711)

Leach, Harry, *The Ship Captain's Medical Guide* (1868)

Leonicen, Niccolo, *Libellus de Morbo Gallico* [A Pamphlet on the French Disease] (1497)

Lind, James, *An Essay on Diseases Incidental to Europeans in Hot Climates* (1768) *On the Most Effectual Means of Preserving the Health of Seamen* (1757) *A Treatise on Scurvy: In Three Parts. Containing an Inquiry into the Nature, Causes, and Cure, of the Disease* (1753)

Lovett, Robert W., *The Treatment of Infantile Paralysis* (1916)

Lowe, Peter, *Discourse of the Whole Art of Chyrurgie* (1612)

Manson, Patrick, *Tropical Disease* (1898)

Manzini, Nicolás, *Pneumopathies Infectieuses Communautaires chez la Personne Agée* [Infectious Lung Disease in the Aged] (1858)

Markowitz, Jacob, *Transfusion of Defibrinated Blood in P.O.W. Camps* (1946)

Martin, Jeanne Parker, *The Person with AIDS: Nursing Perspectives* (1983)

Mather, Cotton, *The Angel of Bethesda* (1722)

Mayerne, Theodore Turquet de, *Epheme-rides Morborum* [Diaries of Diseases] (ca. 1655)

McGrew, Roderick E., *Russia and the Cholera 1823–1832* (1965)

McWilliam, James Ormiston, *Medical History of the Niger Expedition* (1843)

Mociño, José María, *Disertación de la Fiebre Epidémica, Que Padecio Cádiz* [Treatise on the Epidemic Fever That Assailed Cadiz] (1804)

Memoir on the Yellow Fever (ca. 1805)

Monteggia, Giovanni Battista, *Istituzioni di Chirurgiche* [The Profession of Surgery] (1802)

Müller, Moritz Wilhelm, *De Febre in Inflammatoria* [On Fever in Inflammations] (1810)

Noguchi, Hideyo, *Experimental Studies of Yellow Fever in Northern Brazil* (1924)

Ogata Koan, *Korori Chijun* [Treatise on Cholera] (1858)

Ogata Masanori, *Über die Pestepidemie in Formosa* [Concerning the Plague Epidemic in Formosa] (1897)

Olsson, Bengt, *Handbook of the Healing Art* (1578)

Panum, Peter Ludwig, *Observations Made During the Epidemic of Measles on the Faeroe Islands in the Year 1846* (1846)

Pascalis-Ouvière, Felix, *An Account of the Contagious Epidemic Yellow Fever, Which Prevailed in Philadelphia in the Summer and Autumn of 1797* (1798)

Peu, Philippe, *La Pratique des Accouchements* [The Practice of Obstetrics] (1694)

Pfeiffer, Emil, *Drusenfieber* [Glandular Fever] (1889)

Platiensis, Michael, *Account of the Plague* (1357)

Playter, Edward, *Elementary Anatomy, Physiology and Hygiene for the Use of Schools and Families* (1882)

Porcell, Juan Tomás, *Informacion y Curacion dela Peste* [Information and Treatment of Plague] (1565)

Potts, Jonathan, *Utility of Vaccination* (1771)

Pringle, John, *Observations on Diseases of the Army in Camp and Garrison* (1752)

Ramazzini, Bernardino, *De Morbis Artificum Diatriba* [Diseases of Workers] (1700)

Razovsky, Cecilia, *What Every Emigrant Should Know* (1922)

Reed, Walter, Edward O. Shakespeare, and Victor C. Vaughan, *Report on the Origin and Spread of Typhoid Fever in U.S. Military Camps During the Spanish War of 1898* (1898)

Reide, Thomas Dickson, *A View of the Diseases of the Army in Great Britain, America, the West Indies, and on Board of King's Ships and Transports* (1793)

Rhazes of Baghdad, *The Compendium* (ca. 925 C.E.) *De Variolis et Morbillis Commentarius* [On Smallpox and Measles] (910 C.E.)

Rochoux, Jean Andre, *Recherches sur le Fievre Jaune* [Research on Yellow Fever] (1820)

Ruffer, Marc Armand, *Studies in Paleopathology of Egypt* (1921)

Rufus of Ephesus, *Collections of Oribasius Sardianus* (ca. 365 C.E.)

Rutty, John, *A Chronological History of the Weather and Seasons on the Prevailing Diseases in Dublin* (1770)

Sabin, Albert Bruce, *Acute Ascending Myelitis Following a Monkey Bite* (1934)

Sacco, Luigi, *Trattato di Vaccinazione* [Treatise on Vaccination] (1809)

Salk, Jonas, *Poliomyelitis Vaccine in the Fall of 1955* (1956)

Samoilovich, Danilo S., *Description of the Plague Which Attacked Kherson* (1784)

San Bonaventura, Antero Maria Micone da, *Lazaretti della Città, Li* [City Pesthouses] (1658)

Scott, Susan, and Christopher Duncan, *Biology of Plagues* (2001)

Seacole, Mary Jane, *The Wonderful Adventures of Mrs. Seacole in Many Lands* (1857)

Selwyn-Clarke, Percy Selwyn, *Report on the Outbreak of Plague in Kumasi, Ashanti* (1926)

Shilts, Randy, *And the Band Played On: People, Politics and the AIDS Epidemic* (1987)

Shope, Richard, *Swine Influenza* (1931) *Virus Diseases* (1934)

Simond, Paul Louis, *La Propagation de la Peste* [The Spread of Plague] (1898)

Sink, Alice E., *The Grit Behind the Miracle: A True Story of the Determination and Hard Work behind an Emergency Infantile Paralysis Hospital, 1944–1945, Hickory, North Carolina* (1998)

Skeyne, Gilbert, *Brief Description of the Pest* (1568)

Snow, John, *On the Mode of Communication of Cholera* (1855)

Soper, Fred Lowe, *The Organization of Permanent Nation-Wide Anti-Aedes Aegypti Measures in Brazil* (1943)

Soper, George A., *The Air and Ventilation of Subways* (1908)

Sushruta, *Sushruta Samhita* (600 B.C.E.)

Sutton, Daniel, *The Inoculator* (1796) Daniel Sutton

Sydenham, Thomas, *Observationes Medicae* [Clinical Observations] (1676) *Opera Medica* [Medical Works] (1715) *Praxis Medica* [The Practice of Medicine] (1707)

Taché, Joseph-Charles, *Memorandum on Cholera* (1866)

Talbor, Robert, *Pyretologia: A Rational Account of the Cause and Cures of Agues* (1672) *The English Remedy, or Talbor's Wonderful Secret for Curing of Agues and Feavers* (1682)

Tanbo, Yasuhori, *Ishinho* (982 C.E.)

Tenon, Jacques-René, *Mémoires sur les Hôpitaux de Paris* [Memoirs of Paris Hospitals] (1788)

Thacher, Thomas, *A Brief Rule to Guide the Common-People of New England How to Order Themselves & Theirs in the Small-Pocks or Measles* (1678)

Tierney, Patrick, *Darkness in El Dorado* (2000)

Timoni, Emanuel, "An Account or History of the Procuring of the Small Pox by Incision of Inoculation" (1714)

Tulp, Nicholas, *Observationes Medicae* [Clinical Observations] (1652)

Ulsenius, Theodericus, *Cura Mali Francici* [Cure of the French Pox] (1496) *Vaticinium in Epidemicam Scabiem* [Revelation on Epidemic Scabies] (1496)

Underwood, Michael, *Treatise on Diseases of Children* (1789)

Valentyn, François, *Description of the Cape of Good Hope; with Matters Concerning It* (1726)

Villalobos, Francisco López de, *Medical Works* (1498)

Villemin, Jean Antoine, *Études sur la Phthisie* [Studies in Tuberculosis] (1868)

Virchow, Rudolf Ludwig Karl, *Reports on the Typhus Epidemics of Upper Silesia* (1848)

von Economo, Constantin, *Die Encephalitis Lethargica* (1918)

von Ewich, Johann, *De Officio Fidelis* [On the Duty of the Faithful] (1582)

von Pettenkofer, Max, *Handbuch der Hygiene und der Gewerbekrankheiten* [Handbook of Hygiene and Occupational Disease] (1883)

Waterhouse, Benjamin, *A Prospect of Exterminating the Smallpox* (1800)

Webster, Noah, *A Brief History of Epidemic and Pestilential Diseases* (1799)

Wickman, Ivar, *Akute Poliomyelitis* (1911)

Williamson, Anne, *50 Years in Starch* (1948)

Woodall, John, *The Surgeon's Mate, or Military & Domestique Surgery* (1617)

Woodward, Joseph Janvier, *Report on Epidemic Cholera and Yellow Fever in the Army of the United States, During the Year 1867* (1868)

Wu Lien-teh, *History of Chinese Medicine* (1936) *Plague Fighter: The Autobiography of a Modern Chinese Physician* (1959) *Plague: A Manual for Medical and Public Health Workers* (1936)

Wu Youxing, *Wenyilun* [On Pestilence] (1642)

Yu T'ien-Chih, *Miscellaneous Ideas in Medicine* (1643)

Selected Bibliography— General Resources

The following selected sources cover general medical topics or describe two or more diseases or epidemics. The second bibliography lists writings about single diseases under specific topics, e.g., acrodynia, bioterrorism, leishmaniasis.

Amstey, Marvin S., M.D. "The Political History of Syphilis and Its Application to the AIDS Epidemic." *Women's Health Issues* (Spring 1994): 16–19.

Anderson, Warwick. "Immunities of Empire: Race, Disease, and the New Tropical Medicine, 1900–1920." *Bulletin of the History of Medicine* 70: 1 (1996): 94–118.

Armelagos, George J., Kathleen C. Barnes, and James Lin. "Disease in Human Evolution." *National Museum of Natural History Bulletin for Teachers* (Fall 1996).

Arnold, David. *Colonizing the Body: State Medicine and Epidemic Disease in Nineteenth-Century India.* Berkeley: University of California, 1993.

Baker, R. A., and R. A. Bayliss. "William John Ritchie Simpson: Public Health and Tropical Medicine." *Medical History* 31: 4 (1987): 450–465.

Bosch, Xavier. "Old Epidemic Sheds New Light on CJD." *Nature* (6 February 2001).

Bruijn, I. D. R., and G. W. Bruijn. "An Eighteenth-Century Medical Hearing, and the First Observation of Tropical Phagedaena." *Medical History* 35:3 (1991): 295–307.

Burnard, Trevor. "The Countrie Continues Sicklie: White Mortality in Jamaica, 1655–1780." *Social History of Medicine* 12 (1999): 45–72.

Cantor, Norman F., ed. *In the Wake of the Plague: The Black Death and the World It Made.* New York: Harper Perennial, 2001.

Carley, Caroline D. "Historical and Archaeological Evidence of 19th-Century Fever Epidemics and Medicine at Hudson's Bay Company's Fort Vancouver." *Historical Archaeology* 15 (1981): 19–35.

Carr, Ian. "Plagues and Peoples: The Columbian Exchange." http://www.umanitoba.ca/faculties/medicine/history/histories/plagues.html.

Cates, Gerald L. "The Seasoning: Disease and Death among the First Colonists of Georgia." *Georgia Historical Quarterly* 64: 2 (1980): 146–158.

Charbonneau, Pierre. "La Medecine Preventive au Maroc pendant le Protectorat." *Histoire des Sciences Médicales* 26:4 (1992): 301–304.

Chidley, Joe. "Return of an Ancient Killer." *Maclean's* 109:36 (9 September 1996): 45.

Chrysler, Elizabeth. "The Pioneer Catholic Church of Caddo and DeSoto Parishes," *North Louisiana Historical Association Journal* 21: 2–3 (1990): 79–80.

Churchward, W. B. *Blackbirding in the South Pacific.* London: Swan, Sonnenschein & Co., 1888.

Cliff, Andrew, Peter Haggett, and M. R. Smallman-Raynor. *Island Epidemics.* Oxford: Oxford University Press, 2000.

Cockburn, Eve, and Theodore Reyman. *Mummies, Disease, and Ancient Cultures.* London: Cambridge University Press, 1998.

Coelho, Philip R. P., Robert A. McGuire, and Richard H. Steckel. "Diet Versus Diseases: The Anthropometrics of Slave Children." *Journal of Economic History* 60 (2000): 232–246.

Cohen, George C., ed. *Encyclopedia of Plague and Pestilence.* New York: Facts on File, 2001.

Cohen, Israel. *Jewish Life in Modern Times.* London: Methuen, 1914.

Cook, G. C. "Disease in the 19th-Century Merchant Navy: The Seamen's Hospital Society Experience." *Mariner's Mirror* 87:4 (2001): 460–471.

Cook, Noble David. *Born to Die: Disease and New World Conquest, 1492–1650.* New York: Cambridge University Press, 1998.

Crawford, E. Margaret. *The Invisible Enemy: A Natural History of Viruses.* Oxford: Oxford University Press, 2000.

Deeds, Susan. *Defiance and Deference in Mexico's Colonial North: Indians under Spanish Rule in Nueva Vizcaya.* Austin: University of Texas Press, 2003.

Dworkin, Mark S., et al. "Tick-Borne Diseases." *Medical Clinics of North America* 86:2 (March 2002).

Garrett, Laurie. *The Coming Plague.* New York: Farrar, Straus and Giroux, 1994.

Hanson, John H. "Islam, Migration, and the Political Economy of Meaning: *Fergo Nioro* from the Senegal River Valley, 1862–1890." *Journal of African History* 35 (1994): 37–60.

Hollander, Kurt. *Several Ways to Die in Mexico City: An Autobiography of Death in Mexico City.* Port Townsend, WA: Feral House, 2012.

Holmes, Frederick F. "Anne Boleyn, the Sweating Sickness, and the Hantavirus: A Review of an Old Disease with a Modern Interpretation." *Journal of Medical Biography* 6 (1998): 43–48.

Horvath, Stephanie. "Warmer Climate Is Linked to Rise in Epidemics." *Wall Street Journal,* June 21, 2002, B2.

Kelton, Paul. *Epidemics and Enslavement.* Lincoln: University of Nebraska Press, 2009.

Kolata, Gina. *Flu.* New York: Farrar Straus & Giroux, 1999.

Lane, Kris. "Captivity and Redemption: Aspects of Slave Life in Early Colonial Quito and Popayán." *The Americas* 57:2 (October 2000): 225–246.

Leyton, Elliott. *Touched by Fire: Doctors Without Borders in a Third World Crisis.* Toronto: McClelland & Stewart, 1998.

MacLennan, W. J. "The Eleven Plagues of Edinburgh." *Proceedings of the Royal College of Physicians of Edinburgh* (2001): 256–261

Marr, John S., and James B. Kirocofe. "Was the *Huey Cocoliztli* a Hemorrhagic Fever?" *Medical History* (2000): 341–362.

McCann, James C. "Climate and Causation in African History." *International Journal of African Historical Studies* 32:2–3 (1999): 261–279.

Meredith, Howard. "Cultural Conservation and Revival: The Caddo and Hasinai Post-Removal Era 1860–1902." *Chronicles of Oklahoma* (Fall 2001): 278–287.

Moulin, Daniel de. "Paul Barbette, M.D.: A Seventeenth-Century Amsterdam Author of Best-Selling Textbooks." *Bulletin of the History of Medicine* 59, no. 4, 1985, pp. 506–514.

Peck, Douglas T. "Lucas Vasquez de Ayllon's Doomed Colony of San Miguel de Gualdape." *Georgia Historical Quarterly* 84:2 (2001): 183–198.

Perry, Adele. "Hardy Backwoodsmen, Wholesome Women, and Steady Families: Immigration and the Construction of a White Society in Colonial British Columbia, 1849–1871." *Social History* 33: 66 (2000): 343–360.

Petriello, David. *Bacteria and Bayonets: The Impact of Disease in American Military History.* Havertown, PA: Casemate, 2015.

Pickering, L. K., ed. *Red Book 2000: Report of the Committee on Infectious Diseases.* New York: American Academy of Pediatrics, 2000.

Rhoades, Michelle. "'There Are No Safe Women': Prostitution in France During the Great War." *Proceedings of the Annual Meeting of the Western Society for French History* 27 (2001): 43–50.

"Ross and the Discovery That Mosquitoes Transmit the Malaria Parasite." https://www.cdc.gov/malaria/about/history/ross.html.

Suleymanov, Manaf. "Building Infrastructure: Taghiyev's Commitment to the Water Problem." *Azerbaijan International* (Summer 2002): 54–55.

Weiss, Rick. "War on Disease." *National Geographic* (February 2002): 4–31.

Bibliography—
Resources by Disease

Acrodynia

Warkany, Josef, and Donald M. Hubbard. "Acrodynia and Mercury." *Journal of Pediatrics* 42 (1953): 365–386.

AIDS

Muswazi, Paiki. "HIV/AIDS Information Resources and Services: A Swaziland Case Study." *Library Review* 49:1 (2000): 34–39.
"2004 Report on the Global Aids Epidemic." http://files.unaids.org.

Alastrim

Babkin, Igor V., and Irina N. Babkina. "The Origin of the Variola Virus." *Viruses* 7.3 (2015): 1100–1112.

Alkhurma Hemorrhagic Fever

Charrel R.N., et al. "Alkhumra Hemorrhagic Fever Virus in *Ornithodoros savignyi* Ticks." *Emerging Infectious Disease* 13:1 (January 2007): 153–155.

Amoebiasis

Nath, Joyobrato, et al. "Molecular Epidemiology of Amoebiasis." *PLoS Neglected Tropropical Diseases* 9.12 (2015): e0004225.

Anthrax

Remy, Kenneth E., Caitlin Hicks, and Peter Q. Eichacker. "Anthrax Infection." *Human Emerging and Re-emerging Infections: Viral and Parasitic Infections, Volume I* (2016): 773–794.

Argentine Hemorrhagic Fever

Zeitlin, Larry, et al. "Monoclonal Antibody Therapy for Junin Virus Infection." *Proceedings of the National Academy of Sciences* (2016): 201600996.

Arsenic Poisoning

Copping, Matthew. "Death in the Beer Glass." *Journal of the Brewery History Society* 132: 31–57.
Uddin, Riaz, and Naz Hasan Huda. "Arsenic Poisoning in Bangladesh." *Oman Medical Journal* 26:3 (May 2011): 207.

Barmah Forest Virus

Flexman, J. P., et al. "A Comparison of the Diseases Caused by Ross River Virus and Barmah Forest Virus," *Medical Journal of Australia* 169:3 (1998): 159–163.

Bartonellosis

Maguina, Ciro, and Eduardo Gotuzzo. "Emerging and Re-Emerging Diseases in Latin America: Bartonellosis," *Infectious Disease Clinics of North America.* 14:1 March 2000.

Beriberi

Jacobina, R. R., and F. M. Carvalho. "Nina Rodrigues, Epidemiologist: Historical Study of Beriberi Outbreaks in a Mental Illness Asylum in Bahía, Brasil (1897–1904)," *História.* Vol. 8, March–June 2001, pp. 113–132.

Bioterrorism

Regis, Ed. *The Biology of Doom.* New York: Henry Holt, 1999.
Tamanoi, Mariko Asano. "War Responsibility and Japanese Civilian Victims of Japanese Biological Warfare in China." *Bulletin of Concerned Asian Scholars* 32:3 (2000): 13–22.

Black Lung

Derickson, Alan. *Black Lung: Anatomy of a Public Health Disaster.* Ithaca, NY: Cornell University Press, 1998.

Blastomycosis

Castillo, Caroline G., Carol A. Kauffman, and Marisa H. Miceli. "Blastomycosis." *Infectious Disease Clinics of North America* 30:1 (2016): 247–264.

Bornholm Disease

Taylor, Robert B. "Medical Words and Phrases." *White Coat Tales*. New York: Springer International, 2016. 87–110.

Botulism

Yan, Yongyong, et al. "A Combinatory Mucosal Vaccine against Influenza and Botulism." *Journal of Immunology* 196:1 Supplement (2016): 145–213.

Brucellosis

Vassallo, D. J. "The Saga of Brucellosis." *Lancet* (21 September 1996): 804–808.

Buruli Ulcer

Anogwih, Joy. "Buruli Ulcer: A Journey to Unveiling Its Mode of Transmission in Nigeria." *Humanitarian Technology* 159 (2016): 77–81.
Ofori-Adjei, David. "Buruli Ulcer Disease." *Ghana Medical Journal* 45:1 (March 2011): 1.

Campylobacter jejuni

Gill, Carson, Simon Bahrndorff, and Carl Lowenberger. "Campylobacter Jejuni in Musca Domestica." *Insect Science* (2016).

Chagas

Perleth, Matthias. "Historical Aspects of American Trypanosomiasis (Chagas's Disease)." *Memórias do Instituto Oswaldo Cruz* 92:5 (October 1997): 729.

Chicken Pox

Sinha, Arijit, et al. "Acute Disseminated Encephalomyelitis in Chicken Pox." *National Journal of Medical Research* 6:1 (2016): 103–104.

Chikungunya Fever

Rezza, G., et al. "Infection with Chikungunya Virus in Italy." *Lancet* 370:9602 (2007): 1840–1846.

Chlamydia

Gratrix, Jennifer, et al. "Evidence for Increased Chlamydia Case Finding." *Clinical Infectious Diseases* 60:3 (2015): 398–404.

Cholera

Curtis, Bruce. "Social Investment in Medical Forms: The 1866 Cholera Scare and Beyond." *Canadian Historical Review* 83:3 (2000): 347–379.
Ogawa, Mariko. "Uneasy Bedfellows: Science and Politics in the Refutation of Koch's Bacterial Theory of Cholera." *Bulletin of the History of Medicine* 74:4 (2000): 671–707.
Watts, Sheldon. "From Rapid Change to Stasis: Official Responses to Cholera in British-Ruled India and Egypt." *Journal of World History* 12:2 (2001): 321–374.

Conjunctivitis

Maitreyi, R. S., et al. "Acute Hemorrhagic Conjunctivitis Due to Enterovirus 70 in India." *Emerging Infectious Diseases* March-April 1999.
Schrauder, A., et al. "Epidemic Conjunctivitis in Germany, 2004." *Eurosurveillance* 11:7–9 (July-September 2006): 185–87.

Creutzfeldt-Jakob Disease

Bedino, James H. "Creutzfeldt-Jakob Disease and Related Disorders." *Champion Expanding Encyclopedia of Mortuary Practices*. 1995, pp. 2514–2517.
"Cannibalism's Clues to CJD." *BBC News*. March 23, 2001.

Crohn's Disease

Chiodini, Rodrick J. "Crohn's Disease and the Mycobacterioses: A Review and Comparison of Two Disease Entities." *Clinical Microbiology Review* (January 1989).
Greger, Michael. "Paratuberculosis and Crohn's Disease: Got Milk?" *Vegan Outreach* (January 2001).

Cryptosporidiosis

MacKenzie, William R., et al. "A Massive Outbreak in Milwaukee of Cryptosporidium Infection Transmitted through the Public Water Supply." *New England Journal of Medicine* 331 (21 July 1994): 161–167.

Dengue Fever

Packard, Randall M. "'Break-Bone' Fever in Philadelphia, 1780: Reflections on the History of Disease." *Bulletin of the History of Medicine* 90:2 (Summer 2016): 1–50.

Diphtheria

Allam, Ramesh Reddy, et al. "A Case–Control Study of Diphtheria in the High Incidence City of Hyderabad, India." *Pediatric Infectious Disease Journal* 35:3 (2016): 253–256.

Dracunculiasis

Esrey, Steven A., et al. "Effects of Improved Water Supply and Sanitation on Ascariasis, Diarrhoea, Dracunculiasis, Hookworm Infection, Schistosomiasis, and Trachoma." *Bulletin of the World Health Organization* 69:5 (1991): 609.

Dysentery

Xu, X., et al. "Quantifying the Impact of Floods on Bacillary Dysentery in Dalian City, China, from 2004 to 2010." *Disaster Medicine and Public Health Preparedness* (2016): 1–6.

Ebola

Josefson, Deborah. "Vaccine Protects against Ebola Virus." *British Medical Journal* 321: 7274 (9 December 2000): 1433.

Elephantiasis

Molla, Yordanos B., et al. "Modelling Environmental Factors Correlated with Podoconiosis: A Geospatial Study of Non-filarial Elephantiasis." *International Journal of Health Geographics* 13.1 (2014): 1.

Encephalitis

Solomon, T. "Flavivirus Encephalitis and Other Neurological Syndromes." *International Journal of Infectious Diseases* 45 (2016): 24.

Enteritis

Butzler, J. P., and M. B. Skirrow. "Campylobacter Enteritis." *Acta Paediatrica Belgica* 32:2 (1979): 89.

Ergotism

Mundra, Leela S., et al. "The Salem Witch Trials—Bewitchment or Ergotism." *Journal of the American Medical Association Dermatology* 152.5 (2016): 540.

Erysipelas

Vij, Alok, and Kenneth J. Tomecki. "21 Cellulitis and Erysipelas." *Clinical Infectious Disease* (2015): 148.

Escherichia Coli

Rowe, Peter C., et al. "Epidemic Escherichia Coli O157: H7 Gastroenteritis and Hemolytic-Uremic Syndrome in a Canadian Inuit Community." *Journal of Pediatrics* 124:1 (1994): 21–26.

Filiriasis

Michael, Edwin, and D. A. P. Bundy. "Global Mapping of Lymphatic Filariasis." *Parasitology Today* 13:12 (1997): 472–476.

Giardiasis

Thompson, RC Andrew. "Giardiasis as a Re-emerging Infectious Disease and Its Zoonotic Potential." *International Journal for Parasitology* 30:12 (2000): 1259–1267.

Glanders

Srinivasan, Arjun, et al. "Glanders in a Military Research Microbiologist." *New England Journal of Medicine* 345 (26 July 2001): 256–258.

Gonorrhea

Unemo, Magnus, and Robert A. Nicholas. "Emergence of Multidrug-Resistant, Extensively Drug-Resistant, and Untreatable Gonorrhea." *Future Microbiology* 7:12 (2012): 1401–1422.

Guillain-Barré Syndrome

Zhang, X., et al. "Guillain-Barre Syndrome in Six Cities and Provinces of Northern China: Is It a New Entity?" *Chinese Medical Journal* 108:10 (October 1995): 734–738.

Hand, Foot and Mouth Disease

Starr, M., and A. Frydenberg. "Hand, Foot, and Mouth Disease." *Australian Family Physician* 32:8 (2003): 594.

Hantavirus

Johnson, Karl M. "The Discovery of Hantaan Virus: Comparative Biology and Serendipity in a World at War." *Journal of Infectious Diseases* 190 (2004): 1708–1721.

Hemorrhagic Fever

Borio, Luciana, et al. "Hemorrhagic Fever Viruses as Biological Weapons." *Journal of the American Medical Association* 287:18 (2002): 2391–2405.

Hepatitis

Radjef, Nadjia, et al. "Molecular Phylogetntic Analysis Indicate a Wide and Ancient Radiation of African

Hepatitis Delta Virus." *Journal of Virology* 78:5 (March 2004): 2537–2544.

HIV

Aronson, Ian David, et al. "Mobile Interventions to Increase HIV Testing among Underserved: High-Risk Populations." *Proceedings of the 2016 ACM Workshop on Multimedia for Personal Health and Health Care.* ACM, 2016.

Hookworm

Bethony, Jeffrey, et al. "Soil-transmitted Helminth Infections." *Lancet* 367:9521 (2006): 1521–1532.

Influenza

Phillips, Howard, and David Killingray, eds. *The Spanish Influenza Pandemic of 1918–19: New Perspectives.* London: Routledge, 2003.

Itai-Itai

Yoshida, Fumikazu. "*Itai-Itai* Disease and the Countermeasures against Cadmium Pollution by the Kamioka Mine." *Environmental Economics and Policy Studies* 2:3 (1999): 215–229.

Kawasaki Disease

Hsu, Jeremy. "Blowing in the Wind? The Mystery of Kawasaki Disease." https://mosaicscience.com/story/kawasaki, accessed 10/3/16.

Kwashiorkor

Garrett, Wendy S. "Kwashiorkor and the Gut Microbiota." *New England Journal of Medicine* 368:18 (2013): 1746–1747.

Langat Virus

Pletnev, Alexander G., et al. "Tick-Borne Langat/Mosquito-Borne Dengue Flavivirus Chimera." *Journal of Virology* 75:17 (September 2001): 8259–8267.

Lead Poisoning

Burnham, John C. "Biomedical Communication and the Reaction to the Queensland Childhood Lead Poisoning Cases Elsewhere in the World." *Medical History* 43:2 (1999): 155–172.

Lo, Yi-Chun. "Childhood Lead Poisoning Associated with Gold Ore Processing." *Environmental Health Perspectives*, ehp.niehs.nih.gov, accessed 10/4/16.

Legionnaire's Disease

Demirjian, Alicia, et al. "The Importance of Clinical Surveillance in Detecting Legionnaires' Disease Outbreaks." *Clinical Infectious Diseases* 60:11 (2015): 1596–1602.

Leishmaniasis

Hailu, Asrat, Daniel Argaw Dagne, and Marleen Boelaert. "Leishmaniasis." *Neglected Tropical Diseases-Sub-Saharan Africa.* New York: Springer International, 2016, 87–112.

Leprosy

Salgado, Claudio Guedes, et al. "What Do We Actually Know about Leprosy Worldwide?" *Lancet Infectious Diseases* 16:7 (2016): 778.

Leptospirosis

Picardeau, Mathieu. "Leptospirosis: Updating the Global Picture of an Emerging Neglected Disease." *PLoS Neglected Tropical Diseases* 9:9 (2015): e0004039.

Listeria

CDC. "Vital Signs: Listeria Illnesses, Deaths, and Outbreaks—United States, 2009–2011." *Morbidity and Mortality Weekly Report* 62:22 (2013): 448.

Lyme Disease

Kuehn, Bridget M. "CDC Estimates 300,000 US Cases of Lyme Disease Annually." *Journal of the American Medical Association* 310:11 (2013): 1110.

Machupo Virus

Meyerholz, D.K. "Modeling Emergent Diseases." *Veterinary Pathology* 53:3 (2016): 517–518.

Mad Cow Disease

Kim, Keun Soo, et al. "Beef from the United States: Is It Safe?" *Journal of Korean Medical Science* 31:7 (2016): 1009–1010.

Malaria

Cowman, Alan F., et al. "Malaria: Biology and Disease." *Cell* 167:3 (2016): 610–624.

Measles

Hotez, Peter J. "Texas and Its Measles Epidemics." *PLoS Medicine* 13:10 (2016): e1002153.

Meningitis

van de Beek, D., et al. "Community-acquired Bacterial Meningitis." *Nature Reviews* 2 (2016): 16074.

MERS

Feikin, Daniel R., et al. "Association of Higher MERS-CoV Virus Load with Severe Disease and Death, Saudi Arabia, 2014." *Emerging Infectious Diseases* 21:11 (2015): 2029.

Minamata Disease

Iwata, Toyoto, et al. "Characteristics of Hand Tremor and Postural Sway in Patients with Fetal-Type Minamata Disease." *Journal of Toxicological Sciences* 41:6 (2016): 757–763.

Mumps

Gouma, Sigrid, et al. "Severity of Mumps Disease Is Related to MMR Vaccination Status and Viral Shedding." *Vaccine* 34:16 (2016): 1868–1873.

Nipah Virus

Chua, K. B., et al. "Nipah Virus." *Science* 288:5470 (2000): 1432–1435.

Omsk Hemorrhagic Fever

Borio, Luciana, et al. "Hemorrhagic Fever Viruses as Biological Weapons." *Journal of the American Medical Association* 287:18 (8 May 2002): 2391–2405.

Onchocerciasis

Adeoye, G., et al. "Achievements and Challenges of the African Programme for Onchocerciasis Control." *Annals of Tropical Medicine and Parasitology* 96 (2015): S15-S28.

O'nyong-nyong Fever

Tappe, Dennis, et al. "O'nyong-nyong Virus Infection Imported to Europe from Kenya by a Traveler." *Emerging Infectious Diseases* 20:10 (2014): 1766.

Oropouche Fever

da Silva Azevedo, Raimunda do Socorro, et al. "Reemergence of Oropouche Fever, Northern Brazil." *Emerging Infectious Diseases* 13:6 (2007): 912.

Pappataci Fever

Hukić, Mirsada, and I. Salimović-Besić. "Sandfly-Pappataci Fever in Bosnia and Herzegovina: The New-Old Disease." *Bosnian Journal of Basic Medical Sciences* 9:1 (2009): 39–43.

Pellagra

Kelly, Barbara. "Asylum Doctor: James Woods Babcock and the Red Plague of Pellagra." *Southeastern Librarian* 62:4 (2015): 14.

Phagadaena

Bakker, D. J., and A. J. van der Kleij. "Soft Tissue Infections Including Clostridial Myonecrosis." *Handbook on Hyperbaric Medicine* (2012): 343.

Plague

Attar, Naomi. "Bacterial Genomics: Three Centuries of Plague." *Nature Reviews Microbiology* 14:3 (2016): 130.

Schneewind, Olaf. "Classic Spotlight: Studies on the Low-Calcium Response of Yersinia Pestis Reveal the Secrets of Plague Pathogenesis." *Journal of Bacteriology* 198:15 (2016): 2018.

Polio

Luo, Hui-Ming, et al. "Identification and Control of a Poliomyelitis Outbreak in Xinjiang, China." *New England Journal of Medicine* 369:21 (2013): 1981–1990.

Sobti, Deepak, Marcos Cueto, and Yuan He. "A Public Health Achievement under Adversity: The Eradication of Poliomyelitis from Peru, 1991." *American Journal of Public Health* 104:12 (2014): 2298–2305.

Puerperal Fever

Benenson, Shmuel, et al. "Cluster of Puerperal Fever in an Obstetric Ward: A Reminder of Ignaz Semmelweis." *Infection Control & Hospital Epidemiology* 36:12 (2015): 1488–1490.

Puumala Virus

Lundkvist, Åke, et al. "Puumala and Dobrava Viruses Cause Hemorrhagic Fever with Renal Syndrome in Bosnia-Herzegovina." *Journal of Medical Virology* 53:1 (1997): 51–59.

Q Fever

Eldin, Carole, et al. "From Q Fever to Coxiella Burnetii Infection: A Paradigm Change." *Clinical Microbiology Reviews* 30:1 (2017): 115–190.

Pertussis

Winter, Kathleen. "Risk Factors for Pertussis Infection among Infants in California." *2016 CSTE Annual Conference.* Cste, 2016.

Rabies

Hampson, Katie, et al. "Estimating the Global Burden of Endemic Canine Rabies." *PLoS Neglected Tropical Diseases* 9:4 (2015): e0003709.

Relapsing Fever

Ciervo, Alessandra, et al. "Louseborne Relapsing Fever in Young Migrants, Sicily, Italy, July–September 2015." *Emerging Infectious Diseases* 22:1 (2016): 152.

Rheumatic Fever

Guilherme, Luiza, Karen F. Köhler, and Kellen C. Faé. "Editorial: Frontiers in Autoimmune Disease: Rheumatic Fever and Rheumatic Heart Disease." *Frontiers in Pediatrics* 3 (2015): 91.

Rickets

Wheeler, Benjamin, et al. "Incidence and Characteristics of Vitamin D Deficiency Rickets in New Zealand Children." *International Journal of Pediatric Endocrinology* 2015:1 (2015): 1.

Rift Valley Fever

Shawky, Sherine. "Rift Valley Fever." *Saudi Medical Journal* 21:12 (2000): 1109–1115.

Rocky Mountain Spotted Fever

Alvarez-Hernandez, Gerardo, et al. "Clinical Profile and Predictors of Fatal Rocky Mountain Spotted Fever in children from Sonora, Mexico." *Pediatric Infectious Disease Journal* 34:2 (2015): 125–130.

Ross River Fever

Cutcher, Z., et al. "Predictive Modelling of Ross River Virus Notifications in Southeastern Australia." *Epidemiology & Infection* (2016): 1–11.

Rubella

Sutcliffe, Catherine G., et al. "Measles and Rubella Seroprevalence among HIV-Infected and Uninfected Zambian Youth." *Pediatric Infectious Disease Journal* (2016).

Salmonella

Jackson, Brendan R., et al. "Outbreak-associated Salmonella Enterica Serotypes and Food Commodities, United States, 1998–2008." *Emerging Infectious Diseases* 19:8 (2013): 1239–44.

SARS

"Animal Origins of the Severe Acute Respiratory Syndrome Coronavirus." *Journal of Virology* 80:9 (May 2006): 4211–4219.

Knobler, Stacey, et al., eds. *Learning from SARS: Preparing for the Next Disease Outbreak.* Washington, D.C.: National Academies Press, 2004.

"Summary of Probable SARS Cases." http://www.who.int/csr/sars/country/table2004_04_21/en/.

Savill's Disease

Crowther, M. A. "Savill's Disease: A Pauper Epidemic in Britain and Its Implications." *Bulletin of the History of Medicine* 60:4 (1986): 544–558.

Scabies

Micali, Giuseppe, et al. "Scabies: Advances in Noninvasive Diagnosis." *PLoS Neglected Tropical Diseases* 10:6 (2016): e0004691.

Scarlet Fever

Turner, Claire E., et al. "Scarlet Fever Upsurge in England and Molecular-Genetic Analysis in North-West London, 2014." *Emerging Infectious Diseases* 22:6 (2016): 1075.

Schistosomiasis

Gautret, Philippe, et al. "Schistosomiasis Screening of Travelers to Corsica, France." *Emerging Infectious Diseases* 22:1 (2016): 160.

Scrofula

Duarte, G. I., and F. C. Chuaqui. "History of Scrofula: From Humoral Dyscrasia to Consumption." *Revista Médica de Chile* 144:4 (2016): 503.

Scurvy

Armelagos, George J., et al. "Analysis of Nutritional Disease in Prehistory: The Search for Scurvy in Antiquity and Today." *International Journal of Paleopathology* 5 (2014): 9–17.

Shigella

Marteyn, Benoit S., Mark Anderson, and Philippe J. Sansonetti. "Shigella Diversity and Changing Landscape: Insights for the 21st Century." *Frontiers in Cellular and Infection Microbiology* 6 (2016): 45.

Sleeping Sickness

Simarro, Pere P., et al. "Estimating and Mapping the Population at Risk of Sleeping Sickness." *PLoS Neglected Tropical Diseases* 6:10 (2012): e1859.

Smallpox

Fenn, Elizabeth A. *Pox Americana: The Great Smallpox Epidemic of 1775–82.* New York: Hill and Wang, 2001.

Molina, Chai, and David J.D. Earn. "Game Theory of Pre-Emptive Vaccination before Bioterrorism or Accidental Release of Smallpox." *Journal of the Royal Society Interface* 12:107 (2015): 20141387.

Sprue

Samsel, Anthony, and Stephanie Seneff. "Glyphosate, Pathways to Modern Diseases II: Celiac Sprue and Gluten Intolerance." *Interdisciplinary Toxicology* 6:4 (2013): 159–184.

Streptococcus

Madhi, Shabir A., et al. "High Burden of Invasive Streptococcus Agalactiae Disease in South African Infants." *Annals of Tropical Paediatrics: International Child Health* (2013).

Sweating Disease

Heyman, Paul, Leopold Simons, and Christel Cochez. "Were the English Sweating Sickness and the Picardy Sweat Caused by Hantaviruses?" *Viruses* 6:1 (2014): 151–171.

Syphilis

Patton, Monica E., et al. "Primary and Secondary Syphilis—United States, 2005–2013." *Morbidity and Mortality Weekly Report* 63:18 (2014): 402–406.

Tarantism

Corral-Corral, Inigo, and Carlos Corral-Corral. "Tarantism in Spain in the Eighteenth Century: Latrodectism and Suggestion." *Revista de Neurologia* 63:8 (2016): 370.

Tetanus

Jeremijenko, A. "A Tsunami Related Tetanus Epidemic in Aceh, Indonesia." *Asia-Pacific Journal of Public Health* 19:40–44 (January 2007).

Trachoma

Stocks, Meredith E., et al. "Effect of Water, Sanitation, and Hygiene on the Prevention of Trachoma." *PLoS Medicine* 11:2 (2014): e1001605P.

Trench Fever

Raoult, Didier, et al. "Outbreak of Epidemic Typhus Associated with Trench Fever in Burundi." *Lancet* 352:9125 (1998): 353–358.

Trichomoniasis

Secor, W. Evan, et al. "Neglected Parasitic Infections in the United States: Trichomoniasis." *American Journal of Tropical Medicine and Hygiene* 90:5 (2014): 800–804.

Tsutsugamushi

Patil, Navin, et al. "Spot the Dot: Solve the Mystery: Tsutsugamushi Disease." *Research Journal of Pharmaceutical, Biological and Chemical Sciences* 7:1 (2016): 1752–1755.

Tuberculosis

Udwadia, Zarir F., et al. "Totally Drug-Resistant Tuberculosis in India." *Clinical Infectious Diseases* 54:4 (2012): 579–581.

World Health Organization. "Tuberculosis Control in the South-East Asia Region." (2015).

Tularemia

Reintjes, Ralf, et al. "Tularemia Outbreak Investigation in Kosovo." *Emerging Infectious Diseases* 8:1 (January 2002).

Rojko, Tereza, et al. "Cluster of Ulceroglandular Tularemia Cases in Slovenia." *Ticks and Tick-borne Diseases* 7:6 (2016): 1193–1197.

Typhoid

Cronquist, Alicia. "Restaurant-Associated Outbreak of Typhoid Fever Traced to a Chronic Carrier—Colorado, 2015." *2016 CSTE Annual Conference.* 2016.

Khan, Mohammed Imran, et al. "Enteric Fever and Invasive Nontyphoidal Salmonellosis—9th International Conference on Typhoid and Invasive NTS Disease, Bali, Indonesia, April 30–May 3, 2015." *Emerging Infectious Diseases* 22:4 (2016).

Typhus

Blanton, Lucas S., et al. "Reemergence of Murine Typhus in Galveston, Texas, USA, 2013." *Emerging Infectious Diseases* 21:3 (2015): 484–486.

Gastellier, Laura, et al. "Noneruptive Fever Revealing Murine Typhus in a Traveler Returning from Tunisia." *Journal of Travel Medicine* 22:1 (2015): 67–69.

West Nile Fever

Bargaoui, R., S. Lecollinet, and R. Lancelot. "Mapping the Serological Prevalence Rate of West Nile Fever in Equids, Tunisia." *Transboundary and Emerging Diseases* 62:1 (2015): 55–66.

Dinu, S., et al. "West Nile Virus Circulation in Southeastern Romania, 2011 to 2013." *Eurosurveillance* 20:20 (2015): 21130.

Xerophthalmia

Khokhar, Abdul Rehman, Tehseen Iqbal, and Malik-Shah Zaman Latif. "Frequency of Xerophthalmia among Children in Dera Ghazi Khan." *Pakistan Journal of Physiology* 12:2 (2016): 8–10.

Yaws

Revankar, C. R., and K. Asiedu. "Towards Eradication of Yaws Disease by 2020 Using a Single Dose of Azithromycin in Mass Treatment." *International Journal of Infectious Diseases* 21 (2014): 388.

Yellow Fever

Garske, Tini, et al. "Yellow Fever in Africa: Estimating the Burden of Disease and Impact of Mass Vaccination from Outbreak and Serological Data." *PLoS Medicine* 11:5 (2014): e1001638.

Murphy, Jim. *An American Plague: The True and Terrifying Story of the Yellow Fever Epidemic of 1793.* Boston: Houghton Mifflin Harcourt, 2014.

Zika

Mlakar, Jernej, et al. "Zika Virus Associated with Microcephaly." *New England Journal of Medicine* 374:10 (2016): 951–958.

Index

The following names, tribes, historical events, diseases, medicines, treatments, places, and publications appear in the text under the dates listed beside them.